Biological, Physical and Technical Basics of Cell Engineering

Gerhard M. Artmann · Aysegül Artmann
Azhar A. Zhubanova · Ilya Digel
Editors

Biological, Physical and Technical Basics of Cell Engineering

Springer

Editors
Gerhard M. Artmann
Institute for Neurophysiology
University of Cologne
Cologne
Germany

Aysegül Artmann
Institute for Bioengineering,
Medical and Molecular Biology
University of Applied Sciences, Aachen
Campus Jülich
Germany

Azhar A. Zhubanova
Department of Biology and Biotechnology,
Kazakhstan National Academy of Natural Sciences
Al-Farabi Kazakh National University
Almaty
Kazakhstan

Ilya Digel
Institute for Bioengineering,
Cell- and Microbiology
University of Applied Sciences, Aachen
Campus Jülich
Germany

ISBN 978-981-10-7903-0 ISBN 978-981-10-7904-7 (eBook)
https://doi.org/10.1007/978-981-10-7904-7

Library of Congress Control Number: 2017964446

Cover illustration: Cover image courtesy of Dr. Tomo Šarić, Medical Faculty, University of Cologne, Germany. Cardiomyocytes derived from murine iPS cell line. Red indicates the muscle protein alpha-actinin and blue the cell nuclei.

Printed on acid-free paper

This Springer imprint is published by the registered company Springer Nature Singapore Pte Ltd.
part of Springer Nature
The registered company address is: 152 Beach Road, #21-01/04 Gateway East, Singapore 189721, Singapore

Foreword

While writing the preface for Dr. Artmann's book "Biological, Physical and Technical Basics of Cell Engineering" published by Springer Nature, a Confucius quote on learning and friendship came to my mind: "Isn't it a great pleasure to study and practice what you have learned? Isn't it a great delight when friends visit from afar?" I have the privilege of knowing Dr. Artmann and many of the authors of the chapters. In reading this book, I have the great pleasure of renewing my acquaintance with them and reviewing and learning the current states of biomedical sciences and bioengineering. This book was written by outstanding scientists and engineers focusing on the latest developments of bioengineering concepts and technologies in cell and tissue research. The chapters cover advanced frontier technologies and address scientific questions in biology and physics that need analysis and solution in terms of engineering. They provide the readers, including students, faculty and other scientists working in academia and industry, with current information on bioengineering in cell and tissue research. The book aims at promoting readers' understanding of biosciences, physics, engineering and innovative technology and their application to advance human health and well-being. This inspiring book can enhance not only our professional life, but also our understanding of what life really is.

University of California,
San Diego, CA, USA

Shu Chien, M.D., Ph.D.
University Professor of Bioengineering
and Medicine

Contents

Editors and Contributors

About the Editors

Prof. Gerhard M. Artmann (Germany), Book Co-editor, Dr. rer. nat. habil. He received his diploma in physics from the University of Dresden (1974, GDR), his doctorate in physics from the RWTH Aachen (1988, West Germany), and his habilitation from the Technical University of Ilmenau (1998, Germany). In his opinion, modern Western democracy and the freedom of the individual must be absolutely defended including in science. From 1989 to February 2017, he worked at the University of Applied Sciences Aachen, Germany, as Professor of Biophysics and Bioengineering. His research has focused on fundamental biophysical research as well as the development of advanced human cell measurement techniques. He is currently working with Prof. Jürgen Hescheler as Visiting Professor at the University of Cologne in the Institute of Neurophysiology and is researching stem cells with him. E-mail: g.artmann@gmx.net

Mrs. Prof. Aysegül Artmann (Germany), Book Co-editor, M.D., Ph.D. She has been a Full Professor at the University of Applied Sciences Aachen from 2001 to today. Her career is unusual. She attended high school, the TED Ankara College (English speaking), until 1984 and then studied medicine at Ankara University. She earned her medical doctor degree in 1990. Subsequently, she worked in the pharmacological industry. Later, she worked as a practicing physician in various polyclinics, including emergency services. However, she has always had great interest in medical research. She earned her medical certificate in physiology. During her stay at Akdeniz University, Antalya, and at Dokuz Eylül University (DEÜTF), she gained medical laboratory experience. Later on, she not only worked at the Physiology-Biophysics Department at DEÜTF as an assistant professor and specialized academician but also participated in a four-year biochemistry Ph.D. program. Her Ph.D. degree in Biochemistry in 2001 has opened her road to Europe. Although she has been a NATO scholar in 1996 for six months in USA as a researcher at the Department of Physiology and Biophysics at Herb Meiselman's laboratories (USCL), she wanted to continue her science career in Europe. From 2001 through 2003, she taught and researched in Cell Biophysics and Bioengineering as DAAD Sponsored Guest Professor at the University of Applied Sciences Aachen, Germany. Due to her excellent performance and very good feedback from their students and colleagues, she was accepted as a full professor. She successfully integrated her research interests into the interdisciplinary bioengineering/biophysics research team, which she joined, in particular in the development of the CellDrum technology, but also in different medical applications. E-mail: A. Artmann@fh-aachen.de

Prof. Azhar A. Zhubanova (Kazakhstan), Book Co-editor, Ph.D. She is a member of the Kazakhstan's National Academy of Natural Sciences, Doctor of Biology, and Professor of the Department of Biotechnology at the Al-Farabi Kazakh National University. She is well known for her contributions to the fields of microbiology, biochemistry, environmental sciences and biotechnology. She is the author or co-author of over 600 publications, edited 10 books and wrote four monographs. She was invited to numerous guest lectures in the USA, Germany, Russia, India and China.

Prof. Ilya Digel (Germany), Book Co-editor, Ph.D. He received the Dr. degree in Biology from the al-Farabi Kazakh National State University in 1995. Since 2002, he has been researcher and lecturer on cell biology and biophysics at the FH Aachen University of Applied Sciences, Germany. Being the head of the Laboratory of Cell- and Microbiology in the Institute of Bioengineering, he conducts ongoing research on molecular biophysics of proteins and water as well as on development of diagnostic and therapeutic approaches in medicine. E-mail: digel@fh-aachen.de

Contributors

Nuraly Sh. Akimbekov Al-Farabi Kazakh National University, Almaty, Kazakhstan

Ayşegül Artmann Institute for Bioengineering, Medical and Molecular Biology, University of Applied Sciences, Aachen, Jülich, Germany

Gerhard M. Artmann Institute of Neurophysiology, University of Cologne, Cologne, Germany; University of Applied Sciences, Aachen, Jülich, Germany

S. Azat Department of Chemicalphysics and Materialscience, Al-Farabi Kazakh National University, Almaty, Kazakhstan; Institute of Combustion Problems, Almaty, Kazakhstan

Aroj Bhattarai Biomechanics Laboratory, Institute of Bioengineering, University of Applied Sciences, Aachen, Jülich, Germany

Gerd Bungartz Hochschule Döpfer, Cologne, Germany

Kathryn Bungartz K.D. Bungartz, LLC, Riverton, WY, USA

Euisung Chung Graduate Program for Nano Science and Technology, The Graduate School, Yonsei University, Seoul, Korea

Samar Damiati Department of Biochemistry, Faculty of Science, King Abdulaziz University (KAU), Jeddah, Saudi Arabia; Institute for Synthetic Bioarchitectures, Department of Nanobiotechnology, University of Natural Resources and Life Sciences (BOKU), Vienna, Austria

Ilya Digel Institute for Bioengineering, Cell- & Microbiology, University of Applied Sciences Aachen, Campus Jülich, Jülich, Germany

Minh Tuấn Dương University of Applied Sciences, Aachen, Jülich, Germany; University of Erlangen-Nuremberg, Erlangen, Germany; Hanoi University of Science and Technology, Hanoi, Vietnam

Jane Frimodig Department of Bioengineering and Whitaker Institute of Biomedical Engineering, University of California San Diego, La Jolla, CA, USA

Ralf Frotscher Biomechanics Laboratory, Institute of Bioengineering, University of Applied Sciences, Aachen, Jülich, Germany

Oliver Funke German Aerospace Center DLR, Space Administration, Navigation, Bonn, Germany

Bora Garipcan Bio3 Research Laboratory, Institute of Biomedical Engineering, Bogazici University, Istanbul, Turkey

David A. Gough Department of Bioengineering and Whitaker Institute of Biomedical Engineering, University of California San Diego, La Jolla, CA, USA; Sanford Burnham Prebys Medical Discovery Institute, La Jolla, CA, USA

Jürgen Hescheler Institute of Neurophysiology, University of Cologne, Cologne, Germany

Gerda Horneck German Aerospace Center DLR, Institute of Aerospace Medicine, Cologne, Germany

Isgard S. Hueck Department of Bioengineering and Whitaker Institute of Biomedical Engineering, University of California San Diego, La Jolla, CA, USA; Sanford Burnham Prebys Medical Discovery Institute, La Jolla, CA, USA

Gentaro Iribe Okayama University Graduate School of Medicine, Cardiovascular Physiology, Dentistry and Pharmaceutical Sciences, Okayama, Japan

Pamela Itkin-Ansari Department of Bioengineering and Whitaker Institute of Biomedical Engineering, University of California San Diego, La Jolla, CA, USA; Sanford Burnham Prebys Medical Discovery Institute, La Jolla, CA, USA

Igor Jerman Institute Bion, Ljubljana, Slovenia

Ji-Hye Jung Stanford Cardiovascular Institute, Stanford, USA; Division of Cardiovascular Medicine, Center for Clinical Science Research (CCSR); 3115C, Stanford University School of Medicine, Stanford, CA, USA

A. R. Kerimkulova Department of Chemicalphysics and Materialscience, Al-Farabi Kazakh National University, Almaty, Kazakhstan; Institute of Combustion Problems, Almaty, Kazakhstan

Aida Kistaubayeva Al-Farabi Kazakh National University, Almaty, Kazakhstan

Sefa Kizildag Institute for Bioengineering, Medical and Molecular Biology, University of Applied Sciences, Aachen, Jülich, Germany; Department of Medical Biology, Faculty of Medicine, Dokuz Eylul University, Izmir, Turkey

Z. A. Mansurov Department of Chemicalphysics and Materialscience, Al-Farabi Kazakh National University, Almaty, Kazakhstan; Institute of Combustion Problems, Almaty, Kazakhstan

Haritha Meruvu Department of Bioengineering, Gaziosmanpasa University, Tokat, Turkey

Philippe Oger INSA Lyon, CNRS, UMR 5240, Univ Lyon, Villeurbanne Cedex, France

Judith Peters LiPhy, Univ. Grenoble Alpes, Grenoble, France; Institut Laue Langevin, Grenoble, France

Gerald H. Pollack University of Washington, Seattle, WA, USA

Ruth Prassl Institute of Biophysics, Medical University of Graz, Graz, Austria

Victor V. Rodin Institute of Organic Chemistry, Johannes Kepler University Linz, Linz, Austria

Sabra Rostami Bio3 Research Laboratory, Institute of Biomedical Engineering, Bogazici University, Istanbul, Turkey

Volker Seifarth University of Applied Sciences, Aachen, Jülich, Germany

Yong-de Shi Shanghai Medical College, Fudan University, Shanghai, China

Manfred Staat Biomechanics Laboratory, Institute of Bioengineering, University of Applied Sciences, Aachen, Jülich, Germany

Andreas M. Stadler Jülich Centre for Neutron Science, JCNS and Institute for Complex Systems ICS, Forschungszentrum Jülich GmbH, Jülich, Germany

Hwal (Matthew) Suh Graduate Program for Nano Science and Technology, The Graduate School, Yonsei University, Seoul, Korea; Department of Medical Engineering, Yonsei University College of Medicine, Seoul, Korea

Introduction

The publisher Springer Nature approached us and asked if we could prepare a new textbook on cell engineering. We discussed the proposal and formed an editorial team to prepare one. The interdisciplinary character of cell engineering led to the appointment of an editorial team with a very broad background. It consists of a medical doctor and biochemist (Prof. Dr. Aysegül Artmann), a microbiologist (Prof. Dr. Azar Zhubanova), a molecular biophysicist and cell biologist (Prof. Dr. Ilya Digel) and a physicist and bioengineer (Prof. Dr. habil. Gerhard M. Artmann). The team quickly agreed on the title of the book: "Biological, Physical and Technical Basics of Cell Engineering." As an editorial team with deep insights into cell engineering, we therefore have excellent connections to peers worldwide as well. We invited well-known and lesser-known authors and met with an enthusiastic response. This was encouraging and very satisfying for us, in more ways than one. We received positive feedback from almost entirely new members of the cell engineering society, as well as seasoned scientists recognized in the scientific world. And once again, we were surprised at how far scientific advances in medicine and biology have been based on fundamental contributions from other disciplines. These include advances in biophysical research in very basic areas, such as neutron scattering, but also in such fundamental issues as the role of water in biology.

Foreword to the book was written by **Shu Chien, M.D., Ph.D.**, University Professor of Bioengineering and Medicine, University of California, San Diego, USA. He is a major Chinese–American physiologist and engineer (awarded National Medal of Science by President Obama, USA, in 2011). Shu Chien is one of only 11 scholars who are members of all three US national institutes: the National Academy of Sciences, National Academy of Engineering and the Institute of Medicine. His work on blood cells and blood flow is of great importance for the diagnosis and treatment of cardiovascular and microcirculatory diseases. Chien's research has focused on mechanical forces and mechanically induced signal transduction in cells to understand cell behavior in blood vessels.

Bio-membranes are always topologically closed and enclose a space. They form both a separating layer between the inside and the outside of a cell and between cell internal compartments and the cytosol. They are, with all their properties, but especially their selective permeability to lipophobic particles such as water, or large particles such as ions or sugars, a prerequisite for life. Different transport proteins regulate this selectivity. Due to the importance of bio-membranes, the editors placed the chapter by **Mrs. Samar Damiati, Ph.D.**, about bio-membranes right at the beginning of the book. Mrs. Damiati is currently working as an Assistant Professor at the Department of Biochemistry, King Abdulaziz University (KAU), Jeddah, Saudi Arabia. She is interested in applying her knowledge in synthetic biology, biosensors, nanomedicine, lab-on-a-chip techniques and microfluidics. Already after our brief cooperation in the development of her chapter, it seems certain to us that she will be very successful in her job. E-mail: sdamiati@kau.edu.sa.

Mrs. Isgard Hueck, MSc., Germany and USA, wrote the review chapter on encapsulation of stem cells and their application in current clinical cell therapies. Already as an MSc student, Mrs. Hueck surprised her supervisors with publications in the American Journal of Physiology, Cell Physiology and The Microcirculation. She continued working at UC San Diego, USA, together with Profs. David Gough and Geert Schmid-Schoenbein in bioengineering research. At present, and for several years now, she has been intensively involved in immunology research and medical applications of stem cell therapy close to the clinic. Mrs. Hueck's review chapter "Stem Cell Encapsulation for Research and Therapy" is an innovative example of the synergy between stem cell therapy and medical device engineering as a tool to rapidly advance multi-disciplinary development of clinical therapies to treat devastating diseases, like cancer, liver cirrhosis or diabetes mellitus. Mrs. Hueck not only addresses the need for such integrative therapies, but also discusses the multi-leveled approach of bioengineers to overcome technical obstacles, possibly leading to a treatment solution that is very close to be introduced in the clinic. Not only for everyone, who is interested in treatment of chronic liver diseases and diabetes mellitus, but also for engineers, who want to venture out in the world of applied engineering techniques to manipulate biological samples for clinical therapies, this chapter is an eye-opening example of the complexity involved to develop useful therapeutic tools to cure life-threatening illnesses.

Drs. Kathryn and Gerd Bungartz are a scientist couple. Like many of us, they experience and love this varied and, at the same time, restless life. They spent many years of their lives in the USA, mostly working at Harvard University. The scientific hobbyhorse of both of them is stem cells. Their chapter about stem cells in this book can be highly recommended, especially for readers who would like to familiarize themselves with the world of stem cells. Numerous references of review articles and original research papers invite the interested reader for further study. Until today, the couple still works in this field, partly in industry and partly in university teaching. E-mail: gerd.bungartz@gmail.com.

Hwal Suh, DDS, M.SC., Ph.D., was born in Seoul, Korea, in 1953. Hwal Suh is one of the Korean pioneers in cell engineering. Currently, he holds the position of IFMBE chairman of the IFMBE working group cell and stem cell engineering. **Professor Hwal Suh** holds a Master of Divinity degree (M.Div., "magister divinitatis" in Latin) as well. This is the first professional degree of the pastoral profession. Interestingly, Hwal Suh also works as a pastor in his free time! In his book contribution, he reports on a very important chapter in God's creation, the extracellular matrix that surrounds the cells. He reports on porous matrices made from nanofibrous collagen/HA and their interactions with human embryonic stem cells (hESC) and human-induced pluripotent stem cells (hiPSC). E-mail: HWAL@yuhs.ac.

The novel concept of rheology of biological materials is one of the most important biophysical properties of materials which is thoroughly discussed in the chapter "Rheological Properties of Biological Structures, Scaffolds and Their Biomedical Applications" written by **Profs. Sabra Rostami and Bora Garipcan** from the Institute of Biomedical Engineering, Bogazici University, Istanbul. The Bogazici University is renowned worldwide and one of the best in Turkey. The study of the authors gives detailed information about the history of discovery of the concept and, after providing its definition and importance, goes on to a summarized yet comprehensive discussion regarding techniques and methods of measurement. The chapter's main focus is on explaining the rheological properties of biological materials including diverse cell and tissue types in the human body along with various scaffolds and finalizes with their biomedical applications which can definitely be a useful resource for those who are eager to learn about flow properties of biological substances.

The techniques of hysteresis loops and yield stresses are useful biophysical methods to obtain mechanical properties, and reveal the relationship between structure changes and energy transfer for biomaterial movement under force action. In the chapter "Hysteresis loops and yield stresses for cell research" written by **Prof. Yong-de Shi** from Shanghai Medical College, hysteresis loops and yield stresses were elucidated from various points: (1) basic theories, (2) technique requirements and (3) their applications in blood system, in human ejaculate liquefaction course, and during the plant Mimosa pudica movement by a small force. The above-mentioned studies reveal the relationship between structural changes in cytological or molecular biological levels and energy transfers when the biomaterials are subjected to the mechanical shear action, promoting research on how the force acts on cells, during which what relationship will occur between structure and energy.

A typical example of how advances in science and technology can expand biological and medical research opportunities is the CellDrum technique invented by **Prof. Gerhard M. Artmann** back in 1999. It was developed to routinely measure the very small mechanical tensile stress of cell monolayers and its alteration by substances. The measured tensile stress is less than the surface tension of water. Although the CellDrum patent was already published in 2000, the scientific world as well as the pharmaceutical industry has only realized its potential in recent years. Today, in 2017, can only be said that this important technology is gradually entering the research and drug development. The chapter of Artmann shows, more than any other, how strongly a technology development for biology is imprinted by very different scientific/technical disciplines. Incidentally, one of the first who was introduced to the CellDrum principle was the founder of biomechanics, Prof. Y.C. Fung. Professor Artmann was working at that time at his and Prof. Su Chien's laboratories at UC San Diego, USA. Yuan-Cheng Fung (awarded National Medal of Science by President Clinton, USA, in 2000) was surprised by the simplicity of the measuring principle, but smiled when they discussed the hurdles of technical development. E-mail: g.artmann@gmx.net.

Cardiac muscle is continuously exposed to dynamically changing mechanical loads, which is one of the core physiological environments for cardiomyocytes. Chapter "Development in Cell Manipulation Techniques for the Study of Single Cardiomyocyte Mechanics" written by **Prof. Gentaro Iribe** from Okayama University introduces various challenging methods for controlling mechanical environment of single cardiomyocytes in studying cardiac mechanics and underlying subcellular mechanisms. The author also overviews in situ whole heart mechanics to foster better understanding of the link between cardiac physiology and the cell manipulating techniques introduced. This chapter will focus spotlight on the importance of mechanical factors and tools in "real" cardiac physiology. E-mail: iribe@okayama-u.ac.jp.

Computational modeling plays a key role in providing better insights into the bio-physiology of soft tissue. It is mainly discussed in the chapter "Growth Modelling Promoting Mechanical Stimulation of Smooth Muscle Cells of Porcine Tubular Organs in a Fibrin-PVDF Scaffold" written by **Dr. Minh Tuan Duong** from the University of Erlangen Nuremberg and his colleagues from the Aachen University of Applied Sciences. The study indicates that growth modeling can suggest proper mechanical stimulations for smooth muscle cells of biological organs in fibrin-FVDF scaffolds. Furthermore, the authors prove that using external mechanical stimulations shows a significant change in the microstructure of tissues which can be promisingly employed for reconstructive surgery. This chapter can bring great insights and standard guides to scientists who are working in the field of reconstructing tubular organs. E-mail: tuandmhut@yahoo.com.

Chapter "Towards Patient-Specific Computational Modeling of hiPS-derived Cardiomyocytes" written by **Dr. Ralf Frotscher** from the Aachen University of Applied Sciences provokes by implicating that it will be possible to accurately simulate real-life tissue despite the complexity of the biochemical and physiological processes. In the chapter, the authors admit that the modeling of real-world biological processes is not only challenging but merely impossible using unmodified generic deterministic models. While explaining the scientific state of the art of modeling and simulation of cardiac tissues with special focus on the microscopic level, the authors reveal the existing gaps and give an idea how future models of complex human tissues could be formulated and adjusted in order to predict, for instance, patient-specific drug action on cardiac tissue. This chapter shows the comparison of many different experimental results with simulations. The interpretation of the differences between both leads the way toward the necessary experiment-based patient-specific adaption of models in order to be effective and employable in diagnostics. ralffrotscher2@gmail.com.

The chapter "Mechanics of the Soft Tissue Reactions to Different Textile Mesh Implants" presents the mechanical characterization of different monofilament mesh implants that are widely used for the surgical repair of abdominal hernia and pelvic floor disorders. This chapter reports on the biofunctionality of meshs implants. Furthermore, the effectiveness of mesh implants to strengthen the defected area in the body is discussed. The latter was done by comparing the mesh mechanics using the photoelastic uniaxial tensile test, optical strain measurements and biaxial bulge tests on dry meshes and on soft tissue surrogate gelatin-mesh specimens. The authors of the article, **Aroj Bhattarai** and Manfred Staat, also provide an overview on tissue reactions to monofilament mesh implants and mesh related complications such as infection, pain, mesh shrinkage, mesh wrinkling, collagen metabolism, adhesion, seroma, and inflammation. The knowledge will be helpful to predict the functional and biological outcomes after tissue reinforcement with meshes, to optimize the mesh implantation for the specific indications, to avoid mesh rejection or failure, and to improve the success of the surgical treatment. E-mail: arose_bhattarai@yahoo.com.

Professor Zulkhair Mansurov is among those who pioneered design and application of carbonized nanostructured materials originating from agricultural wastes. In his chapter "Extraction of Fusicoccin Like Compounds Using Nanocarbon Sorbents and Study of its Biological and Cytotoxic Activity" written together with his coworkers, Prof. Mansurov extends the frames of nanocarbon sorbents usage for biomedical and pharmacological purposes once again. Fusicoccin-like compounds isolated from fungi and seeds are known for their strong effects on plant cells. Of great importance and interest is the fact that fusicoccin A stimulates growth of axons. Fusicoccin and its derivatives are the main component of the novel biostimulator developed in Aytkhozhin's Institute of Molecular Biology and Biochemistry in Almaty. A critical technological difficulty is to obtain pure biostimulator in preparative quantities. Traditional dextran-based sorbents suffer from quick degradation being easily attacked by microorganisms. Z. Mansurov and his group developed and tested a novel nanostructured carbon sorbent, which is mechanically strong, highly suitable for separation of bio-organic substances and stable for many years. E-mail: ZMansurov@kaznu.kz.

An absolutely "must read" author, a brilliant scientist and a wonderful person is **Prof. Gerald H. Pollack** from the University of Washington in Seattle contributed with his chapter "The Fourth Phase of Water: Implications for Energy, Life and Health." For decades, Dr. Pollack has been studying the most fascinating and "unmanageable" mysteries of nature—the secrets of life and water. Nowadays, for most biophysicists, the role of water in energy storage, transport and transformation in the cell through the quantum physical mechanisms is absolutely unquestionable and yet very controversial. The chapter clearly explains the theory of the fourth phase of water referred to as "exclusion zone" (EZ) water. In the EZ water, water molecules shed their hydrogen ions and organize into orderly lattices of molecules that can grow hundreds of micrometer thick. Dr. Pollack's chapter is, as always, a phenomenal read that is presented in his unique unmistakable style that is easy to understand even for the non-professional. E-mail: ghp@u.washington.edu.

It is his consistent and logically perfect biophysical holism that distinguishes **Prof. Igor Jerman**, one of the world-level stars of water science and a strong proponent of the EZ water theory. In his excellently written chapter "Emergence of

Organisms from Ordered Mesoscopic States of Water (liquids)—Physical Instead of Chemical Origin of Life," Professor Jerman once again draws our attention to the breathtaking prospects and yet not fully discovered potential of the quantum electrodynamic interpretation of the amazing properties of water. This way was paved with the groundbreaking works by Herbert Fröhlich and later Emilio Del Giudice, Giuseppe Vitiello as well as Giuliano Preparata who introduced the idea of water coherent domains. In his excellent study, Prof. Igor Jerman reveals fresh and original evidence for the essential role of the quantum self-organizing phenomena in living processes. E-mail: igor.jerman@bion.si.

Dr. Victor Rodin has gained an excellent expertise in the nuclear magnetic resonance (NMR), both through his endless inquisitiveness and due to his continuous research activities conducted in the world-leading NMR centers in the USA, France, UK and Austria. His research focuses mainly on the (sometimes non-conventional and highly original) practical applications of NMR, and the chapter "Magnetic resonance in studying cells, biotechnology dispersions, fibers and collagen based tissues for biomedical engineering" serves as a bright illustration of the versatility, power and broad spectrum of modern NMR methods. Of particular interest for biomedical engineers will be the NMR-assisted determination of the apparent diffusion coefficient and other parameters related to the growth of engineered tissues. E-mail: victor.rodin@jku.at.

The chapter "Novel MRI contrast from magnetotactic bacteria to evaluate in vivo stem cell engraftment" addresses the MRI (Magnetic Resonance Imaging) visualization of the viability of the implanted stem cells in vivo. The chapter is a particularly good example of the scope of this book. It bridges stem cell biology, MRI imaging technology and physics in an exemplary way. The authors **Ji-Hye Jung, Ph.D.**, et al. report their recent work at Stanford University School of Medicine, USA. E-mail: jihyej@stanford.edu.

One of interesting discoveries in bioengineering and biophysics in the past years made with the help of NMR was the apparently unfolded native state of many functional proteins in the cell. The existence of disordered (also called naturally unfolded) proteins is a mind-boggling mystery and yet a well-established fact that challenges some paradigms of protein biophysics. It is becoming clear that, despite their lack of stable structure, such proteins represent a large and functionally very influential group of proteins involved in most crucial cellular and metabolic events. A large conformational flexibility of such proteins allows, for example, signaling pathway cross talks due to the interaction with various binding partners. Therefore, studies uncovering intimate details of the structure and dynamics of the naturally unfolded proteins are of great theoretical and practical interest. Chapter "Conformational Motions of Disordered Proteins" was written by **Dr. Andreas Stadler**. He is a very talented young protein scientist from Research Center Jülich and good friend of some of the editors of this book. His research confirms that the disordered proteins, being "abnormal" on the one hand, on the other hand obey the fundamental principles of protein molecular dynamics concerning both folding and conformational transitions. E-mail: a.stadler@fz-juelich.de.

Like Dr. Stadler, **Prof. Judith Peters** prefers to use the finest and most sensitive, though somewhat destructive, probes—neutrons—for studying biological objects. “Probing the Structure and Dynamics of Cells, Cell Components and Endogenous Nanoparticles Under Extreme Conditions with Neutrons” clearly demonstrates the huge potential and numerous possibilities for gaining structural and dynamical information about bio-objects through coherent and incoherent neutron scattering. On a high scientific level and yet in a very visual, well-structured and clear fashion, Prof. Peters gives very valuable “insider” information concerning the methodology and equipment used for dynamical studies of bio-systems by neutron scattering techniques, particularly under high-pressure conditions. E-mail: peters@ill.fr.

The crucial role of the fundamental biophysical knowledge in solving actual practical tasks becomes even more evident in chapter “Microbial Sampling from Dry Surfaces: Challenges and Solutions” written by **Prof. Ilya Digel** from the Aachen University of Applied Sciences and his colleagues from Kazakh National State University. In this study, the notorious challenges faced upon microbial sampling from solid surfaces are addressed from a highly fundamental and systematical point of view. The authors not only critically analyze the existing sampling techniques in terms of suitable surfaces, sampling conditions and potential risks, but also give an extensive overview of the physicochemical factors contributing to the strength of microbial attachment to dry solid surfaces. This chapter will definitely be a useful guide and a valuable gift for everyone who deals with environmental or epidemiological microbiology. E-mail: Digel@fh-aachen.de.

The tight and fundamental connection between life and water becomes even more comprehensive and all-embracing through the chapter “The Search for Signatures of Life and Habitability on planets and moons of our Solar system” written by **Dr. Oliver Funke and Dr. Gerda Horneck**—two space research professionals who, together with their long years’ experience in the highly pragmatic space exploration branch, share an absolutely passionate attitude toward the search for life in the universe. Their chapter is a cordial invitation to visit the most amazing sites of our solar system and learn a lot from their peculiar physics, chemistry, biology, geology, atmospheric science and oceanography. In addition, the reader will find the most up-to-date and reliable information about current and future space missions.

The authors and editors of this book of course did not seriously intended to embrace, to cover the gigantic field of modern bioengineering. It was rather an attempt to provide a curious reader (a young professional or maybe an old dilettante?) a slightly zoomed cross-sectional view on some front edges of this extremely versatile and fascinating branch of science and engineering. It was the intention of the editors to address an interested and educated readership with this book and to give an interesting and scientifically relevant insight into the current state of science in cell engineering. If, dear readers, you think we have succeeded, then we see our task as fulfilled. All our thanks, however, go to the authors of the contributions and all those people, woman and men, who have made science possible.

Dear reader, please enjoy this book and be curious. Join us on our flight into the space of cell engineering, a wonderfully diverse field of research. We start with attempts to rebuild cell membranes and at the end of the book we invite you into real space. Please follow the intention of the book and enjoy with the authors the search for signatures of life. Leave your laptop at home and sit in a meadow with wild flowers and read. Look up from time to time—is not nature magical and unbelievably beautiful? Be accompanied by a friend and share such moments. Whether you are a student, scientist or interested reader, take a look at a world that is beautiful and human.

Gerhard M. Artmann
Aysegül Artmann
Azhar A. Zhubanova
Ilya Digel

Part I
Biology

Can We Rebuild the Cell Membrane?

Samar Damiati

Abbreviations

3D	Three-dimensional
AFM	Atomic Force Microscopy
BLM	Black Lipid Membrane
CL	Cardiolipin
EIS	Electrochemical Impedance Spectroscopy
GUVs	Giant Unilamellar Vesicles
hBLM	Hybrid Bilayer Lipid Membrane
HDL	High-Density Lipoprotein
LB	Langmuir-Blodgett
LS	Langmuir-Schaefer
LTA	Lipoteichoic Acids
LUVs	Large Unilamellar Vesicles
MLVs	Multilamellar Lipid Vesicles
MSP	Membrane Scaffolding Protein
PC	Phosphatidylcholine
PE	Phosphatidylethanolamine
PG	Phosphatidylglycerol
PI	Phosphatidylinositol
PS	Phosphatidylserine
QCM-D	Quartz Crystal Microbalance with Dissipation monitoring
S-layer	Surface-layer
SAM	Self-Assembling Monolayer
sLBM	Supported Lipid Bilayer Membrane
SM	Sphingomyelin

S. Damiati (✉)
Department of Biochemistry, Faculty of Science,
King Abdulaziz University (KAU), Jeddah, Saudi Arabia
e-mail: sdamiati@kau.edu.sa

S. Damiati
Institute for Synthetic Bioarchitectures, Department of Nanobiotechnology,
University of Natural Resources and Life Sciences (BOKU), Vienna, Austria

G. M. Artmann et al. (eds.), *Biological, Physical and Technical Basics of Cell Engineering*, https://doi.org/10.1007/978-981-10-7904-7_1

SsLBM	S-layer-supported Lipid Bilayer Membrane
SUVs	Small Unilamellar Vesicles
tBLM	Tethered Bilayer Lipid Membrane

1 Introduction

Synthetic biomimetic membranes that designed to mimic the fundamental architecture of biological membranes are discussed in chapter "Can We Rebuild the Cell Membrane?" written by **Dr. Samar Damiati**. This chapter describes the most common biomimetic membrane models that are fabricated to simplify the complexity of biological membranes by exploiting simple nanostructured materials and the natural molecular mechanisms. Development of biomimetic membranes is a good example for synthetic biology which aims not only to advance our understanding of the biological systems by mimicking existing cellular structure and function, but also to design and engineer semi-natural systems with new functions which do not occur naturally. This chapter can be presented as a synthetic biology toolbox for biomimetic membranes assembly.

Nature is the richest source of inspiration to design and develop man-made molecular machines. Copying nature is a highly demanded skill that can be used to synthesize new biocomponents or to modify existing biocomponents, respectively, to perform novel functions that are not found in nature or to enhance existing systems. The complexity observed in nature arises from the highly organized structures of biomolecules and their complicated interactions. Hence, bioinspired mimics can help build an understanding of how biological systems work. Rapid advances in molecular engineering and high-resolution analytical techniques have allowed scientists to create biomimetic structures with high molecular precision. Understanding and engineering biological systems through mimicry can be defined as synthetic biology. This science is a rapidly growing multidisciplinary field that involves biology, chemistry, physics, mathematics and engineering, and it aims to redesign or construct a semi-synthetic system with the basic properties of cellular life by minimizing the chemical components required and controlling the physical parameters of the system. Although this strategy sounds simple, building an artificial system with the desired characteristics is still difficult and requires many trials to ensure that the system is effective. Although constructing a simple biological system depends on the assembly of basic biological components to perform specific functions, the generated system usually does not behave as expected due to the lack of some biological information. Hence, synthetic biology involves two approaches to building semi-synthetic systems: bottom-up and top-down [53]. In a bottom-up strategy, the basic elements (non-living components) of biological origins, such as nucleic acids, amino acids, lipids, and sugars, are exploited to create artificial life *de novo*. By contrast, the top-down approach starts with a complex living organism and attempts to make it as simple as possible while considering the biological elements as parts of the system.

Synthetic biology has generated interest in biomimetic membranes due to the ubiquitous presence of membranes in biological systems. Moreover, the complex nature of the structure and function of biological membranes is a challenge that can be overcome by constructing biomimetic models. However, membrane technology can be easily scaled up, requires mild conditions, no or low energy consumption, and that presently has a broad range of applications in small- and industrial-scale separations. Biomimetic or bioinspired membranes can be generated through the self-assembly of carefully chosen building materials in a near-natural environment in which several parameters, such as the temperature, pressure, and aqueous surroundings, are controlled. The elementary materials that are used to build the membrane are usually common biocomponents with excellent wetting, adhesive, hydrodynamic, and mechanical properties.

So, can we rebuild a cell membrane? The simple answer is yes, we can. This chapter provides insight into the basic characteristics of biological membranes in terms of membrane composition. The answer to the question above is described by a brief overview of the different biomimetic model membranes and their preparation methods. Furthermore, the advantages and disadvantages of each model are summarized.

2 Biological Membranes

The biological membrane has a critical role in cellular life. It has important functions in cells, such as protection, compartmentalization, and maintaining two-dimensional fluidity, and furthermore, it acts as an effective diffusion barrier that controls membrane permeability. Natural membranes possess unique features such as self-healing, anti-fouling, and selective permeability. These membranes are only a few nanometers thick, they are perfectly organized at the molecular level, and they are composed of three essential components: lipids, proteins, and carbohydrates. Lipids and proteins present a highly complex composition in biological membranes, whereas carbohydrates are located outside the membrane and are attached to lipids or proteins to form glycolipids or glycoproteins, respectively. Hence, the structure and function of the cell membrane are based on the fluid lipid bilayer and the incorporated proteins.

2.1 *Lipid Compositions of Biological Membranes*

The lipid compositions of biological membranes are primarily glycerophospholipids (also known as phospholipids), sphingolipids, and, in eukaryotic cells, sterols. Lipids are small, amphiphilic molecules consisting of a hydrophilic head group

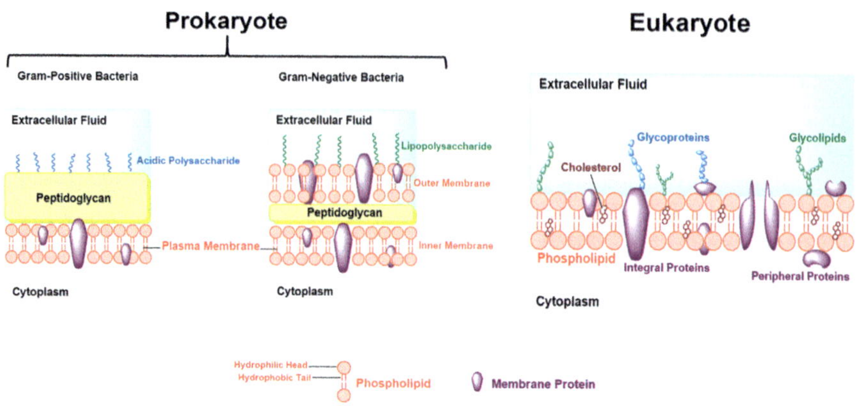

Fig. 1 Scheme of cell membrane structures in bacterial and mammalian cells

and a hydrophobic tail. These two components, and the interface formed between them, define the physical characteristics of any lipid molecule. The cell membrane is formed by a spontaneous self-assembly process driven by the amphipathic nature of phospholipid molecules; in this assembly process, a planar bilayer is formed with the non-polar groups pointed toward the hydrophobic interior of the bilayer and the polar head groups oriented toward the external aqueous phase [24]. Lipid head groups have been classified into several categories, each of which has unique chemical properties. For example, some head groups have a negative charge and exhibit charge-charge repulsion, which forms a larger effective cross-sectional area [26]. Moreover, the changes in the pH values that affect the lipid charge depend on the pKa values of the head group either by imparting or eliminating the charges [12]. However, there are significant differences between bacterial and mammalian membranes (Fig. 1).

Bacteria are classified into two broad groups as follows based on the cell wall structure: Gram-positive bacteria, which have a single cytoplasmic membrane surrounded by a thick layer of peptidoglycan, and Gram-negative bacteria, which have two membranes consisting of cytoplasmic and outer membranes surrounded by a thin layer of peptidoglycan. Both classes of bacteria have different lipopolysaccharides in their membranes. Gram-positive bacteria contain lipoteichoic acids (LTAs) embedded within the cytoplasmic membrane, whereas Gram-negative bacteria contain lipopolysaccharides that form the primary lipid component of the outer leaflet of the outer membrane. Indeed, there is a difference in the lipid compositions of the bacterial cytoplasmic membranes. The predominant zwitterionic phospholipids in most bacteria are phosphatidylethanolamines (PEs), whereas the predominant anionic phospholipids are phosphatidylglycerol (PG) and cardiolipin (CL). Gram-positive bacteria have lower PE contents than Gram-negative bacteria, and there are at least 15% anionic lipids in both bacterial types [25]. Prokaryotic cells lack cholesterol in their membranes; they instead have sterol-like molecules called hopanoids [59]. The exposure of these anionic lipids

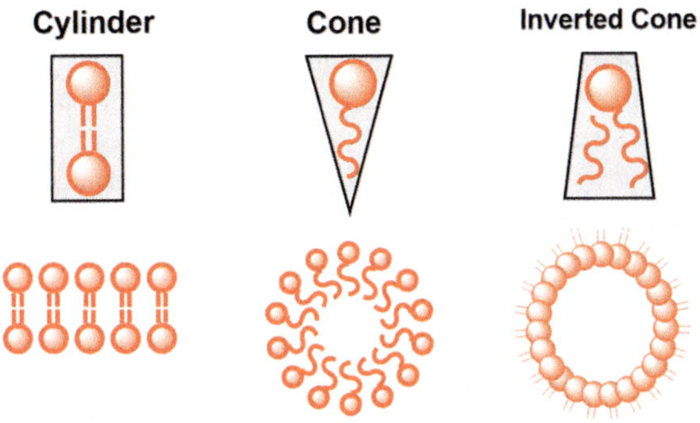

Fig. 2 Shape-structure concept of lipid polymorphism

provides for the selectivity of cationic antimicrobial agents that ensure toxicity against bacterial cells but not against mammalian cells.

Mammalian membranes are primarily composed of different structural lipids, such as phosphatidylcholine (PC), phosphatidylserine (PS), sphingomyelin (SM), phosphatidylinositol (PI), PE and cholesterol. These lipid species are present at different ratios depending on the cell type and function. In general, in most eukaryotic membranes, PC accounts for more than 50% of the phospholipids [75]. Lipids are asymmetrically distributed in the bilayer structure; PE and PS are primarily found in the inner leaflet of the plasma membrane, whereas PC and sphingomyelins are found in the outer leaflet. Cholesterol, the primary steroid in eukaryotic cells, has a significant role in determining membrane fluidity. The polar hydroxyl groups of cholesterol are inserted into the lipid bilayer close to the hydrophilic head groups of phospholipids, whereas the rigid hydrocarbon rings interact with the fatty acid chains. This interaction increases membrane rigidity and decreases its mobility. However, membrane lipids can be present in cylindrical (e.g., PC, and PS), conical (e.g., PE) or inverted conical (e.g., lysophosphatidylcholine) shapes (Fig. 2) [37, 47].

This diversity in shapes (polymorphism) influences the localization of lipids within the biological membranes and is responsible for phase separation within single monolayer leaflets. The shape of a lipid molecule in a membrane depends on the relative sizes of its polar head group and polar tails [69]. Biological membranes contain more lipids than proteins because lipid molecules are smaller than protein molecules and can act as a scaffold for membrane proteins.

2.2 *Membrane Proteins*

Although lipid molecules provide the basic structure of biological membranes, peptides and proteins are the dominant building blocks in biological systems because they perform most of the specific functions of organic tissues. Hence, peptides and proteins provide the cellular membrane with its functional characteristics. The basic building blocks of any protein are 20 amino acids that are encoded by DNA and are composed of two parts: an identical backbone (a carboxyl group and an amino group) and a side chain that varies in size and polarity. The amino acids in any polypeptide chain occur in a specific order (the primary structure) and are folded into a three-dimensional (3D) structure to establish the function of the protein and enhance its stability. Therefore, proteins are proposed to fit into a specific environment, such as an aqueous environment or in a membrane, where they interact with lipids or exist in a complex with other proteins. Protein folding occurs in a complex and crowded molecular environment within the cell and usually requires the assistance of molecular chaperones to prevent protein misfolding. Indeed, these chaperone molecules help the protein to engage in localization, assembly and disassembly without becoming permanent parts of the structure [3, 23].

There are two types of membrane-associated proteins: peripheral and integral proteins. Peripheral proteins are water-soluble molecules that are not inserted into the lipid bilayer but rather are attached to the inner or outer side of the membrane via non-covalent interactions, such as hydrophobic and electrostatic interactions. These proteins can also be attached covalently to the membrane via hydrophobic amino acids. By contrast, integral proteins are water-insoluble molecules that are embedded into a single lipid monolayer or bilayer. Most integral proteins are transmembrane proteins that function as gateways that control the movement of ions and small molecule into or out of the cell. Membrane proteins usually tend to denature once they are isolated from the lipid environment, losing their functionality due to misfolding. Furthermore, the low abundance of these proteins makes their investigation a challenging task due to the difficulty of the isolation and purification processes.

3 Biomimetic Model Membranes

The sophisticated structures and functions of biological membrane components (lipids and proteins) have inspired scientists to design artificial membranes through different chemically and physically controlled mechanisms. Moreover, in biological membranes, the lipid-lipid, protein-lipid, and protein-protein interactions are highly complex, which complicates the investigation of the distinct membrane components that are responsible for specific effects. Hence, biomimetic membranes offer a promising platform to overcome these limitations.

Bioinspired or biomimetic membranes are fabricated with natural or natural-like components by exploiting natural molecular mechanisms and the molecular self-assembly process to generate membranes that possess biologically accurate properties. The most common biomimetic model membranes are described below.

3.1 *Black Lipid Membranes (BLMs)*

The BLM was the earliest artificial membrane derived from extracted brain lipids, and it was introduced by Mueller and colleagues in 1962 [45]. The name of this membrane is attributed to its appearance under optical microscopy, which results from the optical behavior of the membrane when formed. Under reflected light, this very thin lipid film (20–25 Å) reflects little to no light relative to its surroundings and appears "black"; hence, the term "black lipid membrane". The thickness of this membrane fluctuates because it is compressible [52]. There are several methods of forming BLMs, but all of the procedures involve the formation of a free-standing lipid bilayer over a small aperture, which is usually less than 1 mm in diameter and is made in a hydrophobic substrate material, such as Teflon or polyethylene. The aperture is usually part of a wall separating two chambers filled with aqueous solutions containing reference electrodes. The two most popular methods of forming BLMs are described below.

3.1.1 Painting Technique

This method involves the use of a brush to paint a lipid dissolved in an organic solvent, such as n-decane or squalene, directly over an aperture under aqueous conditions (Fig. 3). The deposited lipid mass becomes thinner as it spreads, resulting in the spontaneous formation of a BLM. The primary drawback of the painting BLM technique is the use of hydrocarbon solvents, which are not present in natural biological membranes. Moreover, the resulting membrane retains the solvent, which can change its characteristics and reduce its stability [14].

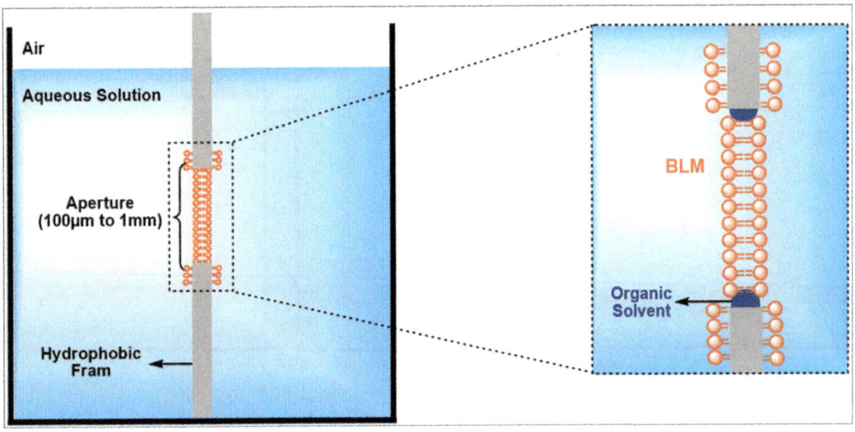

Fig. 3 Schematic of a painted BLM experiment. The lipid membrane spans a 100 μm – 1 mm pin hole in a hydrophobic substrate

3.1.2 Monolayer Folding Painting

This method is a modified form of the previous technique that allows for organic solvent evaporation [44]. Here, it is important to have an experimental cell that controls the solution level independently in each chamber (Fig. 4). The lipid solution is applied on top of one chamber above the aqueous solution and the researcher waits until the volatile solvent evaporates, which results in the formation of a lipid monolayer. The solution level containing the resultant monolayer is slowly lowered below the aperture and then raised again. This step deposits a monolayer on each pass, resulting in the formation of a solvent-free lipid bilayer film.

Both the painting and monolayer folding painting techniques generate a free-standing planar lipid bilayer that allows the transmembrane protein to be fused. Therefore, BLMs are widely used as a model system to study membrane proteins. The suspension of BLMs in solution in the absence of an underlying support allows the membrane proteins to remain fully mobile and active, which makes the BLM similar to a living membrane system. BLMs have been used to examine different biophysical processes, such as the formation of ion channels in the lipid bilayer by pore/channel-forming biomolecules, which include peptides, proteins, and antibiotics [14, 35]. BLMs can be investigated electrochemically or by simple light microscopy. For example, the thickness of a generated BLM may be calculated using the measurements of reflected light intensity [56]. Importantly, BLMs suffer from a limited lifetime (usually less than one hour), poor mechanical stability, and limited detection methods. This poor stability can be addressed by using a supportive substrate. Developing and improving BLM architectures through the use of, for example, chip-based platforms improves the stability, reproducibility, and generation of membranes with different geometries [15]. Indeed, simple imaging and electrochemical techniques are usually used with BLMs, and more advanced optical/fluorescence microscopy or advanced techniques are in high demand.

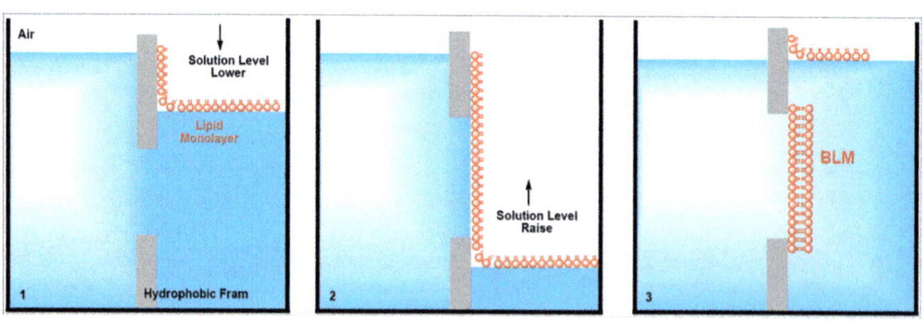

Fig. 4 Illustration scheme for preparation of montal-mueller BLM

3.2 *Langmuir Monolayers*

Another simple model that mimics the biological membrane is the Langmuir monolayer. This two-dimensional system provides a unique model for studying the interactions of small peptides with phospholipids at the air-water interface due to its homogeneity, stability and planar geometry. The Langmuir monolayer is useful in characterizing lipid-lipid interactions and evaluating the insertion of amphipathic molecules, such as drugs or antimicrobial peptides, into the membrane [4, 41]. These monolayers are easily prepared by continuously adding a lipid solution to the air-water interface of a constant area, which subsequently forms an insoluble film via the self-assembly of lipid molecules in a specific orientation. In this self-assembly process, the head groups point toward the aqueous phase, whereas the acyl chains point toward the air; thus, the resulting structure is considered a half-membrane (Fig. 5). This model is suitable for mimicking processes at the membrane surface but is not suitable for investigating transmembrane processes. Active compounds (e.g., peptides, drugs, and biosurfactants) can be injected into the water sub-phase after stabilizing the lipid monolayer at a defined initial surface pressure. The interaction between the active compound and the lipid film can be recorded by monitoring the increases in the surface pressure [22]. However, monolayers present a rich polymorphism that can easily be characterized by several techniques, such as optical microscopy and X-ray diffraction [13]. Even so, Langmuir monolayers have some limitations, such as the absence of bilayer structure similar to that of the biological membrane, the high surface tension of water, and its unsuitableness for membrane protein reconstitution [14, 16].

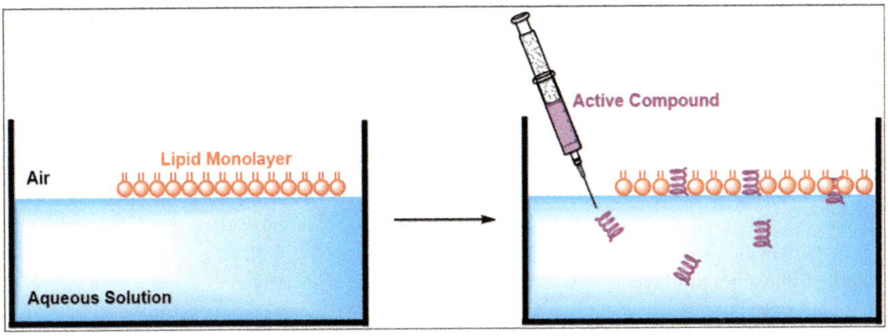

Fig. 5 Schematic view of lipid monolayer in a Langmuir balance (left) and addition of an active compound to the aqueous sub-phase (right)

3.3 *Hybrid Bilayer Lipid Membranes (hBLMs)*

This biomimetic model is a modified lipid monolayer that has been introduced by using alkanethiols (e.g., octadecanethiol) to form self-assembling monolayers (SAMs) on several metals, such as gold (the most widely used), silver, and mercury. Indeed, several polymer films with hydrophobic side chains can also be used to form SAMs (Fig. 6). These SAMs can be formed by incubating alkanethiol in an ethanol solution on an electrode surface, for 12–24 h on gold or silver, and for 5–20 min on mercury. An alternative preparation method involves a Langmuir-Blodgett (LB) transfer (this technique will be described later in this chapter). Both methods produce a tightly packed and well-ordered hydrophobic monolayer. However, the self-assembly of a SAM includes the anchoring of the thiol via the sulfhydryl groups to the metal surface, which leads the hydrocarbon chains to point toward the aqueous solution. There are two general methods for introducing a lipid monolayer (which serves as the upper leaflet) to the SAM that covers the hydrophobic surface. The first method involves the spontaneous fusion of lipid vesicles onto the SAM, whereas the second involves the horizontal transfer of a lipid monolayer that is supported in a Langmuir trough from an air-water interface to a SAM. Both methods result in the generation of a hybrid bilayer membrane (hBLM) that consists of a single lipid layer stabilized onto an alkanethiol SAM [14]. Due to the strong interactions between the SAM layer and the underlying surface, hBLMs are very robust and stable over long time periods. Moreover, some of the properties of a alkanethiol/lipid hybrid bilayer generated at the air-water interface can be preserved without change after the bilayer has been dried and rehydrated.

Different alkanethiols and lipids can alter the physical properties of hBLMs. For example, the membrane capacitance decreases when a thicker membrane is generated due to the increased chain length of the alkanethiol or phospholipids. However, hBLMs can be characterized by a wide range of techniques, such as surface plasmon resonance (SPR), quartz crystal microbalance with dissipation monitoring (QCM-D), and electrochemical measurements that include cyclic voltammetry (CV) and electrochemical impedance spectroscopy (EIS). In fact,

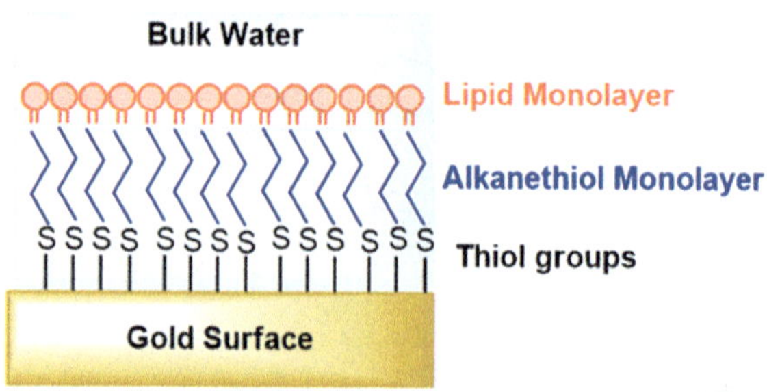

Fig. 6 Schematic of a hBLM consist of phospholipid layer assembled onto an alkanethiol-SAM layer

hBLMs allow for the detection of non-labeled analytes due to the direct coupling of the membrane to the metallic surface. Despite the close packing and rigidity of alkanethiol, this SAM still has some limitations. This supported lipid monolayer model presents a less flexible and fluid environment because its structure (a half-membrane) is more crystalline than a normal lipid bilayer, which affects the protein accommodation and functionality [28].

3.4 *Supported Lipid Bilayer Membranes (sLBMs)*

Since 1958, when Tamm and McConnell were the first to report the formation of a planar supported lipid bilayer membrane (sLBM) in direct contact with a solid surface, this approach has become one of the most popular biomimetic models for developing a surface-confined membrane system (Fig. 7). The simplest technique for obtaining this model is by fusing small unilamellar vesicles onto a smooth, clean solid substrate, such as borosilicate glass, mica, or silicon oxide wafers. Several attempts have been made to deposit a supported lipid bilayer onto single crystals of $SrTiO_2$ and TiO_2 or onto thin films of SiO_2 or $LiNbO_3$ crystals [2, 55]. This one-step fusion procedure results in a flat lipid membrane that can be obtained using different lipid mixtures. When the lipid vesicles (usually liposomes) touch a solid surface, they may adsorb, rupture, or spread to form a lipid bilayer on a hydrophilic surface. Lipid molecules are attached to the substrate surface via electrostatic, hydration, van der Waals and steric forces [51]. In the bilayer model, the polar head groups of one lipid monolayer face toward the solid surface while the hydrocarbon chains are in contact with the other lipid monolayer.

Alternatively, the Langmuir-Blodgett (LB) and Langmuir-Schaefer (LS) techniques can be used to prepare asymmetric bilayers (Fig. 8). In the LB (vertical) technique, dipping the lipid molecules spreads them at the air-water interface to form a lipid monolayer. A solid wafer is then raised or lowered vertically through the generated layer to transfer the lower leaflet of lipids to the solid support. Next, the same support is immersed again through the air-water interface to

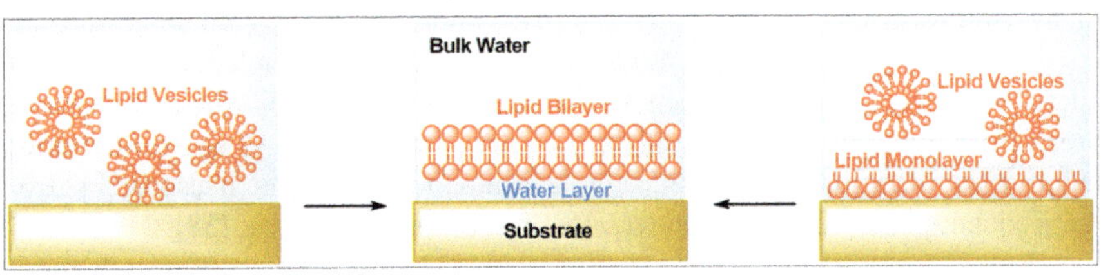

Fig. 7 Two common strategies commonly used for sLBM preparation: fusion of lipid vesicles (left) and combination of the LB technique and vesicle fusion process

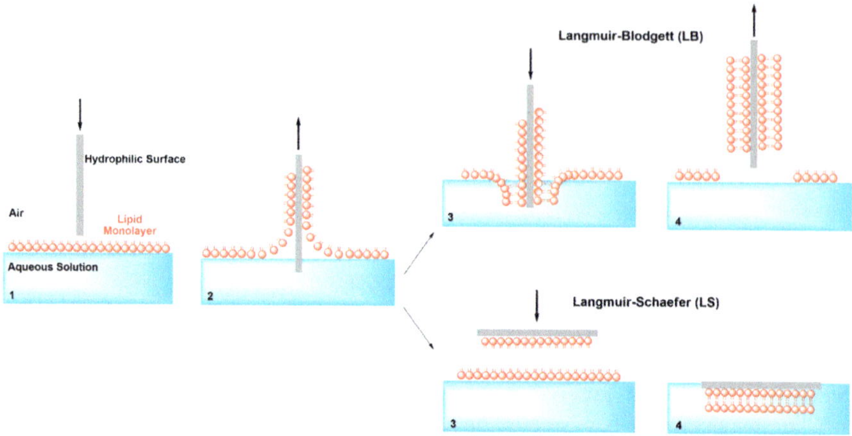

Fig. 8 The LB (vertical) and LS (horizontal) deposition techniques for lipid bilayer formation

generate the upper leaflet, resulting in the formation of a lipid bilayer. In the LS (horizontal) technique, dipping the solid wafer, upon which the first lipid monolayer is adhered, pushes the monolayer horizontally through a second lipid monolayer [27]. A combination of the previous methods can be employed to form the lower and upper leaflets of a lipid membrane; namely, the LB technique is used to form the lower leaflet, and vesicle fusion is subsequently used to form the upper leaflet. The successful generation of the lipid bilayer is dependent on several factors, such as the temperature, vesicle size and composition, and membrane density [38]. In contrast to BLM, sLBM is more stable and robust, and it can be investigated using several surface-specific analytical techniques that include SPR, QCM-D, and atomic force microscopy (AFM).

The fluidity of the sLBM is maintained by a layer of trapped water that separates the lipid membrane from the solid surface at an estimated thickness of 10–20 Å. This hydration layer near the hydrophilic surface is most likely highly ordered, more viscous and has a lower dielectric constant than bulk water. The transmembrane space measures approximately 0.5–2 nm in width, which provides stability and robustness to the generated lipid membrane but is not fully decoupled from the underlying solid surface. Moreover, the space limitation of the membrane-substrate restricts the incorporation and mobility of large integral membrane proteins and allows for direct contact between the surface and the protein, which may lead to protein denaturation and a loss of functionality. It is difficult to control the orientation of the incorporated membrane protein into the sLBM, which also suffers from the frequent presence of a defective area in the lipid membrane [58].

To circumvent these limitations, an auxiliary strategy has been developed in which the lipid membrane is decoupled from the solid substrate by the introduction of an intermediate spacer layer, which is typically less than 100 nm thick and which extends the membrane-substrate distance. A water reservoir increases the thickness of the water layer, which reduces the interaction between the lipid bilayer or incorporated protein and the substrate and minimizes the non-specific adsorption of aqueous protein from the solution. To regulate the lateral diffusivity and function of accommodated membrane proteins into this biomimetic model, it is important to quantitatively control the length and lateral density of the polymer layer by controlling both the membrane-substrate distance and the viscosity of the polymer layer [49, 73].

3.5 *Polymer-Cushioned Lipid Bilayer Membranes*

This model is the modified form of hBLM and sLBM and was pioneered by Sackmann and colleagues [57, 58]. In this biomimetic model, the lipid bilayer membrane is decoupled from the solid surface using soft polymeric molecules, such as chitosan, cellulose, dextran, lipopolymer tethers, and polyelectrolytes, and it presents a model membrane that has greater stability while maintaining the dynamic and structural integrity of a lipid bilayer membrane (Fig. 9). Another alternative to using polymers is to use quasi-crystal bacterial cell surface-layer (S-layer) proteins that can be recrystallized either in suspension or on several solid substrates, including silica, gold, and silicon dioxide. On a solid support, S-layer proteins can form a 5–10-nm-thick, highly porous lattice that effectively organizes the lipid-protein membrane [61, 68]. In general, a cushion layer provides several advantages, such as acting as a supporter that rests on the substrate to support the

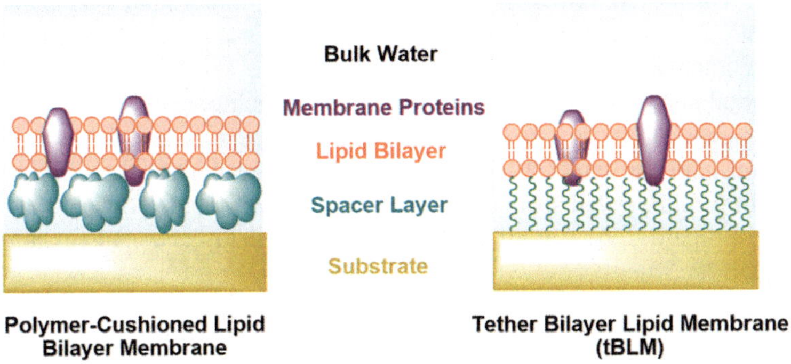

Fig. 9 Schematic of polymer (left) and tethered (right) solid-supported lipid membranes

lipid bilayer without any direct linkage with the substrate. The cushion layer also acts as a lubricant that reduces the interactions between membrane-incorporated proteins and the solid surface (thereby minimizing the risk of protein denaturation), assists in the self-healing of local defects in the membrane, and provides uniform coverage over macroscopically large surfaces (cm^2), which is important because a large number of defect sites contribute to a low signal-to-noise ratio and a high background response. In fact, the polymeric spacer provides a thick (larger than 3 nm) hydrated layer that allows for the incorporation of large transmembrane proteins and prevents the dewetting of the membrane [31, 36, 43, 73, 76]. In general, it is better to choose a polymer support that is soft, hydrophilic, not extensively cross-linked, not too highly charged, and has excellent wetting properties to prevent surface dehydration. All these features lead to the generation of a thermodynamically and mechanically stable supported membrane [58, 72]. To generate polymer-supported membranes, polymer molecules bind to the solid substrate first, and the lipid membrane is subsequently deposited onto the cushion layer. Several techniques have been reported to generate a lipid bilayer on the polymer cushion, such as vesicle fusion and LB/LS transfers.

The fabrication and design of a cushioned membrane can be inspired by nature. For example, the cytoskeleton, which is a protein matrix, supports the lipid bilayer of the cellular membrane in erythrocytes, providing the cell with its distinct shape and support. Thus, a well-designed cushioned membrane should behave similarly to a cytoskeleton and significantly address the balancing of surface forces [57]. The deposition of a lipid membrane on a cushion layer has no chemical link between the lipid molecules and the polymer cushion, but the membrane has attractive and repulsive physical forces that limit potential interactions between the membrane and the solid substrate. However, in a physiological system, an unstable biomimetic system can result from weak interactions between the lipid bilayer and the cushion layer. Hence, to tether the lipid membrane to the cushion layer effectively, the polymer cushions first need to be covalently attached to the underlying substrate, and subsequently, the insertion of anchor lipids or alkyl side chains into the lipid membrane can be employed [72, 73]. Polymer-cushioned lipid membranes are rigid, stable, and allow for the incorporation of large transmembrane proteins in a good orientation and under non-denaturing conditions. This biomimetic model has been reported in many studies for the successful reconstitution of several proteins in their functional forms. It is important to note that the nature and density of the polymer cushion can affect the diffusion of embedded lipids and accommodated proteins. In fact, the polydispersity and swelling behavior of the polymer may affect membrane properties, such as thickness, by producing bilayers containing holes and defects [21, 70, 76].

3.6 Tethered Bilayer Lipid Membranes (tBLMs)

As a polymer-cushioned lipid bilayer membrane, the tBLM is an alternative form of a sLBM that includes a reservoir layer to separate the lipid bilayer and solid support; this structure allows the membrane to behave similarly to the free-standing membrane (Fig. 9). The tBLM minimizes the direct contact between the incorporated membrane protein and the solid surface. Several types of molecules can be used as tethers, such as peptides (e.g., P19), lipopolymers [e.g., poly (2-ethyl-2-oxazoline)] or any self-assembling molecules that contain amino thiols or aminosilane groups [50, 66, 67]. To build a tBLM, tether molecules are self-assembled via thiol chemistry on a mercury substrate, via silane chemistry on a silicon substrate, or via phosphoric acid chemistry on an aluminum substrate [6, 11, 54]. The most commonly used solid substrate is gold, which can easily be functionalized by thiol- or sulfur-based anchor moieties [60]. The final lipid bilayer can be formed by vesicle fusion or via the LB technique. It is important to achieve the correct balance between the chemistry, density, and length of the tether molecules. Having large number of tethers attached to the solid substrate (high density) may reduce the mobility of the lipid bilayer and may change the phase transition temperature. The tether length may affect the membrane integrity; long tethers allow for the homogeneous incorporation of large membrane proteins into the membrane structure, whereas short tethers help in the generation of distinct protein patches in the membrane. However, significant increases in the tether length may lead to a greater number of defective areas in the membrane due to the storing of water molecules in the membrane (high membrane hydration), which reduces the electrical sealing properties [5, 40]. Many studies have used tBLM as a good platform for investigating the incorporation of pore- and channel-forming peptides and proteins [32, 42]. The tBLM, which offers a highly adaptable biosensor that is functionalized by the incorporation of ion channels, has been used to investigate the effects of some nanoparticles on cellular responses [29].

3.7 Lipid Vesicles

In [7], Bangham and colleagues observed the ability of phospholipids to form closed bilayer structures that were hollow and spherical in aqueous solution. These vesicles are versatile biomimetic model membranes that are widely used in biophysical and molecular biology research and, importantly, as a pharmaceutical carrier [34, 74]. Establishing a stable delivery system requires an understanding of lipid polymorphisms. Depending on the type of lipids used and the temperature, many types of assembled vesicles can be generated. However, lipid vesicles are produced by the dispersion of lipid molecules (a single lipid or a mixture of different types of lipids) in aqueous solution, which spontaneously assemble to enclose a small, aqueous compartment [22]. The generated vesicles can be a

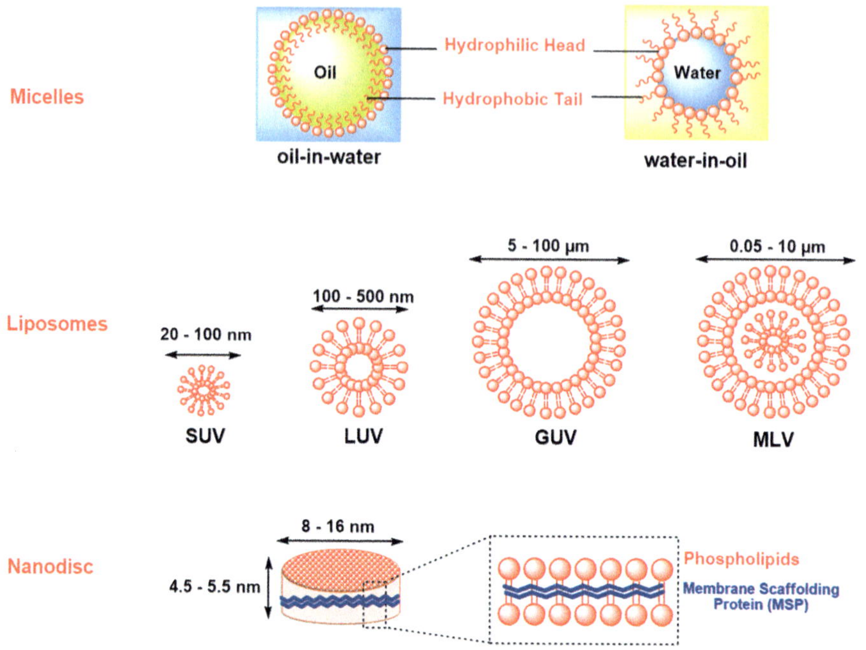

Fig. 10 Schematic drawings of different models of lipid vesicles: micelles, liposomes, and nanodiscs

monolayer (micelles), which presents a simple form of biological membrane structure, or a bilayer (liposomes and nanodiscs), which more closely mimics biological membranes (Fig. 10).

3.7.1 Micelles

Typical micelles are spherical in shape, water soluble, and form spontaneously when surfactants or detergent/lipid mixtures are dispersed in a liquid solution. It is also possible to generate micelles in different shapes, such as cylinders, cones, ellipsoids, and bilayers. Several factors, such as the pH, temperature, ionic strength, surfactant concentration, and molecular geometry of the surfactants, affect the size and shape of the resulting micelles. Two types of micelles can be formed by the micellization process: in a normal-phase micelle (an oil-in-water micelle), the hydrophilic head faces the aqueous exterior, whereas the hydrophobic tail faces the inner core; by contrast, in an inverse micelle (water-in-oil micelle), the head groups face toward the interior side whereas the tail groups face toward the exterior aqueous solution [33, 64]. Therefore, micelles are able to encapsulate non-polar molecules, such as vitamins, anti-oxidants, and antimicrobials.

3.7.2 Liposomes

Liposomes are spherical, self-assembled structures with internal aqueous compartments inside and between the lipid bilayer. These structures vary in size from 10 nm to several hundred micrometers. Liposomes are thermodynamically unstable, and their preparation methods can influence their diameter, size distribution, lamellarity, and encapsulation efficiency. Thus, liposomes are classified by either their structural properties or the method of preparation [30]. Some examples of developed methods presented by Bangham include organic solvent injection and reverse phase evaporation.

The different types of liposomes are as follows:

- **Multilamellar Lipid Vesicles (MLVs)**: These vesicles are characterized by 5–20 concentric lipid bilayers that are separated by water molecules and a bilayer ranging from 0.5–10 μm. MLVs can be formed by the vigorous hydration of a dried lipid film at a temperature above that of the lipid phase transition, and then the size of the generated particles can be reduced by sonication or by performing several freeze-thaw cycles. To form unilamellar vesicles, MLVs can be extruded through a porous membrane, resulting in the generation of either SUVs or LUVs [8, 46].
- **Small Unilamellar Vesicles (SUVs)**: These vesicles are single lipid bilayers varying in size from 20–100 nm, and they can be formed by the hydration or sonication of a lipid film [8].
- **Large Unilamellar Vesicles (LUVs)**: These vesicles are single lipid bilayers varying in size from 100–500 nm and can be formed by extrusion or by reverse-phase evaporation [71].
- **Giant Unilamellar Vesicles (GUVs)**: These vesicles vary in size from 5–100 μm and can be formed using many preparation protocols, which include the hydration of a lipid film at a temperature above the lipid phase transition for a long period of time (up to 36 h), gentle spontaneous swelling, the application of an eternal electrical field (electroformation method) at low or biologically relevant salt concentrations, or electroformation using native membrane extracts purified from biological cells [30, 77].

Similar to micelles, liposomes can encapsulate many functional ingredients. A major advantage is that due to their ability to encapsulate both water- and fat-soluble ingredients, liposomes can encapsulate hydrophilic agents in their internal core and hydrophobic agents into the membrane. Moreover, membrane proteins can be incorporated into the liposome membrane, thereby retaining their activities.

3.7.3 Nanodiscs

The nanodisc is a novel technology that was developed by Sligar and colleagues in the early 2000s. This synthetic model membrane system consists of self-assembled lipid-protein discoid nanoparticles (8–16 nm in diameter) composed of a phospholipid bilayer that is surrounded by a belt of an encircling helical protein called the membrane scaffolding protein (MSP). Each nanodisc is solubilized by two MSP molecules. However, MSPs are amphipathic peptides that are derived from apolipoprotein A-I, an important protein found in the human bloodstream that is fundamental to forming high-density lipoprotein (HDL). The diameter of a nanodisc is determined by the stoichiometry of the given lipid molecules and the length of the MSP. Nanodiscs are formed spontaneously by removing the detergent from a mixture of detergent solubilized-lipids and MSP, and the generated discoid bilayer particles are monodisperse, homogeneous, and can be obtained at high yields. These particles have a hydrophobic core of lipids that is shielded from the solvent by apolipoprotein molecules [9, 20].

Although micelles and liposomes can be used to solubilize and investigate insoluble membrane proteins, nanodiscs are a promising model that exhibit several advantages over other membrane mimetics. Such advantages include the following aspects: (i) their simplicity, which allows for the in vitro investigation of specific interactions between membrane receptors and their native ligands; (ii) their similarity to those of naturally occurring human lipoprotein, thus they can mimic the physiological environment of the plasma membrane, and allow for the incorporated protein to retain its stability in solution; (iii) their small size allows for the incorporation of a limited number of proteins of interest; and (iv) the MSP has the potential to be genetically modified with different tags [10, 19].

4 Applications of Biomimetic Membranes

Synthetic biomimetic membranes can be exploited for many practical applications, including the fabrication of biosensors, the development of nano-devices, and drug delivery systems. Indeed, these membranes help in understanding the molecular functions of biological membranes. Although more than 60% of drugs on the market target membrane proteins, most of their functional and structural properties are still ambiguous because it is difficult to study these proteins in their native environment. Moreover, these proteins tend to denature when isolated from the surrounded lipid membrane. Research on membrane proteins depends on the availability of the protein in its functional form for physical investigations, and it requires a lipid membrane for correct folding. Hence, artificial membranes provide promising strategies that allow for the investigation of membrane proteins in semi-natural environments that mimic biological environments. To create a successful lipid membrane model for embedding large transmembrane proteins, it is important to consider the protein rather than the lipid molecules as the basic building block of the platform. Subsequently, the effective generation of a membrane-incorporated protein can be used for physiological studies or as a sensing platform for diagnostic purposes and drug screening. Many attempts have been made to develop a suitable platform to study membrane proteins in bio-sensing applications. For example, the voltage-dependent anion channel (VDAC) protein, the major protein in the outer mitochondrial membrane, and the pore-forming toxin α-hemolysin, have been chosen for incorporation into the cushion-supported lipid membrane [S-layer-supported lipid bilayer membrane (SsLBM)]. The spontaneous incorporation of the VDAC protein into the SsLBM, which leads to the formation of an open channel in the shape of cylindrical channels that are permeable to many ions, has been successfully observed. The functionality of the formed channels can be detected by monitoring the flow of ions across the membrane, which is recorded as a significant decrease in the membrane resistance. The SsLBM model showed the ability to accommodate the VDAC protein but not α-hemolysin. Perhaps the mushroom-like shape of α-hemolysin could affect the functional incorporation of the protein into the synthetic membrane [17, 18]. Interestingly, α-hemolysin molecules could be functionally incorporated into a free-standing lipid bilayer onto which the S-protein has previously been recrystallized [62, 63].

Artificial membrane technology has attracted more attention by the medical field as an effective green method for the diagnosis, treatment, and prevention of diseases. A major research goal is the development of sensitive biosensors to detect large molecules (such as proteins, DNA, drugs, and toxins), viruses, bacteria, cancer cells, and low molecular weight molecules in samples while directly excluding any preliminary preparation steps. The application of biomimetic membranes in health care centers helps to improve diagnostic tests by producing sensitive (to picomolar concentrations), selective, rapid, easy-to-use, and inexpensive detection systems. Membrane biosensors are versatile and convenient models that can be employed to accurately detect pathogens. For example, a tBLM was developed to detect the influenza virus. This membrane is composed of a lipid bilayer, gramicidin (a peptide antibiotic), and antibody fragments (derived from monoclonal antibodies) that are embedded in the tethered lipid membrane. The functionality of the fabricated sensor is dependent on the recognition of the influenza nucleocapsid protein by the antibody fragments. In brief, chicken cells are infected with influenza virus. The cells are then lysed to extract the protein contents. The resulting lysate contains chicken and viral proteins, including the nucleocapsid protein. In the absence of an analyte, the gramicidin molecules in their monomeric forms are reconstituted into the upper and lower lipid leaflets, and they tend to assemble transiently and spontaneously into the dimeric form. After the analyte is exposed to the tethered lipid membrane, the nucleocapsid protein is captured by the antibody fragments, which prevents the transient formation of gramicidin dimers. These changes are recorded as a drop in admittance, and the signal response changes rapidly (on the order of minutes). This approach is known as the ion channel switch (ICS) biosensor [39].

Another influential use for biomimetic membranes is as a drug carrier. Nanocontainers, such as micelles, liposomes, and nanodiscs, are often used not only for biosensor applications but also for drug delivery. Using recombinant proteins as therapeutic agents has many drawbacks, such as low bioavailability, high toxicity, and rapid clearance [48, 78, 79]. The development of a drug delivery system (DDS) has become an increasingly attractive option for delivering hydrophilic and lipophilic drugs. These drug vehicles are able to increase the drug bioavailability, deliver a drug to a specific target site, control the drug release, and reduce toxicity by using a minimum dose to achieve good therapeutic effects. Due to the high stability, tunability and longevity of these carriers, they offer appropriate environmental conditions for storing drugs while maintaining their activities. In addition to these advantages, lipid nanoparticles have introduced new routes of delivery that were not commonly used in protein drug administration. For example, insulin, a small protein that is composed of two cross-linked peptide chains, is conventionally administered via the subcutaneous injection. As alternatives to this route, oral and nasal modes of administration have been exploited to deliver insulin. Encapsulating insulin in liposomes results in enhanced insulin bioavailability and protects the drug from enzymatic degradation while maintaining the pharmacological effect [1, 65].

5 Conclusion

Biomimetic models are commonly used to explore the biological membrane and, at present, show strong promise in this field. These bioinspired models offer excellent platforms for investigating natural membranes by simplifying their complex structure while maintaining their functionality and quality. This strategy, although simple in concept, is challenging to accomplish, but it can be achieved due to the rapid growth in synthetic biology, particularly via the bottom-up approach. To fabricate a semi-synthetic system, it is important to know what we want to create, which biological materials can be used, which ideas or concepts can be inspired by natural systems, how reorganized biological components can be controlled to perform a desired action, and what alternative solutions are available.

Biomimetic membranes can be built using a well-designed, easily generated system composed of a single lipid or a mixture of lipids to form mono- or bilayer membranes that are free-standing or supported by a solid substrate. As a result, this approach provides a large lipid membrane area that permits the reconstitution of transmembrane proteins. These membrane-based sensors allow the investigation of cellular principles, the functional properties of incorporated proteins, and the effects of proteins on lipid phase behavior. Moreover, biomimetic membrane may provide insight into the pathogenicity of some diseases to which lipids contribute during disease development, such as atherosclerosis. This novel approach will assist in disease diagnosis, treatment, and prevention. Currently, biomimetic membranes are in the phase of case-to-case development, and continuous improvement is revealing a more detailed understanding of cellular life.

References

1. Agrawal, A., Harde, H., Thanki, K., & Jain, S. (2014). Improved stability and antidiabetic potential of insulin containing folic acid functionalized polymer stabilized multilayered liposomes following oral administration. *Biomacromolecules, 5*(1), 350–360.
2. Ajo-Franklin, C., Kam, L., & Boxer, S. (2001). High refractive index substrates for fluorescence microscopy of biological interfaces with high z contrast. *Proceedings of National Academy of Sciences, 98*(24), 13643–13648.
3. Albert, B., Johnson, A., Lewis, J., Morgan, D., Raff, M., Roberts, K., et al. (2014). *Molecular biology of the cell* (6th ed.). New York: Garland Science.
4. Alhakamy, N., Kaviratna, A., Berkland, C., & Dhar, P. (2013). Dynamic measurements of membrane insertion potential of synthetic cell penetrating peptides. *Langmuir, 29*(49), 15336–15349.
5. Andersson, J., & Köper, I. (2016). Tethered and polymer supported bilayer lipid membranes: Structure and function. *Membranes, 6*(2), 30.
6. Atanasov, V., Knorr, N., Duran, R. S., Ingebrandt, S., Offenhäusser, A., Knoll, W., et al. (2005). Membrane on a chip: A functional tethered lipid bilayer membrane on silicon oxide surfaces. *Biophysical Journal, 89*(3), 1780–1788.
7. Bangham, A., Standish, M., & Watkins, J. (1965). Diffusion of univalent ions across the lamellae of swollen phospholipids. *Journal of Molecular Biology, 13,* 238–252.
8. Batzri, S., & Korn, E. (1973). Single bilayer liposomes prepared without sonication. *Biochimica et Biophysica Acta, 298,* 1015–1019.
9. Bayburt, T., & Sligar, S. (2003). Self-assembly of single integral membrane proteins into soluble nanoscale phospholipid bilayers. *Protein Science, 12,* 2476–2481.
10. Bayburt, T. H., & Sligar, S. G. (2010). Membrane protein assembly into nanodiscs. *FEBS Letters, 584,* 1721–1727.
11. Becucci, L., & Guidelli, R. (2014). Mercury-supported biomimetic membranes for the investigation of antimicrobial peptides. *Pharmaceuticals (Basel), 7*(2), 136–168.
12. Bogdanov, M., Dowhan, W., & Vitrac, H. (2014). Lipids and topological rules governing membrane protein assembly. *Biochimica et Biophysica Acta, 1843*(8), 1475–1488.
13. Brezesinski, G., & Möhwald, H. (2003). Langmuir monolayers to study interactions at model membrane surfaces. *Advances in Colloid and Interface Science, 100–102,* 563–584.
14. Castellana, E., & Cremer, P. (2006). Solid supported lipid bilayers: From biophysical studies to sensor design. *Surface Science Reports, 61*(10), 429–444.
15. Chen, T., & Reinhard, B. M. (2013). Characterizing the lateral friction of nanoparticles on on-chip integrated black lipid membranes. *Small (Weinheim an der Bergstrasse, Germany), 9,* 876–884.
16. Costa, A., & Burgess, X. (2012). Langmuir balance investigation of speroxide dimutase interactions with mixed-lipid monolayers. *Langmuir, 28,* 10050–10056.
17. Damiati, S., Zayni, S., Schrems, A., Kiene, E., Sleytr, U.B., Chopineau, J., et al. (2015a). Inspired and stabilized by nature: Ribosomal synthesis of the human voltage gated Ion channel (VDAC) into 2D-protein-tethered lipid interfaces. *Biomaterials Science,* 3, 1406–1413.
18. Damiati, S., Schrems, A., Sinner, E., Sleytr, U.B. & Schuster, B. (2015b). Probing peptide and protein insertion in a biomimetic S-layer supported lipid membranes platform. *International Journal of Molecular Sciences,* 16, 2824–2838.
19. Denisov, I., & Sligar, S. (2016). Nanodiscs for structural and functional studies of membrane proteins. *Nature Structural and Molecular Biology, 23,* 481–486.
20. Denisov, I. G., Grinkova, Y. V., Lazarides, A. A., & Sligar, S. G. (2004). Directed self-assembly of monodisperse phospholipid bilayer nanodiscs with controlled size. *Journal of the American Chemical Society, 126,* 3477–3487.

21. Deverall, M. A., Gindl, E., Sinner, E. K., Besir, H., Ruehe, J., Saxton, M. J., et al. (2005). Membrane lateral mobility obstructed by polymer-tethered lipids studied at the single molecule level. *Biophysical Journal, 88,* 1875–1886.
22. Eeman, M., & Deleu, M. (2010). From biological membranes to biomimetic model membranes. *Biotechnologie, Agronomie, Societe et Environnement, 14*(4), 719–736.
23. Ellis, R. (2005). Chaperone function: The orthodox view. In B. Henderson & A. G. Pockley (Eds.), *Molecular chaperones and cell signalling* (pp. 3–21). Cambridge: Cambridge University Press.
24. Engelmann, D. (2005). Membranes are more mosaic than fluid. *Nature, 438,* 578–580.
25. Epand, R., & Epand, R. (2009). Lipid domains in bacterial membranes and the action of antimicrobial agents. *Biochimica et Biophysica Acta, 1788,* 289–294.
26. Fedyukina, D. V., Jennaro, T. S., & Cavagnero, S. (2014). Charge segregation and low hydrophobicity are key features of ribosomal proteins from different organisms. *Journal of Biological Chemistry, 289*(10), 6740–6750.
27. Girard-Ergot, A. P., & Blum, L. C. (2007). Langmuir-Blodgett technique for synthesis of biomimetic lipid membranes. In D. K. Martin (Ed.), *Nanobiotechnology of biomimetic membranes* (pp. 23–74). New York: Springer.
28. Glazier, S., Vanderah, D., Plant, A., Bayley, H., Valincius, G., & Kasianowicz, J. (2000). Reconstitution of the pore-forming toxin α-hemolysin in phospholipid/18-octadecyl-1-thiahexa (ethylene oxide) and phospholipid/n-octadecanethiol supported bilayer membranes. *Langmuir, 16*(26), 10428–10435.
29. Goreham, R. V., Thompson, V. C., Samura, Y., Gibson, C. T., Shapter, J. G., & Köper, I. (2015). Interaction of silver nanoparticles with tethered bilayer lipid membranes. *Langmuir, 31*(21), 5868–5874.
30. Imura, T., Gotoh, T., Otaka, K., Yoda, S., Takebayashi, Y., Yokoyama, S., et al. (2003). Control of physicochemical properties of liposomes using a supercritical reverse phase evaporation method. *Langmuir, 19*(6), 2021–2025.
31. Ingo, K. (2007). Insulating tethered bilayer lipid membranes to study membrane proteins. *Molecular BioSystems, 3,* 651–657.
32. Jadhav, S. R., Sui, D., Garavito, R. M., & Worden, R. M. (2008). Fabrication of highly insulating tethered bilayer lipid membrane using yeast cell membrane fractions for measuring ion channel activity. *Journal of Colloid and Interface Science, 322*(2), 465–472.
33. Jelinek, R., & Silbert, L. (2009). Biomimetic approaches for studying membrane processes. *Molecular BioSystems, 5,* 811–818.
34. Kamps, J. A. A. M., Scherphof, G. L., Sullivan, S., Gong, Y., & Hughes, J. (2003). In V. P. Torchilin & V. Weissig (Eds.), *Liposomes, a practical approach* (2nd ed., pp. 267–301). New York: Oxford University Press.
35. Khan, M. S., Dosoky, N. S., Berdiev, B. K., & Williams, J. (2016). Electrochemical impedance spectroscopy for black lipid membranes fused with channel protein supported on solid-state nanopore. *European Biophysics Journal, 45,* 843.
36. Kiessling, V., & Tamm, L. (2003). Measuring distances in supported bilayers by fluorescence interference-contrast microscopy: Polymer supports and snare proteins. *Biophysical Journal, 84,* 408–418.
37. Kroon, A., Rijken, P., & De Smet, C. (2013). Checks and balances in membrane phospholipid class and acyl chain homeostasis, the yeast perspective. *Progress in Lipid Research, 52*(4), 374–394.
38. Le Brun, A., Clifton, L., Holt, S., Holden, P., & Lakey, J. (2016). Deuterium labeling strategies for creating contrast in structure-function studies of model bacterial outer membranes using neutron reflectometry. *Methods in Enzymology, 566,* 231–252.
39. Lee, S. K., Cascão-Pereira, L. G., Sala, R. F., Holmes, S. P., Ryan, K. J., & Becker, T. (2005). Ion channel switch array: A biosensor for detecting multiple pathogens. *Industrial Biotechnology, 1,* 26–31.

40. Liu, C., & Faller, R. (2012). Conformational, dynamical and tensional study of tethered bilayer lipid membranes in coarse-grained molecular simulations. *Langmuir, 28*(45), 15907–15915.
41. Maget-Dana, R. (1999). The monolayer technique: A potent tool for studying the interfacial properties of antimicrobial and membrane-lytic peptides and their interactions with lipid membranes. *Biochimica et Biophysica Acta, 1462,* 109–140.
42. McGillivray, D. J., Valincius, G., Heinrich, F., Robertson, J. W., Vanderah, D. J., Febo-Ayala, W., et al. (2009). Structure of functional staphylococcus aureus alpha-hemolysin channels in tethered bilayer lipid membranes. *Biophysical Journal, 96*(4), 1547–1553.
43. Merzlyakov, M., Li, E., Gitsov, I., & Hristova, K. (2006). Surface-supported bilayers with transmembrane proteins: Role of the polymer cushion revisited. *Langmuir, 22*(24), 10145–10151.
44. Montal, M., & Mueller, P. (1972). Formation of bimolecular membranes from lipid monolayers and a study of their electrical properties. *Proceedings of the National Academy of Sciences of the United States of America, 69*(12), 3561–3566.
45. Mueller, P., Rudin, D., Tien, H., & Wscott, W. (1962). Reconstitution of excitable cell membrane structure in vitro. *Circulation, 26,* 1167–1170.
46. Nguyen, T., Tang, Q., Doan, D., & Dang, D. (2016). Micro and nano liposome vesicles containing curcumin for a drug delivery system. *Advances in Natural Sciences: Nanoscience and Nanotechnology, 7*(3), 035003.
47. Osman, C., Voelker, D. R., & Langer, T. (2011). Making heads or tails of phospholipids in mitochondria. *Journal of Cell Biology, 192*(1), 7–16.
48. Petrache, A., Machin, D., Williamson, D., Webb, M., & Beales, P. (2016). Sortase-mediated labelling of lipid nanodiscs for cellular tracing. *Molecular BioSystems, 12,* 1760–1763.
49. Purrucker, P., Förtig, A., Jordan, A., & Tanaka, M. (2004). Supported membranes with well-defined polymer tethers—Incorporation of cell receptors. *Chem PhysChem, 5,* 327–335.
50. Rebaud, S., Maniti, O., & Girard-Egrot, A. P. (2014). Tethered bilayer lipid membranes (tBLMs): Interest and applications for biological membrane investigations. *Biochimie, 107,* 135–142.
51. Reimhult, E., Zach, A., Höök, F., & Kasemo, B. (2006). A multitechnique study of liposome adsorption on Au and lipid bilayer formation on SiO_2. *Langmuir, 22,* 3313–3319.
52. Ries, R., Choi, H., Blunck, R., Bezanilla, F., & Heath, J. (2004). Black lipid membranes: Visualizing the structure, dynamics, and substrate dependence of membranes. *The Journal of Physical Chemistry B, 108,* 16040–16049.
53. Roberts, M. A., Cranenburgh, R. M., Stevens, M. P., & Oyston, P. C. F. (2013). Synthetic biology: Biology by design. *Microbiology, 159*(7), 1219–1220.
54. Roskamp, R. F., Vockenroth, I. K., Eisenmenger, N., Braunagel, J., & Köper, I. (2008). Functional tethered bilayer lipid membranes on aluminum oxide. *ChemPhysChem, 9*(13), 1920–1924.
55. Rossetti, F. F., Bally, M., Michel, R., Textor, M., & Reviakine, I. (2005). Interactions between titanium dioxide and phosphatidyl serine-containing liposomes: Formation and patterning of supported phospholipid bilayers on the surface of a medically relevant material. *Langmuir, 21,* 6443.
56. Ryan, S. R., Hyeon, C., Rikard, B., Francisco, B., & James, R. H. (2004). Black lipid membranes: Visualizing the structure, dynamics, and substrate dependence of membranes. *The Journal of Physical Chemistry B, 108,* 16040–16049.
57. Sackmann, E. (1996). Supported membranes: Scientific and practical applications. *Science, 271,* 43–48.
58. Sackmann, E., & Tanaka, M. (2000). Supported membranes on soft polymer cushions: Fabrication, characterization and applications. *Trends in Biotechnology, 18,* 58–64.
59. Sáenz, J., Grosser, D., Bradley, A., Lagny, T., Lavrynenko, O., Broda, M., et al. (2015). Hopanoids as functional analogues of cholesterol in bacterial membranes. *Proceedings of the National Academy of Sciences, 112*(38), 11971–11976.

60. Schiller, S. M., Naumann, R., Lovejoy, K., Kunz, H., & Knoll, W. (2003). Archaea analogue thiolipids for tethered bilayer lipid membranes on ultrasmooth gold surfaces. *Angewandte Chemie International Edition in English, 42*(2), 208–211.
61. Schrems, A., Larisch, V. D., Stanetty, C., Dutter, K., Damiati, S., Sleytr, U. B., et al. (2011). Liposome fusion on proteinaceous S-layer lattices triggered via β-diketone ligand–europium (III) complex formation. *Soft Matter, 7,* 5514–5518.
62. Schuster, B., & Sleytr, U. B. (2002). Single channel recordings of α-hemolysin reconstituted in S-layer stabilized lipid bilayers. *Bioelectrochemistry, 55,* 5–7.
63. Schuster, B., Pum, D., Sara, M., Braha, O., Bayley, H., & Sleytr, U. B. (2001). S-layer ultrafiltration membranes: A new support for stabilizing functionalized lipid membranes. *Langmuir, 17,* 499–503.
64. Seddon, A., Curnow, P., & Booth, P. (2004). Membrane proteins, lipids and detergents: Not just a soap opera. *Biochimica et Biophysica Acta (BBA)-Biomembranes, 1666*(1–2), 105–117.
65. Sharma, G., Sharma, A. R., Nam, J. S., Doss, G. P. C., Lee, S. S., & Chakraborty, C. (2015). Nanoparticle based insulin delivery system: The next generation efficient therapy for Type 1 diabetes. *Journal of Nanobiotechnology, 13,* 74.
66. Siegel, A. P., Hussain, N. F., Johnson, M., & Naumann, C. A. (2012). Metric between buckling structures and elastic properties in physisorbed polymertethered lipid monolayers. *Soft Matter, 8,* 5873–5880.
67. Sinner, E. K., Reuning, U., Kok, F. N., Sacca, B., Moroder, L., Knoll, W., et al. (2004). Incorporation of integrins into artificial planar lipid membranes: Characterization by plasmon-enhanced fluorescence spectroscopy. *Analytical Biochemistry, 333*(2), 216–224.
68. Sleytr, U. B., Schuster, B., Egelseer, E. M., & Pum, D. (2014). S-layers: Principles and applications. *FEMS Microbiology Reviews, 38*(5), 823–864.
69. Sprong, H., Sluijs, P., & Meer, G. (2001). How proteins move lipids and lipids move proteins. *Nature Reviews Molecular Cell Biology, 2,* 504–513.
70. Su, Z., Jiang, Y., Velázquez-Manzanares, M., Leitch, J. J., Kycia, A., & Lipkowski, J. (2013). Electrochemical and PM-IRRAS studies of floating lipid bilayers assembled at the Au (111) electrode pre-modified with a hydrophilic monolayer. *Journal of Electroanalytical Chemistry, 688,* 76–85.
71. Szoka, F., & Papahadjopoulos, D. (1978). Procedure for preparation of liposomes with large internal aqueous space and high capture by reverse-phase evaporation. *Proceedings of the National Academy of Sciences of the United States of America, 75,* 4194–4198.
72. Tanaka, M. (2006). Polymer-supported membranes: Physical models of cell surfaces. *MRS Bulletin, 31,* 513–520.
73. Tanaka, M., & Sackmann, E. (2005). Polymer-supported membranes as models of the cell surface. *Nature, 437,* 656–663.
74. Torchilin, V. (2005). Recent advances with liposomes as pharmaceutical carriers. *Nature Reviews Drug Discovery, 4,* 145–160.
75. Van Meer, G., Voelker, D., & Feigenson, G. (2008). Membrane lipids: Where they are and how they behave. *Nature Reviews Drug Discovery, 9*(2), 112–124.
76. Wagner, M. L., & Tamm, L. K. (2000). Tethered polymer-supported planar lipid bilayers for reconstitution of integral membrane proteins: Silane-polyethyleneglycol-lipid as a cushion and covalent linker. *Biophysical Journal, 79,* 1400–1414.
77. Walde, P., Cosentino, K., Engel, H., & Stano, P. (2010). Giant vesicles: Preparations and applications. *ChemBioChem, 11,* 848–865.
78. Wu, W., & Jiang, X. (2016). Polymeric micelles for drug delivery. In Y. Zhao & Y. Shen (Eds.), *Biomedical nanomaterials*. Weinheim, Germany: Wiley-VCH Verlag GmbH & Co. KGaA.
79. Yu, X., Trase, I., Ren, M., Duval, K., Guo, X., & Chen, Z. (2016). Design of nanoparticle-based carriers for targeted drug delivery. *Journal of Nanomaterials, 2016*, 1–15.

Author Biography

Dr. Samar Damiati (Saudi Arabia) She is currently an Assistant Professor at Biochemistry Department, King Abdulaziz University (KAU), Jeddah, Saudi Arabia. She received her bachelor's and master's degree in biochemistry from Biochemistry Department, KAU, in 2004 and 2007, respectively. In 2013, she received her Ph.D. from the Institute for Synthetic Bio-architectures, Department of Nano-biotechnology, University of Natural Resources and Life Sciences (BOKU), Vienna, Austria where she focused on the synthetic bio-architectures to develop membrane-based biosensors. She is a member of the European Society for Nanomedicine and the Austrian Association of Molecular Life Sciences and Biotechnology (ÖGMBT). Her current research interests include synthetic biology, biosensors, nanomedicine, lab-on-a-chip and microfluidics.

Encapsulation of Stem Cells in Research and Therapy

Isgard S. Hueck, Jane Frimodig, Pamela Itkin-Ansari and David A. Gough

Abbreviations

ECM Extra Cellular Matrix
hESC Human Embryonic Stem Cell
iPSC Induced Pluripotent Stem Cell
LPC Liver Progenitor Cell
MSC Mesenchymal Stem Cell
PPC Pancreatic Progenitor Cell

1 Introduction

Diseases of the liver and the pancreas are on the rise worldwide (World Health Organization, 2017). The liver and the pancreas are the two organs most responsible for metabolic regulation and, when impaired, have dramatic effects on patients' lifestyles and survival rates [1, 2]. As of 2013, there were more than 30 million Americans [3] and 29 million Europeans suffering from chronic liver diseases [4]. Chronic pancreatitis ranges from 12 (US) up to 120 (Asia) people in each 100,000 [2, 5]. The survival rate of pancreatic cancer is only 8%, which makes it the most lethal cancer worldwide (World Health Organization, 2016). The prevalence of Diabetes cases has risen from 108 million in 1980 to 422 million in 2014 (World Health Organization, 2016). Both, pancreas and liver diseases are increasing to an epidemic level [2, 6], and more than 17,000 people are on the waiting list for liver

I. S. Hueck (✉) · J. Frimodig · P. Itkin-Ansari · D. A. Gough
Department of Bioengineering and Whitaker Institute
of Biomedical Engineering, University of California San Diego,
9500 Gilman Drive, La Jolla, CA 92093-0412, USA
e-mail: ihueck@ucsd.edu

I. S. Hueck · P. Itkin-Ansari · D. A. Gough
Sanford Burnham Prebys Medical Discovery Institute,
10901 North Torrey Pines Road, La Jolla, CA 92037, USA

G. M. Artmann et al. (eds.), *Biological, Physical and Technical Basics of Cell Engineering*, https://doi.org/10.1007/978-981-10-7904-7_2

transplants in the US, of which more than 1500 people die every year while waiting for a transplant organ (American Liver Foundation, 2016). Thus, there is a desperate need for alternative therapies to organ transplantation.

The pancreas and liver develop from the same pool of endodermal progenitor cells [7]. However, expandable long-term primary cell cultures for implantation are not available for either hepatocytes from the liver or endocrine cells from the pancreas. Therefore, another cell type has to be used if cell therapy is to be a feasible alternative to organ transplantation: e.g. stem cells. A therapy that would allow implantation of stem cells to help repair or substitute for organ functions could ultimately reverse the dire prognosis for patients with liver or pancreatic disease. Stem cell therapies could be the key for both short-term approaches until a donor liver is available, or for long-term solutions to mimic the endocrine function of the pancreas by providing in vivo production of insulin.

Since stem cells were discovered, their potential to differentiate into replacement cells for regenerative therapies has been discussed as a cure for many diseases. However, the dynamic play between factors that influence stem cell differentiation, and/or the ability to secrete therapeutic substances in vivo, is very complex and not fully understood. Furthermore, pockets where stem cells can survive, differentiate, or secrete proteins need to be protected in a foreign host environment to avoid the need for chronic immunosuppression. An alternative is cellular encapsulation intended to avoid rejection by the immune system and to control growth of cellular implants in a defined space. Encapsulation techniques for cellular implants, ranging from single cells to complex structures of organoid tissues, are engineered as an immunoprotective packaging tool for stem cell survival in vivo. This chapter will summarize and examine the advances of stem cell encapsulation approaches in research and therapy. Selected examples for stem cell encapsulation devices will be discussed with focus on the treatment of chronic and acute liver diseases and Diabetes Mellitus.

2 History of Cell Encapsulation

By the early 1900s, the idea of isolating cells via encapsulation was an appealing scientific method for research. In 1912, Murphy and Rous developed a method of implanting tumor cells in a chicken embryo inside a chicken egg [8]. After carefully patching the egg's shell using paraffin under sterile conditions, the chicken egg resembled a perfect encapsulation device to keep the implanted cells alive. The cellular environment was contained inside the eggshell, yet the implanted cells were protected from mechanical destruction or immune responses. About 20 years later, in 1933, the scientist Vincenzo Bisceglie, at the Royal University of Bari in Italy, followed up on the encapsulation technique using chicken eggs. He discovered that an immune response is lacking in chicken eggs until the 18th day of embryo development [9]. Therefore, his implanted cells, encapsulated in chicken eggs, grew rapidly for about 18 days before an immune response arose. In his experiments,

Dr. Bisceglie used sarcoma and carcinoma cells, and a tumor mass was able to develop for 2½ weeks of protected growth in the egg capsule. In further cell encapsulation approaches, Dr. Bisceglie used a collodion bag to hold cells for implantation in vivo [10]. The collodion bag was prepared of material synthesized of gelatin and nitrocellulose treated carefully in a procedure with alcohol, ether, and hot and cold water. In these studies, Dr. Bisceglie used mouse tumor cells contained in the collodion bag, which functioned as a protective semipermeable membrane, for implantation into the abdominal cavities of guinea pigs. It was reported that the tumor cells survived for 12 days in the guinea pig's abdomen [10]. This was the birth, not only of cellular encapsulation techniques, but also of modern cancer research with tumor studies in foreign hosts. However, the implanted cells were originated from aggressive growing tumors and did not have any regenerative effects on the host. Another 30 years passed before cell encapsulation was considered as a tool for tissue engineering in regenerative therapy.

In 1964, Thomas Chang introduced the first polymer membrane encapsulation of aqueous solutions of proteins. He developed stable capsules up to 100 μm in diameter that were protected from immunoreactions, and were capable of delivering therapeutic enzymes in vivo [11]. However, until a more biocompatible material was developed and other issues were addressed, cell and tissue encapsulation was not ready for clinical applications in humans. The idea of encapsulated cell replacement therapies for humans was driven by Diabetes Mellitus research, which was highly impacted by advances of Dr. Ernst Pfeiffer in the 20th century [12]. In the 1970s and 1980s, several research groups working in the field of diabetes were pioneers in experiments with the transplantation of pancreatic *Islet of Langerhans* cells, the endocrine tissue that releases insulin [13]. Furthermore, Australian Dr. Kevin Lafferty laid the groundwork for encapsulation of cell therapies by exploring immunoreactions to implants [14]. American Dr. William Chick used primary islet cells for encapsulation techniques in pre-clinical studies with small and large animals [15]. Drs. Lim and Sun were the first to do microencapsulation of primary islet cells in diabetic rats [16]. At the same time, Dr. Robert Langer was recognized as the first researcher to integrate biology and material sciences for cellular encapsulation techniques [17, 18]. Not until 1994, however, were encapsulated vital islet cell implants successfully used in humans. A study conducted by Drs. Soon-Shiong and Heinz et al. is recognized as the first clinical trial using alginate encapsulated human islets in one diabetic patient [19]. The diabetic and end-stage kidney diseased patient was reported to be insulin-free for 9 months after cell implantation [20]. Although this was a great first success, later clinical studies using this approach, did not confirm insulin-independence in patients with diabetes [21]. In 1999, a German scientist group, led by Dr. Mathias Löhr in Berlin, used cellulose sulfate polymers in clinical trials as an innovative strategy to embed cells in micro-capsules for treatment of patients with end-stage pancreatic cancer [22–24]. Since then, many variations of embedding materials, either of natural and biocompatible components, synthetized polymers, or a mixture of thereof have been tested to encapsulate cells for regenerative therapies.

In recent decades, the discovery of stem cells for cellular replacement therapies made the encapsulation approach a crucial engineering tool. Stem cell populations can be uniquely sensitive in their need for a specialized environment in order to survive and differentiate to therapeutically useful cells in vivo. Encapsulation materials with specific characteristics for filtering molecules that may pass through the encapsulation layer have been developed in the past 20 years [25–31].

This chapter will review how the combination of two developments has advanced the prospect for cell therapy. Encapsulation materials and designs, used as an engineering tool for device development, and the understanding of stem cell manipulation, may provide a promising approach to the treatment of liver diseases and Diabetes Mellitus.

3 Background

3.1 Stem Cells for Therapeutic Use

Currently, expandable long-term primary cell cultures are not possible for either hepatocytes from the liver or endocrine cells from the pancreas. Therefore, another cell type has to be found if cell therapy for the liver and pancreas is to be scalable and feasible. One possibility is stem cells. There are several different types of stem cells that can be utilized for cellular implantation devices to treat chronic diseases [32]. Depending on the clinical application, stem cells for use in regenerative therapies include *embryonic stem cells (ESC), induced pluripotent stem cells* (iPSC), adult *mesenchymal stem cells* (MSC), and *progenitor cells* (PC). Although stem cells can develop into various functional mature cells, this chapter is focused on examples of liver and pancreas cell development.

Embryonic stem cells (*ESCs*) originate in the embryonic blastocyst, are pluripotent, carrying the potential to differentiate into any cell of the body, and can proliferate in vitro. They are usually cultivated on feeder cell layers or matrigel, so that a sufficient quantity of the proper cell type is produced prior to implantation [33]. However, this process is still limited in scale. There also may be a risk of introducing murine viruses from matrigel [33]. Human embryonic stem cells (hESC) can be manipulated to differentiate into hepatocyte-like cells, as well as beta-cell-like cells that produce insulin in vitro and in vivo. HESCs have been used in cellular implantation devices, as will be described later in this chapter. However, the use of embryonic stem cells has been a subject of ethical debate ever since their discovery. Furthermore, immune responses from the implant host need to be addressed in transplantation approaches that include hESCs.

Another stem cell type used in encapsulation devices is *induced pluripotent stem cells, or iPSCs.* These cells are usually derived from adult terminally differentiated cells that have been genetically altered to introduce pluripotency. The first human iPSCs were originated from adult human fibroblasts genetically reprogrammed with viral vectors [34]. Since then, other easily accessible cells have been used for induction of pluripotency, such as keratinocytes from a plucked hair, peripheral blood cells, and epithelial cells excreted in urine [35–37]. Although these cells are readily available, the process of reprogramming creates challenges for use in humans. Disadvantages of iPSCs include low efficiency in the rate of successful conversions, risk of inserting genomic mutations into the cells, and high tumorigenecity [38]. However, iPSCs are free from immune and ethical issues that are associated with ESCs. IPSCs have shown promise insofar as they can be differentiated to hepatocytes and hold potential for autologous transplantation in liver disease treatments [39, 40]. Still, use of iPSCs entails worries about the use of viral vectors, changes in cell cycle proteins, and the origination of teratomas. Development of efficient non-viral reprogramming techniques, better biological understanding of reprogramming effects on the epigenetic state, and certified stocks of iPSC clones, will all be critical to clinical use of iPSCs in regenerative therapies [41, 42].

Adult stem cells of mesodermal origin *(MSC)* can be isolated from bone marrow, adipose tissue, and other sources [43]. They remain undifferentiated throughout most of their life, resting in adult tissue until needed. However, in cell culture, they lose their proliferative capacity over time. Although they are *adult* stem cells, MSCs can be stimulated to differentiate into hormone-secreting cells as they are found to express human insulin, glucagon, and somatostatin in immunocytochemistry experiments [44]. MSCs are shown to exhibit insulin-production in vivo in diabetic mice [45] and seem to improve insulin sensitivity in peripheral tissues. MSCs can be genetically modified to express a variety of therapeutic factors; therefore they can serve as a promising platform for cell-based microencapsulation. MSCs are known to be hypoimmunogenic. In vitro and in vivo studies revealed that encapsulated human MSCs showed a significant decrease in cytokine expression, compared to other encapsulated stem cells, because they do not express certain surface receptors involved in immune responses [46]. Encapsulated hMSCs were shown to maintain their mesenchymal surface markers and can differentiate to all typical mesoderm lineages. However, MSCs lack the high proliferative potential that ESC's and iPSCs ensure [47].

Progenitor cells (*PC*) are adult stem cells that are generally able to differentiate, but are limited to the cell types of their resident organ. They are still in the process of being defined for both the liver and the pancreas. Generally, use of ESC or iPSC involves a differentiation step to committed liver progenitor (LPC) or pancreas progenitor cells (PPC) before they can be implanted. Terminal differentiation to the desired mature cell type then takes place in vivo. Figures 1 and 2 show a schematic overview of stem cells used in encapsulation therapies for the liver and pancreas, as well as their differentiation potential.

In summary, disadvantages of iPSCs and ESCs include ethical issues and possible mutations. However, advantages of these stem cell therapies are the strong proliferative potential of iPSCs and ESCs, as well as their capacity to differentiate into any desired cell [47]. On the other hand, MSCs are widely available and have excellent hypoimmunogenicity to keep immune responses low. PCs are specific to

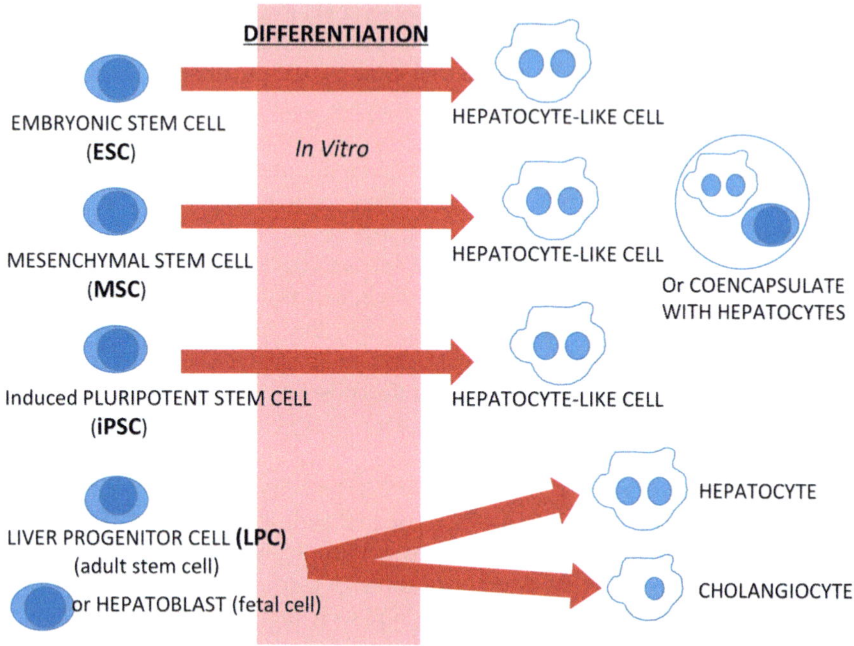

Fig. 1 Stem Cell Differentiation in the Liver. Embryonic Stem Cells (ESCs), Mesenchymal Stem Cells (MSCs), and induced Pluripotent Stem Cells (iPSCs) can differentiate to Hepato-Like Cells in vitro. Co-encapsulation with Hepatocytes enhances the effects of stem cell regenerative therapies. Liver Progenitor Cells (LPCs) and Hepatoblasts are bipotent. LPSs are generally a differentiation step from stem cells to committed organ-specific cells. Hepatoblasts are human fetal cells from early gestation and are extremely limited as a cell source

their organ of origin. The choice of the cell type used in encapsulation devices is affected by the clinical application and the desired therapeutic outcome. The beneficial effects of encapsulated stem cell therapies may not be restricted to cell restoration, but stem cells may also provide therapeutic use due to their delivery of proteins in vivo, as they actively contribute to their environment by secreting cytokines, growth factors and extracellular matrix (ECM) molecules that act either on themselves (autocrine actions), or on neighboring cells (paracrine actions) [48].

3.2 *Types of Cell Encapsulation*

Many approaches have been made to cure disease by implantation of whole donor organs, organ parts, or functional units of organs (organoids), down to single cell implantations. Depending on the desired function of the implant, the implant size, the choice of implanted cells, and the encapsulation material, may vary.

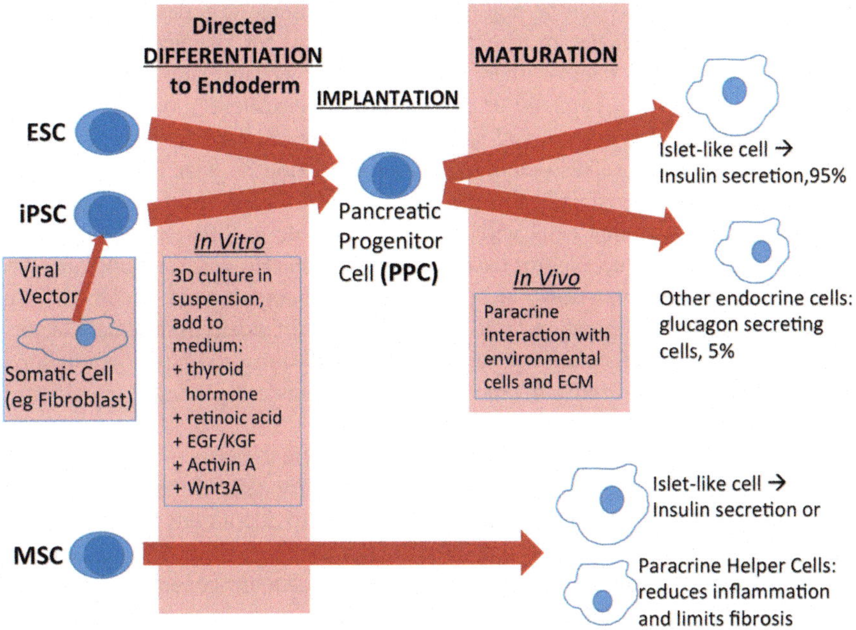

Fig. 2 Stem Cell Differentiation in the Pancreas. Embryonic Stem Cells (ESCs) or induced Pluripotent Stem Cells (iPSCs), after they were transduced with a viral vector, can differentiate to Pancreatic Progenitor Cells (PPCs) in vitro. PPCs then mature into endocrine islet-like cells in vivo under certain circumstances. Mesenchymal Stem Cells (MSCs) can develop into paracrine helper cells if encapsulated

The choice of cell type used in an implantation device depends on many factors, such as the availability of cells or clinical need. Generally, there are 3 types of cells used for cellular therapy: *autologous* cells, *allogeneic* cells, and *xenogeneic* cells [49]. *Autologous* cells are derived from the same patient and eliminate any immune reactions. However, depending on the nature of the specific cells, these may be limited in availability since a lack of healthy cells might cause the disease in the first place. New approaches, such as developing cellular therapies from frozen umbilical stem cells, collected at birth, may fill the gap of healthy autologous cells in the future. *Allogeneic* cells are derived from another donor and are controversial due to the transmission of viral diseases and immunoreaction concerns [50]. *Xenogeneic* cells are derived from another species and are known to trigger strong immunoreactions due to inter-species differences [51]. However, transgenic pig islet cells have been used in experiments to treat Diabetes Mellitus [52–54]. Though xenogeneic cells are easily accessible, the danger of viral disease transmission still remains, as the example of porcine endogenous retrovirus infections has shown [55]. In the age of gene manipulation, genetically engineered cells, either from allogeneic or xenogeneic origin, have become promising encapsulation cells [56].

Encapsulated stem cells can be engineered to differentiate in a desired manner, to secrete a specific substance, or to express fluorophores for visualization in vivo.

Another choice in cellular therapies is the cell encapsulation method. Cell encapsulations can be *permanent* or *degradable*. If permanent, most likely the cell is being implanted for purposes of secretion of some required factor, e.g. insulin. The implanted cells are restricted by its encapsulation so that cell-cell contact cannot be made with host cells. On the other hand, if the encapsulation material is degradable, the implanted cells will eventually be able to differentiate and incorporate with its host tissue, e.g. for tissue repair therapies. However, all implants face challenges of immunoreactions, mechanical damage, containment of cells, control of growth and differentiation, and mismatch of microenvironmental factors. Numerous engineering techniques of the encapsulation method have been tailored to help achieve the desired outcome. These techniques are generally classified as *microencapsulation* and *macroencapsulation* [57].

Microencapsulation usually refers to small spherical capsules or conformally coated vehicles of single cells or small cell clusters. The size of microcapsules can vary from 0.3 to 1.5 mm. The small capsule sizes provide an advantage for mass transport considerations due to the large surface-area-to-volume ratio, and the spherical form is a relative stable shape. Microcapsules are generally highly biocompatible as a consequence of hydrophilic properties of the encapsulating material [57]. It has been shown that smaller capsule sizes seem to trigger less cellular immune response in foreign hosts [58]. Traditional microcapsules contain cells in a hydrogel core surrounded by a membranous structure for stability. Microbeads without a distinct membrane have been successfully used for certain applications. For example, the encapsulation of three-dimensional aggregates of pluripotent stem cells in soft alginate micro-beads (so called embryoid-body cells) have shown promising results in regenerative therapies for liver cell re-growth [59].

Macroencapsulation devices deal with larger living cell populations or part of tissues that are physically isolated from directly interacting with the surrounding host tissue. The implanted cells are typically encapsulated by a 2-dimensional membrane of natural or synthesized material that varies in structural, functional, and mechanical properties [57]. While the microenvironment inside the encapsulated compartment can be varied, these devices still rely on the host's homeostatic mechanisms for control of pH, metabolic waste removal, and nutrients (extravascular supply). Therefore, the association with the host's vasculature plays an important role in the success of implanting macroencapsulation devices. To sustain the viability of the encapsulated cell mass, extravascular devices rely on the formation of new blood vessels (angiogenesis) at the host-device interface following implantation. Angiogenic stimulation in the host, applied in combination with the use of encapsulated stem cell populations, have made it possible to keep functional cell organoids alive for prolonged periods of time following implantation into various body cavities or tissues. While preventing direct contact with host cells, the encapsulation material can also be engineered to minimize adhesion properties on the implant's surface, or to selectively filter certain sizes of molecules for transport through the encapsulation layer. By engineering specific properties of the encapsulation material (e.g. chemical affinity to certain molecules or pore size), immune responses to the implant can be minimized, while at the same time, angiogenesis or tissue regrowth through paracrine factors can be stimulated around the implant site. An example of a macroencapsulation device using cellular therapy for treating Diabetes Mellitus is discussed in the pancreas section (part 5) of this book chapter.

3.3 Therapeutic Reasons for Encapsulation of Stem Cells

(a) Immunoprotection and Paracrine Effects

The foremost reason for cell encapsulation is immunoprotection, designed to isolate the implanted cells from attacks by host immune cells. The encapsulation layer prevents the direct cell-cell contact of implanted cells with host cells. However, paracrine effects, as a form of neighboring cell-to-cell communication, still need to be transmitted through the encapsulation layer in order to alter the behavior of the cells inside the encapsulation layer, as well as cells of the surrounding host tissue. With careful design, this effect can be therapeutically used to induce changes in cells, either on the encapsulated cells to secrete certain substances, or on host cells surrounding the implant to stimulate re-growth of tissue or new vascularization (angiogenesis). In order to allow transport of oxygen, water, and necessary nutrients to enter, and metabolic waste products to exit the encapsulation chamber, the encapsulation layer must be semipermeable [60]. However, circulating larger host molecules (e.g. certain immunoglobins like IgM) or immune cells can be prevented from contact with the cells inside the implantation chamber [61]. Figure 3 shows a model of necessary barrier properties and molecules traveling in and out of an encapsulated chamber containing implanted cells.

It is essential that the encapsulation layer controls the rate of therapeutic products exiting the encapsulation chamber. It has been shown that different cell types have unique metabolic requirements; therefore the encapsulation layer has to be optimized for each cell type according to pore size or other membrane properties [57].

Furthermore, if a specific secretion product from implanted cells is desired (e.g. insulin), the secreted protein must be able to enter the blood system of the host and be distributed throughout the host's body quickly. Therefore, blood vessel growth towards the implant is essential to provide short diffusion distances necessary for secreted proteins to enter the blood system of the host. Growth of new vascularization surrounding the cellular implant is initiated by paracrine effects of angiogenic growth factors, either produced by the implanted cells, or as a result of

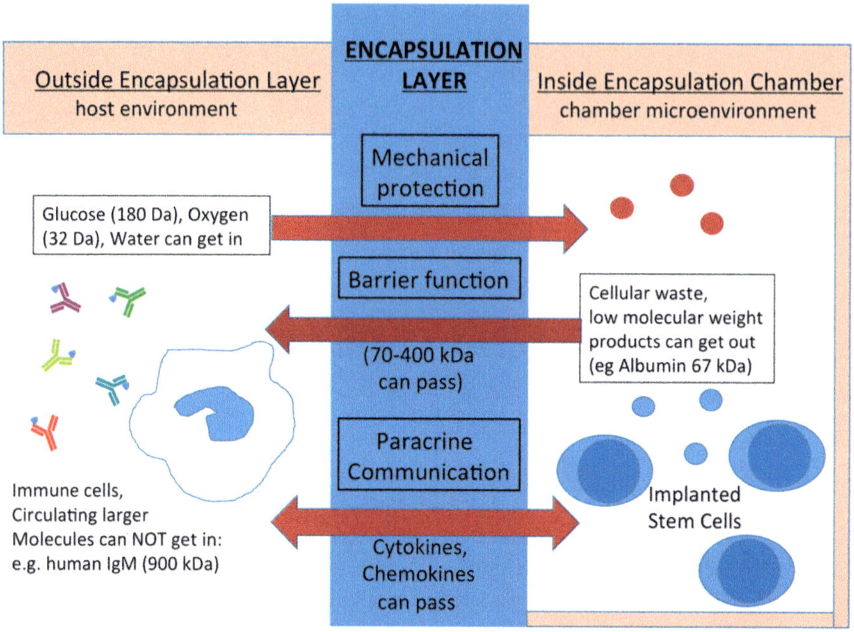

Fig. 3 Barrier Functions of the Encapsulation Layer. The semipermeable membrane of the device encapsulation layer controls the traffic in and out of the encapsulation chamber. Necessary nutrients for cell survival, like glucose, oxygen, and water, as well as cellular waste products are small enough to pass, whereas immune response cells of the host are prevented from immunogenic sampling of the implanted cells due to their larger size. Paracrine communication between the encapsulated cells and the host is still attained through the layer

engineered properties of the encapsulation layer [62]. An example of angiogenic growth around a cellular implantation device post explantation is shown in the pancreas section of this book chapter (part 5).

(b) Microenvironment

As the significance of the microenvironmental role in cellular survival has become more obvious in recent years of research, designing specific characteristics of the environment around stem cells has become an essential tool in the engineering of encapsulation devices. Since the extracellular matrix (ECM) around encapsulated cells in the chamber can be engineered in regard to the nature of the biomimetic scaffold, it's architecture, and the presence of growth factors, as well as recognition motifs, implanted cells can be enabled to exhibit specific functionality in vitro and in vivo [63]. Furthermore, encapsulation material can be manipulated to replicate cell-cell interactions, cell-matrix interactions, and chemical gradients. Approaches to engineering the cellular microenvironment for stem cells have been explored by (1) providing a 3D environment for cell growth, (2) co-culturing of cells (usually with non-parenchymal cells), (3) using cellular genetic modifications, (4) modifying the ECM in encapsulation chambers, and (5) adding growth factors and other signaling moieties [63–65].

The ECM represents an essential player in the stem cell niche environment, since it directly influences the maintenance, proliferation, self-renewal, and differentiation of stem cells. During human embryonic development, the ECM acts on embryonic stem cells to alter their gene expression and induce their proliferation or differentiation needed for the development of tissue. In adult humans, adult stem cells are maintained in a quiescent state until the surrounding microenvironment actively signals the need for self-renewal or differentiation to form new tissues [66].

The physiochemical composition of the ECM is critical to the regulation of stem cell fate; this includes paracrine factors, cytokines, chemokines, glycoproteins, and oxygen tension. These factors will affect important properties, such as mechanotransduction, the effect of mechanical forces on the cell. Furthermore, the ECM can promote interactions between stem cells, neighboring differentiated cells, and/or adhesion molecules [67]. To mimic certain conditions of the ECM in encapsulation devices, biophysiochemical factors, such as composition, shape, topography, stiffness, and mechanical strength, can be modified to control stem cell behavior. Various studies involving different types of scaffolds that regulate stem cell fate by mimicking and altering ECM properties have been done [68]. The complex interactions between stem cells and their niche creates the dynamic system necessary for repairing tissues, and for the ultimate design of stem-cell therapeutics [69].

The main advantage of implanting encapsulated stem cells is that cues, not only from the direct microenvironment around the stem cells, but also from the host tissue and/or the encapsulation material, can serve to direct the differentiation of the stem cells into precisely the kind of cells desired, in order to optimize therapeutic effect. Cues can be designed to work either to maintain proliferation and stem cell phenotype, or to direct differentiation to mature cells. Cues can be added in vitro, before cells are implanted, incorporated during the encapsulation process, or initiated by the implant microenvironment. For example, ways to control behavior-initiating signals to stem cells are enhanced by 3D cell culturing, cell-cell signaling as in co-cultures, use of stimulants from the ECM, by growth and diffusible factors, and by genetic engineering [63]. Therefore, engineering tools, like cell manipulation and the design of encapsulation material, are critical in determining cell fate of stem cells living in a foreign host.

Another important factor to consider is the geometry of the implant. Many years of in vitro cell culturing, most commonly performed in a 2D environment in flasks or plates, demonstrated that growth and behavior of cells outside their natural 3-dimensional environment is not ideal in 2D layers. The optimal shape for *microencapsulation* implants is a sphere, due to maximal surface area and to complement diffusion limits of oxygen (about 200 μm through tissue). For *macroencapsulation* approaches, the optimal implant device contains a 3D interior growth chamber to provide a microenvironment that is specific to the needs of the implanted cell population. The effects of 3D cultures and co-cultures seem to be particularly important in regenerative therapies for liver diseases, which is further discussed in the liver section (part 4) of this book chapter.

(c) Cell Containment and Control of Growth and Differentiation

Cell encapsulation naturally provides mechanical protection of cell populations during the implantation process, as well as from contractile forces from adjacent tissues during movement of the host's body, when implanted. Cell therapy devices are usually implanted in body cavities or soft tissue areas; however everyday movement can shift the implants around. Furthermore, cells themselves have the tendency to migrate through tissue and may try to escape from the implantation device with amoeboid movement, driven by motogenic signals from cytokines, or cell-cell interactions of their environment [70]. The mechanisms that govern the movement of eukaryotic cells are still not fully understood because of the dynamic nature of their regulation through biochemical and mechanical interaction that control cell motility [71]. Physical properties of the ECM and the encapsulation layer, such as rigidity, porosity, topography and insolubility, are factors that influence various mobility-related cell functions, like cell division, tissue polarity, and cell migration [72]. Therefore, the encapsulation layer of a cell therapy device needs to be designed to prevent cellular protrusion, in order to prohibit cell escape from the chamber and to keep host immune cells from antigenic sampling of the implanted cells. The encapsulation layer, in combination with microenvironmental factors, needs to protect the encapsulation chamber from cellular depletion, while also preventing over-exuberant stem-cell proliferation [72]. A carefully designed microenvironment, as well as the encapsulation layer itself, both play a specific role in the proliferation rate and overall fate of the implanted cells. Encapsulation material can be engineered in such a way as to either help retain stem cells in an undifferentiated pluripotent state, or to direct their differentiation in a desired manner. Modifications of alginate encapsulation materials, using peptides on the surface of the encapsulation wall, were observed to control the proliferation and rate of differentiation of certain encapsulated cells [73]. Several studies demonstrate that surface modifications containing nano-particles in the cell-encapsulation material allow control over growth and cellular differentiation [74, 75]. Cell-to-cell communication, as initiated in co-encapsulation with different companion cells, is also influential in directing the stem cell's differentiation. A specific example, discussing beneficial versus unwanted cell behavior due to co-culturing of liver cells with MSCs, is discussed in the liver section (part 4) of this book chapter.

3.4 Materials for Cell Encapsulation

Encapsulation material can be either a *naturally occurring* substance or a *synthetic* polymer. Both types can be modified further to improve desired properties of the encapsulation layer. Important properties of encapsulation material include biocompatibility, pore size, material density, stiffness, and material degradability [63]. For manufacturability considerations, the cost and reproducibility, as well as the ease of use in clinical settings, are also essential. Engineering techniques in the design of implantable biomaterials is a large field of research by itself. For the purpose of this chapter, only the most widely used materials for cellular encapsulation therapies will be discussed.

Naturally occurring encapsulation materials include alginate, chitosan, collagen, gelatin, agarose, and combinations of cellulose sulfate. Synthetic materials include PEG (PolyEthyleneGlycol), PLGA (PolyLacticGlycolicAcid), PVA (PolyVinylAlcohol), PU (PolyUrethane), and PTFE (PolyTetraFluoroEthylene); all are polymers with elastic properties and variable biodegradability [76]. For example, PTFE is not biodegradable and, as many synthetic materials, lack recognition motifs in a cell environment, but can be modified by means of chemical engineering and nano-engineering techniques [63, 76]. Most cellular encapsulation approaches use a combination of a natural and a synthesized component to achieve the desired encapsulation properties and cellular responses in vivo. The speed and sterility of the process, and the quality and purity of the encapsulation agents, are critical for success and reproducibility of the product. Use of non-toxic manufacturing procedures contribute further to the viability of the cells, when implanted in humans or animals. Impurities of the encapsulation material may influence cell viability and function, including the rate of production of the therapeutically relevant molecules secreted by the encapsulated cells [77].

Alginate is by far the most common material used for stem cell encapsulation due to its early discovery, excellent biocompatibility, ease of availability, and low cost. Alginate, extracted from seaweed, can either be used by itself, or in combination with a synthetic polymer such as Poly-L-Lysine. Alginate has a high permeability, but is fairly weak in physical strength and stability. Its softness is advantageous for hydrogel-based encapsulation of cells. For some applications, such as the embryoid-body cells, the flexibility of pure alginate is a beneficial characteristic, as small membrane-less microcapsules are filled with 3-dimensional cell clusters [59]. This is known as conformal encapsulation. Figure 4 shows an example of experimental alginate microbeads used to determine optimal cell density for implants. Encapsulated cells are engineered to express fluorescent signals for visualization purposes. The production of uniformly sized microcapsules, and the use of biocompatible cross-linking reagents for alginate capsule stabilization, are critical factors for successful microbead production (see Fig. 4).

Alginate capsules are often coated with Poly-L-Lysine (PLL) in order to improve mechanical strength of the encapsulation. However, PLL lacks biocompatibility [78]. This can be corrected by adding another thin alginate layer on the outside of the PLL-alginate capsule. The triple layer construct is called APA (Alginate-Poly-L-Lysine-Alginate). Unfortunately for use in long-term applications, alginate exhibits issues with low durability over time. The stability of alginate decays relatively fast, which can lead to cracking of the alginate layer and increased porosity of the capsule after a short implantation time. Consequences of a porous alginate layer are the exposure of the encapsulated cells to immune responses of the host, and may result in the failure of a long-term implant [79]. Furthermore, some studies suggest that even ultrapure alginate may contain endotoxins and polyphenols, which could produce an inflammatory response in vivo [80, 81].

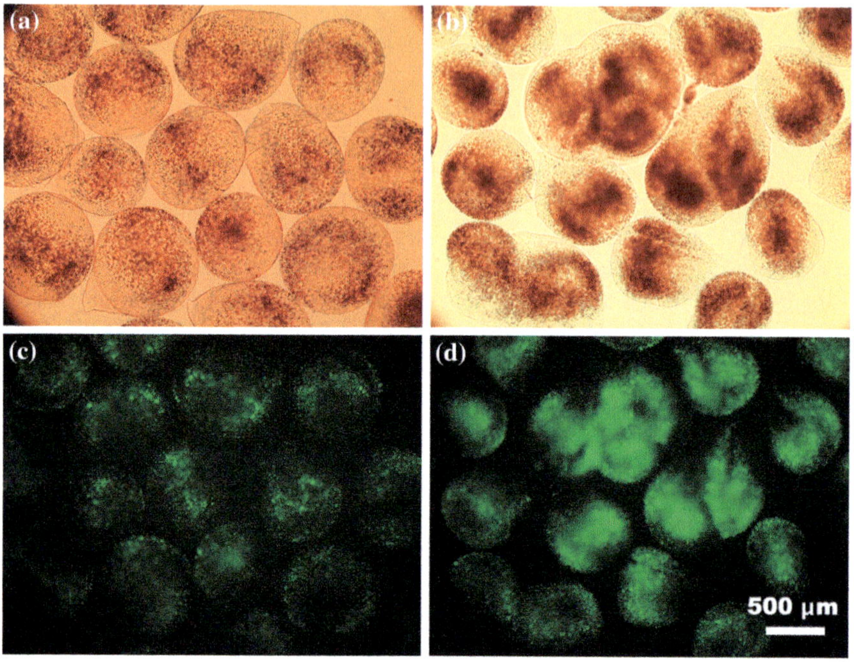

Fig. 4 Alginate Microbead visualization for optimization of cell density in beads. In pre-clinical studies, cells were encapsulated in 2% alginate beads. Encapsulated cells were engineered to express fluorescent signals (eGFP-green), which were used to determine cell count for optimization of cell density in beads. Images **a** and **c** (top row) show bright field, while images B and D (bottom row) show fluorescent images of the same beads. Beads on the left (image **a** and **b**) have a density of approximately 1.5×10^6 cells/ml alginate, while the beads on the right (image **c** and **c**) have a density of about 10×10^6 cells/ml alginate. The production of uniformly sized microcapsules is critical for the success of the microbeads in vivo

Other natural materials for cell encapsulation include *chitosan*, a polysaccharide derived from chitin. It has been used in application for encapsulations of drug delivery [82], and wound dressings [83]. In some cell encapsulations, a combination of alginate-chitosan (AC) was used for cell delivery applications [28, 84]. However, as with other alginate constructs, the encapsulation capsules have limited stability [85, 86]. Therefore, chitosan is more useful in combination with other materials, such as collagen, to form a layer with stronger mechanical properties, as needed in cell encapsulation devices [87].

Collagen, or its denatured form, *gelatin,* is a major protein component of the ECM (extracellular matrix) in tissues. It naturally provides support to tissues and is found in most building blocks of the body, like skin, cartilage, bones, blood vessels, and ligaments. In tissue engineering, collagen is used as a model scaffold due to its perfect biocompatibility, biodegradability, and the ability to promote cell binding [88]. Collagen can be obtained from animal tissues, is commercially available, and is used in many encapsulation applications for drug delivery [89].

Agarose is a polysaccharide derived from seaweed and used for mircroencapsulation of cells in microbeads or suspensions. However, a polysaccharide wall is not suited to keep the cells localized, since they can protrude through the matrix of the encapsulation layer after the formation of the capsules.

Suspensions of *cellulose sulfate*, originally derived from cotton, can be processed to form a semipermeable membrane around suspended cells. Both mammalian cells, as well as bacterial cells, have been shown to remain viable and continue to replicate within a cellulose sulfate capsule membrane [90]. Cellulose sulfate capsules filled with antibody-producing cells have been used for gene therapies and antibody treatments [25].

In the development of cellular macroencapsulation devices for humans, very few materials have been shown to be suitable for long-term implants. According to the FDA, alginate is the only approved material safe for use in humans [91]. Nevertheless, engineered synthetic polymers (PTFE) have been established as an encapsulation layer between implanted insulin-producing cells and host cells for macroencapsulation devices approved for clinical trials in humans. An example of a macroencapsulation device intended as an 'artificial pancreas' is further discussed in the pancreas section (part 5) of this book chapter.

4 Applications of Stem Cell Micro-encapsulation for Liver Disease

4.1 The Need for Liver Cell Therapies

The liver is the largest organ in the body, and the only one with the ability to regenerate from as little as 40% of its original mass. Rat livers, for example, can regenerate two thirds of their mass within a few weeks following acute injury [92]. Even though the liver is one of the organs best able to regenerate, it is also the organ most in demand for transplantation. As a blood filter, the liver is a target of toxic substances, including alcohol and viruses (hepatitis B/C) that compromise its function and ability to regenerate. Metabolic dysfunction, resulting in fat accumulation and aggravated fibrosis in the liver, is also destructive. Repeated adverse events in the liver ultimately result in damage and loss of function [93]. Such damage may also contribute to the development of liver cancer, which is on the rise globally [94]. More than 6000 liver transplants are performed every year. There are currently 17,000 people on the waiting list for a liver in America. The current mean wait time for a liver is 149 days for adults and 86 days for children, and these wait times are increasing. Every year, more than 1500 people die while waiting for a liver (American Liver Foundation, 2016). Therefore, there is a desperate need for livers. A therapy that would allow implantation of encapsulated stem cells to help restore liver function by replacing dysfunctional cells could ultimately mitigate this demand. Stem cell therapy could be used to keep the patient alive while they are waiting for a liver transplant (so called 'bridging'). Advantages of cell micro-transplantation over using an entire organ include: more patients can be treated from a single donor organ; cells can be frozen and used later; surgery is less invasive; the procedure can be repeated, if needed, with low morbidity; and the procedure is less expensive [95].

4.2 Cells for Liver Therapeutics

Currently, primary hepatocytes do not survive or maintain their phenotype in vitro. Hepatocytes embedded in their natural microenvironment in vivo are long-lived and survive up to a year [96], but when removed from their microenvironment and placed in vitro, they lose their phenotype within a day, and perish, usually within a week [97]. Therefore, there is a need for stem cells to be used in liver cell therapy. Stem cells have several potential advantages over primary hepatocytes: they are self-replicating and capable of differentiating into other types of cells, by definition. Stem cells are capable of differentiating into the types of cells that may be required for liver regeneration.

Liver progenitor cells (LPC) have also been called oval cells, liver hepatic stem cells, atypical ductular cells, ductular hepatocytes, and intermediate hepatobililary cells [98]. LPCs are found in human, mice and rat livers [99]. LPCs are quiescent stem cells in the adult liver; they are activated to proliferate during liver injury, and can differentiate to both hepatocytes and cholangiocytes (also known as biliary epithelial cells, BEC), which line the bile ducts. LPCs are thought to be a heterogeneous population of cells in regard to their differentiation capacity and stage of differentiation [96]. These cells originate from the canals of Hering, the interface between hepatocytes and cholangiocytes. Their proliferation and differentiation are regulated by chemokines and cytokines produced by immune cells that are activated at an injury site [100–104]. Nonetheless, in the majority of cases, hepatocytes that assume proliferative capability, and NOT LPCs, are responsible for the bulk of liver regeneration [105, 106]. LPCs can become important during injury, but their role in this process is still not well understood. It is believed that LPCs become active when hepatocyte proliferation is impaired, but even this is being debated [106]. Recently, it was found that hepatocytes can be reprogrammed to LPCs in a mouse model of chronic liver injury [106] and that they can differentiate back to hepatocytes when transplanted into recipient, non-injured mice [107]. Their low frequency can make them difficult to isolate, thereby limiting the supply of LPCs. Additionally, there are concerns about carcinogenesis, because LPCs also occur in hepatocellular carcinoma [108].

Hepatoblasts are bipotent progenitor cells and can differentiate to hepatocytes or cholangiocytes similar to LPCs (see Fig. 1). They are isolated from human fetal livers during early gestation. Hepatoblasts proliferate well in vitro and are able to integrate and proliferate in injured livers of a host. However, due to their source, availability is extremely limited [105].

Non-liver specific stem cells, for possible use in liver therapy, include MSCs, ESCs and iPSCs (see Fig. 1). So far, MSCs, transplanted in a host with liver injury, do not, to any great extent, differentiate to functional hepatocytes (range 0.05–0.23% of implanted cells). To the contrary, the natural ability of MSCs to differentiate into myofibroblasts for tissue repair purposes at an injury site does increase unwanted liver fibrosis in vivo, and is particularly of concern in liver regenerative applications [109]. However, if implanted MSCs are encapsulated, differentiation input from the environment can be more rigorously controlled and stimuli towards fibrogenic differentiation could possibly be limited, leaving only beneficial paracrine effects of MSCs.

ESCs, on the other hand, have been differentiated to hepatocyte-like cells [110–112]. However this process is still limited in scale. Also, iPSCs have shown promise insofar as they differentiated to hepatocytes in vitro [40]. In summary, the identification of suitable cells to be used in encapsulated implantation devices for liver regeneration is still in need of further investigation.

4.3 Micro-encapsulation Designs

To find the best-suited encapsulation material specific for the needs of liver cells, alginate has been explored. Depending on the concentration, alginate is reported to have varying effects on liver cell differentiation. High concentrations maintain stemness and lower concentrations allow differentiation of liver cells [113]. The concentration most used in studies seems to be 2% (weight/volume). Combining alginate with other polymers, such as PLL and PEG, can strengthen the construct. Alginate, itself, is subject to degradation. Cho and Elazar found that viruses could infect encapsulated LPCs through cracks in the coating of alginate beads [114]. On the other hand, use of alginate for liver stem cell encapsulation can influence stem cell differentiation and improve function of mature hepatocytes. When ESCs were encapsulated in alginate microbeads and grown in vitro, the ESCs differentiated towards hepatocyte-like cells, as shown by Fang et al. Hepatocyte function was assessed by secretion of urea and albumin, glycogen storage, and cytochrome P450 transcription factor activity. A cell density of 5 million cells/ml in a 2% alginate concentration was determined to be optimal for stem cells to exhibit liver cell function. After about two weeks in vitro, ESCs differentiated into hepatocyte-like cells, as indicated by albumin secretion and positive staining for glycogen [59].

Some in vivo studies used MSCs for liver regeneration experiments. Meier et al. designed encapsulation material for microbeads that contained a combination of alginate and synthetic PEG for mechanical strength and stability. Beads were made in one step with interactions of calcium alginate molecules and covalent crosslinking from vinyl sulfone-terminated multi-arm PEG molecules [115, 116]. This material combination allowed adjustment of the permeability and swelling of the cell-carrying microbeads, making them more stable. In Meier's experiments, the new material was then injected in the peritoneal cavities of mice, with and without encapsulated MSCs, to treat induced liver fibrosis. Experimental results showed that microcapsules loaded with MSCs had lower inflammatory effects on the mice compared to empty beads (negative control) or beads loaded with human fibroblasts. Furthermore, encapsulated MSCs had positive paracrine effects on the host by secreting interleukin-10 (Il-10), hepatocyte growth factor (HGF), vascular endothelial growth factor (VGF), insulin-like growth factor binding protein (IGFBP), and matrix metalloproteinase (MMP). The in vivo secreted proteins helped remodel the fibrotic tissue and decreased liver fibrosis overall. In comparison, in another study, injected MSCs WITHOUT encapsulation differentiated to collagen-producing fibroblasts after contact with host cells, which actually worsened the *chronic* liver problem by increasing fibrosis [117]. These two findings emphasize the fact that a microencapsulation device loaded with encapsulated cells can exhibit beneficial *paracrine effects*, where as, in the contrary, the same cells can contribute to unwanted fibrosis when not encapsulated. The results indicate that the avoidance of contact between host and implanted cells through an encapsulation layer is absolutely crucial to benefit the host in MSC derived liver regenerative therapies. A careful design of the encapsulation layer as a permeable, non-degradable casing for implanted cells, which will allow paracrine drug delivery in vivo, seems to be the key for implanted cellular drug delivery therapies, particularly for liver diseases.

4.4 Co-encapsulation

The last decade of research in tissue engineering has made it very apparent that for appropriate tissue regeneration, stem cells require two things: companion cells (e.g. stromal cells), and a 3-dimensional geometry of a cell culture growth environment. Cell encapsulation is inherently 3-dimensional and that has been found to improve hepatocyte-like function on its own [118].

The most investigated co-encapsulation for hepatic cellular therapy is MSCs and hepatocytes. Hepatocytes might be the best suited cells to treat acute liver disease, when there is no time for differentiation; while MSCs, with their paracrine ability to secrete beneficial proteins over a relatively long time period in vivo, might be good for treatment of chronic liver ailments. It makes sense that a combination of both, MSCs and hepatocytes encapsulated together, could provide an efficient long-term cellular therapy design for liver regeneration.

Liu and Chang designed in vivo experiments to co-encapsulate MSCs, derived from bone marrow, with primary rat hepatocytes. Microcapsules were loaded with both cell types and implanted in the peritoneum of Wistar rats [79]. Liu and Chang found that co-encapsulation largely increased hepatocyte viability. The ratio of co-encapsulated cells was 2:1 of hepatocytes to MSCs. To ensure immunoisolation by avoiding entrapment of cells in the surface of the microcapsule, a two-step encapsulation method, combining alginate and PLL material, was used. The length of cell survival in vivo and retrievability of viable co-cultures was assessed.

In comparison to prior observations, the enhanced encapsulation method resulted in a significant survival increase of the implants and successful retrieval of more than 90% of encapsulated co-cultures after 1 month of implantation. Furthermore, the two-step encapsulation method produced lasting microcapsules with therapeutically effective co-encapsulated cells eliminating most of the fibrosis and inflammatory responses in the host. Some vital hepatocyte clusters were retained in vivo for up to 3 months, indicating that this method could be useful for bridging patients waiting for a liver transplantation. However, in this study, function of encapsulated hepatocytes was not quantitatively assessed.

In another study, hepatocytes and MSC's were co-encapsulated 2:1 and implanted in the peritoneum of rats, in which *acute* liver failure had been chemically initiated [65]. The co-encapsulated cells dramatically improved survival rate and liver function in the treated rats. PLL-coated alginate was used to encapsulate the co-cultures, but the core of the microbeads was dissolved using sodium citrate. With this encapsulation design, it was demonstrated that MSCs were able to minimize inflammatory and fibrotic processes after acute liver failure through their paracrine effects, while hepatocytes could help replace lost liver tissue. MSC/ hepatocyte treated rats had a 60% survival rate after one week, compared to 30% for rats treated with encapsulated hepatocytes only, and to 10% for untreated rats. Furthermore, less damage was done to the livers of rats treated with the co-encapsulated cells, as assessed by liver enzymes, plasma ammonia, and total bilirubin measurements. The improvement in survival rates in rats with acute liver injury indicates that treatment with co-encapsulated MSCs and hepatocytes holds potential for a cellular encapsulation therapy after acute liver damage.

4.5 The Future of Cellular Liver Regenerative Therapies

Stem cell encapsulation therapies for liver diseases hold great potential for both chronic and acute liver injuries. The use of stem cells can overcome disadvantages of hepatocytes, such as limited availability and inability to maintain proliferation and function in vitro. The microencapsulation of implanted stem cells for liver applications needs to be carefully designed, in order to direct differentiation to hepatocyte-like cells that benefit the host via paracrine effects while avoiding cell-to-cell contact with host cells. Co-culture of adult liver cells with stem cells provides a strong advantage for hepatocyte survival rate in vivo. Microencapsulated cellular implants have effectively increased survival rate after liver damage in rodent models. For *short-term* solutions, implants of encapsulated co-cultures of hepatocytes and MSCs carry the potential as a bridge therapy to liver transplantation. However many problems still need to be resolved before stem cell encapsulation for liver disease treatment is clinically feasible.

A major bottleneck in stem cell therapy is reproducibility. In vitro systems need to be controllable and scalable [119]. Protocols for stem cells must be robust and completely efficient. Cellular signaling must be better understood in order to enable optimal differentiation of stem cells into functional hepatocytes and maintenance of functional hepatocytes in vivo. To date, stem cells have not been differentiated to fully functional hepatocytes, only hepatocyte-like cells [120]. Scale-up is still an issue. Also, differentiation requires expensive growth factors and other supplements. Synthetic compounds that will support these processes would be beneficial in regard to reducing variability and cost. Further development of sophisticated synthetic materials that solve the problems of biocompatibility, localization, and differentiation of stem cells will be a major step in the advancement of therapeutic stem cell encapsulation.

5 Application of Stem Cell Macro-encapsulation for Diabetes Mellitus

5.1 *The Need for Islet Cell Therapies*

In 2012, more than 371 million people had diabetes, and it is predicted that by 2030, this number will increase to 552 million people worldwide (International Diabetes Foundation, 2016). Type I diabetics, and about 25% of Type II diabetics, rely on exogenous insulin for survival [121]. In the past 10 years, insulin-dependent diabetes cases have been successfully treated by transplanting donor *Islet of Langerhans* cells in the approach to replace endocrine tissue. However, with only 10,000 organ donors per year, the major obstacle of islet transplantation is the limited human donor supply [122]. Furthermore, in order to avoid transplant rejection, immune responses against the foreign islet tissue in the recipient need to be suppressed. Life-long administration of immunosuppressant medications are known to have harmful side effects and favor the manifestation of other unrelated infections [123]. Together these limitations justify the search for alternative sources of insulin-producing cells. In xenotransplantation approaches, swine islet cells were used for implantation, since swine pancreases were readily available, and the human and pig insulin differ only by one amino acid [53]. However, even with immunosuppressant medications, the swine cells were rejected by both humoral and cellular immune reactions, due to immunologic species barriers [124].

5.2 *Islet Cell Encapsulation*

In search of immunoprotective methods for islet cell implantation, sophisticated encapsulation techniques for cells are explored as an alternative to immunosuppressants. Initially, therapeutic approaches included encapsulation of small numbers of individual islets coated by semipermeable membranes that were intended to prevent contact between implanted cells and host cells [16]. In subsequent rodent and primate experiments, some microencapsulation approaches were successful in increasing cell survival in vivo [125, 126]. Early-phase human trials with mature porcine islets, embedded in alginate microcapsules and injected intraperitoneally in diabetic patients, were conducted by a New Zealand company named 'Living Cell Technologies' [127]. Despite initial promising achievements in the field of encapsulated islet transplantation, pericapsular fibrosis led to increased failure rates of the implanted microcapsules [128]. Several studies confirmed that the choice of encapsulation material is responsible for the activation of foreign body responses in the host, as certain compositions of alginate stimulates monocytes and macrophages to produce pro-inflammatory cytokines [81, 129, 130]. The development of more refined encapsulation material was crucial to further advance encapsulated islet cell implantation technology [78, 131–133]. Barium crosslinking methods of alginate materials, for example, have been demonstrated to reduce antibody- or cytokine-mediated islet-injury in vivo [19]. Other in vivo experiments revealed complications with the injection site of the microcapsules [123]. For example, APA microcapsules were loaded with 1–3 islets and injected into the liver of diabetic patients via the portal vein [134]. However, the injection of the microcapsules into the portal vein led to complications, such as hypertension and portal thrombosis [135]. Alternative injection sites, like the peritoneum, the kidney capsule, and subcutaneous tissues, were tested in mice and primates [136–138]. Generally, with advances in encapsulation materials, immune and foreign body responses by the host have been reduced, in comparison to un-encapsulated islet cells. However, even encapsulated islets have not been proven viable in vivo for long periods of time. Due to the porosity of the cell-enclosing microcapsules, missing microenvironmental factors, and the choice of the implantation site, only short-term effects of mature islet cell implantations were reported to be successful [139, 140]. Without sufficient survival rates of the implanted islets, the *micro*encapsulation approach of single islets is not ready to be suitable for long-term application in humans.

To provide durability and retrievability of transplanted islet cells as well prevent cell escape, approaches with a larger encapsulation chamber led to the first *macro*encapsulation device with living pancreas islet cells [141]. In order to supply the larger number of implanted cells inside the encapsulation chamber with nutrients, diffusion distances to blood vessels in close proximity of the extravascular device surface need to be kept minimal. Thus, the original designs of a macroencapsulation device were focused on cellular monolayers, encapsulated in a flat-sheet double membrane chamber system that incorporated 2-dimensional monolayers of cells in a planar islet sheet device [142]. Only later, it was established that a 3-dimensional growth chamber does encourage cell growth and survival, as it better mimics the cell's natural environment [143]. Furthermore, the initial choices of encapsulation material (nitrocellulose, crude alginate, acrylonitrile, and agarose) with little biocompatibility or initiators of inflammatory side effects made the early macroencapsulation device less successful. Toxicity of the encapsulation layer and other factors, e.g. device surface irregularities, triggered fibroblast attachment and activated nonspecific foreign body responses, resulting in subsequent necrosis of the encapsulated tissue and device failure [144]. When *adult* islet cells were co-encapsulated with mesenchymal *stem cells* (*MSCs*), fibrotic growth could be limited due to positive anti-fibrotic paracrine effects of encapsulated MSCs, and the macroencapsulation devices exhibited prolonged survival rates in vivo [145, 146].

5.3 Stem Cells for Treatment of Diabetes Mellitus

A more recent development is moving away from the implantation of *mature* islet cells with limited life span. The new focus is on the use of *stem cells* in macroencapsulation devices. However, before macroencapsulated stem cell devices can become a possibility for treatment of diabetic patients, stem cell treatment protocols are in need of being refined and standardized, so that the differentiation of human ESCs or iPSCs into insulin-producing islet-like cells in vitro and in vivo can be repeatedly documented. The pioneering discovery was that human primary pancreatic islet progenitor cells, derived from human embryonic stem cells, will continue to differentiate in vivo into insulin-producing cells inside the encapsulation chamber. Studies showed that glucose-stimulated insulin responses started to increase after a few weeks post-implantation, suggesting that implanted encapsulated primary islet progenitor cells had matured to insulin-producing cells in vivo [147]. In some instances, however, mass transport in and out of the implanted device chamber was compromised by wound healing and fibrotic host reaction processes, i.e. the foreign body response, leading to the shut-down of paracrine communication between cells, and ultimately resulting in insufficient insulin secretion [148]. The device encapsulation layer's chemical and mechanical cues seemed to stimulate fibrotic processes in the host due to its biocompatibility, its architecture, and the oxygen permeability of the material itself [149]. Also, it has been demonstrated that the stiffness of the alginate substrate can affect stem cell differentiation by triggering transcriptional responses [150]. Direct effects of the encasing material on surrounding tissues and implanted cells show the significance of the need to carefully choose the design and material for an encapsulation device, in order to ensure long-term implant survival and to achieve intended differentiation of cells in vivo.

In summary, two major obstacles of encapsulated stem cell transplantation, intended for in vivo insulin production devices, need to be addressed: (1) the stimulation of hESCs or iPSCs to differentiate into insulin-producing cells in vivo, while the inherent ability of hESC/iPSCs to proliferate rapidly in an unpredictable manner is blocked, and (2) the risk of graft failure due to immune response attacks by the host or fibrotic reactions resulting from foreign body and wound repair processes [19]. To provide a successful macroencapsulation device for the treatment of diabetes, device engineering tools have to be optimized to provide a biocompatible, immunoprotective, semi-permeable growth chamber with ideal conditions for stem differentiation in a foreign host, while promoting host vascularization for nutrient exchange and insulin distribution.

5.4 Stem Cell Differentiation

Stem cell engineering refers to the ability to manipulate and control the powerful potential of stem cells to self-renew and/or to differentiate into specialized cells. Particularly, in the manipulation of human stem cells to specialize into insulin-secreting pancreatic cells in long-term implantation devices, detailed sequential steps in the differentiation pathway have to be identified, identically repeated, standardized, controlled, and they have to provide the ability to be scaled–up for manufacturing purposes [151]. The process has to be efficient and homogenous to ensure repeatable functionality of the implantable cells. This can be achieved with standardized protocols that promote precise sequential changes in the cellular microenvironment in vitro and in vivo. Stem cells need to be trained to respond to signals (glucose) and secrete proteins (insulin) required for glucose control. This can be executed by means of specialized cell media with added growth and other factors in vitro and/or via ECM or cell-cell signals around implanted cells in vivo. In order to allow for the desired differentiation steps to occur at specific time points, in vitro and in vivo steps need to be optimized with precision. In 2014 and 2015, three independent research groups developed refined stem cell treatment protocols to create insulin-secreting cells [152–154]. The commonalities of the protocols were used to optimize the cell differentiation protocol [155] for use in cellular macroencapsulation devices, which potentially could function as an 'artificial pancreas' device.

Possible stem cells used in macroencapsulation devices for treatment of diabetes are iPSCs and human ESCs (hESC) (see Fig. 2). IPSCs, originated from human fibroblasts, have shown the ability to differentiate into functional β-cells in vitro, which responded to glucose stimulation [156]. To generate iPSCs, reprogramming of the adult fibroblasts was achieved by the activation of three transcription factors (OCT4, SOX2, KLF4) [34, 156, 157]. Also, adult pancreatic ductal or epithelial cells, and even α-cells, have been genetically reprogrammed to iPSCs using adenoviral vectors and certain combinations of transcription factors [158]. After iPSCs are successfully engineered, the cells are then initiated to differentiate into β-cell-like cells that respond with insulin production upon a glucose challenge [159]. Although approaches to use iPSCs for the treatment of insulin-dependent diabetes are promising, clinical use is not yet feasible due to targeting of reprogramming transcription factors and viral mutagenicity [160]. Non-viral mediated introduction of reprogramming factors however, could enable safe clinical use.

ESCs can be directed to differentiate into functional endocrine pancreatic cells without the use of viral vectors, but instead by means of reproducible differentiation protocols, additions to growth media, selected hydrogels, and co-culturing approaches [161]. However, large populations of original obtained endocrine cells from ESC differentiation were poly-hormonal, meaning single cells are expressing simultaneously somatostatin, glucagon, and insulin. Cells expressing multiple hormone markers within a single cell are found to have poor responses to a glucose challenge in terms of insulin release [160]. Instrumental changes in the development of a protocol that could differentiate hESCs into islet-like cell populations containing single-hormone-producing, insulin-secreting pancreatic endoderm cells, were based on studies of D'Amour et al. [162] and Kroon et al. [163]. The differentiation protocols were subsequently optimized for in vivo implantation by Agulnick et al. in 2015 [155]. All three research groups used a 2-step procedure with differentiation of hESCs to pancreatic progenitor cells (PPC) in vitro, and implantation of the encapsulated PPCs for maturation into islet-like cells in vivo (see Fig. 2). In this protocol, hESCs are initiated on a desired pathway to differentiate into committed pancreatic progenitor cells (PPC) in vitro with 7 distinct steps in the use of growth media, containing Activin A and Wnt3A at different time points. With these steps, complex signaling pathways are suppressed or promoted during the initiation of the differentiation process. Wnt signaling has been shown to promote proliferation of PPCs in the developing mouse, and Activin family signaling has been implicated in pancreatic specification [155]. Unique to this treatment protocol is the combination of adherent and suspension culturing during the in vitro differentiation step. 15 days of adherent cell culturing in flasks is followed by a 1-day aggregation step that is finalized by 12 days of cell differentiation in suspension, allowing 3-dimensional growth of cell clusters. HESCs develop into pancreatic endoderm cells (PPCs) during this step (see Fig. 2). After the differentiation to PPCs in suspension is completed, a 3-day preparation step, using various suspension formulations, prepares the cells for the implantation device. The implant preparation step enables progress to final maturation of the implanted cells, which does take place in vivo, post implantation, and produces insulin-producing islet-like cells inside the implant chamber [151]. Resulting cell populations from hESCs treated with this differentiation protocol exhibited about 80% of matured cells in the implant with endoderm characteristics and many expressed insulin [155, 163]. Main factors required to generate glucose-responsive, insulin-producing cells were culturing in suspension (3D culture), and the temporary activation or deactivation of certain signaling pathways during cell differentiation via substances in the microenvironment of the hHSCs and PPCs [152, 154]. Specifically, the presence of retinoic acid to dampen the sonic hedgehog signaling pathway (a known inhibitor of pancreas development) [164], and the presence of the thyroid hormone tri-iodothyronine (T3), known to be required for liver development [165], were demonstrated to be influential in the cell differentiation steps. It was also found that high retinoic acid levels during the beginning of the hESC differentiation, followed by treatment with specific growth factors (EGF/KGF), reduced the number of poly-hormonal cells [154].

Furthermore, procedures to preserve and store the implantable cells were established, which is necessary to generate a consistent and robust population of implantable cells suitable for cellular therapeutic manufacturing. Experiments to cryopreserve and re-aggregate treated hESCs showed no significant difference in functionality and insulin secretion after implantation in vivo [166].

Itkin-Ansari et al. demonstrated that encapsulated hESC derived islet-PCs acquire glucose-responsiveness in vivo without a significant change in biomass or any evidence of cell escape from the encapsulation chamber [121]. In her experiments, macroencapsulated cellular devices were implanted in mice, and located and imaged with bioluminescent imaging techniques in vivo (Fig. 5). In immunohistochemical images of explanted devices (Fig. 6a), encapsulated matured islet-like cells are visible inside an implantation device (device encapsulation membrane layers appear red due to autofluorescence), and did not reveal an increase in cellular biomass over time. This finding may indicate that the matured PPCs lost the proliferation and self-renewal capacity of hESCs. The finding of constant cellular biomass in a cellular implant device eliminates the general concern with stem cells to uncontrollably self-renew in vivo, possibly leading to overpopulation of implanted stem cells. Furthermore, as visible from immunohistochemistry images of explanted devices (Fig. 6b), the encapsulated cells show a high percentage of insulin-expressing cells (green), as well as a minor amount of glucagon-expressing cells (red). It was noted that no single cells in the graft were positive for both hormones (see Fig. 6b). These experiments showed that hESC derived, implanted, and matured endocrine cells either expressed insulin or glucagon, but not both, confirming that progenitor cells in the device had matured along the desired distinct endocrine lineages [121].

The combination of several recent advances in diabetes-related stem cell research may finally bring bench-top research to the clinic, and reinstates hope to manufacture a therapeutic encapsulation stem cell device, adequate to treat diabetic

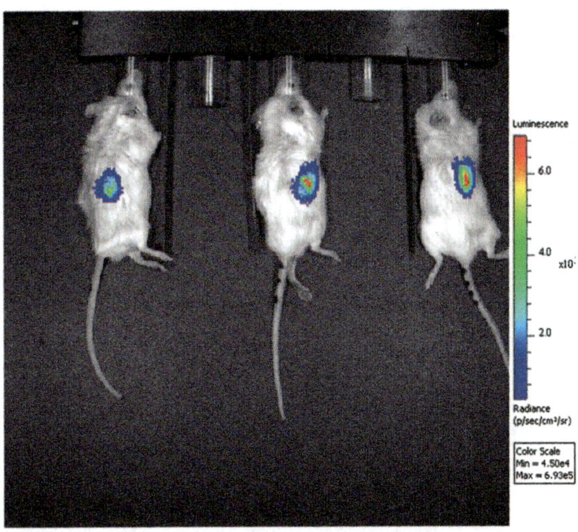

Fig. 5 Whole mice imaging with bioluminescent cellular device implants. *First published by Elsevier* [121]. Mice were implanted with macro-encapsulation devices containing pancreatic progenitor cells transducted with a self-inactivating lentivirus carrying firefly luciferase for visualization. Live, anesthetized, implant-carrying mice were scanned in vivo using an IVIS Spectrum imaging platform (Perkin-Elmer) to quantify luciferase expression of implanted cells. No significant change in biomass or any evidence of cell escape from the encapsulation chamber was detected

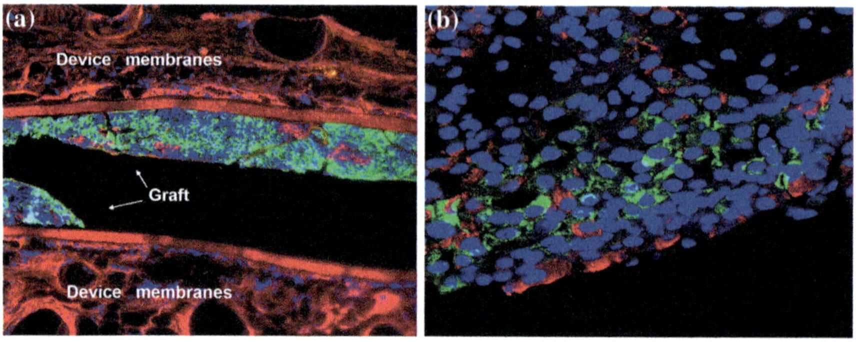

Fig. 6 Immunohistochemical images of a cell graft from an explanted encapsulation chamber in mice experiments. *First published by Elsevier* [121]. Cross-sections through an encapsulated cell graft show matured hormone-producing islet-like cells in vivo. **a**: Device cross-section. A vital cell graft is visible inside the encapsulation chamber (device membrane appears red due to autofluorescence). Vital graft cells are immunostained for insulin (green) and glucagon (red). Cell nuclei appear blue. Human ESC-derived pancreatic cells, encapsulated in the device, secrete human insulin (green) in vivo. Magnification: 100x. **b**: High power view of cellular graft. The cell graft is compromised of a high percentage of insulin positive cells (green) and a smaller amount of glucagon positive cells (red). Cell nuclei appear blue. It was noted that no single cell in the graft was positive for both hormones, confirming that hESC-derived pancreas progenitor cells had matured into distinct endocrine single-hormone-producing pancreatic cells inside the encapsulation device in vivo. Magnification: (1200x)

humans. In these recent experiments, optimized hESC growth and differentiation protocols were used to successfully obtain implantable islet-like cell clusters.

In summary, milestones have been met to enable hESCs to form endoderm cells with subsequent development through pancreatic PCs into mature endocrine cells, capable of synthesizing pancreatic hormones in vivo. Recent refinement in the treatment protocols of hESCs has generated a consistent, transplantable cell population that differentiates in vivo, producing insulin-secreting cells. Furthermore, bioprocessing strategies to enable controlled manufacturing of these cells, have opened the path for possible clinical use in humans.

5.5 *Macroencapsulation-Device Engineering*

Maintaining long-term graft viability, while eliminating the need for immunosuppressive medication, can only be accomplished through encapsulation [167]. The biomaterial used for the encapsulation layers of the device has to provide specific properties that resonate with the implanted cells, as well as protect them from the host, yet the layer must allow transportation of paracrine and endocrine products. Biomaterial research and chemical advances provide continuous opportunities to design, synthesize, test, and optimize desired polymers for encapsulation purposes [168]. Modern encapsulation device engineering for clinical applications in humans includes the testing of biocompatible materials, risk management, concept and feasibility assessment, design controls, and process validation controls. For clinical purposes, the device also needs to be accessible for monitoring with common clinical imaging systems and allow for easy retrieval. With the advancement of sophisticated micro- and nano-manufacturing techniques, it has become possible to 'engineer' a membrane for encapsulation of live cellular grafts with precise morphologies, surface characteristics, and immunoisolation properties [169]. In recent experiments, a synthesized polymer PTFE (PolyTetraFluoroEthylene) was coated with alginate and engineered for an encapsulation device to be implanted under the skin of humans [169]. PTFE has been shown to limit the induction of fibrotic repair processes and to exhibit good vascularization-stimulating qualities, allowing for better cell viability in implants [170, 171]. An important factor for these qualities is the pore size of the engineered PTFE. Smaller pore sizes in the sub-micron range have been demonstrated to provide good immunoprotection for transplanted cells in vivo [121, 172]. Immunological cells, such as macrophages and leukocytes, about 6–10 microns in diameter, cannot pass through pores sized in the sub-micron scale. On the other hand, larger pore sizes in the host-interface layer of PTFE material, starting at about 5 μm, initiated a substantial increase in vascularization on the membrane-tissue surface. This observation remained, even when the larger pore size membrane was laminated to an inner smaller pore size membrane to prevent cell protrusion [169]. Silicon micro-machining allowed the production of the macroencapsulation material with uniform and well-controlled pore sizes, channel lengths, and surface properties [169, 173].

Based on the ability to engineer precise pore sizes in synthetic polymers and the consequential tissue reactions that are triggered by different pore sizes, an American company named 'Theracyte' developed an advanced material for a durable, planar macroencapsulation device. The innovation was an encapsulation membrane consisting of a laminated PTFE bilayer with 2 different pore sizes. The new design of the encapsulation membrane combined a PTFE outer layer with larger pore size (5 μm) and a PTFE inner layer with sub-micron pore size (0.45 μm) [169]. This specialized PTFE material was another step toward an implantation prototype device, combining angiogenic stimulation properties, initiated by larger pores at the surface with host cells, and crucial immunoisolation properties, provided by the smaller pores of the inside layer. Immunoisolation characteristics could be further enhanced by other nano-techniques, such as precise surface alignment of nano-fibers and coating with peptide-modified alginate hydrogels that inhibit inflammatory cell surface receptors [173]. The Theracyte encapsulation device was unbreakable, retrievable, supported neovascularization near the implant, kept implanted cells contained, and avoided host cell contact via the small pore inner membrane layer. In subsequent studies, this device was used for implantation of mature islet cells in pre-clinical rodent experiments [174–176]. One research group reported survival of cell implants in vivo in rats for up to 6 months using the Theracyte macroencapsulation device [177]. However, human clinical trials using mature islet cells, macroencapsulated within the Theracyte device, remained unsuccessful for long-term implantation due to fibrotic host reactions around the implant. Fibroblast attachment on the polyester mesh that was attached to the outer 5 μm encapsulation layer of the device hindered long-term device performance in vivo [170].

The development of new multi-layered encapsulation material becomes even more crucial for implantation of stem cells. Synthetic membranes of macroencapsulation devices for stem cells do not only need to provide immunobarrier properties (see Fig. 3) and neovascularization stimuli, but also have to ensure an environment for differentiation and maturation of implanted stem cells within the encapsulation chamber. The design and material of devices like the Theracyte one, holds great potential for the implantation of stem cells, since oxygenation and effective nutrient and waste exchange through enhanced angiogenesis around the implant was achieved. Neovascularization with ample blood supply to the implanted stem cell mass is absolutely crucial for survival of the implant [149]. Once stem cells mature to insulin-producing endocrine cells in vivo, the distribution of secreted therapeutic substances (e.g. insulin) is also ensured by proximity to the host's vasculature. Furthermore, desired insulin secretion of the implanted cells can be regulated by the host's blood glucose levels, as initiated through paracrine communication from close-by blood vessels through the device encapsulation layer [169].

A biotechnology company in California USA, 'ViaCyte LLC', has pioneered in the area of encapsulated stem cell transplants. ViaCyte has demonstrated the feasibility of encapsulated hESCs using a Theracyte-like device. In Fig. 7, an example of neovascularization around the macroencapsulation device from clinical studies is shown. The cellular implant device is carrying insulin-secreting islet-like cells, which were derived from hESCs, differentiated into islet progenitors prior to implantation, and which matured in vivo. The device was explanted after 18 weeks in vivo and shows sufficient neovascularization on the implant's surface (see Fig. 7).

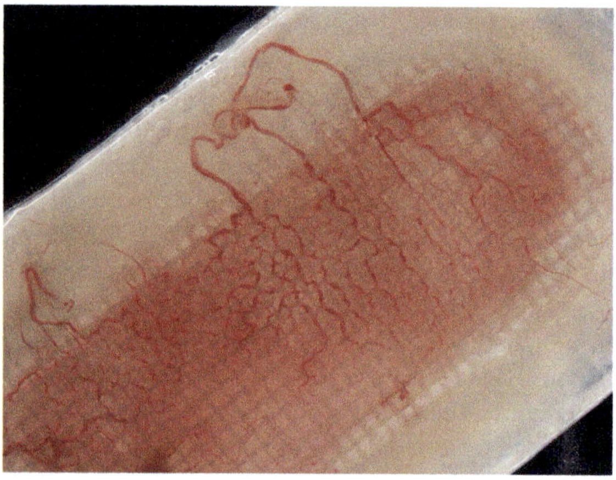

Fig. 7 Neovascularization around a micro-encapsulation device for treatment of Diabetes Mellitus. *Image provided by ViaCyte.* The micro-encapsulation device becomes densely vascularized by the host after 18 weeks of implantation. Vascularization is critical for robust engraftment and function of a cell-carrying device

5.6 *The Synergy Between Stem Cell Therapy and Device Engineering*

As a true Bioengineering approach, combining expertise across disciplines, ranging from nano-engineering, biomaterials, biotechnology, and tissue engineering, into the fields of immunology and clinical medicine, we can begin to address the multiple challenges that are involved in translating encapsulated stem cell therapy from the laboratory to the clinic. Recent improvements in cellular graft viability, due to encapsulation methods with improved biomaterial manufacturing and purification techniques, as well as the identification of safe, reliable, scalable tissue sources with refinement in stem cell isolation and culturing techniques, may make it possible to introduce a successful cellular implantation device into the clinic [19].

With the hope of providing prolonged periods of insulin independence for diabetic patients, recent research advances have utilized many modern technologies to develop a cellular macroencapsulation device, using un-differentiated human ESCs to replace dysfunctional islet cells in diabetic patients. It has been shown that encapsulated hESCs can be successfully directed down an endocrine lineage, implanted without direct contact to the host environment, and that the hESC derived differentiated cells can deliver optimal insulin production after 3–5 months of maturation in vivo, as determined by improved glucose-stimulated insulin responses and C-peptide levels [163]. This finding, as similar to previous findings with primary human islet progenitor cells, suggests that stem cell differentiation continued after encapsulation and reached its optimum performance in insulin delivery after 3–5 months of implantation [147]. The example of the San Diego Biotechnology company ViaCyte demonstrates how stem cell therapies and device engineering are utilized to develop an implantable stem cell macroencapsulation device that shows promise as an in vivo cellular therapeutic tool to treat Diabetes Mellitus diseases. In Fig. 8, a histological image of a cross-section through a ViaCyte encapsulation device, after 18 weeks of implantation in a clinical trial, is shown. In the cross-section through the device chamber, encapsulated islet-like cells show viability (see Fig. 8a H&E histological staining) and development of insulin-producing cells in vivo (see Fig. 8b immunohistochemical stain for insulin, insulin appears red).

Starting with the original idea of implanting stem cells in a patient that then differentiate into therapeutic cells and deliver insulin in vivo, many challenges have been successfully addressed with bioengineering techniques and an engineered encapsulation device that provides immunoisolation from host cells, while allowing paracrine information, as well as nutrients, to pass. Successful preclinical results with a stem cell implantation device that could produce and deliver insulin in vivo, have lead to clinical trials. ViaCyte is currently carrying out a 3-year, human phase I/II clinical trial to assess safety and efficacy of the implant system, using stem cell-derived cell sources for an encapsulated cell replacement therapy in Diabetes

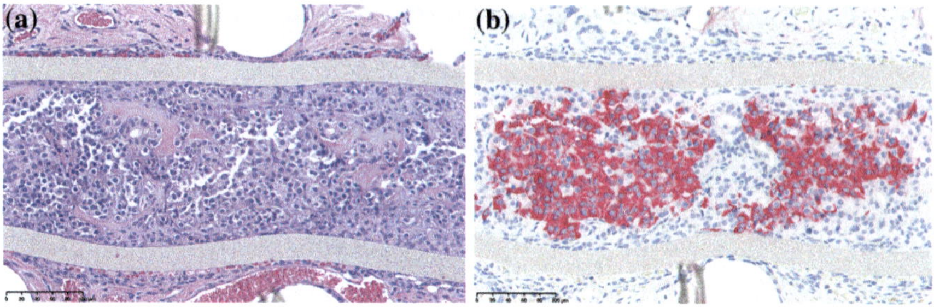

Fig. 8 Cross-section through an encapsulation implant. *Image provided by ViaCyte.* Histological images of cross-sections through a cell-carrying encapsulation device show vital, hormone-producing cells inside the encapsulation chamber (gray layers on top and bottom) after 18 weeks of implantation in clinical trials. **a**: H&E stain. Implanted encapsulated multipotent pancreatic progenitor cells matured into vital hormone-producing pancreatic cell types in vivo. **b**: Immunohistochemical stain for insulin (red) illustrates the differentiation into functional insulin-producing pancreatic cells

Mellitus [178]. Human subjects are Type I Diabetes Mellitus patients. The device is implanted subcutaneous and can be easily removed. Looking into the future, ViaCyte is also developing a platform for cryopreserving undifferentiated hESCs to provide the necessary foundation for scale-up and manufacturing needs of the cellular implantation device, if clinical trials prove to be successful in long-term settings.

In summary, many challenges were faced to develop an encapsulated stem cell device to treat Diabetes Mellitus. Issues like host vascularization and immune rejection of the implant seem to have been successfully addressed. However, long-term stability of cellular implants seems to be the bottleneck. Even if initial wound healing host reaction to an implant is successfully suppressed, long-term host tissue reconstruction via tissue repair mechanisms may trigger scar tissue and fibrotic host responses around the implant over time, which could negatively influence the long-term implant performance. To create a therapeutic cellular implant device that continuously provides insulin-independence for diabetic patients, long-term interface reactions between host and implant still need to be further investigated. Nevertheless, in the clinic, a cellular implantation device using encapsulated stem cell therapy for insulin-dependent Diabetes Mellitus patients has never been closer.

6 Conclusion

The synergy of stem cell therapy and medical device engineering for liver disease and Diabetes Mellitus has led to great progress in the development of encapsulation techniques with stem cells for human implantation therapies. The first clinical trials are underway. However, diabetic human subjects will not be able to benefit from these achievements immediately, as reproducible results from multi-centered, multidisciplinary, randomized, controlled clinical trials will be required to establish treatment modalities for patients with diabetes. Nevertheless, the hope for life-quality improving treatments and the elimination of the need for daily insulin injections is a driving force in research and clinical therapy. The life-threatening shortage of human donor livers and pancreases is rationale to push forward with stem cells therapies and micro-and macroencapsulation techniques for therapeutic applications in humans. The idea of a bio-artificial pancreas or a cell-based liver replacement therapy has gained strong interest in recent research and therapy through encapsulation devices. Due to new bioengineering advancements and the better understanding of stem cell biology, the synergy between stem cell therapy and device engineering may provide promising options for new treatments of human disease in the future.

Acknowledgements We thank Dr. Kevin D'Amour for providing images from pre-clinical and clinical work at ViaCyte, San Diego, USA.

References

1. Everhart, J. E., & Ruhl, C. E. (2009). Burden of digestive diseases in the United States part III: Liver, biliary tract, and pancreas. *Gastroenterology, 136*(4), 1134–1144.
2. Yadav, D., & Lowenfels, A. B. (2013). The epidemiology of pancreatitis and pancreatic cancer. *Gastroenterology, 144*(6), 1252–1261.
3. Congress, I. L. (2016). Fast facts about liver disease—The international Liver Congress 2016, April 13–17. The Home of Hepatology. http://2016.ilc-congress.eu/wp-content/uploads/2016/04/Liver-disease-backgrounder.pdf.
4. Blachier, M., Leleu, H., Peck-Radosavljevic, M., Valla, D. C., & Roudot-Thoraval, F. (2013). The burden of liver disease in Europe: A review of available epidemiological data. *Journal of hepatology, 58*(3), 593–608.
5. Lévy, P., Domínguez-Muñoz, E., Imrie, C., Löhr, M., & Maisonneuve, P. (2014) Epidemiology of chronic pancreatitis: burden of the disease and consequences. *United European Gastroenterology Journal, 2*(5), 345–354.
6. Peery, A. F., Crockett, S. D., Barritt, A. S., Dellon, E. S., Eluri, S., Gangarosa, L. M., … Schmidt, M. (2015). Burden of Gastrointestinal, Liver, and Pancreatic Diseases in the United States. *Gastroenterology, 149*(7), 1731–1741 e3.
7. Zaret, K. S. (2008). Genetic programming of liver and pancreas progenitors: Lessons for stem-cell differentiation. *Nature Reviews Genetics, 9*(5), 329–340.
8. Rous, P., & Murphy, J. B. (1912). The histological signs of resistance to a transmissible sarcoma of the fowl. *Journal of Experimental Medicine, 15*(3), 270–286.
9. Bisceglie, V. (1933). Uber die antineoplastische Immunitat: (I) Heterologe Einpflanzung von Tumoren in Huhnerembryonen. *Zeitschrift für Krebsforschung, 40,* 122–140.
10. Bisceglie, V. (1933). Uber die antineoplastische Immunität; (II) Die Wachstumfahigkeit von heterologen Geschwulsten in erwachsenen Tieren nach Einpflanzung in Kollodiumsackchen. *Zeitschrift für Krebsforschung, 40,* 141–158.
11. Chang, T. M. (1964). Semipermeable microcapsules. *Science, 146*(3643), 524–525.

12. Federlin, K., & Raptis, S. (2001). Ernst F. Pfeiffer: His impact on diabetes research and treatment over the last 45 years. *Experimental and Clinical Endocrinology & Diabetes, 109* (2), S83–S85.
13. Bowen, K. M., Andrus, L., & Lafferty, K. J. (1980). Successful allotransplantation of mouse pancreatic islets to nonimmunosuppressed recipients. *Diabetes, 29*(1), 98–104.
14. Lafferty, K. J., Andrus, L., & Prowse, S. J. (1980). Role of lymphokine and antigen in the control of specific T cell responses. *Immunological Reviews, 51,* 279–314.
15. Chick, W. L., Perna, J. J., Lauris, V., Low, D., Galletti, P. M., Panol, G., et al. (1977). Artificial pancreas using living beta cells: Effects on glucose homeostasis in diabetic rats. *Science, 197*(4305), 4780–4782.
16. Lim, F., & Sun, A. M. (1980). Microencapsulated islets as bioartificial endocrine pancreas. *Science, 210*(4472), 908–910.
17. Cima, L. G., Vacanti, J. P., Vacanti, C., Ingber, D., Mooney, D., & Langer, R. (1991). Tissue engineering by cell transplantation using degradable polymer substrates. *Journal of Biomechanical Engineering, 113*(2), 143–151.
18. https://ki.mit.edu/people/faculty/langer. *Robert S. Langer*. 2017.
19. Krishnan, R., Alexander, M., Robles, L., Foster III, C. E., & Lakey, J. R. (2014). Islet and stem cell encapsulation for clinical transplantation. *The Review of Diabetic Studies, 11*(1), 84–101.
20. Soon-Shiong, P., Heintz, R. E., Merideth, N., Yao, Q. X., Yao, Z., Zheng, T. I. A. N. L. I., … Mendez, R. (1994). Insulin independence in a type 1 diabetic patient after encapsulated islet transplantation. *Lancet, 343*(8903), 950–951.
21. Calafiore, R., Basta, G., Luca, G., Lemmi, A., Montanucci, M. P., Calabrese, G., … Brunetti, P. (2006). Microencapsulated pancreatic islet allografts into nonimmunosuppressed patients with type 1 diabetes: first two cases. *Diabetes Care, 29*(1), 137–138.
22. Löhr, M., Bago, Z. T., Bergmeister, H., Ceijna, M., Freund, M., Gelbmann, W., … Henninger, W. (1999). Cell therapy using microencapsulated 293 cells transfected with a gene construct expressing CYP2B1, an ifosfamide converting enzyme, instilled intra-arterially in patients with advanced-stage pancreatic carcinoma: a phase I/II study. *Journal of Molecular Medicine, 77*(4), 393–398.
23. Löhr, M., Hoffmeyer, A., Kröger, J. C., Freund, M., Hain, J., Holle, A., … Nizze, H. (2001). Microencapsulated cell-mediated treatment of inoperable pancreatic carcinoma. *Lancet, 357* (9268), 1591–1592.
24. Löhr, M., Kroger, J. C., Hoffmeyer, A., Freund, M., Hain, J., Holle, A., … Saller, R. (2003). Safety, feasibility and clinical benefit of localized chemotherapy using microencapsulated cells for inoperable pancreatic carcinoma in a phase I/II trial. *Cancer Therapy, 1,* 121–131.
25. Pelegrin, M., Marin, M., Noel, D., Del Rio, M., Saller, R., Stange, J., … Piechaczyk, M. (1998). Systemic long-term delivery of antibodies in immunocompetent animals using cellulose sulphate capsules containing antibody-producing cells. *Gene Therapy, 5*(6), 828–834.
26. Hartgerink, J. D., Beniash, E., & Stupp, S. I. (2001). Self-assembly and mineralization of peptide-amphiphile nanofibers. *Science, 294*(5547), 1684–1688.
27. Kobayashi, T., Aomatsu, Y., Iwata, H., Kin, T., Kanehiro, H., Hisanaga, M., … Nakajima, Y. (2003). Indefinite islet protection from autoimmune destruction in nonobese diabetic mice by agarose microencapsulation without immunosuppression. *Transplantation, 75*(5), 619–625.
28. Haque, T., Chen, H., Ouyang, W., Martoni, C., Lawuyi, B., Urbanska, A. M., & Prakash, S. (2005). In vitro study of alginate-chitosan microcapsules: an alternative to liver cell transplants for the treatment of liver failure. *Biotechnology Letters, 27*(5), 317–322.
29. Young, S., Wong, M., Tabata, Y., & Mikos, A. G. (2005). Gelatin as a delivery vehicle for the controlled release of bioactive molecules. *Journal of Controlled Release, 109*(1–3), 256–274.
30. Dong, H., Fahmy, T. M., Metcalfe, S. M., Morton, S. L., Dong, X., Inverardi, L., … Wang, H. (2012). Immuno-isolation of pancreatic islet allografts using pegylated nanotherapy leads

to long-term normoglycemia in full MHC mismatch recipient mice. *PLoS One*, *7*(12), e50265.
31. Kim, A. R., Hwang, J. H., Kim, H. M., Kim, H. N., Song, J. E., Yang, Y. I., … Khang, G. (2013). Reduction of inflammatory reaction in the use of purified alginate microcapsules. *Journal of Biomaterials Science, Polymer Edition*, *24*(9), 1084–1098.
32. Hueck, I. S., Haas, M., Finones, R., Frimodig, J., & Gough, D. A. (2011). The potential of selectively cultured adult stem cells re-implanted in tissues. In Artmann (Ed.), *Stem cell engineering: Principles and applications* (pp. 79–117). Berlin: Springer.
33. Seale, N. M., & Varghese, S. (2016). Biomaterials for pluripotent stem cell engineering: from fate determination to vascularization. *Journal of Materials Chemistry B*, *4*(20), 3454–3463.
34. Takahashi, K., Tanabe, K., Ohnuki, M., Narita, M., Ichisaka, T., Tomoda, K., & Yamanaka, S. (2007). Induction of pluripotent stem cells from adult human fibroblasts by defined factors. *Cell*, *131*(5), 861–872.
35. Aasen, T., Raya, A., Barrero, M. J., Garreta, E., Consiglio, A., Gonzalez, F., … Edel, M. (2008). Efficient and rapid generation of induced pluripotent stem cells from human keratinocytes. *Nature Biotechnology*, *26*(11), 1276–1284.
36. Loh, Y. H., Hartung, O., Li, H., Guo, C., Sahalie, J. M., Manos, P. D., … Lensch, M. W. (2010). Reprogramming of T cells from human peripheral blood. *Cell Stem Cell*, *7*(1), 15–19.
37. Zhou, T., Benda, C., Dunzinger, S., Huang, Y., Ho, J. C., Yang, J., … Bao, X. (2012). Generation of human induced pluripotent stem cells from urine samples. *Nature Protocols*, *7* (12), 2080–2089.
38. Selvaraj, V., Plane, J. M., Williams, A. J., & Deng, W. (2010). Switching cell fate: the remarkable rise of induced pluripotent stem cells and lineage reprogramming technologies. *Trends in Biotechnology*, *28*(4), 214–223.
39. Liu, H., Ye, Z., Kim, Y., Sharkis, S., & Jang, Y. Y. (2010). Generation of endoderm-derived human induced pluripotent stem cells from primary hepatocytes. *Hepatology*, *51*(5), 1810–1819.
40. Sullivan, G. J., Hay, D. C., Park, I. H., Fletcher, J., Hannoun, Z., Payne, C. M., … Wang, G. (2010). Generation of functional human hepatic endoderm from human induced pluripotent stem cells. *Hepatology*, *51*(1), 329–335.
41. Wu, S., Xu, H., Chen, B., Wen, Y., Ikusika, O. M., Ocker, A., … Ildstad, S. T. (2011). Sensitized recipients exhibit accelerated but not hyperacute rejection of vascularized composite tissue allografts. *Transplantation*, *92*(6), 627–633.
42. Yamanaka, S. (2012). Induced pluripotent stem cells: Past, present, and future. *Cell Stem Cell*, *10*(6), 678–684.
43. Pittenger, M. F., Mackay, A. M., Beck, S. C., Jaiswal, R. K., Douglas, R., Mosca, J. D., … Marshak, D. R. (1999). Multilineage potential of adult human mesenchymal stem cells. *Science*, *284*(5411), 143–147.
44. Kadam, S., Muthyala, S., Nair, P., & Bhonde, R. (2010). Human placenta-derived mesenchymal stem cells and islet-like cell clusters generated from these cells as a novel source for stem cell therapy in diabetes. *The Review of Diabetic Studies*, *7*(2), 168–182.
45. Ngoc, P. K., Van Phuc, P., Nhung, T. H., Thuy, D. T., & Nguyet, N. T. M. (2011). Improving the efficacy of type 1 diabetes therapy by transplantation of immunoisolated insulin-producing cells. *Human Cell*, *24*(2), 86–95.
46. Goren, A., Dahan, N., Goren, E., Baruch, L., & Machluf, M. (2010). Encapsulated human mesenchymal stem cells: a unique hypoimmunogenic platform for long-term cellular therapy. *FASEB J*, *24*(1), 22–31.
47. Wilson, J. L., & McDevitt, T. C. (2013). Stem cell microencapsulation for phenotypic control, bioprocessing, and transplantation. *Biotechnology and Bioengineering*, *110*(3), 667–682.
48. Baraniak, P. R., & McDevitt, T. C. (2010). Stem cell paracrine actions and tissue regeneration. *Regenerative Medicine*, *5*(1), 121–143.

49. Orive, G., Gascón, A. R., Hernández, R. M., Igartua, M., & Pedraz, J. L. (2003). Cell microencapsulation technology for biomedical purposes: Novel insights and challenges. *Trends in Pharmacological Sciences, 24*(5), 207–210.
50. Hunkeler, D. (2001). Allo transplants xeno: As bioartificial organs move to the clinic. Introduction. *Annals of the New York Academy of Sciences, 944,* 1–6.
51. Kang, H. K., Wang, S., Dangi, A., Zhang, X., Singh, A., Zhang, L., … Thorp, E. B. (2017). Differential role of B cells and IL-17 versus IFN-gamma during early and late rejection of pig islet xenografts in mice. *Transplantation.*
52. Dufrane, D., & Gianello, P. (2012). Pig islet for xenotransplantation in human: Structural and physiological compatibility for human clinical application. *Transplantation Reviews, 26* (3), 183–188.
53. Dufrane, D., & Gianello, P. (2012). Macro- or microencapsulation of pig islets to cure type 1 diabetes. *World Journal of Gastroenterology, 18*(47), 6885–6893.
54. Zhu, H. T., Lu, L., Liu, X. Y., Yu, L., Lyu, Y., & Wang, B. (2015). Treatment of diabetes with encapsulated pig islets: An update on current developments. *Journal of Zhejiang University-Science B, 16*(5), 329–343.
55. Gunzburg, W. H., & Salmons, B. (2000). Xenotransplantation: Is the risk of viral infection as great as we thought? *Molecular Medicine Today, 6*(5), 199–208.
56. Bowie, K. M., & Chang, P. L. (1998). Development of engineered cells for implantation in gene therapy. *Advanced Drug Delivery Reviews, 33*(1–2), 31–43.
57. Uludag, H., De Vos, P., & Tresco, P. A. (2000). Technology of mammalian cell encapsulation. *Advanced Drug Delivery Reviews, 42*(1–2), 29–64.
58. Sakai, S., Mu, C., Kawabata, K., Hashimoto, I., & Kawakami, K. (2006). Biocompatibility of subsieve-size capsules versus conventional-size microcapsules. *Journal of Biomedical Materials Research Part A, 78*(2), 394–398.
59. Fang, S., & MAO, L. (2007). Differentiation of embryoid-body cells derived from embryonic stem cells into hepatocytes in alginate microbeads in vitro. *Acta Pharmacologica Sinica, 28*(12), 1924–1930.
60. Orive, G., Hernández, R. M., Gascón, A. R., Calafiore, R., Chang, T. M. S., de Vos, P., … Pedraz, J. L. (2004). History, challenges and perspectives of cell microencapsulation. *Trends in Biotechnology, 22*(2), 87–92.
61. Rabanel, J. M., Banquy, X., Zouaoui, H., Mokhtar, M., & Hildgen, P. (2009). Progress technology in microencapsulation methods for cell therapy. *Biotechnology Progress, 25*(4), 946–963.
62. Kidd, K. R., Nagle, R. B., & Williams, S. K. (2002). Angiogenesis and neovascularization associated with extracellular matrix-modified porous implants. *Journal of Biomedical Materials Research, 59*(2), 366–377.
63. Vasanthan, K. S., Subramanian, A., Krishnan, U. M., & Sethuraman, S. (2012). Role of biomaterials, therapeutic molecules and cells for hepatic tissue engineering. *Biotechnology Advances, 30*(3), 742–752.
64. Maguire, T., Davidovich, A. E., Wallenstein, E. J., Novik, E., Sharma, N., Pedersen, H., … Yarmush, M. (2007). Control of hepatic differentiation via cellular aggregation in an alginate microenvironment. *Biotechnology and Bioengineering, 98*(3), 631–644.
65. Shi, X. L., Zhang, Y., Gu, J. Y., & Ding, Y. T. (2009). Coencapsulation of hepatocytes with bone marrow mesenchymal stem cells improves hepatocyte-specific functions. *Transplantation, 88*(10), 1178–1185.
66. Gattazzo, F., Urciuolo, A., & Bonaldo, P. (2014). Extracellular matrix: A dynamic microenvironment for stem cell niche. *Biochimica et Biophysica Acta, 1840*(8), 2506–2519.
67. Vishwakarma, A., & Karp, J. M. (Eds.) (2017). *Biology and engineering of stem cell niches.* London: Elsevier Academic Press.
68. Jhala, D., & Vasita, R. (2015). A review on extracellular matrix mimicking strategies for an artificial stem cell niche. *Polymer Reviews, 55*(4), 561–595.
69. Scadden, D. T. (2006). The stem-cell niche as an entity of action. *Nature, 441*(7097), 1075–1079.

70. Stoker, M., & Gherardi, E. (1991). Regulation of cell movement: The motogenic cytokines. *Biochimica et Biophysica Acta, 1072*(1), 81–102.
71. Welf, E. S., & Haugh, J. M. (2011). Signaling pathways that control cell migration: Models and analysis. *Wiley Interdisciplinary Reviews: Systems Biology and Medicine, 3*(2), 231–240.
72. Watt, F. M., & Huck, W. T. (2013). Role of the extracellular matrix in regulating stem cell fate. *Nature Reviews Molecular Cell Biology, 14*(8), 467–473.
73. Rowley, J. A., & Mooney, D. J. (2002). Alginate type and RGD density control myoblast phenotype. *Journal of Biomedical Materials Research, 60*(2), 217–223.
74. Benoit, D. S., Schwartz, M. P., Durney, A. R., & Anseth, K. S. (2008). Small functional groups for controlled differentiation of hydrogel-encapsulated human mesenchymal stem cells. *Nature Materials, 7*(10), 816–823.
75. Chokkalingam, V., Tel, J., Wimmers, F., Liu, X., Semenov, S., Thiele, J., … Huck, W. T. (2013). Probing cellular heterogeneity in cytokine-secreting immune cells using droplet-based microfluidics. *Lab on a Chip, 13*(24), 4740–4744.
76. Severian Dumitriu, V. P. (2013) *Medical and pharmaceutical applications, Vol. 2, 3rd ed.* ISBN 978-1-4200-9468-8.
77. Brandtner, D. (2013). *Bioencapsulation of living cells for diverse medical applications.* Bentham books. ISBN: 978-1-60805-721-4.
78. Strand, B. L., Ryan, L., Veld, P. I. T., Kulseng, B., Rokstad, A. M., Skjåk-Bræk, G., & Espevik, T. (2001). Poly-L-Lysine induces fibrosis on alginate microcapsules via the induction of cytokines. *Cell Transplant, 10*(3), 263–275.
79. Liu, Z. C., & Chang, T. M. (2002). Increased viability of transplanted hepatocytes when hepatocytes are co-encapsulated with bone marrow stem cells using a novel method. *Artificial Cells, Blood Substitutes, and Immobilization Biotechnology, 30*(2), 99–112.
80. Otterlei, M., Østgaard, K., Skjåk-Bræk, G., Smidsrød, O., Soon-Shiong, P., & Espevik, T. (1991). Induction of cytokine production from human monocytes stimulated with alginate. *Journal of Immunotherapy, 10*(4), 286–291.
81. Espevik, T., Otterlei, M., Skjåk-Braek, G., Ryan, L., Wright, S. D., & Sundan, A. (1993). The involvement of CD14 in stimulation of cytokine production by uronic acid polymers. *European Journal of Immunology, 23*(1), 255–261.
82. Aiedehe, K., Gianasii, E., Orienti, I., & Zecchi, V. (1997). Chitosan microcapsules as controlled release systems for insulin. *Journal of Microencapsulation, 14*(5), 567–576.
83. Altiok, D., Altiok, E., & Tihminlioglu, F. (2010). Physical, antibacterial and antioxidant properties of chitosan films incorporated with thyme oil for potential wound healing applications. *Journal of Materials Science. Materials in Medicine, 21*(7), 2227–2236.
84. Green, D. W., Leveque, I., Walsh, D., Howard, D., Yang, X., Partridge, K., … Oreffo, R. O. (2005). Polysaccharide capsules for encapsulation, organization, and delivery of human cell types and growth factors. *Advanced Funtional Materials, 15*(6), 917–923.
85. Krasaekoopt, W., Bhandari, B., & Deeth, H. (2004). The influence of coating materials on some properties of alginate beads and survivability of microencapsulated probiotic bacteria. *International Dairy Journal, 14*(8), 737–743.
86. Chen, H., Ouyang, W., Jones, M., Metz, T., Martoni, C., Haque, T., … Prakash, S. (2007). Preparation and characterization of novel polymeric microcapsules for live cell encapsulation and therapy. *Cell Biochemistry and Biophysics, 47*(1), 159–168.
87. Tan, W., Krishnaraj, R., & Desai, T. A. (2001). Evaluation of nanostructured composite collagen–chitosan matrices for tissue engineering. *Tissue Engineering, 7*(2), 203–210.
88. Chevallay, B., & Herbage, D. (2000). Collagen-based biomaterials as 3D scaffold for cell cultures: Applications for tissue engineering and gene therapy. *Medical and Biological Engineering and Computing, 38*(2), 211–218.

89. Malafaya, P. B., Silva, G. A., & Reis, R. L. (2007). Natural-origin polymers as carriers and scaffolds for biomolecules and cell delivery in tissue engineering applications. *Advanced Drug Delivery Reviews, 59*(4–5), 207–233.
90. Dautzenberg, H., Schuldt, U. T. E., Grasnick, G., Karle, P., MÜller, P., LÖhr, M., … Salmons, B. (1999). Development of cellulose sulfate-based polyelectrolyte complex microcapsules for medical applications. *Annals of the New York Academy of Sciences, 875*, 46–63.
91. Swioklo, S., & Connon, C. J. (2016). Keeping cells in their place: The future of stem cell encapsulation. *Expert Opinion on Biological Therapy, 16*(10), 1181–1183.
92. Higgins, C. C. (1931). Solitary cysts of the kidney. *Annals of Surgery, 93*(4), 868–879.
93. Byass, P. (2014). The global burden of liver disease: A challenge for methods and for public health. *BMC Medicine, 12,* 159.
94. Hindley, C. J., Cordero-Espinoza, L., & Huch, M. (2016). Organoids from adult liver and pancreas: Stem cell biology and biomedical utility. *Development Biology, 420*(2), 251–261.
95. Fox, I. J., & Chowdhury, J. R. (2004). Hepatocyte transplantation. *American Journal of Transplantation, 4*(6), 7–13.
96. Sell, S. (2001). Heterogeneity and plasticity of hepatocyte lineage cells. *Hepatology, 33*(3), 738–750.
97. Miyajima, A., Tanaka, M., & Itoh, T. (2014). Stem/progenitor cells in liver development, homeostasis, regeneration, and reprogramming. *Cell Stem Cell, 14*(5), 561–574.
98. Itoh (2015). *Tissue-specific stem cell niche, stem cell biology and regeneration* ed. K. Turksen. Switzerland: Springer.
99. Conigliaro, A., Brenner, D. A., & Kisseleva, T. (2010). Hepatic progenitors for liver disease: Current position. *Stem Cells Cloning, 3,* 39–47.
100. Yang, W., Yan, H. X., Chen, L., Liu, Q., He, Y. Q., Yu, L. X., …. Chen, C. (2008). Wnt/beta-catenin signaling contributes to activation of normal and tumorigenic liver progenitor cells. *Cancer Research, 68*(11), 4287–4295.
101. de Lau, W., Barker, N., Low, T. Y., Koo, B. K., Li, V. S., Teunissen, H., … Stange, D. E. (2011). Lgr5 homologues associate with Wnt receptors and mediate R-spondin signalling. *Nature, 476*(7360). 293–297.
102. Boulter, L., Govaere, O., Bird, T. G., Radulescu, S., Ramachandran, P., Pellicoro, A., … Sansom, O. J. (2012). Macrophage-derived Wnt opposes notch signaling to specify hepatic progenitor cell fate in chronic liver disease. *Nature Medicine, 18*(4), 572–579.
103. Huch, M., Dorrell, C., Boj, S. F., Van Es, J. H., Li, V. S., Van De Wetering, M., … Haft, A. (2013). In vitro expansion of single Lgr5 + liver stem cells induced by Wnt-driven regeneration. *Nature, 494*(7436), 247–250.
104. Xinguang, Y., Huixing, Y., Xiaowei, W., Xiaojun, W., & Linghua, Y. (2015). R-spondin1 arguments hepatic fibrogenesis in vivo and in vitro. *Journal of Surgical Research, 193*(2), 598–605.
105. Kopp, J. L., Grompe, M., & Sander, M. (2016). Stem cells versus plasticity in liver and pancreas regeneration. *Nature Cell Biology, 18*(3), 238–245.
106. Yanger, K., Knigin, D., Zong, Y., Maggs, L., Gu, G., Akiyama, H., … Stanger, B. Z. (2014). Adult hepatocytes are generated by self-duplication rather than stem cell differentiation. *Cell Stem Cell, 15*(3), 340–349.
107. Tarlow, B. D., Pelz, C., Naugler, W. E., Wakefield, L., Wilson, E. M., Finegold, M. J., & Grompe, M. (2014). Bipotential adult liver progenitors are derived from chronically injured mature hepatocytes. *Cell Stem Cell, 15*(5), 605–618.
108. Ji, J., & Wang, X. W. (2012). Clinical implications of cancer stem cell biology in hepatocellular carcinoma. *Seminars in Oncology, 39*(4), 461–472.
109. Di Bonzo, L. V., Ferrero, I., Cravanzola, C., Mareschi, K., Rustichell, D., Novo, E., … Davit, A. (2008). Human mesenchymal stem cells as a two-edged sword in hepatic

regenerative medicine: Engraftment and hepatocyte differentiation versus profibrogenic potential. *Gut, 57*(2), 223–231.

110. Cai, J., Zhao, Y., Liu, Y., Ye, F., Song, Z., Qin, H., … Guo, Y. (2007). Directed differentiation of human embryonic stem cells into functional hepatic cells. *Hepatology, 45*(5), 1229–1239.
111. Basma, H., Soto–Gutiérrez, A., Yannam, G. R., Liu, L., Ito, R., Yamamoto, T., … Muirhead, D. (2009). Differentiation and transplantation of human embryonic stem cell-derived hepatocytes. *Gastroenterology, 136*(3), 990–999.
112. Touboul, T., Hannan, N. R., Corbineau, S., Martinez, A., Martinet, C., Branchereau, S., … Weber, A. (2010). Generation of functional hepatocytes from human embryonic stem cells under chemically defined conditions that recapitulate liver development. *Hepatology, 51*(5), 1754–1765.
113. Magyar, J. P., Nemir, M., Ehler, E., Suter, N., Perriard, J. C., & Eppenberger, H. M. (2001). Mass production of embryoid bodies in microbeads. *Annals of the New York Academy of Sciences, 944*(1), 135–143.
114. Cho, N. J., Elazar, M., Xiong, A., Lee, W., Chiao, E., Baker, J., … Glenn, J. S. (2008). Viral infection of human progenitor and liver-derived cells encapsulated in three-dimensional PEG-based hydrogel. *Biomedical materials, 4*(1), 011001.
115. Meier, R. P., Mahou, R., Morel, P., Meyer, J., Montanari, E., Muller, Y. D., … Bühler, L. H. (2015). Microencapsulated human mesenchymal stem cells decrease liver fibrosis in mice. *Journal of Hepatology, 62*(3), 634–641.
116. Meier, R. P., Montanari, E., Morel, P., Pimenta, J., Schuurman, H. J., Wandrey, C., … Bühler, L. H. (2017). Microencapsulation of hepatocytes and mesenchymal stem cells for therapeutic applications. *Hepatocyte Transplantation: Methods and Protocols,* 259–271.
117. Kisseleva, T., Uchinami, H., Feirt, N., Quintana-Bustamante, O., Segovia, J. C., Schwabe, R. F., & Brenner, D. A. (2006). Bone marrow-derived fibrocytes participate in pathogenesis of liver fibrosis. *Journal of hepatology, 45*(3), 429–438.
118. Maguire, T., Novik, E., Schloss, R., & Yarmush, M. (2006). Alginate-PLL microencapsulation: Effect on the differentiation of embryonic stem cells into hepatocytes. *Biotechnology and Bioengineering, 93*(3). 581–591.
119. Fang, S., & MAO, L. (2007). Differentiation of embryoid-body cells derived from embryonic stem cells into hepatocytes in alginate microbeads in vitro. *Acta Pharmacologica Sinica, 28*(12), 1924–1930.
120. Palakkan, A. A., Hay, D. C., TV, K., & Ross, J. A. (2013). Liver tissue engineering and cell sources: Issues and challenges. *Liver International, 33*(5), 666–676.
121. Kirk, K., Hao, E., Lahmy, R., & Itkin-Ansari, P. (2014). Human embryonic stem cell derived islet progenitors mature inside an encapsulation device without evidence of increased biomass or cell escape. *Stem Cell Research, 12*(3), 807–814.
122. Scharp, D. W., & Marchetti, P. (2014). Encapsulated islets for diabetes therapy: History, current progress, and critical issues requiring solution. *Advanced Drug Delivery Reviews, 67–68,* 35–73.
123. Sakata, N., Sumi, S., Yoshimatsu, G., Goto, M., Egawa, S., & Unno, M. (2012). Encapsulated islets transplantation: Past, present and future. *World Journal of Gastrointestinal Pathophysiology, 3*(1), 19–26.
124. Kirchhof, N., Shibata, S., Wijkstrom, M., Kulick, D. M., Salerno, C. T., Clemmings, S. M., … Hering, B. J. (2004). Reversal of diabetes in non-immunosuppressed rhesus macaques by intraportal porcine islet xenografts precedes acute cellular rejection. *Xenotransplantation, 11*(5), 396–407.
125. Elliott, R. B., Escobar, L., Tan, P. L. J., Garkavenko, O., Calafiore, R., Basta, P., … Bambra, C. (2005, October). Intraperitoneal alginate-encapsulated neonatal porcine islets in a

placebo-controlled study with 16 diabetic cynomolgus primates. *Transplantation Proceedings, 37*(8), 3505–3508.
126. Duvivier-Kali, V. F., Omer, A., Lopez-Avalos, M. D., O'neil, J. J., & Weir, G. C. (2004). Survival of microencapsulated adult pig islets in mice in spite of an antibody response. *American Journal of Transplantation, 4*(12), 1991–2000.
127. Elliott, R. B., Escobar, L., Tan, P. L., Muzina, M., Zwain, S., & Buchanan, C. (2007). Live encapsulated porcine islets from a type 1 diabetic patient 9.5 yr after xenotransplantation. *Xenotransplantation, 14*(2), 157–161.
128. Tuch, B. E., Keogh, G. W., Williams, L. J., Wu, W., Foster, J. L., Vaithilingam, V., & Philips, R. (2009). Safety and viability of microencapsulated human islets transplanted into diabetic humans. *Diabetes Care, 32*(10), 1887–1889.
129. King, A., Lau, J., Nordin, A., Sandler, S., & Andersson, A. (2003). The effect of capsule composition in the reversal of hyperglycemia in diabetic mice transplanted with microencapsulated allogeneic islets. *Diabetes Technology & Therapeutics, 5*(4), 653–663.
130. King, A., Andersson, A., Strand, B. L., Lau, J., Skjåk-Bræk, G., & Sandler, S. (2003). The role of capsule composition and biologic responses in the function of transplanted microencapsulated islets of Langerhans. *Transplantation, 76*(2), 275–279.
131. Kulseng, B., Thu, B., Espevik, T., & Skjåk-Bræk, G. (1997). Alginate polylysine microcapsules as immune barrier: Permeability of cytokines and immunoglobulins over the capsule membrane. *Cell Transplant, 6*(4), 387–394.
132. Juste, S., Lessard, M., Henley, N., Ménard, M., & Hallé, J. P. (2005). Effect of poly-L-lysine coating on macrophage activation by alginate-based microcapsules: Assessment using a new in vitro method. *Journal of Biomedical Materials Research Part A, 72*(4), 389–398.
133. Pueyo, M. E., Darquy, S., Capron, F., & Reach, G. (1994). In vitro activation of human macrophages by alginate-polylysine microcapsules. *Journal of Biomaterials Science, Polymer Edition, 5*(3), 197–203.
134. Brennan, D. C., Shannon, M. B., Koch, M. J., Polonsky, K. S., Desai, N., & Shapiro, J. (2004). Portal vein thrombosis complicating islet transplantation in a recipient with the Factor V Leiden mutation. *Transplantation, 78*(1), 172–173.
135. Kawahara, T., Kin, T., Kashkoush, S., Gala-Lopez, B., Bigam, D. L., Kneteman, N. M., … Shapiro, A. J. (2011). Portal vein thrombosis is a potentially preventable complication in clinical islet transplantation. *American Journal of Transplantation, 11*(12), 2700-2707.
136. Sakata, N., Tan, A., Chan, N., Obenaus, A., Mace, J., Peverini, R., … Hathout, E. (2009, February). Efficacy comparison between intraportal and subcapsular islet transplants in a murine diabetic model. *Transplantation Proceedings, 41*(1), 346–349.
137. Christoffersson, G., Carlsson, P. O., & Phillipson, M. (2011). Intramuscular islet transplantation promotes restored islet vascularity. *Islets, 3*(2), 69–71.
138. McQuilling, J. P., Arenas-Herrera, J., Childers, C., Pareta, R. A., Khanna, O., Jiang, B., … Opara, E. C. (2011, November). New alginate microcapsule system for angiogenic protein delivery and immunoisolation of islets for transplantation in the rat omentum pouch. *Transplantation Proceedings, 43*(9), 3262–3264.
139. Elliott, R. B., Escobar, L., Calafiore, R., Basta, G., Garkavenko, O., Vasconcellos, A., et al. (2005). Transplantation of micro- and macroencapsulated piglet islets into mice and monkeys. *Transplantation Proceedings, 37*(1), 466–469.
140. Dufrane, D., van Steenberghe, M., Goebbels, R. M., Saliez, A., Guiot, Y., & Gianello, P. (2006). The influence of implantation site on the biocompatibility and survival of alginate encapsulated pig islets in rats. *Biomaterials, 27*(17), 3201–3208.
141. Algire, G. (1943). An adaptation of the transparent-chamber technique to the mouse. *Journal of the National Cancer Institute, 1943*(4), 1–11.
142. Storrs, R., Dorian, R., King, S. R., Lakey, J., & Rilo, H. (2001). Preclinical development of the Islet sheet. *Annals of the New York Academy of Sciences, 2001*(944), 252–266.

143. Nyitray, C. E., Chavez, M. G., & Desai, T. A. (2014). Compliant 3D microenvironment improves beta-cell cluster insulin expression through mechanosensing and beta-catenin signaling. *Tissue Engineering Part A, 20*(13–14), 1888–1895.
144. de Vos, P., Hamel, A. F., & Tatarkiewicz, K. (2002). Considerations for successful transplantation of encapsulated pancreatic islets. *Diabetologia, 45*(2), 159–173.
145. Veriter, S., Aouassar, N., Xhema, D., Beaurin, G., Adnet, P. Y., Igarashi, Y., … Dufrane, D. (2011). Islets and mesenchymal stem cells co-encapsulation can improve subcutaneous bioartificial pancreas survival in diabetic primates. The Transplantation Society: CTS-IXA. Parallel Session 5—ISLET xenotransplantation—Preclinical models.
146. Davis, N. E., Beenken-Rothkopf, L. N., Mirsoian, A., Kojic, N., Kaplan, D. L., Barron, A. E., & Fontaine, M. J. (2012). Enhanced function of pancreatic islets co-encapsulated with ECM proteins and mesenchymal stromal cells in a silk hydrogel. *Biomaterials, 33*(28), 6691–6697.
147. Lee, S. H., Hao, E., Savinov, A. Y., Geron, I., Strongin, A. Y., & Itkin-Ansari, P. (2009). Human beta-cell precursors mature into functional insulin-producing cells in an immunoisolation device: Implications for diabetes cell therapies. *Transplantation, 87*(7), 983–991.
148. Matveyenko, A. V., Georgia, S., Bhushan, A., & Butler, P. C. (2010). Inconsistent formation and nonfunction of insulin-positive cells from pancreatic endoderm derived from human embryonic stem cells in athymic nude rats. *American Journal of Physiology-Endocrinology and Metabolism, 299*(5), E713–E720.
149. Colton, C. K. (2014). Oxygen supply to encapsulated therapeutic cells. *Advanced Drug Delivery Reviews, 67–68,* 93–110.
150. Candiello, J., Singh, S. S., Task, K., Kumta, P. N., & Banerjee, I. (2013). Early differentiation patterning of mouse embryonic stem cells in response to variations in alginate substrate stiffness. *Journal of Biological Engineering*, *7*(1), 9.
151. Schulz, T. C. (2015). Concise review: Manufacturing of pancreatic endoderm cells for clinical trials in Type 1 diabetes. *Stem Cells Translational Medicine, 4*(8), 927–931.
152. Pagliuca, F. W., Millman, J. R., Gürtler, M., Segel, M., Van Dervort, A., Ryu, J. H., … Melton, D. A. (2014). Generation of functional human pancreatic beta cells in vitro. *Cell, 159*(2), 428–439.
153. Rezania, A., Bruin, J. E., Arora, P., Rubin, A., Batushansky, I., Asadi, A., … Yang, Y. H. C. (2014). Reversal of diabetes with insulin-producing cells derived in vitro from human pluripotent stem cells. *Nature Biotechnology*, *32*(11), 1121–1133.
154. Russ, H. A., Parent, A. V., Ringler, J. J., Hennings, T. G., Nair, G. G., Shveygert, M., … Blelloch, R. (2015). Controlled induction of human pancreatic progenitors produces functional beta-like cells in vitro. *The EMBO Journal*, *34*(13), 1759–1772.
155. Agulnick, A. D., Ambruzs, D. M., Moorman, M. A., Bhoumik, A., Cesario, R. M., Payne, J. K., …Kerr, J. (2015). Insulin-producing endocrine cells differentiated in vitro from human embryonic stem cells function in macroencapsulation devices in vivo. *Stem Cells Translational Medicine*, *4*(10), 1214–1222.
156. Tateishi, K., He, J., Taranova, O., Liang, G., D'Alessio, A. C., & Zhang, Y. (2008). Generation of insulin-secreting islet-like clusters from human skin fibroblasts. *Journal of Biological Chemistry*, *283*(46), 31601–31607.
157. Maehr, R., Chen, S., Snitow, M., Ludwig, T., Yagasaki, L., Goland, R., … Melton, D. A. (2009). Generation of pluripotent stem cells from patients with type 1 diabetes. *Proceedings of the National Academy of Sciences*, *106*(37), 15768–15773.
158. Lysy, P. A., Weir, G. C., & Bonner-Weir, S. (2012). Concise review: pancreas regeneration: Recent advances and perspectives. *Stem Cells Translational Medicine, 1*(2), 150–159.
159. Noguchi, H., Xu, G., Matsumoto, S., Kaneto, H., Kobayashi, N., Bonner-Weir, S., & Hayashi, S. (2006). Induction of pancreatic stem/progenitor cells into insulin-producing cells by adenoviral-mediated gene transfer technology. *Cell transplantation*, *15*(10), 929–938.
160. Alberto Hayek, C.K., Brief review: Cell replacement therapies to treat type 1 diabetes mellitus. *Clinical Diabetes and Endocrinology, 2*(4).

161. Mason, M. N., & Mahoney, M. J. (2009). Selective beta-cell differentiation of dissociated embryonic pancreatic precursor cells cultured in synthetic polyethylene glycol hydrogels. *Tissue Engineering Part A, 15*(6), 1343–1352.
162. D'Amour, K. A., Bang, A. G., Eliazer, S., Kelly, O. G., Agulnick, A. D., Smart, N. G., … Baetge, E. E. (2006). Production of pancreatic hormone-expressing endocrine cells from human embryonic stem cells. *Nature Biotechnology, 24*(11), 1392–1401.
163. Kroon, E., Martinson, L. A., Kadoya, K., Bang, A. G., Kelly, O. G., Eliazer, S., … Agulnick, A. D. (2008). Pancreatic endoderm derived from human embryonic stem cells generates glucose-responsive insulin-secreting cells in vivo. *Nature Biotechnology, 26*(4), 443–452.
164. Chen, Y., Pan, F. C., Brandes, N., Afelik, S., Sölter, M., & Pieler, T. (2004). Retinoic acid signaling is essential for pancreas development and promotes endocrine at the expense of exocrine cell differentiation in Xenopus. *Developmental Biology, 271*(1), 144–160.
165. Gomes, L. F., Lorente, S., Simon-Giavarotti, K. A., Areco, K. N., Araújo-Peres, C., & Videla, L. A. (2004). Tri-iodothyronine differentially induces Kupffer cell ED1/ED2 subpopulations. *Molecular Aspects of Medicine, 25*(1–2), 183–190.
166. Schulz, T. C., Young, H. Y., Agulnick, A. D., Babin, M. J., Baetge, E. E., Bang, A. G., … Kadoya, K. (2012). A scalable system for production of functional pancreatic progenitors from human embryonic stem cells. *PLoS One, 7*(5), e37004.
167. Shapiro, A. M., Nanji, S. A., & Lakey, J. R. (2003). Clinical islet transplant: Current and future directions towards tolerance. *Immunological Reviews, 196,* 219–236.
168. Wang, T., Lacík, I., Brissová, M., Anilkumar, A. V., Prokop, A., Hunkeler, D., … Powers, A. C. (1997). An encapsulation system for the immunoisolation of pancreatic islets. *Nature Biotechnology, 15*(4), 358–362.
169. Desai, T., & Shea, L. D. (2017). Advances in islet encapsulation technologies. *Nature Reviews Drug Discovery, 16*(5), 367.
170. Brauker, J. H., Carr-Brendel, V. E., Martinson, L. A., Crudele, J., Johnston, W. D., & Johnson, R. C. (1995). Neovascularization of synthetic membranes directed by membrane microarchitecture. *Journal of Biomedical Materials Research Part A, 29*(12), 1517–1524.
171. Geller, R. L., Loudovaris, T., Neuenfeldt, S., Johnson, R. C., & Brauker, J. H. (1997). Use of an immunoisolation device for cell transplantation and tumor immunotherapy. *Annals of the New York Academy of Sciences, 1997*(831), 438–451.
172. Colton, C. K., & Avgoustiniatos, E. S. (1991). Bioengineering in development of the hybrid artificial pancreas. *Journal of Biomechanical Engineering, 113*(2), 152–170.
173. Su, J., Hu, B. H., Lowe, W. L., Kaufman, D. B., & Messersmith, P. B. (2010). Anti-inflammatory peptide-functionalized hydrogels for insulin-secreting cell encapsulation. *Biomaterials, 31*(2), 308–314.
174. Rafael, E., Wu, G. S., Hultenby, K., Tibell, A., & Wernerson, A. (2003). Improved survival of macroencapsulated islets of Langerhans by preimplantation of the immunoisolating device: A morphometric study. *Cell Transplantation, 12*(4), 407–412.
175. Yakhnenko, I., Wong, W. K., Katkov, I. I., & Itkin-Ansari, P. (2012). Cryopreservation of human insulin expressing cells macro-encapsulated in a durable therapeutic immunoisolating device theracyte. *Cryo Letters, 33*(6), 518–531.
176. Rafael, E., Wernerson, A., Arner, P., Wu, G. S., & Tibell, A. (1999). In vivo evaluation of glucose permeability of an immunoisolation device intended for islet transplantation: A novel application of the microdialysis technique. *Cell Transplantation, 8*(3), 317–326.
177. Kumagai-Braesch, M., Jacobson, S., Mori, H., Jia, X., Takahashi, T., Wernerson, A., … Tibell, A. (2013). The TheraCyte device protects against islet allograft rejection in immunized hosts. *Cell Transplantation, 22*(7), 1137–1146.
178. ClinicalTrials.gov/beta/. (2016). ViaCyte: Three Year Follow-up Safety Study in Subjects Previously Implanted With VC-01™.

Author Biography

Isgard S. Hueck (Germany, USA), MSc. Mrs. Hueck has studied biology at the Westfälische Wilhelms University of Münster, Germany. She received her license as Cyto-pathologist in Cologne in 1987. After many years of clinical work in hematology, cancer diagnosis, and bone marrow transplantation centers, she studied Biomedical Engineering at the University of Applied Sciences, Aachen. She received her Master of Science degree in Bioengineering in 1998 in conjunction with the University of California San Diego (UCSD), California, USA. Based on her extensive expertise in the areas of cellular engineering, cytohisto-pathology, and clinical cancer diagnostic, Isgard continued to conduct cell-based research in cancer treatment and Diabetes Mellitus at UCSD. She has published in scientific journals like Microcirculation, American J. Physiology, Cell Physiol., and Springer books in Stem Cell Engineering. Her main research interests are immunology and medical applications based on the synergy between engineering and clinical cell therapy. Stem cells are still Isgard's most fascinating topic.

Stem Cells: Cellular and Extracellular Requirements for Generation and Use

Gerd Bungartz and Kathryn Bungartz

1 Stem Cells—Definition

Stem cells are defined by their capacity to self-renew and their ability to give rise to one or more type of differentiated progeny [4, 34]. Stem cells can be isolated from numerous sites and from numerous stages of development, each site and stage providing specific characteristics to the stem cell.

Cells that can differentiate into all three germ layers of a human (endoderm, mesoderm, and ectoderm) and into extra-embryonic (placental) cell types are called totipotent or omnipotent stem cells. These cells are the immediate descendants of a zygote, which is produced from the fusion of an egg cell and a sperm cell [42].

Embryonic stem (ES) cells are derived from the inner cell mass of a blastocyst, which represents a pre-implantation stage of embryogenesis. ES cells can differentiate into all three germ layers of an organism as well as their descendant cells. However, the potential of an ES cell is confined to the embryo proper. Hence, ES cells are described as being pluripotent [51].

During the development of an organism, ES cells lose their pluripotency to differentiate along different cellular lineages. In other words, these pluripotent cells lose their stem cell properties and develop into more specialized cells. Within the developing embryo, lineage-commitment is guided by extracellular cues, such as signals from the extracellular matrix (ECM), which are dependent upon the position of a certain ES cell within the developing blastocyst. Typically, a miniscule fraction of tissue-specific cells persist that do not fully complete maturation and remain in an undifferentiated, yet lineage-committed, developmental state throughout the life-

G. Bungartz (✉)
Hochschule Döpfer, Cologne, Germany
e-mail: gerd.bungartz@gmail.com

K. Bungartz
K.D. Bungartz, LLC, Riverton, WY, USA

G. M. Artmann et al. (eds.), *Biological, Physical and Technical Basics of Cell Engineering*, https://doi.org/10.1007/978-981-10-7904-7_3

time of an organism. These cells are referred to as adult stem cells and are considered multipotent.

2 History of Stem Cell Discoveries and Milestones

In 1981, two research groups, one led by Martin Evans and Matthew Kaufman and another led by Gail R. Martin, published the isolation and in vitro culture of pluripotent cells; the term embryonic stem cell was coined at this time by Dr. Martin [18, 32]. These ES cells were isolated from the inner cell mass of developing mouse embryos at the blastocyst stage. Special measures were taken to maintain the ES cells in an undifferentiated state in the petri dish. Combined with gene targeting, this was the first step toward generating a genetically altered animal.

In 1989, just eight years after their isolation ES cells were used to create the first genetically engineered mouse through gene targeting [13, 49]. The capability to isolate, grow, and genetically modify mouse ES cells and subsequently generate genetically distinct animals sparked an entire field of research [insert reference here]. In the following years, ES cell culture and the generation of genetically altered mice using DNA recombination became a standardized technique in countless labs around the world ([23] and references therein).

The research stemming from these early studies revolutionized the life sciences as it was used to understand gene function in development and disease within a mammalian system; mice were developed to model human conditions including cancer, heart disease, diabetes, hearing loss, and countless more. For these reasons, the 2007 Nobel Prize in Physiology or Medicine was awarded to Capecchi, Evans, and Smithies for the development of genetically engineered mice using ES cells.

Embryonic stem cell research indeed led to a better understanding of gene and protein function and promoted knowledge in the basic sciences. While these animal models could suggest pathways for new therapeutic approaches and these therapies could even be tested in the mice, the translation of mouse studies to human conditions lagged behind for a variety of reasons, with one major reason being the unanticipated difficulty of isolating and culturing human ES cells.

But in 1998, James Thomson spearheaded a research group was able for the first time to isolate and grow human ES cells in culture [50]. This accomplishment represented a tremendous advance toward tissue engineering for medical purposes as the ability to culture and sustain human ES cells in vitro provided a theoretically unlimited source of these precious cells. However, despite further technical improvements to establish human ES cell lines in culture, ethical difficulties regarding the isolation cells from pre-implantation stage human embryos left its use problematic. Additionally, while ES cells bear the potential to differentiate into all types of tissue-specific cells, in practice, the challenge to induce and control directed cell-type specific differentiation for therapeutic use remained unmet.

Meanwhile, another approach to cellular programming was being used. In 1996, the concepts and innovative work of Ian Wilmut, Keith Campbell, and colleagues came to fruition with the birth of Dolly [11, 55]. This famous sheep stemmed from a second line of research in regenerative medicine that discovered the DNA from a differentiated cell contains all of the information necessary to give rise to a new organism. Dolly was created from the contents of a single cell isolated from the udder of an adult sheep inserted to the ovum of another sheep and transplanted to a third sheep for surrogate gestation and birth. This approach, called somatic cell nuclear transfer (SCNT), utilized a series of epigenetic changes of the DNA induced upon transfer of the DNA into an enucleated oocyte, an oocyte with it's nuclear content removed. This method resulted in reprogramming of donor cell nucleus. The cloning of Dolly sparked a hot ethical debate about the possibility and use of cloning human beings. (The assembly of recombinant DNA molecules and their replication within host organisms, often *E. coli*, using techniques and methods of molecular biology is also referred to as 'cloning', which is a technique ubiquitous in research labs around the world.) Ethical concerns surrounding cloning and stem cell use are discussed later in this chapter. While Dolly proved the principal that cloning a mammal was possible, SCNT was found to be very inefficient and instead of developing into a widely used method, sparked the field of cellular reprogramming.

In 2006, induced pluripotent stem (iPS) cells were generated and published for the first time by Shinya Yamanaka, presenting a major milestone in stem cell research and opening up the field of regenerative medicine [48]. The use of iPS cells has the potential to circumvent the major ethical hurdles faced by ES cell research and cloning attempts, as mature (adult) cells are induced to de-differentiate or are reprogrammed into a pluripotent, ES cell-like state without SCNT. In order to accomplish this, Yamanaka and his colleague Takahashi expressed different combinations of candidate transcription factors in a cellular system designed to test for pluripotency. Starting with 24 candidate genes, the researches pared it down to just four genes, Oct3/4, Sox2, Klf4, and c-Myc, that were determined to be indispensable reprogramming factors. These four factors are commonly referred to as the 'Yamanaka factors'. The explosion of studies that immediately followed this publication highlights the impact of this discovery.

Within the following years, numerous research groups, including those led by Yamanaka [47], Jaenisch [33], Hochedlinger [31], Zhou [58], Gao [26] and Thomson [57] adopted and improved this approach, resulting in the generation of iPS cell-derived mice and human iPS cells.

With these accomplishments, iPS technology appeared to overcome principal ethical concerns and possible immunological barriers of ES cells and SCNT, making the clinical use of stem cells appear attainable for the first time since their discovery more than four decades earlier. Ideas about clinical uses of iPS cells, including tissue repair, in vitro tissue and organ generation for subsequent transplantation, and the possibility of generating disease or patient-specific iPS cells for drug testing and therapy development were emerging. However, these prospective goals were hindered by the current state of the technique, which required the DNA sequence of the reprograming Yamanaka factors to be inserted into the host genome. This presented two major issues. First, the integration sites could potentially disturb endogenous gene function, as integration was random. Second, the inserted genetic sequence could transcribe a protein known to drive cancer, an oncogene, like c-Myc, required for iPS cell generation. Additionally, the efficiency of the reprogramming process proved low and sometimes incomplete, i.e. did not fully erase epigenetic imprints.

One emphasis of the last decade's stem cell research was increasing iPS reprogramming efficiency and developing alternative delivery methods of the reprogramming factors. To date, integration-free vector delivery methods have proven successful, as have protein delivery systems requiring no integration of genetic material. In 2013, the research group surrounding Hongkui Deng achieved a successful mouse somatic cell reprogramming solely by small molecule compounds. This approach utilized seven small compounds and achieved a reprogramming frequency of 0.2% [56]. Today, stem cells arising from different sources in the human body are being tested in clinical studies (Tables 1 and 2).

Table 1 Sampling of clinical trials using HSCs as intervention including the disease/condition and the clinical developmental stage

Disease/purpose/condition	Intervention	Status
Alpha thalassemia major; hemoglobinopathy	In utero HSC transplantation	Phase I
Stiff person syndrome	Autologous HSC transplantation	Phase I/II
Inflammatory bowel diseases	Autologous HSC transplantation	Phase I/II
Neuromyelitis optica (Devic's Disease)	HSC transplantation	Phase I/II
End stage renal disease	Kidney and HSC transplantation	Pilot
Multiple sclerosis	Autologous HSC transplantation	Phase I
Crohn's disease	HSC transplantation	Pilot
Pancreatic adenocarcinoma	Allogeneic HSC transplantation	Phase I/II
Beta-thalassemia	Autologous HSC genetically modified with lentiviral vector encoding for the human beta-globin gene	Phase I/II
Triple-negative invasive breast carcinoma	Pharmacologic and autologous HSC transplantation	Phase II
Sickle cell anemia	Fludarabine and HSC transplantation	Phase I/II
Fanconi anemia	Pharmacologic and HSC transplantation	Phase II
Mitochondrial neurogastrointestinal encephalomyopathy (MNGIE)	Allogeneic HSC	Phase I
Immunodeficiency	HSC transplantation	Phase II
Systemic lupus erythematosus	Immunoablation and autologous HSC transplantation	Phase II
Pancreatic adenocarcinoma	Allogeneic HSC transplantation	Phase I/II
Wiskott-Aldrich syndrome	Autologous CD34 positive cells transduced with a lentiviral vector containing human WAS gene	Phase I/II

Table 2 Sampling of clinical trials using non-HSC stem cells populations as intervention including the disease/condition and the clinical developmental stage

Disease/purpose/condition	Intervention	Status
HSC transplantation	MSC infusion, cyclophosphamid	Phase III
Facial rejuvenation	Adipose-derived stem cells	Pilot
Urticaria	Autologous MSC therapy	Phase I
Hair restoration therapy for androgenetic alopecia	Adipose tissue derived autologous MSC	Phase II
Non-obstructive azoospermia male infertility	Autologous MSC transplantation	Phase I/II
Diabetes mellitus type 1	Autologous CD34+, CD133+ MSC transplantation	Phase I/II
Respiratory distress syndrome	MSC therapy	Phase II
Limbal stem cell deficiency	Limbal epithelial stem cell graft	Phase II
Amyotrophic lateral sclerosis	Injection of adipose derived MSCs	Phase I
Chronic myocardial ischemia	Autologous Bone Marrow-derived MSC Administration	Phase I
Type 1 diabetes with diabetic ketoacidosis	umbilical cord MSC transplantation	Phase II
Liver cirrhosis	MSC transplantation	Phase I/II
Peripheral vascular disease; ischemia; diabetic foot	Adipose-derived stem cell transplantation	Phase I
COPD	Adipose derived stem cell transplantation	Pilot
Full thickness rotator cuff tear	MSC augmentation in rotator cuff repair	Pilot
Cystic fibrosis	Allogenic MSC infusion	Phase I
Stroke	MSC transplantation	Phase I
Age related macular degeneration	Human embryonic stem cell derived retinal pigmented epithelium	Phase I/II
Spinal cord injury	Human spinal cord-derived neural stem cell transplantation	Phase I/II
Parkinson disease	Human neural stem cell injection	pilot
Heart failure	Intracoronary injection of autologous cardiac stem cells	Phase II
Glioma	Neural stem cells loaded with an oncolytic adenovirus	Phase I
Muscle dystrophy	Intramuscular injection of muscle derived stem cell and adipose derived MSC	Phase I
Hepatic cirrhosis	Human umbilical cord-MSC transplantation	Phase I

3 Types of Stem Cells

3.1 Embryonic Stem (ES) Cells

ES cells arise from the cell division of a fertilized egg in the inner cell mass of a developing embryo. During normal development, ES cells give rise to the embryo proper and differentiate into three germ layers, endoderm, mesoderm, and ectoderm. Later in development, cells from the endoderm give rise to the gastrointestinal and respiratory tracts and tissues forming the liver and thyroid. Mesodermal cells contribute to the development of organs such as the heart, blood vessels, lymphoid tissues, and blood. Furthermore, the kidneys, skeletal muscle, connective tissue, and bone are derived from the mesoderm. Finally, the ectoderm differentiates to form the skin and the neuronal system. Due to the pluripotency of ES cells, their potential for scientific research and eventually for therapeutic purposes is tremendous.

As outlined above, the capability of expanding ES cells in culture dishes without the cells differentiating paved the way for the experimental genetic alteration of these cells by homologous recombination, called 'gene targeting', and the subsequent selection of individual (targeted) ES cell clones, resulting in the generation of genetically modified mice. More recently, the workflow of cellular gene editing has been accelerated considerably by the development of the CRISPRCas9-System [16]. These technological developments revolutionized the entire field of life sciences and boosted the knowledge of gene function. Translating these findings from murine models to the human system has proven difficult and it was not until 1998 that human ES cell culture techniques were established to maintain these cells in culture in an undifferentiated state. As ES cells are derived from the pre-implantation stage embryo, the development of patient-matched stem cell lines to support precision medicine is currently out of reach using ES cells. Additionally, as pre-implantation stage embryos are repurposed to develop ES cell cultures, their use remains highly controversial.

3.2 Induced Pluripotent Stem (iPS) Cells

iPS cells are cells that are either de-differentiated or reprogrammed from adult somatic cells to harbor the characteristics of ES cells, having the ability to then differentiate into numerous cell types; they are pluripotent. In order to reprogram adult cells, certain transcription factors, the Yamanaka factors, need to be expressed. Expression vectors, a plasmid or virus designed to express a certain gene or set of genes, were first used to introduce these factors into the adult cells. This created a hurdle as these gene delivery systems (expression vectors) remained in the genome of the reprogrammed cells. Because the expression vectors could theoretically insert anywhere in the host genome, deleterious mutations could be induced.

With this in mind, several research groups have developed methods of generating iPS cells without the use of stably introduced expression vectors, in a transgene-free manner [44]. Around the same time, it was shown using a tratraploid complementation assay, the most rigorous assay available, that iPS cells are comparable to ES cells in their potential to contribute to all cells of an organism including germ cells [26, 58].

This technique has the potential to address ethical concerns about human ES cell isolation and provided ways to generate unlimited amounts of patient specific pluripotent cells. These cells have been used with great advantage for studying mechanisms of human disease or drug effects.

3.3 Adult Stem Cells

Adult stem cells are generally multipotent cells and are present in numerous adult tissues. Unlike the pluripotent ES and iPS cells, adult stem cells have the potential to develop into a restricted line or a family of closely related cells. Examples of adult stem cells include hematopoietic stem cells (HSCs), which give rise to all lineages of the blood system, mesenchymal stem cells (MSCs), which are capable of differentiating into bone, cartilage, and adipose tissue, hepatic stem cells, which can differentiate into cells of the liver, satellite cells of the muscle, and neuronal stem cells, which serve as precursors of neurons, astrocytes, and glia. Techniques have been developed to isolate certain stem cell populations, sorting out the cells of interest from the rest of the cells in the tissue. Of the many types of adult stem cells, MSCs including muse cells and HSCs are the best characterized.

3.4 Mesechymal Stem Cells (MSCs)

MSCs have first been described as colony forming units (CFUs) from ex vivo mouse bone marrow stroma cells [5]. These cells are defined by their ability to differentiate into bone, cartilage, and adipose tissue. MSCs have also been suggested to have the potential to differentiate into muscle [38]. MSC populations are accessible from bone marrow and adipose tissue and can be isolated using the cell surface markers CD34 and CD133, which are receptor proteins expressed specifically by these cells. MSCs have immuno-modulatory properties and avoid immune rejection upon allo-transplantation, meaning they themselves are unlikely to be recognized as foreign cells and when transplanted together with HSCs, make the other cells less likely to be rejected as foreign cells. These properties make them an exquisite candidate for medical use in numerous transplantation settings. A number of clinical trials are currently testing MSC use in autoimmune and other diseases (Table 1).

Muse cells

In 2010, research guided by Mari Dezawa led to the discovery and isolation of a rare subpopulation of MSCs that was termed Muse (multilineage differentiating stress enduring) cells. These cells have been shown to express pluripotency markers including SSEA3, TRA1-60, Nanog, Oct3/4, and Sox2 at low levels and to self-renew. They are able to differentiate into cells from all three germ layers and have the capacity to home to damaged tissue and differentiate into the tissue at the site of damage, contributing to functional tissue repair [28]. Implications of this discovery are discussed below in the transdifferentiation section.

3.4.1 Hematopoietic Stem Cells (HSCs)

HSCs describe a small population of cells within the bone marrow. HSCs are also found in and isolated from umbilical cord blood, which is the blood that remains in the placenta and umbilical cord after a child is born. All blood cells derive from HSCs, which have the potential to self-renew and to differentiate into all hematopoietic cell lineages, which can be distinguished by their expression of characteristic cell surface markers (Fig. 1). During normal development, HSCs are generated outside the embryo proper in the yolk sac and within the embryo in the para-aortic splanchnopleura (PAS)/aorta-gonad mesonephros (AGM) region. HSCs give rise to lymphatic, myeloid, and erythroid precursor cells. The stem cells themselves can be sorted out from the milieu of other bone marrow cells in a laboratory with a combination of surface proteins using the technique of flow cytometry. In addition to cell surface markers, HSCs are functionally defined by their ability to give rise to long-term multi-lineage reconstitution in lethally irradiated mice. In other words, a single HSC can be injected into mice that have had their bone marrow cells completely eliminated by radiation and, if the cell has been properly sorted, this single cells is capable of reconstituting the entire immune system of the mouse, demonstrating both self-renewal capacity and multipotency of HSCs.

Much research has been invested in determining the molecular cues guiding hematopoietic and other stem cells to either differentiate or to self-renew [19, 22, 27, 53]. Our own research identified the transcription factor, nuclear factor Y (NFY), to act upstream of a signaling cascade critical to these decisions in HSCs [8]. This work helped decipher the nature of these molecular pathways and to better understand how this network is altered during diseases such as leukemia and how it can be manipulated for therapeutic purposes. We took this work one step further and demonstrated that NFY is instrumental in expanding CD34+ cord blood cells in vitro [14]. This is an important discovery as cord blood cells, like isolated HSCs, are used to treat cancers of the blood system and different forms of anemia; methods to expand their number in culture promote their clinical utility.

The comparatively easy accessibility and the established clinical use of HSCs, which will be discussed later in this chapter, propelled their investigation and

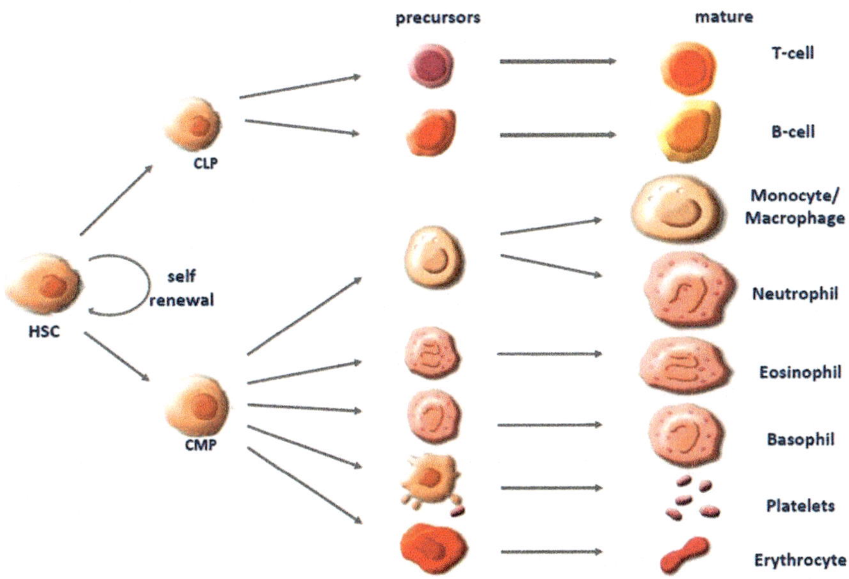

Fig. 1 Simplified depiction of the hematopoietic hierarchy. HSCs give rise to common lymphoid precursors (CLPs) and common myeloid precursors (CMPs), which further differentiate into different mature cell types. While lineage committed cells can be described upon their expression of certain markers on the cell surface, HSCs and precursor cells are usually identified by a combination of several surface markers [12]

additionally made them a model of stem cell research. In fact, concepts about cellular differentiation hierarchies and cellular self-renewal stem from HSC research.

3.4.2 Cancer Stem Cells

The concept that a small number of stem cells are able to support a tissue comprised of millions of different cells and perhaps dozens of different cell types can be translated from HSCs to other stem cell populations. When properly controlled, this exquisite system is beneficial for a developing or injured organism. However, unbalanced cell proliferation and differentiation can result in metaplastic or dysplastic growth and cancer. Metaplasia refers to the replacement of one differentiated cell type with that of another differentiated cell type. Dysplasia refers to abnormal development, often meaning an abnormally large number of immature cells in a tissue.

In 1994 and 1997, John E. Dick and colleagues published research demonstrating that the same principles of self-renewal and differentiation discovered for HSCs hold true for cancer cells [6, 29]. The underlying experiments evaluated the number and properties of leukemic cells necessary to transplant leukemia from one mouse to another. Interestingly, only a small specific population of cancer cells was necessary and sufficient to transmit leukemia. These cells were termed leukemic stem cells. In perfect analogy to HSCs, cancer stem cells are capable of self-renewal and differentiation, thus feeding the tumor without being exhausted. Like HSCs, cancer stem cells likely divide slowly and give rise to rapidly dividing cells. Thus, cancer therapies targeting rapidly dividing cells may temporarily ease the symptoms of cancer, but are doomed to fail as the cancer stem cells continue to slowly divide. This concept necessitates new strategies for the treatment of cancer, many of which are currently being tested in clinical studies (Table 1).

4 Extracellular Matrix and the Stem Cell Niche

When cells are removed from their natural environments, they behave differently than when they remain within their natural environments. Much effort has been concentrated on developing external environments for cells, including stem cells, that simulate the natural environment.

Stem cells thrive in particular microenvironments referred to as niches. The stem cell niche is an environment that fosters proper growth, proliferation, and differentiation of stem cells.

The idea of a specific niche for stem cells was proposed four decades ago by Schofield [41], who proposed the stem cell niche [10].

It is now known that stem cell niches do properties of hematopoietic stem cells are due to their association with other cells within a particular niche. Since this time, niches have been proposed for stem cells of numerous different origins, including skin, hair follicle, and intestine, to name a few [1, 2, 20, 30, 36, 46]. Cancer not only consists of cells, but the extracellular matrix (ECM) is a critical component contributing to uncontrolled cellular proliferation [39]. In addition to providing structure and support to tissue and anchorage to cells, the ECM is involved in receptor signaling and plays a dynamic role in establishing growth factor gradients and other secreted factors that serve as modulating cues that determine cell fate [15, 35]. Only the most prominent ECM molecules which we also reviewed earlier [40] will be discussed below, including collagen, proteoglycans/glycosaminoglycans, laminins, and fibronectin.

As the name implies, ECM proteins exist outside of the cell. They are secreted molecules that are highly modified, meaning they undergo post-translational processing in which certain residues are covalently phosphorylated, acetylated, glycosylated, or otherwise modified. These modifications impart a tremendous degree of diversity and allow for exquisite spatial and temporal functional regulation of these proteins. The ECM proteins will be discussed below all undergo such post-translational modification and, while classified as glycoproteins, possess a number of other covalent modifications that are dependent on their tissue location and the particular stage of cellular development.

The most abundant ECM molecule is also the most abundant protein in the body, collagen. Collagen provides a large degree of structural support and contains epitopes found to influence cellular processes such as angiogenesis, or blood vessel growth. Like other ECM molecules, collagen binds to transmembrane receptors such as integrins and helps to mediate both inside-out and outside-in signaling between the cell proper and its extracellular environment [9].

Heparan sulphate proteoglycans (HSPGs) stand out as additional key players in the development and maintenance of the stem cell niche [21, 35]. HSPGs consist of specific core proteins with a variable number of polysaccharide chains, called glycosaminoglycans (GAGs) attached [7, 25, 37]. Unique combinations of core proteins and GAG side chains imparts a tremendous degree of diversity to the ECM and provides a single proteoglycan the ability to play diverse roles within a tissue in time-dependent manner [7].

Laminin and fibronectin are additional ECM components that provide structure and individuality to specific stem cell niches. Laminins comprise a family of glycoproteins consisting of alpha, beta, and gamma chains that combine to form various functional heterotrimers. Fibronectin is another large glycoprotein found within the ECM. Like collagens, laminins and fibronectins bind integrin receptors as one means of regulating cellular behavior.

4.1 Three-Dimensional Scaffolds

In the organ system, cells reside in highly specialized niches consisting of supportive cells and ECM. These components provide mechanical stability, generate and transduce biophysical signals via cellular receptor binding to intracellular and extracellular ligands, and sequester and release soluble growth factors to promote growth and differentiation. These specific and interdependent cell-matrix interactions provide instructive cues for stem cells that determine their behavior. For these reasons, growing, expanding, and differentiating cells outside of the body has proved challenging. A flat plastic petri dish cannot provide the optimal conditions for stem cell growth and expansion.

Three-dimensional scaffolds represent an artificial microenvironment aimed at imitating the natural niche and come in as many varieties as do stem cells themselves. The emergence of three-dimensional (3D) scaffolds to mimic the natural stem cell environment has enriched the field of stem cell research and advanced the field of regenerative medicine. Most scaffolds consist of biodegradable materials that can be loaded with cells and supplied with nutrients and morphogens within a bioreactor, an engineered system used to support a biological environment, providing a home away from home in order to obtain the desired behavior from the cultured stem cells 24,570,851, 27,851,739. Thus, the desired cellular behavior dictates the design and properties of the scaffold.

Research led by Dennis Discher provided the first evidence that the nature of the matrix strongly influences the lineage into which stem cells differentiate [17]. Using identical culture serum conditions, this group found soft matrices to support neurogenic differentiation, stiffer matrices myogenic differentiation, and solid matrices osteogenic differentiation. The results of these studies demonstrate how physical factors influence stem cell differentiation.

Work stemming from our own research contributed to show that physical factors affecting stem cell behavior are not limited to the stiffness of culture matrices. Mechanical stimulation using extracorporeal shock waves additionally has the potential to modify migration and proliferation of stem cell populations [45]. In addition to the stiffness of the matrix and physical stimulation, ligand-receptor interactions, autocrine and paracrine signals, as well as oxygen, cytokine, and nutrient concentrations need to be considered synergistically in the design of a bioreactor [52].

In order to be clinically relevant, biologically relevant culture of stem cells and even culture of entire organ grafts is necessary. One avenue taken in rodent models and a few single human cases has been to de-cellularize a donor organ, meaning to take out all cells and small molecules, leaving only the natural three-dimensional extracellular matrix behind. Recipient cells seeded onto this matrix have been successfully grown into transplantable organs. Studies of such organ transplant research are ongoing, including studies of trachea, lung, kidney, and heart. Advances and risks including immunological concerns surrounding this line of research have been reviewed [54].

4.2 *Stem Cell Therapies in Clinical Development*

HSC culture has been established for the treatment of blood cancer, anemia, and autoimmune diseases for decades. However, techniques are constantly improving. The number of studies and trials involving HSCs highlights the interest and potential clinical impact of research in this field. As of today, more than 700 interventional clinical studies involving HSCs are listed at clinicaltrials.gov (access at clinicaltrials.gov, Apr. 26th, 2017). Table 1 lists a representation of ongoing interventional clinical trials involving HSC, with therapies targeting blood cancers omitted from the list.

In recent years, treatment options beyond HSCs have been designed and are being tested clinically using new approaches that employ different stem cell populations. Currently, there are also around 700 such trials listed (access at clinicaltrials.gov, Apr. 26th, 2017). Table 2 provides examples of ongoing stem cell based interventional approaches utilizing stem cells other than HSCs that are currently in clinical trials for a variety of disease.

4.2.1 Transdifferentiation

For many stem cell populations, the central limitation hindering clinical use remains the accessibility of large numbers of differentiated cell types. While ES cells but not differentiated/mature cells can be expanded infinitesimally in culture, this shortage can only be overcome if patient-specific pluripotent cells could be differentiated into the target cells. This entails, first, the accessibility of such cells and second, ways of controlled, directed differentiation.

The first challenge has been overcome by the generation of induced pluripotent cells (iPSC) from basically any differentiated cell type [48].

The second obstacle, i.e. directed differentiation of these cells in vitro, remains challenging. The intracellular factors and extracellular stimuli are not yet sufficiently defined to deliver these factors or design 3D-culture systems that adequately mimic an appropriate microenvironment (see above, [52]). In 2008, a group led by Douglas Melton achieved to convert pancreatic exocrine cells in adult mice into cells indistinguishable from endogenous β-cells [59]. Besides the obvious importance for diabetes research, this study proved that a differentiated cell can be transdifferentiated into another differentiated cell type without being de-differentiated into the state of pluripotency. However, it does not seem entirely clear what happens during this process. As to the best of our knowledge, it cannot be excluded at that time that differentiated cells during the process of lineage conversion resemble a common progenitor or even a pluripotent cell. However, while this is an important question in stem cell research, it appears less relevant for clinical purposes as long as the lineage converted cells take over the desired functions.

From the medical point of view, the use of Muse cells (discussed above) might outperform trans-differentiated cells because of their relative accessibility, expandability, and applicability. It remains to be seen if these cells hold their promise in clinical trials.

4.3 Ethics in Stem Cell Research

Given the sensitivity of the topic of stem cell research, ethical considerations are addressed in this chapter. While oocytes are not fertilized for the specific use of research, there are numerous fertilized oocytes that remain unutilized from in vitro fertilization (IVF). ES cell cultures have been obtained from supernumerary (excess) products of IVF. Discussions about IVF, a preceding scientific innovation, include arguments similar to those raised in the debate about ES cell research. Likewise, abortion the cloned sheep, Dolly, and genetically modified crops caused a fundamental debate over the limits of human intervention in all matters of life [3]. Interestingly, today the majority of the population in Western societies is in favor of IVF, given its undisputable medical usefulness. Discussions over GM crops endure with different outcomes concerning their acceptance in different parts of the world despite a scientific consensus for their safety [3]. This highlights the complexity of ethics in the field of research, which is further complicated by political, cultural, and religious confounders.

The major ethical position against ES cell research is the isolation of ES cells from supernumerary pre-implantation stage embryos from fertility clinics. The argument is that using these embryos for research purposes prevents the development of a human being. Follow-up questions include how ethical the unlimited storage of these embryos is or what should be done with the supernumerary fertilized oocytes. The position in favor of using ES cells for research argues that failing to use this powerful tool that has the potential cure serious diseases would be unethical. Different political and governmental administrations in different countries have adopted different positions concerning the use of ES cells for research, highlighting that a global agreement is not in sight.

As a bright spot, while there is a need for research on human ES cell lines [24] their widespread use for clinical purposes can be omitted as the iPS technology has provided an elegant way around this ethical dilemma. Furthermore, the discovery of additional adult stem cell populations may provide plentiful opportunities for the development of cellular therapies and tissue grafts.

4.4 Summary

The field of stem cell research continues to advance at a remarkable pace and holds tremendous promise for clinical applications. From the discovery of embryonic stem cells to the creation of Dolly to the ever-increasing use of induced pluripotent stem cells, this exciting field continues to surprise and challenge our understanding of biology and tissue repair as well as our beliefs regarding autonomy and life.

Stem cell research has now entered a new phase; one that includes clinical studies in human subjects where this wealth of knowledge gained through years of innovative research can be tested for therapeutic utility. Reaching this lofty goal required the combined efforts of diverse groups of scientists, including cell biologists, extracellular matrix experts, and bioengineers.

Additional possibilities for cell based patient specific therapies may arise once combined with contemporary technologies of gene editing, such as CRISPR/Cas9, which has itself progressed into the clinical development stage [43].

As stem cell research continues, it will be important to continue an open dialogue concerning ethical matters surrounding the techniques used in this filed. Discussions about gene editing and the possibility of creating germline changes that would affect future generations will certainly join those concerning the use, origin, development, and disposing of various types of stem cells. With the potential to effectively treat countless medical conditions on an individual bases, so called precision medicine, comes the responsibility to define and follow the developing code of ethics.

References

1. Adams, G. B., & Scadden, D. T. (2006). The hematopoietic stem cell in its place. *Nature Immunology*. https://doi.org/10.1038/ni1331. [pii]\r10.1038/ni1331.
2. Aguila, H. L., & Rowe, D. W. (2005). Skeletal development, bone remodeling, and hematopoiesis. *Immunological Reviews*. https://doi.org/10.1111/j.0105-2896.2005.00333.x.
3. Allum, N., Allansdottir, A., Gaskell, G., Jackson, J., Moldovan, A., Priest, S., et al. (2017). Religion and the public ethics of stem-cell research: Attitudes in Europe, Canada and the United States, 1–14.
4. Becker, A. J., McCulloch, E. A, & Till, J. E. (1963a). Cytological demonstration of the clonal nature of spleen colonies derived from transplanted mouse marrow cells. *Nature, 197*(4866), 452–454. https://doi.org/10.1038/197452a0.
5. Becker, A. J., McCulloch, E. A., & Till, J. E. (1963b). Cytological demonstration of the clonal nature of spleen colonies derived from transplanted mouse marrow cells. *Nature*. https://doi.org/10.1038/197452a0.
6. Bonnet, D., & Dick, J. (1997). Human acute myeloid leukemia is organized as a hierarchy that originates from a primitive hematopoietic cell. *Nature Medicine, 3*(7), 730–737.
7. Bülow, H. E., & Hobert, O. (2006). The molecular diversity of glycosaminoglycans shapes animal development. *Annual Review of Cell and Developmental Biology*. https://doi.org/10.1146/annurev.cellbio.22.010605.093433.
8. Bungartz, G., Land, H., Scadden, D. T., & Emerson, S. G. (2012). NF-Y is necessary for hematopoietic stem cell proliferation and survival. *Blood, 119*(6), 1380–1389. https://doi.org/10.1182/blood-2011-06-359406.
9. Bungartz, G., Stiller, S., Bauer, M., Müller, W., Schippers, A., Wagner, N., et al. (2006). Adult murine hematopoiesis can proceed without beta1 and beta7 integrins. *Blood, 108*(6), 1857–1864. Retrieved from http://bloodjournal.hematologylibrary.org/cgi/content/full/108/6/1857.
10. Cabarcas, S. M., Mathews, L. A., & Farrar, W. L. (2011). The cancer stem cell niche-there goes the neighborhood? *International Journal of Cancer, 129*(10), 2315–2327. https://doi.org/10.1002/ijc.26312.
11. Campbell, K. H., McWhir, J., Ritchie, W. A., & Wilmut, I. (1996). Sheep cloned by nuclear transfer from a cultured cell line. *Nature, 380*(6569), 64–66. https://doi.org/10.1038/380064a0.
12. Chao, M. P., Seita, J., & Weissman, I. L. (2008). Establishment of a normal hematopoietic and leukemia stem cell hierarchy establishment of a normal hematopoietic and leukemia stem cell hierarchy, *LXXIII* (Ogawa 1993), pp. 439–449. https://doi.org/10.1101/sqb.2008.73.031.
13. Doetschman, T., Gregg, R. G., Maeda, N., Hooper, M. L., Melton, D. W., Thompson, S., et al. (1987). Targetted correction of a mutant HPRT gene in mouse embryonic stem cells. *Nature*. https://doi.org/10.1038/330576a0.
14. Domashenko, A. D., Danet-Desnoyers, G., Aron, A., Carroll, M. P., & Emerson, S. G. (2010). TAT-mediated transduction of NF-Ya peptide induces the ex vivo proliferation and engraftment potential of human hematopoietic progenitor cells. *Blood, 116*(15), 2676–83. https://doi.org/10.1182/blood-2010-03-273441.
15. Domingues, M. J., Cao, H., Heazlewood, S. Y., Cao, B., & Nilsson, S. K. (2017). Niche extracellular matrix components and their influence on HSC. *Journal of Cellular Biochemistry*. https://doi.org/10.1002/jcb.25905.
16. Doudna, J. A., & Charpentier, E. (2014). Genome editing. The new frontier of genome engineering with CRISPR-Cas9. TL–346. *Science (New York, N.Y.)*. https://doi.org/10.1126/science.1258096.
17. Engler, A. J., Sen, S., Sweeney, H. L., & Discher, D. E. (2006). Matrix elasticity directs stem cell lineage specification. *Cell, 126*(4), 677–689. https://doi.org/10.1016/j.cell.2006.06.044.
18. Evans, M. J., & Kaufman, M. H. (1981). Establishment in culture of pluripotential cells from mouse embryos. *Nature*. https://doi.org/10.1038/292154a0.

19. Fuchs, E., & Chen, T. (2012). A matter of life and death: Self-renewal in stem cells. *EMBO Reports*. https://doi.org/10.1038/embor.2012.197.
20. Fuchs, E., Tumbar, T., & Guasch, G. (2004). Socializing with the neighbors: Stem cells and their niche. *Cell*. https://doi.org/10.1016/S0092-8674(04)00255-7.
21. Gupta, P., Oegema, T. R., Brazil, J. J., Dudek, A. Z., Slungaard, A., & Verfaillie, C. M. (1998). Structurally specific heparan sulfates support primitive human hematopoiesis by formation of a multimolecular stem cell niche. *Blood, 92*(12), 4641–4651. Retrieved from http://www.ncbi.nlm.nih.gov/pubmed/9845530.
22. Hadjimichael, C., Chanoumidou, K., Papadopoulou, N., Arampatzi, P., Papamatheakis, J., & Kretsovali, A. (2015). Common stemness regulators of embryonic and cancer stem cells. *World Journal of Stem Cells*. https://doi.org/10.4252/wjsc.v7.i9.1150.
23. Hall, B., Limaye, A., & Kulkarni, A. B. (2009). Overview: Generation of gene knockout mice. *Current Protocols in Cell Biology*. https://doi.org/10.1002/0471143030.cb1912s44.
24. Hyun, I., Hochedlinger, K., Jaenisch, R., & Yamanaka, S. (2007). New advances in iPS cell research do not obviate the need for human embryonic stem cells. *Cell Stem Cell*. https://doi.org/10.1016/j.stem.2007.09.006.
25. Iozzo, R. V. (1998). Matrix proteoglycans: from molecular design to cellular function. *Annual Review of Biochemistry*. https://doi.org/10.1146/annurev.biochem.67.1.609.
26. Kang, L., Wang, J., Zhang, Y., Kou, Z., & Gao, S. (2009). iPS cells can support full-term development of tetraploid blastocyst-complemented embryos. *Cell Stem Cell, 5*(2), 135–138. https://doi.org/10.1016/j.stem.2009.07.001.
27. Kanji, S., Pompili, V. J., & Das, H. (2011). Plasticity and maintenance of hematopoietic stem cells during development. *Recent Patents on Biotechnology*. https://doi.org/10.2174/187220811795655896.
28. Kuroda, Y., & Dezawa, M. (2014). Mesenchymal stem cells and their subpopulation, pluripotent muse cells, in basic research and regenerative medicine. *Anatomical Record, 297* (1), 98–110. https://doi.org/10.1002/ar.22798.
29. Lapidot, T., Sirard, C., Vormoor, J., Murdoch, B., Hoang, T., Caceres-Cortes, J., et al. (1994). A cell initiating human acute myeloid leukaemia after transplantation into SCID mice. *Nature*. https://doi.org/10.1038/367645a0.
30. Lin, H. (2002). The stem-cell niche theory: lessons from flies. *Nature Reviews. Genetics*. https://doi.org/10.1038/nrg952.
31. Maherali, N., Sridharan, R., Xie, W., Utikal, J., Eminli, S., Arnold, K., et al. (2007). Directly reprogrammed fibroblasts show global epigenetic remodeling and widespread tissue contribution. *Cell Stem Cell, 1*(1), 55–70. https://doi.org/10.1016/j.stem.2007.05.014.
32. Martin, G. R. (1981). Isolation of a pluripotent cell line from early mouse embryos cultured in medium conditioned by teratocarcinoma stem cells. *Proceedings of the National Academy of Sciences of the United States of America, 78*(12), 7634–7638. https://doi.org/10.1073/pnas.78.12.7634.
33. Meissner, A., Wernig, M., & Jaenisch, R. (2007). Direct reprogramming of genetically unmodified fibroblasts into pluripotent stem cells. *Nature Biotechnology*. https://doi.org/10.1038/nbt1335.
34. Metcalf, D., Moore, M. A. S., & Shortman, K. (1971). Adherence column and buoyant density separation o f bone marrow stem cells and more. *Journal Cell Physiology, 78*(3), 441–449.
35. Nurcombe, V., & Cool, S. M. (2007). Heparan sulfate control of proliferation and differentiation in the stem cell niche. *Critical Reviews in Eukaryotic Gene Expression, 17* (2):159–171.
36. Ohlstein, B., Kai, T., Decotto, E., & Spradling, A. (2004). The stem cell niche: Theme and variations. *Current Opinion in Cell Biology*. https://doi.org/10.1016/j.ceb.2004.09.003.
37. Oldberg, A., Antonsson, P., Hedbom, E., & Heinegard, D. (1990). Structure and function of extracellular matrix proteoglycans. *Biochemical Society Transactions, 18*(5), 789–792.

38. Pittenger, M. F., Mackay, A. M., Beck, S. C., Jaiswal, R. K., Douglas, R., Mosca, J. D., et al. (1999). Multilineage potential of adult human mesenchymal stem cells. *Science, 284*(5411), 143–147. Retrieved from http://www.ncbi.nlm.nih.gov/pubmed/10102814.
39. Raaijmakers, M. H. G. P., Mukherjee, S., Guo, S., Zhang, S., Kobayashi, T., Schoonmaker, J. A., et al. (2010). Bone progenitor dysfunction induces myelodysplasia and secondary leukaemia. *Nature*. https://doi.org/10.1038/nature08851.
40. Rodgers, K. D., San Antonio, J. D., & Jacenko, O. (2008). Heparan sulfate proteoglycans: A GAGgle of skeletal-hematopoietic regulators. *Developmental Dynamics, 237*(10), 2622–2642. https://doi.org/10.1002/dvdy.21593.
41. Schofield, R. (1978). The relationship between the spleen colony-forming cell and the haemopoietic stem cell. *Blood Cells, 4*(1–2), 7–25 https://Chronicischaemic mitralregurgitation.Currenttreatmentresultsandnewmechanism-basedsurgicalapproaches.
42. Seydoux, G., & Braun, R. E. (2006). Pathway to totipotency: Lessons from germ cells. *Cell, 127*(5), 891–904. https://doi.org/10.1016/j.cell.2006.11.016.
43. Sheridan, C. (2017). CRISPR therapeutics push into human testing. *Nature Biotechnology, 35*, 3. Retrieved from http://dx.doi.org/10.1038/nbt0117-3.
44. Stadtfeld, M., Hochedlinger, K., Stadtfeld, M., & Hochedlinger, K. (2010). Induced pluripotency: History, mechanisms, and applications. *Genes & Development*. https://doi.org/10.1101/gad.1963910.
45. Suhr, F., Delhasse, Y., Bungartz, G., Schmidt, A., Pfannkuche, K., & Bloch, W. (2013). Cell biological effects of mechanical stimulations generated by focused extracorporeal shock wave applications on cultured human bone marrow stromal cells. *Stem Cell Res, 11*(2), 951–964. https://doi.org/10.1016/j.scr.2013.05.010. [pii]\r10.1016/j.scr.2013.05.010.
46. Taichman, R. S., Dc, W., & Taichman, R. S. (2011). Blood and bone : Two tissues whose fates are intertwined to create the hematopoietic stem-cell niche. *Blood, 105*(7), 2631–2639. https://doi.org/10.1182/blood-2004-06-2480.
47. Takahashi, K., Tanabe, K., Ohnuki, M., Narita, M., Ichisaka, T., Tomoda, K., et al. (2007). Induction of pluripotent stem cells from adult human fibroblasts by defined factors. *Cell, 131* (5), 861–872. https://doi.org/10.1016/j.cell.2007.11.019.
48. Takahashi, K., & Yamanaka, S. (2006). Induction of pluripotent stem cells from mouse embryonic and adult fibroblast cultures by defined factors. *Cell, 126*(4), 663–676. https://doi.org/10.1016/j.cell.2006.07.024.
49. Thomas, K. R., & Capecchi, M. R. (1987). Site-directed mutagenesis by gene targeting in mouse embryo-derived stem cells. *Cell*. https://doi.org/10.1016/0092-8674(87)90646-5.
50. Thoms, J. A., Joseph, I.-E., Sander, S. S., Waknitz, M. A., Swiergiel, J. J., et al. (1998). Embryonic stem cell lines derived from human blastocysts. *Science, 282*(2391), 1145–1147.
51. Thomson, J. A., Itskovitz-eldor, J., Shapiro, S. S., Waknitz, M. A., Swiergiel, J. J., Marshall, V. S., et al. (1998). Embryonic stem cell lines derived from human blastocysts. *Advancement Of Science, 282*(5391), 1145–1147. https://doi.org/10.1126/science.282.5391.1145.
52. Vunjak-Novakovic, G., & Scadden, D. T. D. T. (2011). Biomimetic platforms for human stem cell research. *Cell Stem Cell*. https://doi.org/10.1016/j.stem.2011.02.014.
53. Wagers, A. J., & Weissman, I. L. (2004). Plasticity of adult stem cells. *Cell 116*(Icm), 639–648.
54. Wiles, K., Fishman, J. M., De Coppi, P., & Birchall, M. A. (2016). The host immune response to tissue-engineered organs: Current problems and future directions. *Tissue Engineering Part B: Reviews, 22*(3), 208–219. https://doi.org/10.1089/ten.teb.2015.0376.
55. Wilmut, I., Schnieke, A. E., McWhir, J., Kind, A. J., & Campbell, K. H. (1997). Viable offspring derived from fetal and adult mammalian cells. *Nature*. https://doi.org/10.1038/385810a0.
56. Ye, J., Ge, J., Zhang, X., Cheng, L., Zhang, Z., He, S., et al. (2015). Pluripotent stem cells induced from mouse neural stem cells and small intestinal epithelial cells by small molecule compounds. *Cell Research, 26*(1), 34–45. https://doi.org/10.1038/cr.2015.142.

57. Yu, J., Vodyanik, M. A., Smuga-Otto, K., Antosiewicz-Bourget, J., Frane, J. L., Tian, S., et al. (2007). Induced pluripotent stem cell lines derived from human somatic cells. *Science (New York, N.Y.)*. https://doi.org/10.1126/science.1151526.
58. Zhao, X., Li, W., Lv, Z., Liu, L., Tong, M., Hai, T., et al. (2009). iPS cells produce viable mice through tetraploid complementation. *Nature*. https://doi.org/10.1038/nature08267.
59. Zhou, Q., Brown, J., Kanarek, A., Rajagopal, J., & Melton, D. A. (2008). In vivo reprogramming of adult pancreatic exocrine cells to beta-cells. *Nature*. https://doi.org/10.1038/nature07314.

Author Biography

Dr. Gerd Bungartz (Germany) Dr. Bungartz has a diverse and international career in science and research. After his studies at the University of Cologne and the German Sport University, he focused his research on adhesion molecules, at Lund University in Sweden. He then continued this work at the Max Planck Institute for Biochemistry where he expanded his interests to include research on hematopoietic (stem) cells. After receiving his Ph.D. with honors in 2005, Dr. Bungartz relocated to work with Professor Emerson at the University of Pennsylvania and later to Massachusetts General Hospital/Harvard Medical School to work with Professor Scadden of the Harvard Stem Cell Institute. In 2012, he moved to the German Sport University in Cologne and later into industry working for Pfizer. Dr. Bungartz' research was awarded several grants and published in high-ranking journals. Currently, he holds positions at 'Hochschule Döpfer' and Merck Serono.

A Collagen-Hyaluronic Acid Matrix for Stem Cell Culture

Euisung Chung, Geajin Yoon and Hwal (Matthew) Suh

1 Introduction

Recently, the regenerative medicine has been introduced to promote self-restoration, replacement and regeneration of impaired bodily function of tissues or organs by delivering viable cells which are genetically appropriate to patients [1].

As cells are living by being attached on extracellular matrix (ECM) which is released from the cells and contains variable biological molecules necessary to provide favorable environment for cellular metabolism and structurally support tissue, mimicking the natural mammalian tissue by artificially hybridizing viable cells with biocompatible structural scaffold materials has been introduced [2, 3].

For scaffold, various biocompatible materials have been studied, as synthesized polymers or natural ECM components, according to the specific target tissue structure. An ideal artificial scaffold shall contain porous network that permits cell proliferation and entrapment, and transportation of oxygen, carbon dioxide, and nutritional substances [4–6].

By applying the ECM substances such as proteins and polysaccharides as scaffold components, as these are naturally synthesized, secreted and decomposed by enzymes in the body, less immunological events are expected after implantation [6–9]. Utilization of the signal transduction system in ECM is also possible. Additional vascular endothelial growth factor (VEGF) combined scaffold promotes

E. Chung · G. Yoon · H. (Matthew) Suh (✉)
Graduate Program for Nano Science and Technology,
The Graduate School, Yonsei University, Seoul, Korea
e-mail: hwal@yuhs.ac

H. (Matthew) Suh
Department of Medical Engineering,
Yonsei University College of Medicine, Seoul, Korea

G. M. Artmann et al. (eds.), *Biological, Physical and Technical Basics of Cell Engineering*, https://doi.org/10.1007/978-981-10-7904-7_4

neovascularization and infiltration of blood vessels, which are essential to provide vitality to the transferred cells, throughout the porous scaffolds after implantation [6].

In the ECM substances, collagen and hyaluronic acid (HA) have taken attention from numerous researchers for the scaffold materials [10–15].

Collagen occupies about 25% of the total body proteins and plays decisive role as structural scaffold for every tissue and organ, and the peptide sequences of RGD (Arg-Gly-Asp) in their molecules conduct cell attachment. Atelocollagen is an immune-free collagen molecule produced by eliminating telopeptides which exist in the both extremities of type I atelocollagen molecule that induce immune reactions, and has been recommended as the firstly chosen ECM substances for scaffold fabrications [2, 5, 9, 16–18].

In nature, glycosaminoglycan (GAG) is a polysaccharide existing in connective tissues as hydrophilic viscoelastic gel. Researches on various GAGs, such as chondroitin sulfate, heparin or HA, have reported that stiffness, chemical composition, and reaction against the growth factors of GAGs are affecting cell differentiation and proliferation [19–21].

HA consists of alternatively repeating β-D-glucuronic acid and β-N-acetyl-D-glucosamine molecules as one of the GAG. On HA matrices, cell attachment, migration and proliferation are promoted by the CD44 in molecules, and hydrophilic property affects mechanical strength and shape or structure of the artificial matrices made of HA [10, 14, 22].

Scaffolds designed to apply biological advantages of the collagen and HA revealed conduction of less immune reactions and wound healing through recognizing their ingredients dissolved in serum [20, 21]. Electrospinning method has been introduced to produce ultra-micro scaled fibers, and suggested as an appropriate tool to model ECM fibers with ranges about 50–500 nm in diameter. Nano-fibers have less diameter in comparison with length, and be easily fabricated to porous structure by controlling concentration of solute, as the diameter of the fiber increases as the solute increases, therefore, electrospun fiber is recommendable to model structural ECM [23–25].

For collagen-HA composites, collagen used to be dissolved by 1,1,1,3,3,3,-hexafluoro-2-propanol (HFP) through hydrolysis by disintegrating hydrogen bonds [26, 27], while HA is dissolved by formic acid or NaOH and N, N-dimethylformamide (DMF), and each solution is adulterated to produce electrospun nano-fibrous scaffold. For these methods produced collagen-HA scaffold, the residual organic solvents such as HFP, formic acid and DMF remained in body may have possibility to occur cytotoxicity [14, 15], high concentration of phosphate buffered salt (PBS) and ethanol are recommended as solvents for collagen to avoid organic solvents [28].

Also, to produce scaffold made of atelocollagen and HA, cross-linking agents are necessary to control the biodegradable property which is induced by enzymes in body. In a previous study reported by Park and Suh [11], 1-ethyl-3-(3-dimethylaminopropyl) carbodiimide (EDC) demonstrates reasonable result as such a carbodiimide becomes a hydrolytic residuum and extracted from body without cytotoxicity [10–12].

In this study, porous matrices of nano fibrous collagen-HA were produced, and their affinity to human bone marrow derived mesenchymal stem cells (hBM-MSCs), human embryonic stem cells (hESC) and human induced pluripotnent stem cells (hiPSC) were evaluated through observing expression of specific proteins, DNA or RNA, to investigate possibility of using the matrix in regenerative medicine.

2 Materials and Methods

2.1 *Materials*

Type I atelocollagen (collagen) was extracted by pepsin treatment from porcine dermal skins and lyophilized under −40 °C [29]. Molecular weight of HA was from 120 to 150 k. Collagen and HA were stored under −20 °C before use.

2.2 *Fabrication of Collagen-HA Nano-Fibrous Matrices*

HA was dissolved in 20X PBS solution at pH 7.3, that contains NaCl 160.0 g/L, KCl 4.0 g/L, Na2HPO4 28.4 g/L, and KH2PO4 4.8 g/L in DW [28].

Collagen-HA solution with concentrations of 5, 10, and 15% (w/v) atelocollagen and 0, 0.25, 0.5, 0.75, and 1% (w/v) HA were prepared, and a self-designed electro spinner using 18 gauge nozzle was applied to produce the collagen-HA nano-fibers. Figure 1 shows procedures making collagen-HA solutions and a self-designed electro spinner, and Table 1 is the condition of electro spinning.

To crosslink the nano-fibrous matrices, 50 mM EDC was dissolved in 95% (v/v) ethanol. Each matrix was immersed in EDC solution for 6, 12, 18, and 24 h respectively at room temperature. To remove the residual EDC, matrices were washed by gradually diluted 20X, 10X, 5X PBS and DW in ultrasound chambers. Nano-fibrous matrices were collected on either cover glasses or aluminum foils.

2.3 *Characterizations of Collagen-HA Nano-Fibrous Matrices*

Morphological characterization of the fabricated matrices were observed by using a Field Emission Scanning Electron Microscope(FE-SEM) and diameter of nano-fibers was measured for more than 100 fibers randomly selected from each

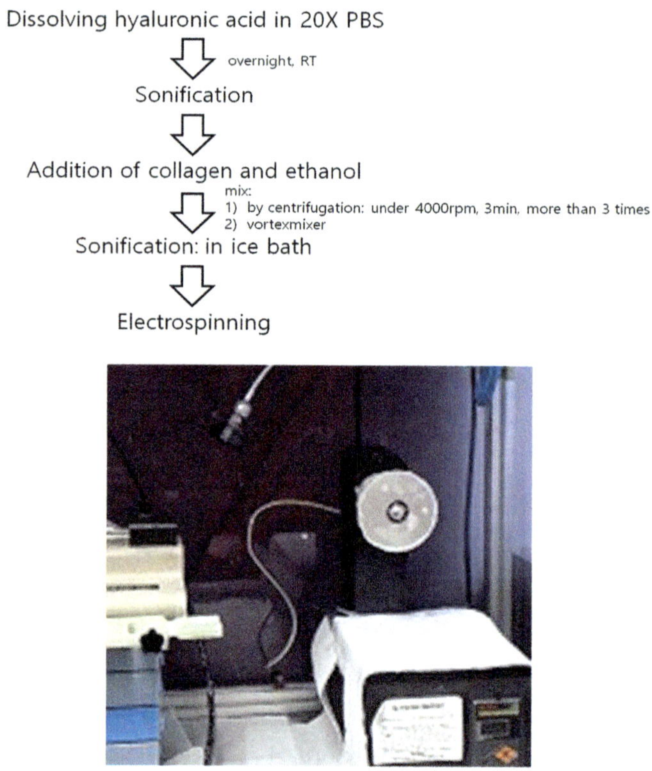

Fig. 1 Process of producing the collagen-HA solution and photo of the self-designed electro spinner

Table 1 Experimental condition of electrospinning

Voltage	15, 20, 25 kV
Nozzle size	18 gauge
Rotation speed of collector	300 rpm
Distance from nozzle to collector	5, 10, 15 cm
Injection speed of solution	0.5 ml/h

group, and the mean diameter was obtained depending on the conditions. Pore structure was defined as the space surrounded by more than 3 fibers, and diameter of pore was measured from more than 20 spots randomly selected in each group.

Alteration of chemical bonds in matrix by treatment of EDC was observed by a Fourier-transformed Infrared Spectroscope (FT-IR) at the wave length between 500 and 4000 cm^{-1}.

2.4 Cell Cultures

2.4.1 hBM-MSCs

Human bone marrow was provided by the department of Orthopedics in Yonsei University Severance hospital and approved by the institutional review board of the Yonsei University Health System in Seoul, Korea. hBM-MSCs was isolated from bone marrow by conventional methods. hBM-MSCs was cultured using Dulbecco's Modified Eagle's Medium-low glucose (DMEM-lg) containing 1% penicillin/streptomycin/amphotericin B and 10% FBS. Cells were passaged when it reached confluence more than 90% by trypsin-EDTA treatment. Cell culture was conducted under an incubator under 37 °C, 5% CO_2, 99% humidity conditions. Medium was changed every 3–4 days. Cells with passage number between 3 and 5 were seeded onto nano-fibrous matrices and checked the attachment, morphology and viability.

2.4.2 hiPSC, hESC

For feeder layer, STO cells were cultured in 10% FBS and 1% penicillin/streptomycin/amphotericin B added DMEM-high glucose (DMEM-hg) medium, and mitomycin C was treated to prevent STO cells proliferation.

Two types of cells were cultured on STO cell line. hiPSC and hESC were cultured using Dulbecco's Modified Eagle Medium Nutrient Mixture F-2(Ham) (1:1) (DMEM/F12) medium. 20% Knockout Serum Replacement, 1% L-glutamine, 0.1% Non-essential amino acids, 1% β-mercaptoethanol 0.5% penicillin/streptomycin, and 4 ng/ml human recombinant basic fibroblast growth factor (bFGF) were supplemented in DMEM/F12 medium.

hiPSC and hESC were purchased from American Type Culture Collection (ATCC, University Boulevard, Manassas, VA, USA).

After dissociating the undifferentiated cells, cells were suspended in cryopreservation solvents and stored in LN_2 tank before use.

2.5 Cell Seeding

Nano-fibrous matrices with dimensions of 2 cm × 2 cm collected on cover glasses were sterilized under 100, 70% ethanol for 30 min respectively. After sterilization, matrices were dried and washed with DW and 1X PBS and stored in medium before cell seeding. Matrices were placed upon a sterilized uncoated cover glasses.

2.5.1 hBM-MSCs

5×10^3 cells/cm^2 were seeded on sterilized matrices. Firstly, small amount of cells suspension was seeded on the surface of matrices. After keeping matrices to be soaked in the cell suspension in incubator for 6 h, medium was added to fill the matrix for full immersion. Cell attachment efficiency was confirmed 24 h after seeding.

2.5.2 hiPSC and hESC

Matrices soaked in medium were washed with DMEM/F12 medium and hiPSC and hESC were seeded onto each matrix. Cells were seeded at the middle of the matrices which immersed in medium. 48 h after seeding, 4 ng/ml bFGF was added to the medium. Cell attachment was observed under light microscope on 4 days after seeding. Medium was changed every 5 days, and same volume of new medium was added to the previous one.

2.6 Evaluation of the Attached Stem Cells onto Collagen-HA Matrices

For morphological characterization, cells were fixed with 4% formaldehyde solution for 20 min at RT and observed by a light microscope.

CCK8 analysis was performed to investigate the viability of cells attached on each matrix. Optical density was measured by Enzyme-Linked Immunosorbent Assay (ELISA), and observed the effect of bFGF in culture media,

3 Results

3.1 Fabrication and Characterization of Collagen-HA Nano-Fibrous Matrices

To select the proper concentrations for fabrication of collagen-HA matrices, HA concentration was set 0, 0.25, 0.5, 0.75, and 1% (w/v). By increasing collagen concentrations, morphological and chemical characterizations were compared with.

In all collagen concentrations, nano-fibrous structure were observed by FE-SEM. In 5 and 10% (w/v) collagen specimens, nodes were randomly distributed, and diameters of fibers was measured including the nodes (Figs. 2, 3, 4 and 5; Tables 2, 3 and 4).

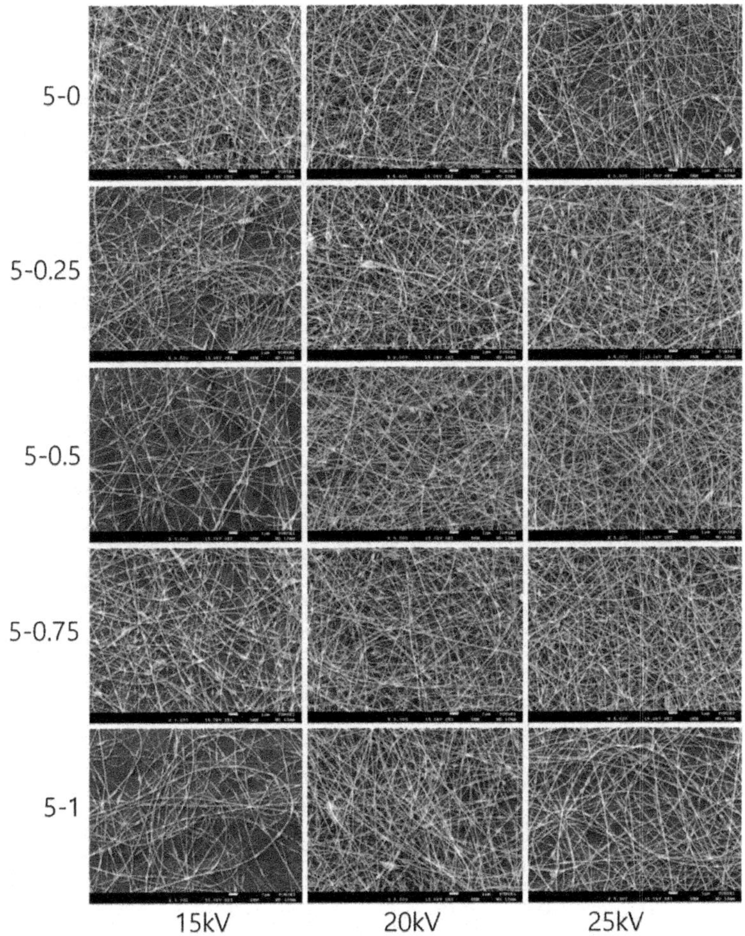

Fig. 2 Morphologies of the Collagen-HA nano-fibrous matrices of 5% (w/v) collagen containing 0, 0.25, 0.5, 0.75 and 1% (w/v) HA, produced by electrospinner at 15, 20 and 25 kV. (×5000, scale bar: 1 μm, FE-SEM)

Diameter of porous structures was measured in 15% (w/v) collagen nano-fibrous matrices. Maximum pore size was 3.55038 μm in 0% (w/v) HA mixed group spun at 25 kV, and minimum was 1.9266 μm in 1% (w/v) HA mixed group spun at 20 kV (Fig. 6).

In FT-IR analysis, type I atalocollagen, HA dissolved in 20X PBS, Ethanol dissolved 15% (w/v) collagen containing 0, 0.25, 0.5, 0.75 and 1% (w/v) HA were surveyed. Amide peaks of collagen were appeared at 3300, 1600 and 600 cm^{-1}, while HA were appeared at 3400 and 1600 cm^{-1}. Peak of proteoglycan ring of HA was detectable from 1140 to 980 cm^{-1}. Expressions of each collagen and HA peaks were revealed in entire concentrations [12, 28, 30] (Fig. 7).

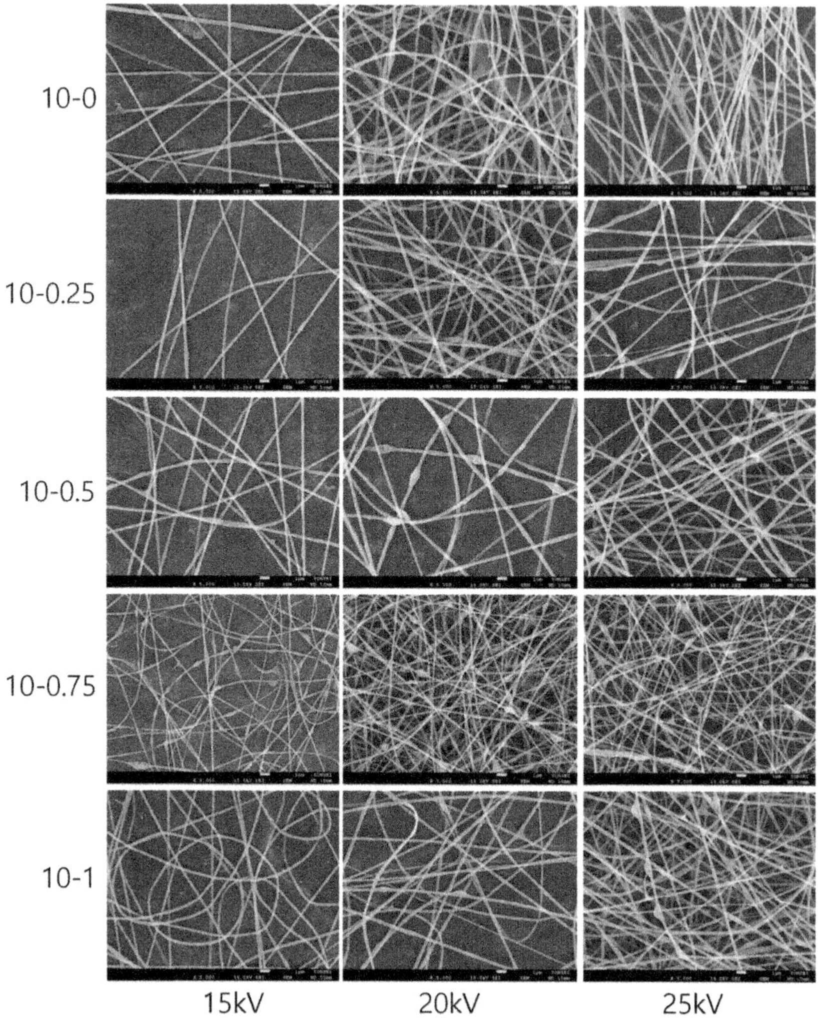

Fig. 3 Morphologies of the Collagen-HA nano-fibrous matrices of 10% (w/v) collagen containing 0, 0.25, 0.5, 0.75 and 1% (w/v) HA, produced by electrospinner at 15, 20 and 25 kV. (×5000, scale bar: 1 μm, FE-SEM)

Formation of amide bond by the EDC crosslinking was confirmed by appearance of the peaks near 1650–1400 cm^{-1}. Also peak of ester bond was appeared at 1100 cm^{-1}. At 1400 cm^{-1}, peaks related to carboxylate salt were shown. Morphological characteristics were verified by the EDC crosslinking time through FE-SEM. Nanofibers were adhered each other after EDC crosslinking (Figs. 8, 9, 10, 11, 12, 13 and 14).

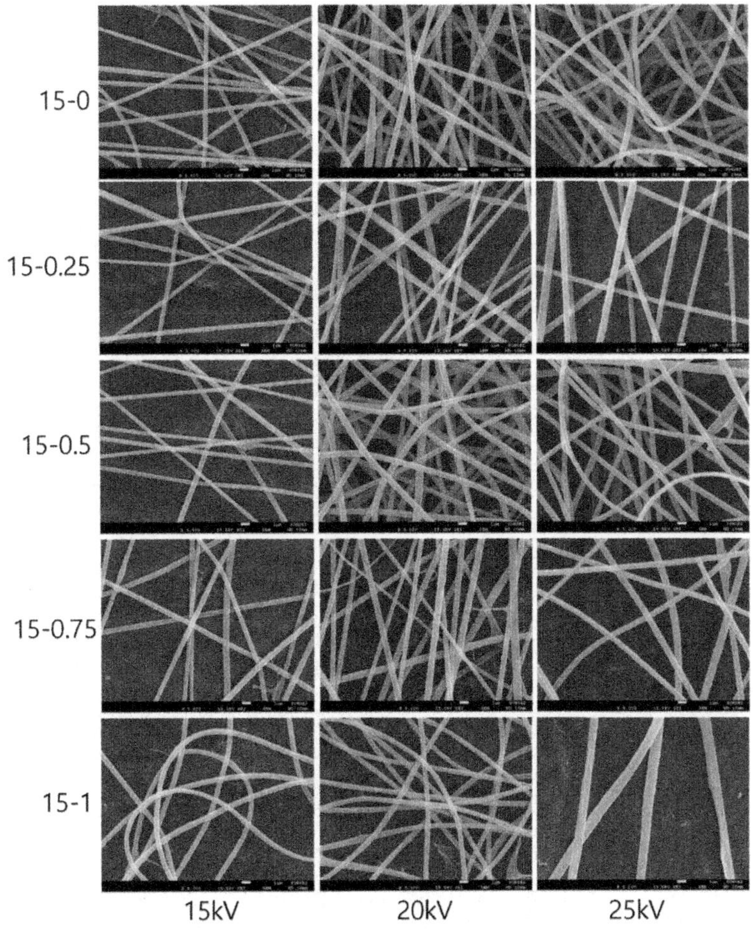

Fig. 4 Morphologies of the Collagen-HA nano-fibrous matrices of 15% (w/v) collagen containing 0, 0.25, 0.5, 0.75 and 1% (w/v) HA, produced by electrospinner at 15, 20 and 25 kV. (×5000, scale bar: 1 μm, FE-SEM)

3.2 Characterization of Cells After Seeding on Matrix

Analysis of cell viability was measured by CCK8 kit. hBM-MSC cells were seeded on 15% (w/v) collagen-0, 0.25, 0.5, 0.75 and 1% (w/v) HA matrices and compared with the control groups of culture flask. 4 days after cell seeding, 0.25% (w/v) HA containing 15% (w/v) collagen matrices demonstrated the highest value of cell viability, and cell viability decreased as the concentration of HA increased (Fig. 15).

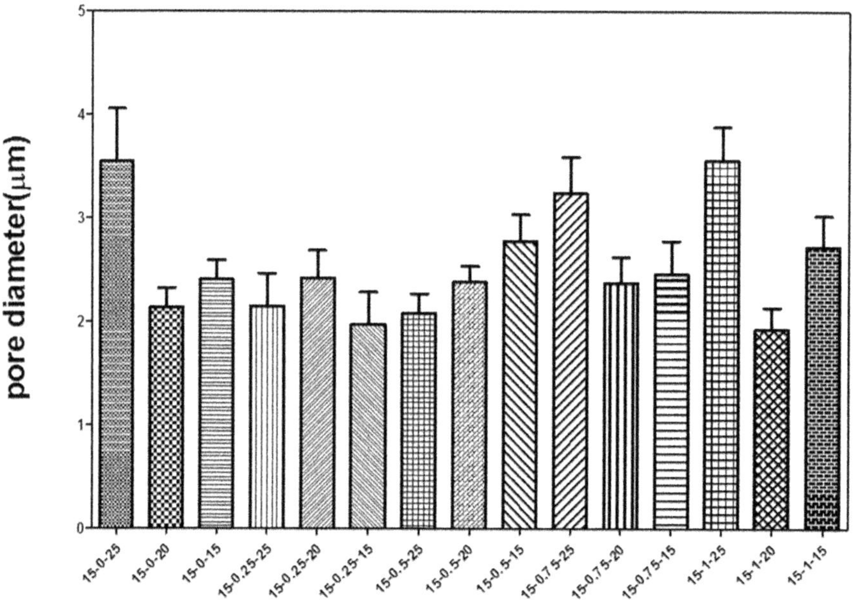

Fig. 5 Pore diameters of collagen-HA nano-fibrous matrices by 15% (w/v) collagen and 0, 0.25, 0.5. 0.75 and 1% (w/v) HA at 25, 20 and 15 kV. (Collagen concentration-HA concentration-Voltage)

Table 2 Diameters of Collagen-HA nano-fiberes (including nodes)

		5–0	5–0.25	5–0.5	5–0.75	5–1
15 kV	Mean	75.825	74.0	94.2625	81.4375	103.637
	SD	21.4717	23.8662	28.7145	26.679	30.2811
	SE	2.4006	2.66832	3.21038	2.9828	3.38553
20 kV	Mean	89.4375	83.675	96.2	80.975	104.537
	SD	22.9622	22.4261	28.8586	22.0511	28.2596
	SE	2.56725	2.50732	3.22649	2.46539	3.15952
25 kV	Mean	89.075	89.175	101.613	99.4875	98.6375
	SD	24.4643	22.0744	23.9358	23.9361	26.79
	SE	2.73519	2.46762	2.67611	2.67614	2.99521

5% (w/v) collagen containing 0, 0.25, 0.5, 0.75 and 1% (w/v) HA, produced by electrospinner at 15, 20 and 25 kV

Morphologies of hBM-MSC, hESC, and hiPSC seeded upon 15% (w/v) collagen-0, 0.25, 0.5, 0.75, 1% (w/v) HA matrices were observed 4, 10, 16, 22, 28 days after seeding (Figs. 16, 17, 18, 19, 20 and 21).

In control group of culture flask, hBM-MSCs reached 90% confluence after 4 days, and it reached 100% confluence after 10 days at the site of cell seeded.

Table 3 Diameters of Collagen-HA nano-fiberes (including nodes)

		10–0	10–0.25	10–0.5	10–0.75	10–1
15 kV	Mean	248.113	251.726	256.97	143.682	225.7
	SD	36.071	40.7983	37.0524	31.0127	32.8593
	SE	4.03286	5.18138	4.52667	3.8174	3.67378
20 kV	Mean	267.237	270.2	289.288	167.237	269.515
	SD	58.5617	40.7078	67.0893	32.3204	55.8218
	SE	6.5474	4.55127	8.25813	3.70741	6.76939
25 kV	Mean	268.075	279.567	293.625	169.365	242.279
	SD	54.5947	48.2164	48.451	40.447	49.5818
	SE	6.10387	5.89058	5.41699	4.70187	6.01267

10% (w/v) collagen containing 0, 0.25, 0.5, 0.75 and 1% (w/v) HA, produced by electrospinner at 15, 20 and 25 kV

Table 4 Diameters of Collagen-HA nano-fiberes (including nodes)

		15–0	15–0.25	15–0.5	15–0.75	15–1
15 kV	Mean	490.152	517.104	527.808	631.44	576.263
	SD	86.0754	98.8448	181.322	128.905	104.978
	SE	7.4919	8.84094	17.7801	11.9686	10.7705
20 kV	Mean	612.821	620.137	722.01	620.875	607.687
	SD	124.119	137.837	269.153	195.279	107.139
	SE	11.1915	12.743	26.2667	17.8264	9.99076
25 kV	Mean	694.566	711.611	736.719	753.986	768.537
	SD	166.038	202.704	184.011	155.902	226.267
	SE	14.2377	20.797	18.7805	18.3732	27.6429

15% (w/v) collagen containing 0, 0.25, 0.5, 0.75 and 1% (w/v) HA, produced by electrospinner at 15, 20 and 25 kV

Control had larger growth area than matrices, and showed cells keep spreading across the surface of flasks.

hBM-MSCs flatly distributed throughout the surface, therefore, morphology of hBM-MSCs was hard to define by light microscope. On the other hand, colonies of hiPSC and hESC protruded from the surfaces of matrices. Spherically formed colony could be seen in both hiPSC and hESC on 10 days after seeding. But on the 4th day, hiPSC demonstrated flatly spreading differentiated cells. Particularly, 0.5% (w/v) HA contained matrices showed large sized colonies, and cells grew out of the colony and spread as time courses. This phenomenon was not seen in the other groups. Cell viability was confirmed by the CCK8 analysis. There were no evidences of proliferation of hiPSC but hESC showed signs of proliferation by increasing of optical density in comparison with the results at day 10 and 28. Morphological characteristic was not altered as time courses. And CCK8 results were not affected by the inclusion of growth factors (Figs. 19, 20 and 21).

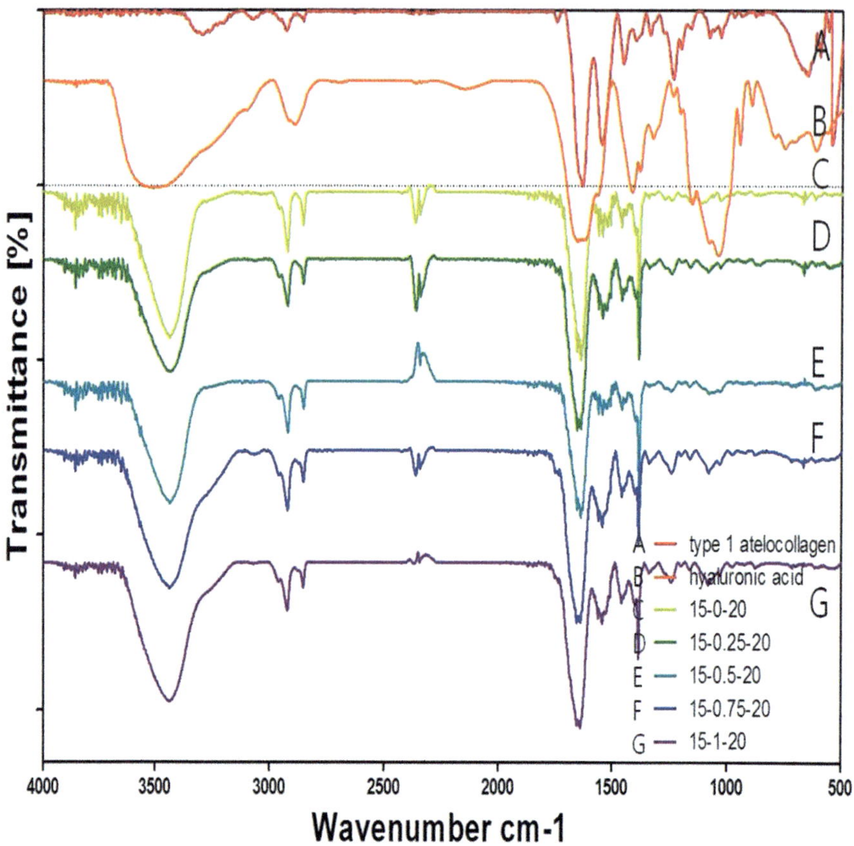

Fig. 6 FT-IR analysis. (A) non-treated type I atelocollagen, (B) non-treated HA, (C) 15% (w/v) collagen-0% HA at 20 kV, (D) 15% (w/v) collagen-0.25% (w/v) HA at 20 kV, (E) 15% (w/v) collagen- 0.5% (w/v) HA at 20 kV, (F) 15% (w/v) collagen-0.75% (w/v) HA at 20 kV, (G) 15% (w/v) collagen-1% (w/v) HA at 20 kV

4 Discussions

To produce optimal solvent condition for fabrication of collagen-HA nano-fibrous matrices, conditions suggested by Park and Suh were adopted [11]. This condition was used to solve the problem that HA has low solubility in organic solvents. Conventionally, binary solvents like NaOH/N,N-dimethylformamide(DMF) or HFP/formic acid were used to make collagen-HA nano- fibrous matrix. But, these works had some limitations because of the use of toxic solvents like formic acid, DMF, and HFP [8, 14, 27]. Therefore, benign solvents like PBS and ethanol were applied for this work to select the conditions to produce proper collagen-HA nano-fibrous matrix structures.

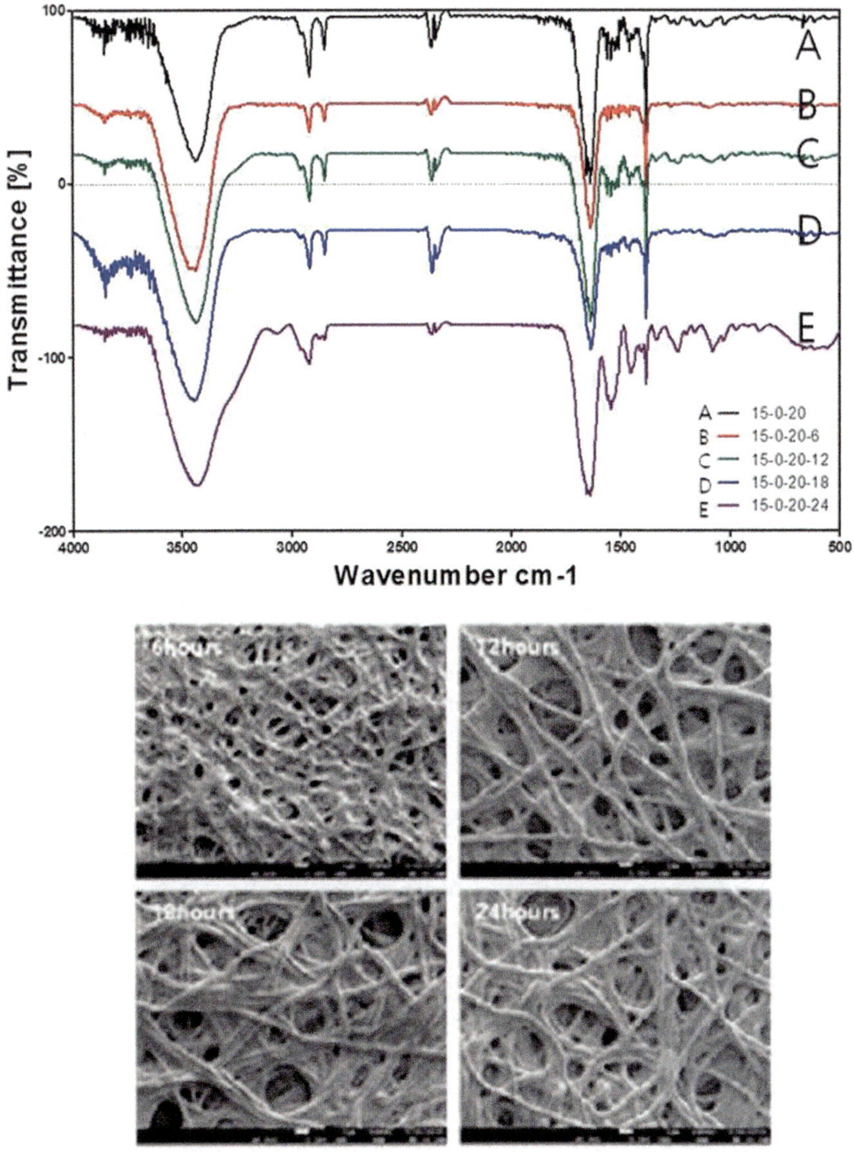

Fig. 7 15% (w/c) collagen-0% (w/v) HA at 20 kV nano-fibrous matrices after crosslinking by EDC. Upper: FT-IR analysis after (A) before crosslinking, (B) crosslinking for 6 h, (C) crosslinking for 12 h, (D) crosslinking for 18 h, (E) crosslinking for 24 h. Lower: FE-SEM view of nano-fibrous matrices after crosslinking by EDC. (×5000, scale bar: 1 μm)

HA concentrations were set 0, 0.25, 0.05, 0.75, and 1% (w/v) in PBS and ethanol conditions. Gel was formed during solution making processes due to its hydrophilic property. And to solve the low solubility of HA in ethanol, HA was dissolved in

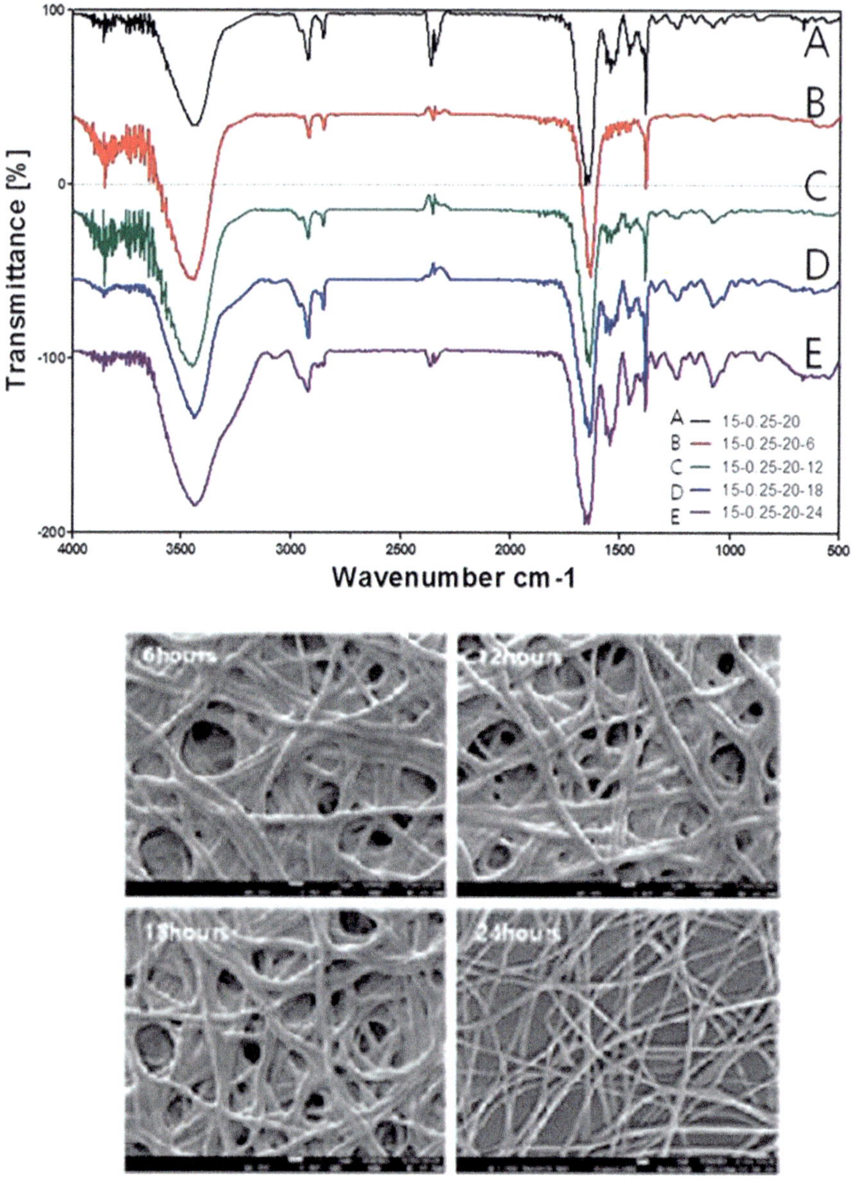

Fig. 8 15% (w/c) collagen-0.25% (w/v) HA at 20 kV nano-fibrous matrices after crosslinking by EDC. Upper: FT-IR analysis after (A) before crosslinking, (B) crosslinking for 6 h, (C) crosslinking for 12 h, (D) crosslinking for 18 h, (E) crosslinking for 24 h. Lower: FE-SEM view of nano-fibrous matrices after crosslinking by EDC. (×5000, scale bar: 1 μm)

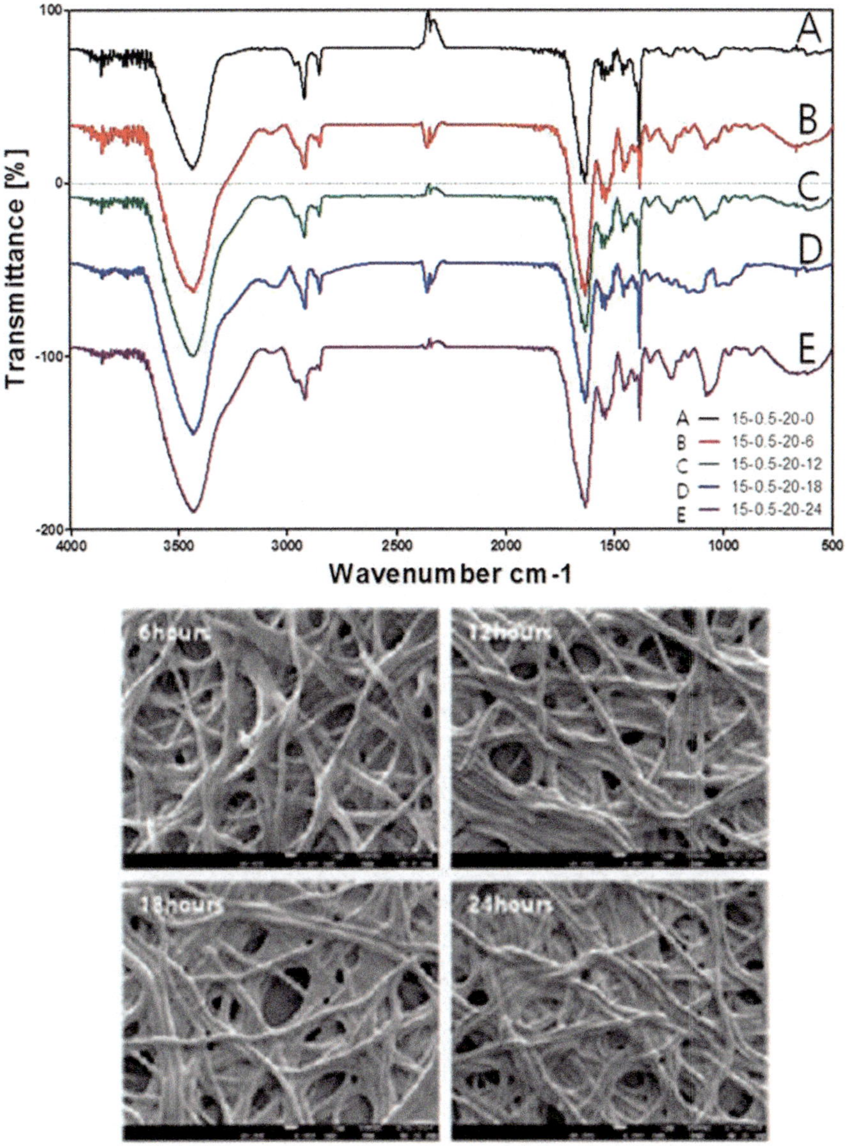

Fig. 9 15% (w/c) collagen-0.5% (w/v) HA at 20 kV nano-fibrous matrices after crosslinking by EDC. Upper: FT-IR analysis after (A) before crosslinking, (B) crosslinking for 6 h, (C) crosslinking for 12 h, (D) crosslinking for 18 h, (E) crosslinking for 24 h. Lower: FE-SEM view of nano-fibrous matrices after crosslinking by EDC. (×5000, scale bar: 1 μm)

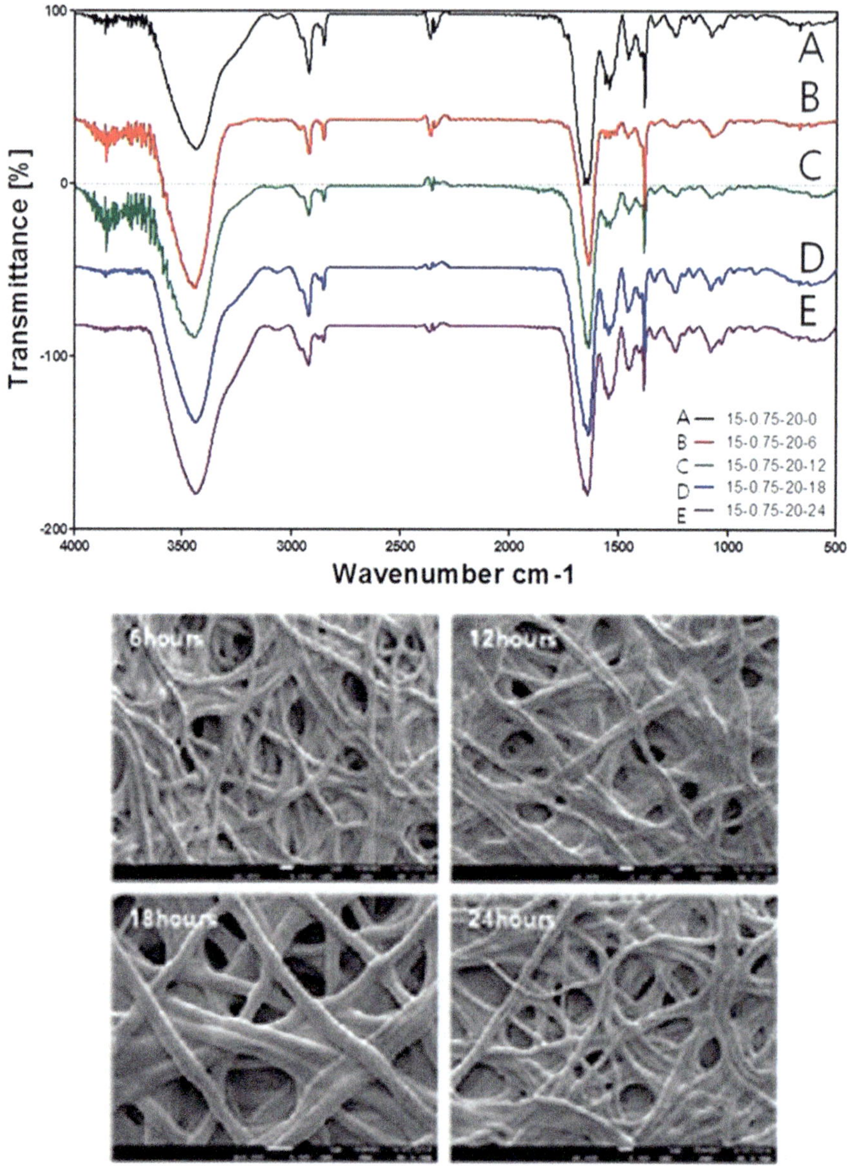

Fig. 10 15% (w/c) collagen-0.75% (w/v) HA at 20 kV nano-fibrous matrices after crosslinking by EDC. Upper: FT-IR analysis after (A) before crosslinking, (B) crosslinking for 6 h, (C) crosslinking for 12 h, (D) crosslinking for 18 h, (E) crosslinking for 24 h. Lower: FE-SEM view of nano-fibrous matrices after crosslinking by EDC. (×5000, scale bar: 1 μm)

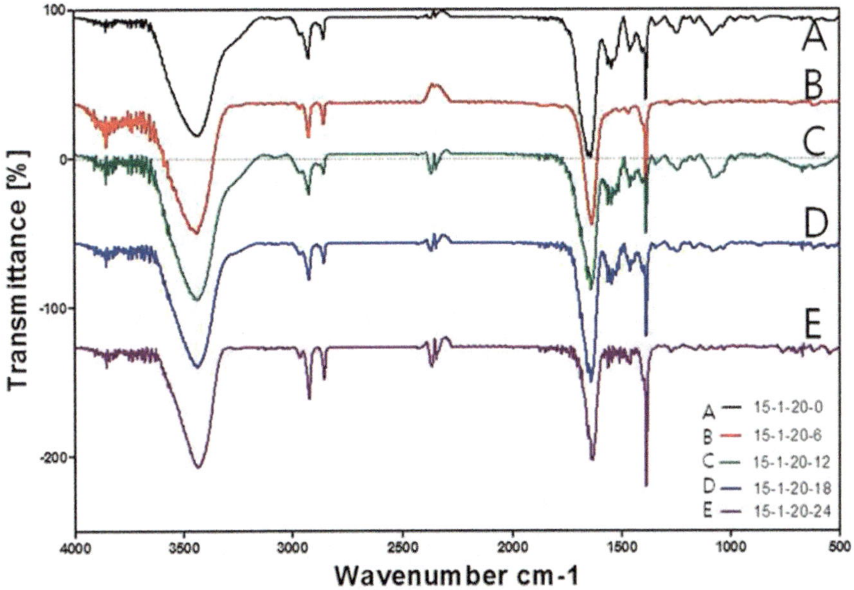

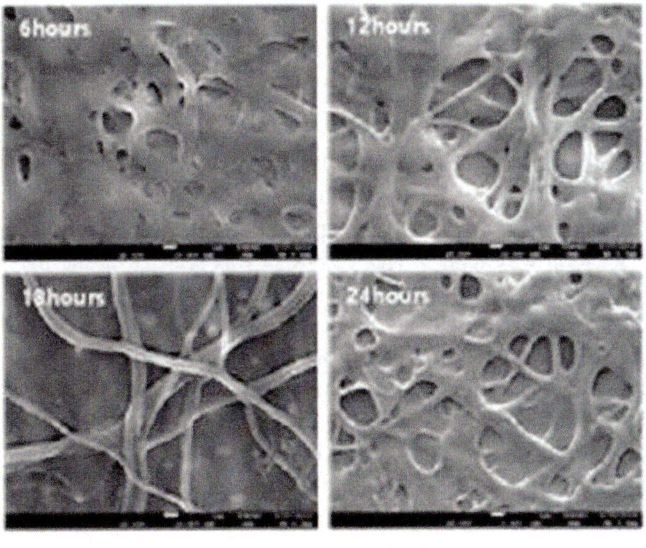

Fig. 11 15% (w/c) collagen-1% (w/v) HA at 20 kV nano-fibrous matrices after crosslinking by EDC. Upper: FT-IR analysis after (A) before crosslinking, (B) crosslinking for 6 h, (C) crosslinking for 12 h, (D) crosslinking for 18 h, (E) crosslinking for 24 h. Lower: FE-SEM view of nano-fibrous matrices after crosslinking by EDC. (×5000, scale bar: 1 μm)

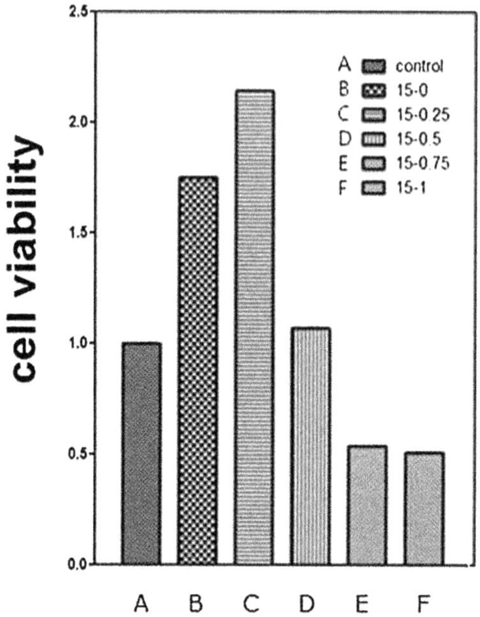

Fig. 12 Cell viability of hBM-MSC cultured for 4 days on collagen-HA nano-matrices (relative values vs. control group)

PBS 24 h at room temperature before processing. Solutions were mixed by either vortex mixer or centrifuge to dissolve collagen and HA in solvents. After that, remained solutes were dissociated by ultrasonicator to remove the lump. When concentration of HA reached more than 1% (w/v), gel lumps could be remained in the solution even though sonication was applied, and 1% (w/v) was selected as the maximum concentration of HA. Collagen-HA solution was produced as the Fig. 1 and electro spun with the conditions of Table 1. In the pilot test to set the collagen concentrations, same collagen concentration of Dong et al. [28] was used by addition of HA. By comparing the structures of nano-fibers with each group, fixing HA concentrations were chosen, and collagen concentrations were set at 5, 10, 15% (w/v).

In 5% (w/v) collagen-HA matrices, there were fine nano-fibrous structures in all conditions, but it has some nodal structures. To figure out the effect of formation of nodal structures, distribution of diameter was observed. With including nodal structures, maximum value of diameter was 137.41 nm whereas minimum was 91.2794 nm. If nodal structures were excluded, every diameter was less than 100 nm. The standard deviations were also changed by addition of nodal structures in the distribution. Randomly generated nodal structures affected the diameter and standard deviation. But there was no significant relation of the effect of nodal structures to cells. Collagen concentration was increased to remove the nodal structures from the matrices.

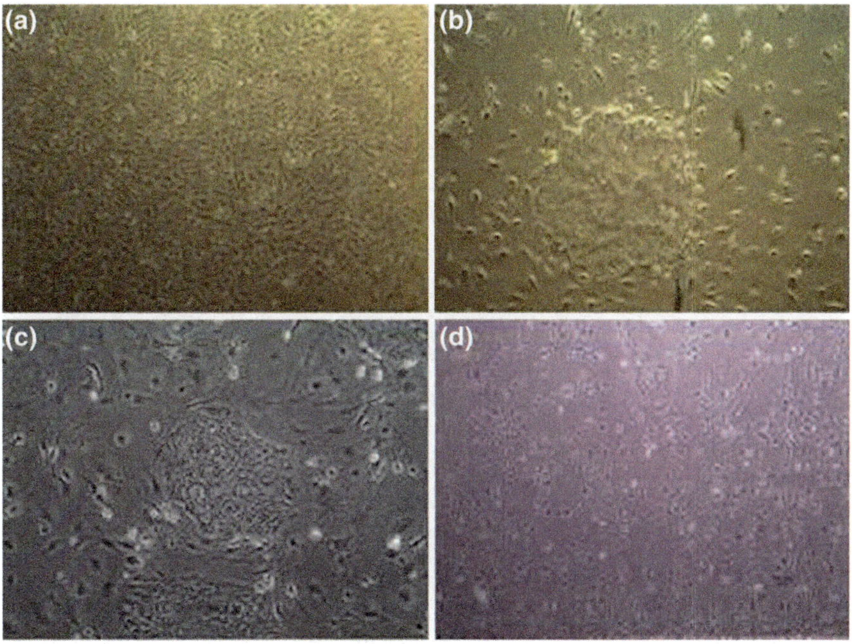

Fig. 13 Morphologies of stem cells attached onto STO feeder layer 5 days after seeding. **a** STO, **b** hESC, **c** hiPS, **d** hBM-MSC (light microscope, ×100)

In 10% (w/v) collagen-HA nano-fibrous matrices, occurrence of nodules decreased as collagen concentration increased. Also the surface of nano-fiber became smoothened. Maximum diameter was 302.575 nm (with nodes) and 293.625 nm (without nodes). And the minimum value was 172.049 nm (with nodes) and 143.682 nm (without nodes). Specifically, 0.75% (w/v) HA contained matrices showed decrease of diameters in every voltage, and it did not affected by the formation of the nodal structures.

There were no nodes in 15% (w/v) collagen-HA nano-fibrous matrices. Formation of nodal structures was affected by the concentrations of solutes. Since HA concentration was fixed, collagen played a more important part than HA in nodal formation. Maximum diameter was 768.537 nm and minimum was 490.152 nm. By increasing the collagen concentrations, diameter increased and that is referred to characteristics of electro spinning [31, 32]. And attachment of the dissolved salts on nano-fibers in PBS also increased. 15% (w/v) collagen was proper concentration to fabricate fine and stable nano-fibers. Porous structure was affected by the concentration of HA. 1% (w/v) HA contained matrices fabricated at 20 kV showed minimum sized pores.

FT-IR analysis was applied to characterize the deformation or degradation of collagenous structure by the variously applied voltage during electro spinning process. There were difference in intensity, but there was no shift of each collagen peaks. It showed that there was no structural difference by the voltage. 20 kV was

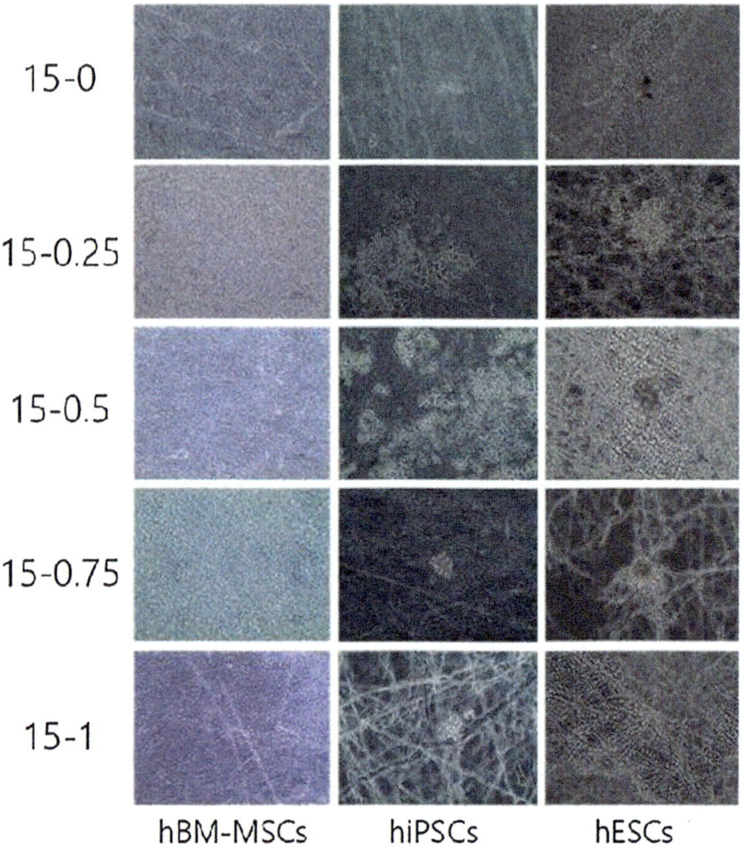

Fig. 14 Morphologies of stem cells seeded upon 15% (w/v) collagen-0, 0.25, 0.5, 0.75, 1% (w/v) HA matrices after 4 days (light microscope, ×200)

ideal voltage to fabricate the matrix based on the results of FT-IR and FE-SEM. Because there was no difference in FT-IR peaks. At 20 kV, collected large amount of nano-fibers were observed by FE-SEM. According to the overall results, 15% (w/v) collagen and 20 kV was optimal to fabricate the collagen-HA nano-fibrous matrices.

In previous works, collagen-HA matrices were fabricated using 50 mM EDC solutions for 24 h [11, 12, 27]. But structural differences and concentration of collagen and HA could affect the EDC crosslinking time. Therefore, EDC crosslinking time was varied for 6, 12, 18, and 24 h without changing EDC concentration, and compared the FT-IR with FE-SEM results as time course. After EDC crosslinking, each matrix was washed by using ultrasonicator. Washing order was 20X PBS, 10X PBS, 5X PBS and DW to remove the salts derivate from the electro spinning solvents. Alteration of amide bond, ester bond, and carboxylate salt peak were investigated by FT-IR. Each peak appeared after crosslinking for 12 h. Particularly, these peaks' intensity or appearance was diminished in 1% (w/v) HA added matrices. Therefore, 12 h was set as the proper

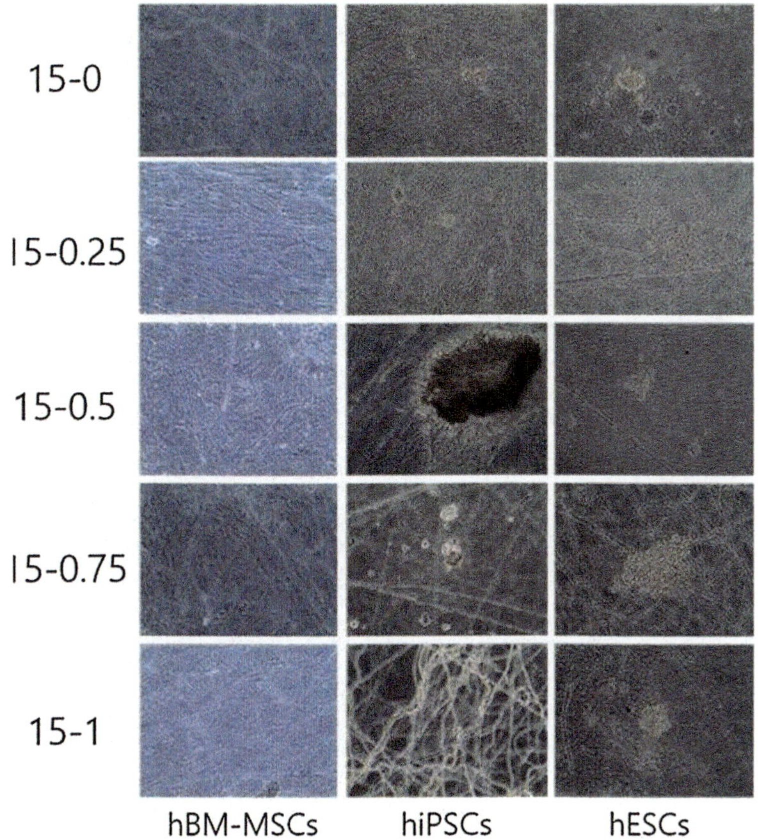

Fig. 15 Morphologies of stem cells seeded upon 15% (w/v) collagen-0, 0.25, 0.5, 0.75, 1% (w/v) HA matrices after 10 days (light microscope, ×200)

time of EDC crosslinking. The morphological change was observed by FE-SEM, and nano-fibers were adhered each other. This was the same results of Dong et al. [28] in which 0.2 M EDC/N-hydroxysuccinimide(NHS) was used. Although EDC crosslinking was conducted for 24 h, there was no signs of adhesion of nano-fibers in 0.5% (w/v) HA contained matrices. These results could be interpreted as that crosslinking would be occurred by the adhesion between the adjacent nano-fibers. But it could reduce the pore sizes and also trigger a negative effect of the migration of materials through the pores.

Before seeding the cells on collagen-HA nanofibrous matrices, sterilization was done using 100 and 70% ethanol for 30 min respectively. After each step, ethanol was fully dried under clean bench, washed with 1X PBS and medium, and immersed in medium before use.

To find out the characteristics of cells after the short-term culture, hBM-MSCS were cultured for 4 days. Control was cultured on culture flasks to compare the cell attachment efficiency. Cell density was set as 5×10^3 cells/cm^2. 6 h after cell seeding, hBM-MSCs were grown in the seeding spot with radially spreading by the

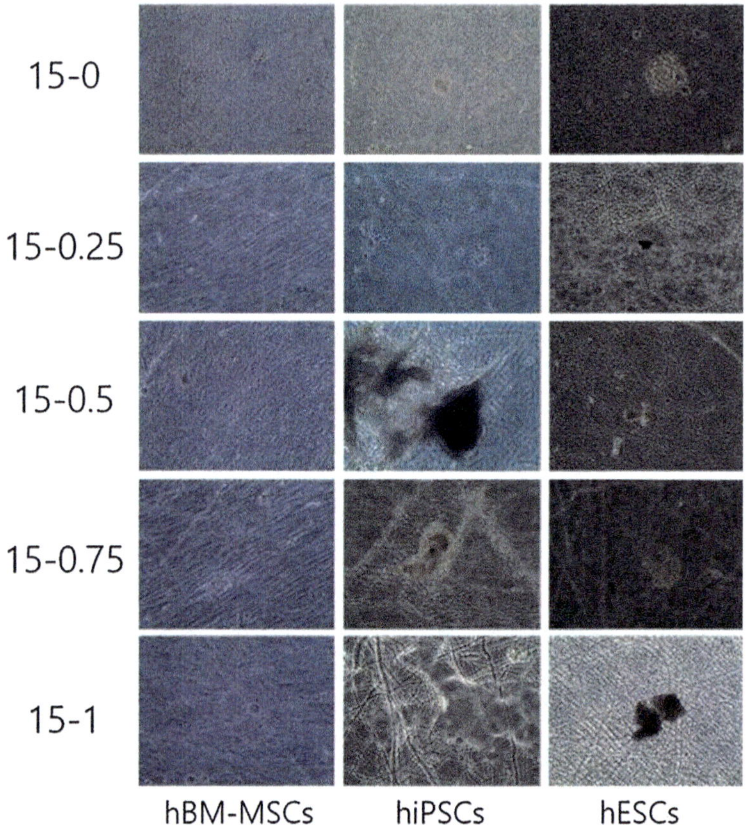

Fig. 16 Morphologies of stem cells seeded upon 15% (w/v) collagen-0, 0.25, 0.5, 0.75, 1% (w/v) HA matrices after 16 days (light microscope, ×200)

permeation of medium in matrices in control groups. After observing the morphology, medium was added to the each matrix to fill it completely. Each matrix was transferred to new flask to calculate the efficiency of cell attachment 24 h after seeding. At the same time, cell seeded matrices were washed with 1X PBS to remove the dead or unattached cells. And the previous flasks were trypsinized to detach the cells from the surface of flasks, and the cells were calculated by hemocytometer. Cell attachment was highest at 0% (w/v) HA added matrices, and it decreased as increasing HA contents but increased again in 1% (w/v) HA contained matrix. Cellular distribution and morphological images were observed by a light microscope. And CCK8 analysis was also performed to verify the cell viability as time courses. After removing medium, matrices were washed with PBS and added new medium. CCK8 solution was added into each matrix, incubated for 2 h in CO_2 incubator, and measured the optical density at 450 nm.

The cell viability of 0.75% (w/v) HA containing matrices demonstrated lower than the control, and that is assumed to be affected by the initial cell attachment

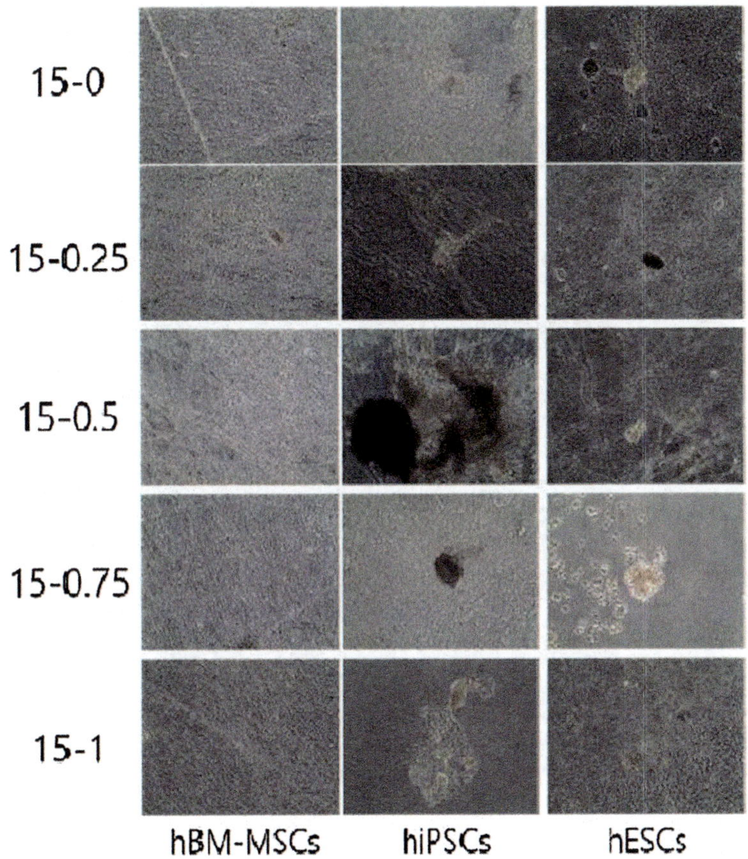

Fig. 17 Morphologies of stem cells seeded upon 15% (w/v) collagen-0, 0.25, 0.5, 0.75, 1% (w/v) HA matrices after 22 days (light microscope, ×200)

efficacy. Meanwhile, cell viability of 1% (w/v) HA containing matrices revealed relatively low value whereas cell attachment was maintained. It would be the results of the higher HA content in the matrices, which could negatively affect cell proliferation at early stage that requires adaptation period to the high HA content. But in the short- term experiments, cell passage was 5 and cultured for only 4 days, and it would be difficult to define the influence of matrices on cells. Therefore, to assure the effect of matrices, experiments were conducted in long-term period, and behaviors of hBM-MSC, hESC and hiPSC on matrices were observed.

Long-term culture of hBM-MSCs was conducted same as short-term culture. In case of hESC and hiPSC, colonies are easily destroyed and damaged by ethanol or PBS. To completely remove ethanol, matrices were immersed in DMEM-hg medium for overnight. After washing with DMEM/F12, cells were seeded at the middle of the matrices which have been fully immersed in DMEM/F12. Since colony formation of hiPSC and hESC took more than 4 days on STO feeder layer,

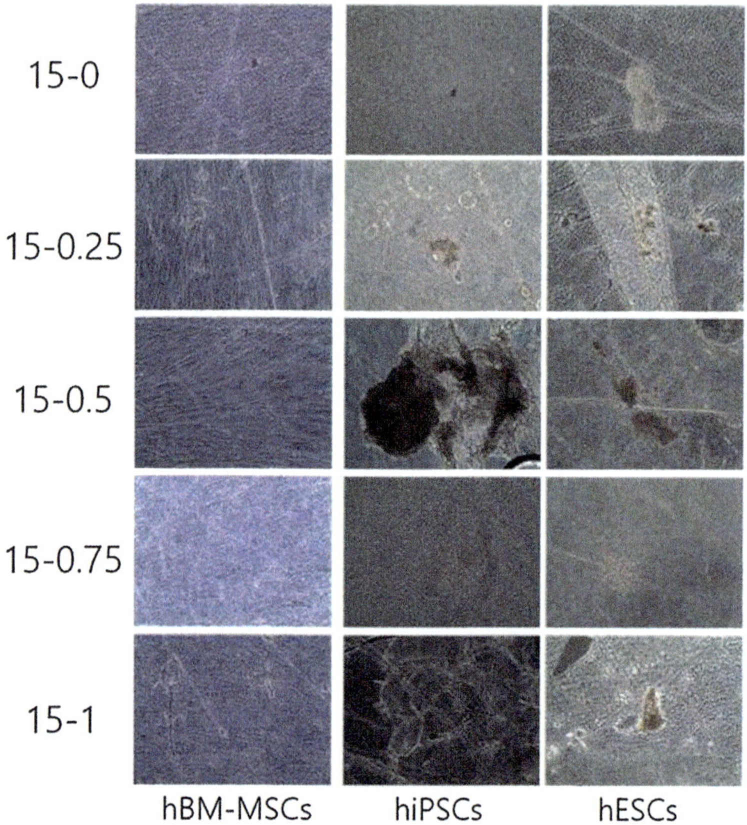

Fig. 18 Morphologies of stem cells seeded upon 15% (w/v) collagen-0, 0.25, 0.5, 0.75, 1% (w/v) HA matrices after 28 days (light microscope, ×200)

cells were checked under microscope 4 days after seeding. And to improve the colony formation, 4 ng/ml bFGF was added 48 h after seeding the cells on matrices. Culture media of hiPSC and hESC were changed every 5 days. Suspended cells were centrifuged and reseeded on each matrix during changing medium. To reduce the adaptation time, same volume of new and used medium was mixed. In the pilot study, hiPSC and hESC easily escaped from the matrices because the size of these cells was smaller than hBM-MSCs. To solve this problem, matrices were covered with sterilized cover glasses. Matrices were flipped every 6 days to find out the migration between cover glasses and matrices. Characteristics of every cell types were verified in period of 4, 10, 16, 22, and 28 days. bFGF was added into the hESC and hiPSC medium in contrary to the hBM-MSCs medium. In case of these cells, cell viability was measured whether the growth factor was added or not. Cells were seeded directly after thawing the cells from LN_2. This method was used to maintain the cell condition without adapting the cells on flasks for

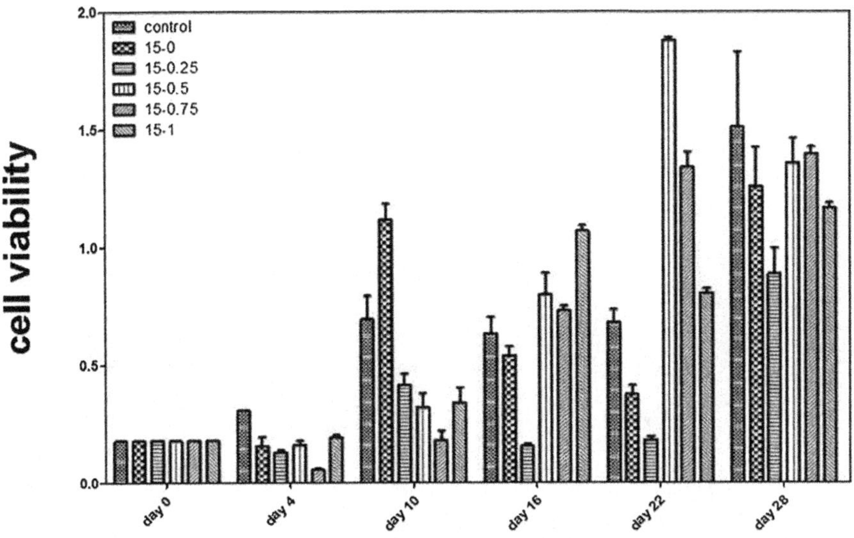

Fig. 19 Cell viability of hBM-MSC seeded onto 15% (w/v) collagen and 0, 0.25, 0.5, 0.75. 1% (w/v) HA matrices for 0, 4, 10, 16, 22 and 28 days by CCK8 analysis

further in vivo tests. The control group of hBM-MSCs were cultured on culture flask. Since hiPSC and hESC needed the feeder layer, feeder cells could affect the results of cell viability.

Colonies of hiPSC and hESC were grown under space between the matrix and glass cover by penetrating matrix, and over grown or differentiated cells were loosened and detached from the colony. Differentiated cells were flatly covered the surface and grew. Meanwhile, hBM-MSCs were grown within the matrices, and the over grown cells spread out from the matrix as increase in population as time courses.

In CCK8 analysis, cell viability of hBM-MSCs within 0.75% (w/v) HA containing matrices showed lowest values, likewise the short-term results. But values were increased as time courses. Therefore 0.75% (w/v) HA including matrices were proper for long-term culture, while the results of 1% (w/v) HA included matrix had the average values in comparison with the other matrices. This would be resulted from the difference of passage. FE-SEM results revealed there was no nano-structural deformation as time courses. But the 3D structures constructed by the concentration of matrices could affect the proliferation, and cell viability were different according to the matrices.

Microscopic images of hiPSC demonstrated maintenance of colony morphology in 0, 1% (w/v) HA containing matrices, while differentiation and spreading over the matrix was observed in 0.25, 0.75% (w/v) HA specimens. Particularly, large sized colony could found 10 days after seeding in 0.5% (w/v) HA contained matrices. And cells were grown coming out from the colony and spread as time courses. But these aspects did not appeared in the other matrices. Further confirmation is

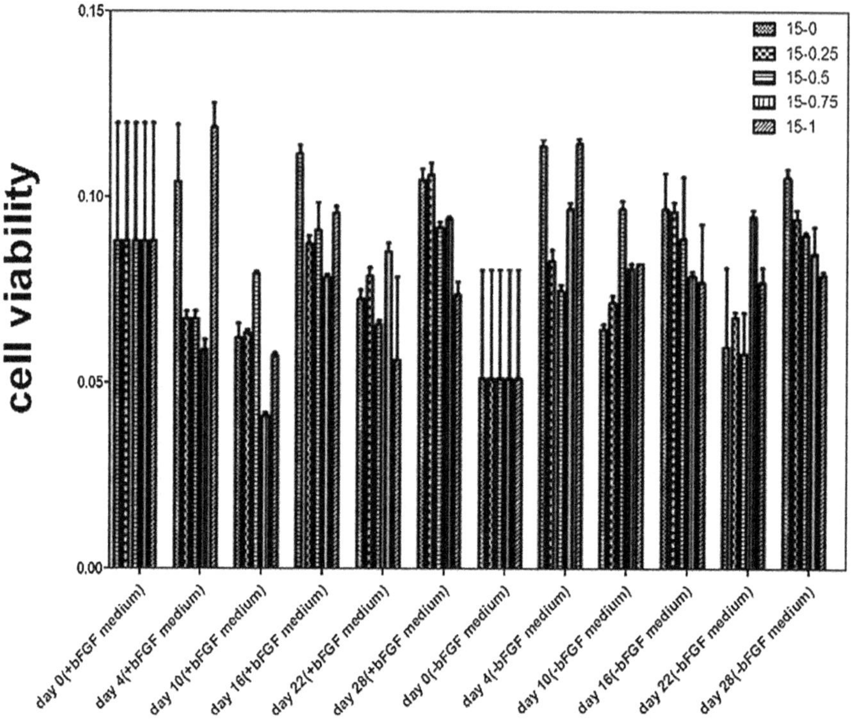

Fig. 20 Cell viability of hiPSC seeded onto 15% (w/v) collagen and 0, 0.25, 0.5, 0.75. 1% (w/v) HA matrices for 0, 4, 10, 16, 22 and 28 days in growth factor added culture media (+bFGF) and growth factor free media (−bFGF) by CCK8 analysis

necessary to define whether it was affected by the initial size of colony, cell number or concentration of HA, and ihPSC viability in collagen matrix was not affected by additional growth factor

According to the CCK8 results, 0.25 and 0.5% (w/v) HA containing matrices were proper for long- term culture, as they revealed the highest value at 28 day. Whereas other types of matrix could be proper for short-term culture since the maximum value appeared at early and middle stage of culture. In case of hESC, colony spreads and differentiates in 0, 0.25, 0.5, and 0.75% (w/v) HA containing matrices, since oval shaped colonies were observed 4 days after seeding. In most matrices, spreading and differentiating oval shaped colony appeared earlier than hiPSC. CCK8 values on 10th day and 28th day increased as time courses disregarding additional growth factor was or not. In comparison of the results of day 10 with day 28, cells attached on the surface of cover glasses without directly attaching on matrices. This assumed to be resulted from the pressure of matrices weight. A loose or pressured biomechanical environment provided by matrix fibers assumed to influence cell viability. In consideration of 1.6 times larger amount of hiPSCs and hESCs than the general cell numbers are necessary to produce colony

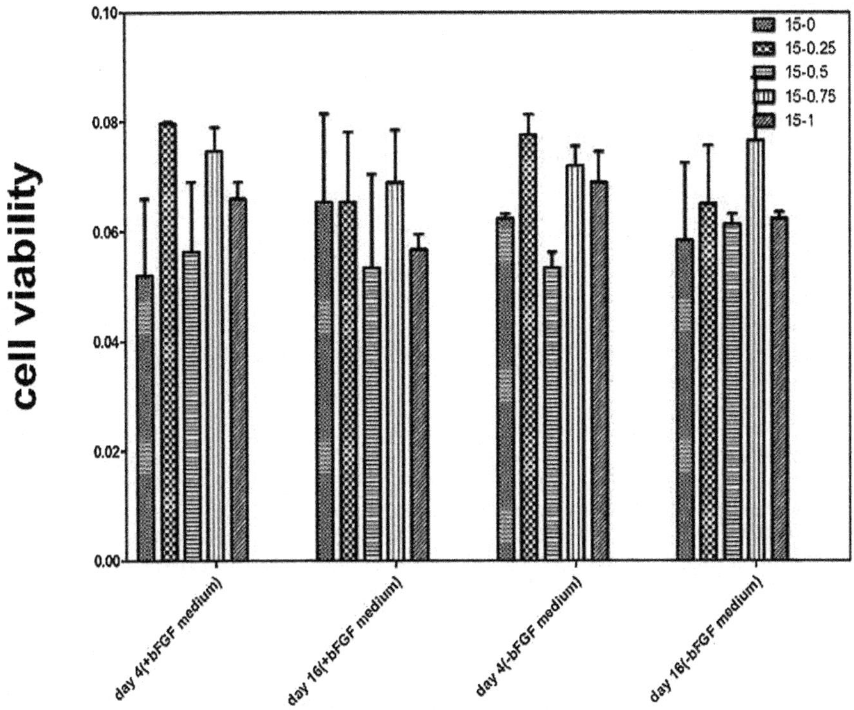

Fig. 21 Cell viability of hESC seeded onto 15% (w/v) collagen and 0, 0.25, 0.5, 0.75. 1% (w/v) HA matrices for 4, and 16 days in growth factor added culture media (+bFGF) and growth factor free media (−bFGF) by CCK8 analysis

for STO cell line feeder layers, respectively, the CCK8 analysis resulted no significant cell proliferation in relation with viability. And the sizes of colonies and the numbers of hiPSC and hESC were difficult to quantify, and it was necessary to increase the initial cell numbers on matrix to subdue the effect of collagen-HA matrices which govern cell adaptable attachment. It is recommended to use cells dissociated from the cells cultured in matrix. Therefore, methods to uniformly distribute cells in matrix should be established.

However, electro spun 15% (w/v) collagen containing 0.5% (w/v) HA porous nano-fibrous matrices provided favorable environment for cell hBM-MSC, hiPSC and hESC attachment and proliferation with maintaining cell viability. But proper cell amount, proliferation rate, differentiation and viability shall be established for each stem cell type.

References

1. Suh, H., Chung, E., & Jae Hyung, K. (2013). Global safety guidance for the cell-based implants. *Tissue Science & Engineering, 4*.
2. Suh, H. (1998). Fundamental concepts for the tissue engineering. *Biomaterials Research, 2*, 1–7.
3. Langer, R., & Vacanti, J. P. (1993). Tissue engineering. *Science, 260*, 920–926.
4. Langer, R., & Tirrell, D. A. (2004). Designing materials for biology and medicine. *Nature, 428*, 487–492.
5. Murphy, S. V., & Atala, A. (2013). Organ engineering—combining stem cells, biomaterials, and bioreactors to produce bioengineered organs for transplantation. *BioEssays, 35*, 163–172.
6. Naderi, H., Matin, M. M., Bahrami, A. R. (2011). Review article: Critical issues in tissue engineering: Biomaterials, cell sources, angiogenesis, and drug delivery systems. *Journal of Biomaterials Applications*. https://doi.org/10.1177/0885328211408946.
7. Matthews, J. A., Wnek, G. E., Simpson, D. G., & Bowlin, G. L. (2002). Electrospinning of collagen nanofibers. *Biomacromolecules, 3*, 232–238.
8. Hsu, F. Y., Hung, Y. S., Liou, H. M., & Shen, C. H. (2010). Electrospun hyaluronate-collagen nanofibrous matrix and the effects of varying the concentration of hyaluronate on the characteristics of foreskin fibroblast cells. *Acta Biomaterialia, 6*, 2140–2147.
9. Sell, S. A., Wolfe, P. S., Garg, K., McCool, J. M., Rodriguez, I. A., & Bowlin, G. L. (2010). The use of natural polymers in tissue engineering: A focus on electrospun extracellular matrix analogues. *Polymers, 2*, 522.
10. Suh, H., & Lee, J. E. (2002). Behavior of fibroblasts on a porous hyaluronic acid incorporated collagen matrix. *Yonsei Medical Journal, 43*, 193–202.
11. Park, S. N., Lee, H. J., Lee, K. H., & Suh, H. (2003). Biological characterization of EDC-crosslinked collagen- hyaluronic acid matrix in dermal tissue restoration. *Biomaterials, 24*, 1631–1641.
12. Park, S. N., Park, J. C., Kim, H. O., Song, M. J., & Suh, H. (2002). Characterization of porous collagen/hyaluronic acid scaffold modified by 1-ethyl-3-(3-dimethylaminopropyl)carbodiimide cross-linking. *Biomaterials, 23*, 1205–1212.
13. Lee, J. M., Edwards, H. H. L., Pereira, C. A., Samii, S. I. Crosslinking of tissue-derived biomaterials in 1-ethyl-3-(3-dimethylaminopropyl)-carbodiimide (EDC). *Journal of Materials Science: Materials in Medicine, 7*, 531–541.
14. Fischer, R. L., McCoy, M. G., & Grant, S. A. (2012). Electrospinning collagen and hyaluronic acid nanofiber meshes. *Journal of Materials Science Materials in Medicine, 23*, 1645–1654.
15. Kim, T. G., Chung, H. J., & Park, T. G. (2008). Macroporous and nanofibrous hyaluronic acid/collagen hybrid scaffold fabricated by concurrent electrospinning and deposition/leaching of salt particles. *Acta Biomaterialia, 4*, 1611–1619.
16. Glowacki, J., & Mizuno, S. (2008). Collagen scaffolds for tissue engineering. *Biopolymers, 89*, 338–344.
17. Parenteau-Bareil, R., Gauvin, R., & Berthod, F. (2010). Collagen-based biomaterials for tissue engineering applications. *Materials, 3*, 1863.
18. Furthmayr, H., & Timpl, R. (1976). Immunochemistry of collagens and procollagens. *International Review Connective Tissue Research, 7*, 61–99.
19. Her, G. J., Wu, H. C., Chen, M. H., Chen, M. Y., Chang, S. C., & Wang, T. W. (2013). Control of three-dimensional substrate stiffness to manipulate mesenchymal stem cell fate toward neuronal or glial lineages. *Acta Biomaterialia, 9*, 5170–5180.
20. Murphy, C. M., Matsiko, A., Haugh, M. G., Gleeson, J. P., & O'Brien, F. J. (2012). Mesenchymal stem cell fate is regulated by the composition and mechanical properties of collagen-glycosaminoglycan scaffolds. *Journal of the Mechanical Behavior of Biomedical Materials, 11*, 53–62.

21. Hortensius, R. A., & Harley, B. A. (2013). The use of bioinspired alterations in the glycosaminoglycan content of collagen-GAG scaffolds to regulate cell activity. *Biomaterials, 34,* 7645–7652.
22. Knudson, C. B. (2003). Hyaluronan and CD44: Strategic players for cell-matrix interactions during chondrogenesis and matrix assembly. *Birth Defects Res C Embryo Today, 69,* 174–196.
23. Wang, X., Ding, B., & Li, B. (2013). Biomimetic electrospun nanofibrous structures for tissue engineering. *Materials Today, 16,* 229–241.
24. Gupta, K. C., Haider, A., Y, Choi, & I, Kang. (2014). Nanofibrous scaffolds in biomedical applications. *Biomaterials Research, 18,* 1–11.
25. Lu, T., Li, Y., & Chen, T. (2013). Techniques for fabrication and construction of three-dimensional scaffolds for tissue engineering. *International Journal of Nanomedicine, 8,* 337–350.
26. Shih, Y. R. V., Chen, C. N., Tsai, S. W., Wang, Y. J., & Lee, O. K. (2006). Growth of mesenchymal stem cells on electrospun type I collagen nanofibers. *Stem Cells, 24,* 2391–2397.
27. Park, S. N., Kim, J. H., Kim, I., Seol, A., & Suh, H. (2006). Electrospun nanofibrous matrices for the engineering of cultured skin substitute. *Biomaterials Research, 10,* 7.
28. Dong, B., Arnoult, O., Smith, M. E., & Wnek, G. E. (2009). Electrospinning of collagen nanofiber scaffolds from benign solvents. *Macromolecular Rapid Communications, 30,* 539–542.
29. Suh, H., Suh, S., & Min, B. (1994). Anti-infection treatment of a transcutaneous device by a collagen- rifampicine composite. *ASAIO Journal, 40,* M406–M411.
30. Wang, X., Um, I. C., Fang, D., Okamoto, A., Hsiao, B. S., & Chu, B. (2005). Formation of water-resistant hyaluronic acid nanofibers by blowing-assisted electro-spinning and non-toxic post treatments. *Polymer, 46,* 4853–4867.
31. Bhardwaj, N., & Kundu, S. C. (2010). Electrospinning: A fascinating fiber fabrication technique. *Biotechnology Advances, 28,* 325–347.
32. Barnes, C. P., Sell, S. A., Boland, E. D., Simpson, D. G., & Bowlin, G. L. (2007). Nanofiber technology: Designing the next generation of tissue engineering scaffolds. *Advanced Drug Delivery Reviews, 59,* 1413–1433.

Author Biography

Professor Hwal (Matthew) Suh, (South Korea), DDS, MSC, PhD. He was born in Seoul, Korea in 1953. He has studied on mesenchymal stem cell manipulation and hybridization of cells with extracellular matrix for therapeutic purpose. Working at Laboratory of Tissue Regenerative Medicine, Department of Medical Engineering, Yonsei University College of Medicine in Seoul, Korea as Professor, his current research is to develop multilayered therapeutic cells for implantation.

Rheological Properties of Biological Structures, Scaffolds and Their Biomedical Applications

Sabra Rostami and Bora Garipcan

1 What Is Rheology?

The term rheology is derived from the Greek words rheo, "to flow," and logos, "science." Rheology, therefore, is "the scientific study of the deformation and flow properties of matter" [79].

The discipline of rheology was first introduced formally on April 29, 1929 by Professor E. C. Bingham of Lafayette College and G. W. Scott Blair [82]. The importance of this concept was so rapidly understood that in the same year at the Third Plasticity Symposium in 1929 a permanent organization with the aim of study and development of the new discipline of rheology under the name of "The Society of Rheology" was founded.

To this day, the role of rheology in materials both of natural origin and synthetic along with the extent of its efficacy has being investigated and during past century with the advancement of measurement techniques, studying rheology in small scale has become possible. Hence the hot topic of principles and influence of rheology on living materials has become an appealing and amusing research subject resulting in attainment of fascinating results which so far have proven to be greatly helpful in unlocking wide range of mysteries of life itself.

S. Rostami · B. Garipcan (✉)
Bio3 Research Laboratory, Institute of Biomedical Engineering,
Bogazici University, Istanbul, Turkey
e-mail: bora.garipcan@boun.edu.tr

G. M. Artmann et al. (eds.), *Biological, Physical and Technical Basics of Cell Engineering*, https://doi.org/10.1007/978-981-10-7904-7_5

2 Role of Rheology in Life

Biological materials including cells and tissue, are complex structures made of diverse unites scaling from nano-scale to micron. Cells, as building blocks of living things, have various types of structural units and subunits namely nucleus, mitochondria and bilayer cell membrane each with their specific physicochemical and mechanical properties. Combination and accumulation of numerous of these sub-unites inside cytosol, the gel-like aqueous component of cytoplasm which constitutes about 70% of the total volume of a normal cell, gives rise to a non-homogeneous entity [22]. Despite the fact that cells are made of solid and fluidic materials, the nonhomogeneous nature of this complex leads to unique characteristics unlike solid and liquid such as flow properties.

In material science mechanical properties of materials are usually classified in three main categories of elastic, plastic and viscous. Elasticity is a constant value calculated by scaling between stress and strain of a solid material also called the Young's modulus which determines the ability of solid materials to sustain their original shape when mechanical stress is applied. Plasticity on the other hand, refers to very large deformation of a material as it is subjected to stress levels beyond its certain yield limits [79]. In contrast to solids, if stress is applied to fluids, elastic energy cannot be stored and the flow pattern of the fluid changes. The concept of changes of the rate of flow under applied load is known as viscosity [51]. Deformation of a material can be measured while a force is applied to it. Figure 1 is an illustration of various ways a force that can be applied to a material.

Cells, being constructed of mostly fluidic and partially solid materials show neither elastic nor viscous characteristics. Hence cells can be categorized as

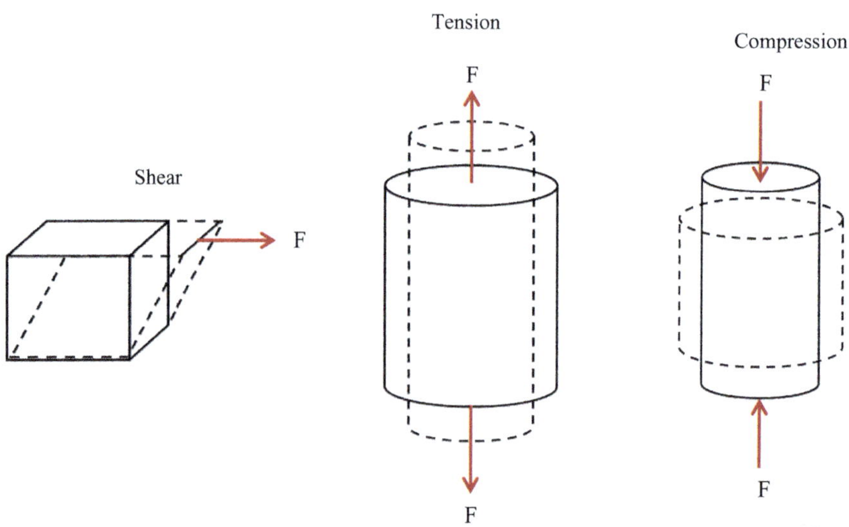

Fig. 1 Schematic illustration of forces that can be applied to a material

viscoelastic [79]. Unlike soft materials, biological materials do not just react to a certain mechanical stimuli; they behave via generating an active response by the means of mechanotransduction.

Mechanotransduction refers to cell's ability of generating biochemical signals in response to external mechanical cues with the aim of transporting information from nucleus to other organelles inside the cell and vice versa in order to trigger series of particular responses. These responses usually results in functional behaviors like adhesion, migration, proliferation and differentiation as well as apoptosis or necrosis [17, 34, 78].

Activation of mechanosensitive signaling pathways is one of the most basic and vital physiological mechanisms through which cells are able to react to environmental conditions and physical cues. These reactions occur by reorganization and force generation of cytoskeletal structure. It has also been found that any disruption in cytoskeletal architecture caused by genetic mutations or pathogens can lead to changes in cell mechanical properties such as elasticity, adhesiveness, and viscosity [34, 78]. Hence studying the mechanical behavior of cells and therefore tissues has attracted considerable interest over the past century.

Since majority of cell volume contains fluidic substance, among mechanical properties, study of flow properties of cells has attracted significant attention resulting in the massive amount of research regarding the concept of cellular rheology [34, 49].

In material science, during the past century, data obtained from rheological measurements of nonliving complex materials has provided vast range of information regarding conformational changes or rearrangements of material's constituents in micro and nano-scale when they undergo mechanical force [49]. These findings shed light on the potential effects and importance of rheological properties of cells on their function. Thus, over the past decade there have been many studies concerning the role of rheological properties on cellular functions namely cell crawling, wound healing, protein regulation, cell malignancy and how they may cause deficiency in cell's life [73].

As a result, today there exist a myriad of experimental techniques and devices for measuring flow properties of biological materials such as Atomic Force Microscopy (AFM) that can provide massive amount of information (in the form of interdisciplinary concepts of combined molecular biology and advanced cell mechanics) which ease our understanding of cellular mechanisms and effects of mechanical forces on their physiology, metabolism, development, function and disease [17].

3 Techniques and Methods of Measurement

When it comes to measuring quantitative values of a property, the scale of measurement plays the most important role of all. In case of cells, almost all sorts of measurements need to be done at micron scale [34]. Thus devices and instruments of choice must be in submicron scale and precision. Since over the past decades, techniques of micron and nano-scale measurements have been developed extensively and thoroughly, it has become quite easy and simple to measure flow properties of cells of different origin in diverse conditions, namely Atomic force Microscopy (AFM), Optical Tweezers, Magnetic Tweezers or Magnetic Twisting Cytometry, Particle Tracking Microrheology, Microplate Rheometry or Whole Cell Stretching, Micropipette Aspiration, Traction Force Microrheology and Optical Cell Stretcher [34, 51, 68]. Each of these methods is briefly mentioned in Table 1, and Fig. 2 represents a schematic of each technique.

Table 1 Short summary of measurement techniques for cell rheology measurement and their applications

Techniques	Measurement techniques	Applications	References
Atomic force microscopy (AFM)	Tip of the cantilever is placed on and then pressed against cell membrane. By moving the cantilever over the surface of the cell, the force and the imposed cellular deformations are recorded	High spatial resolution, measures large range of forces, more precise and easy to control cantilever position	[28, 44, 46, 48, 56, 60, 67]
Optical tweezers	Laser beams are used to exert forces onto a small dielectric particle attached to the cell membrane. Force application occurs in the form of pulling the cell membrane away from the cell using optical tweezers and membrane tether	High timescale resolution, force and position can be controlled more accurately, well suited to sub cellular measurements such as the pulling of membrane tethers	[5, 26, 36, 60, 64, 71]
Magnetic tweezers or magnetic twisting cytometry	A magnetic bead is placed on the surface of the cell membrane and twisted via external magnetic force. The bead displacement and resulting cell deformations are detected by video microscopy	Large range of frequencies [0.01–1.000 Hz], parallel measurements of large number of cells possible, high timescale resolution	[2, 41, 81]
Particle tracking microrheology	Via injection of fluorescent microbeads of less than 1 μm in diameter into the living cells, mechanical changes inside the cells can be studied due to the simple and random diffusion of beads in the cytoplasm	Quantitative measurements of shear modulus possible, can be used under physiological conditions	[8, 16, 29, 77]

(continued)

Table 1 (continued)

Techniques	Measurement techniques	Applications	References
Micropipette aspiration	Includes a micro-scale glass pipette with tip diameter of about 2–50 μm, able to apply a force of suction to the cell membrane. The applied force onto the cell can be as small as 0.1–0.2 $pN/\mu m^2$. Negative pressure causes the cell membrane to be aspirated into the micropipette. The process of aspiration is closely monitored under light microscope to an accuracy of ±25 nm	Measurements of non-linear deformations with high accuracy, both soft and rigid cells can be used	[4, 14, 21, 33, 38, 45, 55, 57, 71]
Microplate rheometry or whole cell stretching	A two parallel micron-sized flat plates system functioning as attachment surfaces for the cell, act as parallel clamps that hold the cell and can apply both stretching and compressing forces in nN-scale to the cell. Deformations of cells under force can be observed by an inverted microscope	Several manipulations in the same instrument are possible, a large range of forces are measurable (1 nN − 1 μN), control of cellular pre-stress	[34, 68, 74]
Traction force microscopy (TFM)	Two distinct testing methods: microbead embedded gel substrate and micropillar substrate. Force applied by the cell to the substrate is calculated by displacement of beads or pillars measured by optical microscope	Measuring relation between cell and ECM	[24, 42]
Optical cell stretcher	Two non-focused laser beams are used in opposed position, the cell is placed in between these beams. When sum of forces of the laser beams equals zero, the cell is steadily suspended i.e. optically trapped. Solely internally generated forces deform cells which are detected via optical microscopy	Mechanical characterization of individual suspended cells	[11, 69]

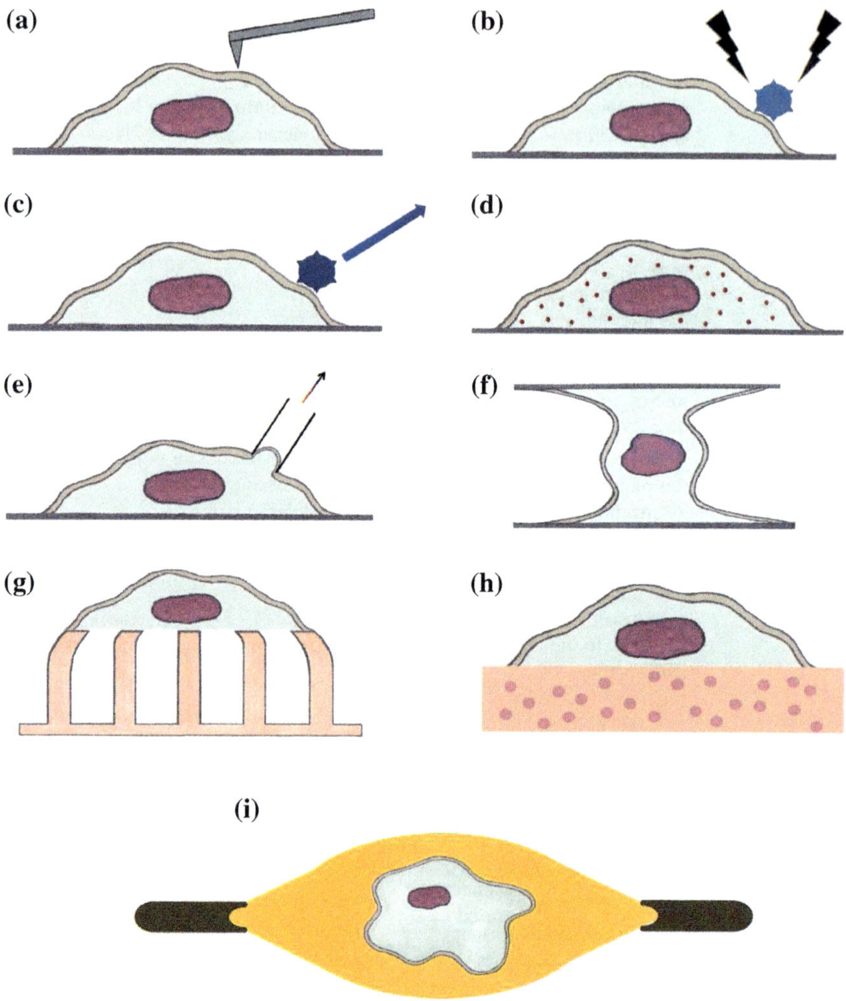

Fig. 2 Schematic representation of microrheology measurement techniques: **a** AFM, **b** optical tweezers, **c** magnetic tweezers, **d** particle tracking microrheology, **e** micropipette rheometry, **f** microplate aspiration, **g** traction force microrheology (embedded microbeads), **h** traction force microrheology (micropillars), **i** optical cell stretcher

Unfortunately all of the measurements techniques developed so far show some disadvantages. For instance AMF provides high resolution cell images, quantitative information about the cell surface properties, the Young modulus, the viscosity or the relaxation times but the measurements are too slow. In optical tweezers, local heating and prototoxicity occurs during measurements but it is a highly sensitive technique. Magnetic tweezers can only probe microenvironment inside the cell whereas Traction force microscopy can only measure relation between cell and ECM.

Micropipette Aspiration technique as a quantitative method, relies heavily on theoretical models, thus results may not be much reliable. Overall it can be said that in order to characterize rheological properties of cells, a combination of these methods must be used so as to obtain realistic and reliable results [3, 5, 41, 43, 56, 60, 75].

4 Rheological Properties of Cells: Principles, Discoveries and Applications

4.1 Principles of Cell Mobility: Mechanotransduction

In human body there is a wide and diverse range of mechanical stimuli in form of mechanical stress being applied to almost all cells with both external and internal origins [1]. In case of occurrence of a mechanical stimulus, the cells are equipped with sensing mechanisms which are called mechanoreceptors. These receptors help cells feel their surrounding environment in terms of mechanical stress. Since the main goal of dynamic nature of cells is to promote homeostasis, including maintenance of internal structural integrity of cell while interacting with its ECM, which is vital for cell's survival, the fundamental importance of motion regulators, mechanoreceptors and mechanotransduction process are well recognized and appreciated [35].

Hence mechanotransduction was defined as sequential processes of understanding how mechanical forces are sensed by cells, how sensed forces are transduced into biochemical signals and finally how they affect cellular metabolism and function [50]. Mechanotransduction involves two main procedures: mechanosensing and mechanoregulation. In the former, mechanical stresses applied to the cell are detected and identified via mechanoreceptors and in the latter cells regulate series of functions so as to respond to the sensed force in the most appropriate way [35].

Over the years many studies have been conducted on revealing unknown facts of mechanosensing and so far several molecules and structures have been identified as to be involved in this process namely ion channels, cell adhesion receptors, cytoskeletal and extracellular matrix molecules [30]. Some of the cytoskeletal molecules that are immensely involved in cellular motion are actin and myosin. Actin is a protein that can shape microfilaments and can be found in almost all eukaryotic cells. It is found in two forms of actin stress fibers and actin microfilaments. Actin structures can also be found in the form of thin filaments in contractile apparatus of muscle tissue in Z disc [19].

Just like actin, myosin is one of the major components of cell cytoskeleton as well and it is a motor protein that can be found in most eukaryotic cells specially muscle cells [78]. Myosin proteins are categorized in two groups: myosin I and myosin II. Myosin I is in close coordination with actin proteins and since it has contractile properties by itself, plays a vital role in contractility of muscle cells. On the other hand myosin II, having a massive amino acidic side, is a molecule which is divided in two main parts: head domain which merges with actin and initiates force and tail domain that acts as a coordination tool between other myosin subunits and communication facilitator with cargo muscle [54, 78].

Actin and myosin together coordinate in a systematic manner to control cellular mobility and deformation along with many other functions that require structural deformations of cell. This systematic regulation is called actomyosin machinery [30, 35, 54].

As mentioned previously, cell movement mainly starts with cell adhesion which in turn starts with formation of binding sites between cell membrane and underlying substrate called focal adhesion sites. Integrins are one of the most important adhesive membrane receptors that bind the cell to the surrounding environment mechanically. Integrins are structures inside the cells that are in direct contact with focal adhesion sites [1, 25].

Many mechanical stresses that cells sense, are inflicted by ECM. There are also many forces that are generated within the contractile cytoskeleton of cell itself [50]. Convergence of both of these forces occurs on membrane integrin receptors that assemble together within the focal adhesion sites. Focal adhesions are the linking point of cytoskeleton to ECM [50]. The forces that are sensed by the cell enforces to generate a stress-induces strengthening reaction as the result of adaptation to cell generated applied forces to integrin or rigidity changes of the ECM. Therefore the mechanical stiffness of the cell increases as the overall applied force rises. The ability of cells to change their stiffness is an indication of a level of tolerance towards membrane tearing [83].

Microbead microrheology and optical tweezers method have been used in order to understand the mechanism by which cells respond to mechanical cues and it was revealed that cells need to create focal adhesion at the site of applied force as a vital step on stiffening the cell. In short, stiffening (in response to external stimulus) starts by changes in focal adhesion size, composition, and position which simultaneously occur within the first few seconds to minutes after external stresses are sensed by mechanoreceptors [65].

When the cell is subjected to internal mechanical stress which is applied to integrins, focal adhesions are again assembled and the cell stiffness changes due to activation of actin and myosin. In this process myosin light chain is phosphorylated causing increase in cell contractility along with polymerization of actin. These procedures occur in a very complex and precise manner in the matter of seconds to minutes based on the intensity of applied stress [65, 80].

In addition to activation of actin-myosin network, mechanosensitive calcium ion channels can also be activated due to the stress imposed on integrins resulting in calcium influx into the cell which can cause contractility and therefore changing cell stiffness [53].

In addition to internal mechanical stimuli, there are external mechanical forces which apply sheer stress on the cells. For instance upon some cells, these forces are applied continuously such as sheer and stretch forces applied to Endothelial Cells (ECs) in blood vessels via blood flow which are called hemodynamic forces. These forces have the ability to change the physical shape and function of the ECs. These changes occur due to the ability of ECs to sense and adapt to the pressure fluctuations in order to maintain in an optimal stage of homeostasis and therefore function [13]. In general it can be said that mechanosensing is a vital regulatory mechanism of the cells with the main purpose of keeping the cell alive and at highest functional performance.

4.2 Rheological Properties of Eukaryotic Cells

Cells are composed of mostly fluid trapped inside a lipid membrane with various types of channels embedded inside this membrane allowing transportation of all sorts of molecules. The fluid inside the cell, cytoplasm, contains solid-like tiny structures called organelles. Existence of these organelles inside the cell along with constant chemical composition changes of the cytoplasm result in a unique flow characteristic which is referred to as viscoelastic. This viscoelastic fluid trapped in a lipid sack has extraordinary mobility abilities such as proliferation, migration, differentiation, along with contract, stiffen, stretch, fluidize, reinforce, crawl, intravasate, extravasate, invade, engulf, divide, swell, shrink or remodel [6].

Over the past century many studies have been conducted so as to measure flow properties of different eukaryotic cells in various stages of cell's life. As a result many techniques have been developed and extremely valuable insights have been provided in this regard. It has been found that in eukaryotic cells, for approximately 1% of applied strain, cell-generated contractile stress which result in cell deformation, is described by weak power law worldwide and is strongly related to cell contents in terms of composition and number of organelles [72, 84]. Hence cellular rheological behaviors cannot be measured and defined in time-dependent manners.

So far most rheological studies of cells have focused on red blood cells, lung alveolar epithelium, and brain cells along with human mesenchymal-derived stem cells, primary osteoblasts, chondrocytes, and adipocytes. In addition some studies focused on mechanical properties of cell nucleus [27]. AFM, micropipette aspiration, tweezers and optical measurement methods are the most preferred methods of microrheometry [62].

Starting from stem cells in 2007 Pajerowski et al. conducted a study on physical plasticity of nucleus of primary human embryonic stem cells (ESCs) also known as naive stem cells, proving the fact that structural reorganizations due to fundamental and massive amount of changes in gene expression necessary for differentiation, affect the flexibility of cell nucleus noticeably [62]. These internal structural deformations occur by condensation of chromatin and immobilization of nucleoprotein. The results of this study indicate that after sixfold through terminal differentiation, the nuclei becomes so stiffen that it loses the ability of going back to the original shape. Lamin A/C was identified to be the nucleo-skeletal component behind this phenomenon. By knocking down lamin A/C in human epithelial cells and bearing in mind that this component does not express in stem cells, deformability of epithelial cells were measured at close values to that of adult hematopoietic stem cells [62, 72, 84].

One of the major outcomes of this study was proving the flow, distend, and reorganization of chromatin whilst the lamina stretches, meaning that despite the fact that rheological behavior of nucleus is mainly regulated by nucleoplasm/ chromatin interactions, lamina is the modulator of extent of nucleus deformations [58].

Blood as a tissue contains many cell types such as red blood cell (erythrocytes), white blood cells (leucocytes) and platelets. Each of these go through extensive deformations through out their life, erythrocytes change their physical form in order to be able to move in and out of narrow capillaries. This function requires the cells to have the capability to regulate cellular stiffness and elasticity imperative for desired actions. Hence studying mechanical and rheological properties of red blood cells are of great importance [63].

Like other eukaryotic cells, blood cells have a lipid bilayer attached to the two dimensional elastic spectrin networks. This network is the main regulator of morphological and mechanical changes of the cell [59]. Studying mechanical deformations and therefore rheological properties of red blood cells are mostly done in two aspects: deformability and aggregability [63].

Aggregability occurs when erythrocytes attached to one another which can impair blood circulation at the areas where shear rate is low (Matthias Brust 2013) [15, 63]. Investigating rheological properties of red blood cells via micropipette aspiration has revealed that shape deformations of erythrocytes occur due to changes in volume of the cell rather than membrane disfigurement and structural changes of cell membrane. By measuring surface area and cell volume of aspirated cells verses relaxed cells, it was discovered that while the total surface area of the cell remains constant, as the cell is being aspirated, the volume of the cell changes in order to fit in the micropipette. This result proved the fact that, cell membrane of red blood cells has resistance properties against mechanical forces [38]. Erythrocyte stiffness and relaxation time are the fundamental factors that affect blood flow and viscosity to a great extent by controlling deformation and aggregation. These factors have been investigated thoroughly over that past century and plenty of methods have been developed ever since namely Microscopic Photometric Monolayer Technique. In this method erythrocytes' resistance to deformation (elongation) is measured along with relaxation time when cells are subjected to various physicochemical conditions [3].

Despite the fact that majority of the cells found in blood are erythrocytes and white blood cells, also known as leucocytes, can only negligibly affect blood flow and viscosity in main blood vessels, they contribute significantly to the hemodynamic forces at the tissue level. In fact in microcirculation leucocytes have major effects on flow rate and viscosity since the diameter of capillaries is close to diameter of a single cell or even smaller. There are many types of white blood cells such as Granulocytes which have the ability to alter its morphology when an external mechanical force is sensed such as the resistance for passing applied by the internal wall of narrow capillaries. The capability of leucocytes for shape alterations (mainly dependent on type) is considered to be greater than erythrocytes but the transient time necessary for passing a leucocyte of any kind is much longer than that of erythrocytes. In cases of severe infection, due to increase in the rigidity of leucocytes, this phenomenon can cause blockages in the microcirculation network and lead to serious problems [7].

One type of cell in human body that is always subjected to considerable amount of physical tension is alveolar epithelial lung cells. These cells constantly undergo stretching during inhale and exhale. This high amount of mechanical force changes cell viscoelastic properties. In all eukaryotic cells, stabilized cytoskeletal structure occurs mainly due to overall tensile stresses endured by filamentous structures. Consequently, cell stiffness increases relatively with the level of the tensile stress, which is called the prestress [83].

Alveolar cells like other cells are under pre-stress. In a detailed study regarding viscoelastic properties of these cells under stretch, it was found that cytoskeletal prestress is the main regulator of the frictional and elastic properties of lung alveolar cells. This result was obtained via experiments on human epithelial lung cells under equibiaxial stretch forces. It was observed by Trepat et al. that the applied force in form of stretch, increased cellular viscoelasticity but inhibited by latrunculin A via sequestering G-actin and preventing F-actin assembly [83].

This process once again proves the vital role of actin network in cell stiffness. Another important outcome of this study was the idea which suggested the cell detachment from the substrate and increase in cell stiffness coincide in high stretch forces revealing the highest limit to which alveolar cells can withstand the mechanical strain forces. This discovery shed some light on the disruption of alveolar barrier in injured lung by explaining the unbalanced forces both in cell-cell and cell-ECM interaction caused by cell stiffening induced by stretching forces [76].

Another organ in the body under constant shear stress is brain. Human brain tissue is a quite elastic and plastic tissue consisting of diverse types of cells called neurons. This flexible organ floats in a liquid called cerebrospinal fluid. Cerebrospinal fluid is the recycling system of the central nervous system since it circulates nutritious and chemicals which were filtered from the blood along with removing waste products of cells. Additionally it acts as a shield for brain. Constant circulation of this fluid causes a slight shear force on the brain tissue [18].

Apart from this, some neurological activities also change the physical form of brain tissue at specific regions namely, playing a musical instrument for a long time. Physical deformation of brain tissue starts with shape change of neurons. In human brain there are diverse types of nerve cells categorized in two groups of primary signaling cells or neurons and supporting cells or glial cell. Glial cells were discovered nearly 150 years ago as supporting cells for neurons. The name glial comes from a Greek word meaning glue. In the early days of discovery of glial cells, they were thought to be either supporting cells that glued primary signaling neurons together or act as scaffolds for them [76]. Over the past century it has been discovered that these cells have individual vital functions apart from protecting primary neuron. Glial cells can only be found in central nervous system.

The shape changing characteristics of human brain has made it an appealing research subject in terms of mechanical properties [47]. Most preferred method in studying viscoelastic characteristics of human brain cells are bulk rheology, AFM, Scanning Force Microscopy (SFM), and optical measurement methods. Primary findings of these experiments indicated that all cells of central nervous system behave more like elastic materials rather than viscous materials [47].

Secondly, unlike most eukaryotic cells, both primary signaling neurons and glial cells are very soft biological structures, but when comparing primary neurons with glial cells, neurons are considerably stiffer than glial cells. Thirdly, and most interestingly, different part of nerve cells showed different mechanical properties. This noticeable local diversity of viscoelastic values was assumed to be due to distribution of organelles. These findings revealed that initial roles assigned to glial cells cannot be true and they serve neither as support cells nor as scaffolds for primary neurons [47].

Taking another step forward towards discerning mysteries of mechanical properties of eukaryotic cells, mesenchymal-derived stem cells, primary osteoblasts, chondrocytes, and adipocytes attracted particular attention among mechanobiologists. The aim of investigating viscoelastic properties of these cells simultaneously by Darling et al. [17] was to understand the differences in mechanical behavior between stem cells and primary stem cells of mesenchymal lineage. They chose primary osteoblasts, chondrocytes, and adipocytes cells for this purpose. Their findings demonstrated that viscoelastic characteristics of cells not only affect cell-cell and cell-ECM interactions, but also greatly influences the production of biomarkers on the cells. Biomarkers are biological markers that demonstrate phenotypical changes that cells experienced during cellular transformation, differentiation and changes which were caused by diseases. A group of these biomarkers can provide precise and detailed information regarding phenotypic characteristics of different types of stem cells. Over the past years, focus of mechanobiology has been on the identification of these biomarkers via cells' mechanical characteristics with the aim of using the results for identification of different stem cells from mesenchymal stem cells. According to results of this study, when comparing adipocytes, osteoblast and chondrocyte with mesenchymal stem cells, the three cell types were all found to be stiffer than mesenchymal stem cells. A comparison among adipocytes, osteoblast and chondrocyte showed that adipocytes are the softest of them all followed by osteoblast and then chondrocyte as the stiffest cell type of this group. Interestingly, adipose-derived stem cells showed similar viscoelastic values to that of mesenchymal stem cells, but the two were mechanically distinct from primary differentiated stem cells [17].

5 Rheological Properties of Tissues

After detailed investigation on rheological properties of cells, it is also necessary to study rheological properties of tissues as well. Since tissue are biological structures made of considerable number of cells bound together by ECM, and almost all types of cells in human body demonstrate viscoelastic behaviors, it would be rational to assume that tissues would show similar mechanical characteristics as well. Hence a noticeable number of scientists in the fields of biology, mechanics and biomedical science have focused on studying the mechanical properties of tissues of human body, in particular their rheological behaviors [79].

Among all tissues in the body some have been more investigated than others including blood and brain tissue (white and grey matters). There are indeed many studied conducted over the years to investigate rheological properties of tissues but here some major examples are given and discussed with the aim of introducing the concepts and demonstrating their importance.

Blood as discussed before consists of three major cell types. Despite the common belief that blood is a viscous fluid, accumulation of these cells each with their unique mechanical characterizations, result in blood's behavior to be far from viscous material. Instead, it can be considered as a two phase liquid or in more simple term as a solid-liquid suspension.

In case of blood as a tissue, the mechanical behavior of blood in terms of flow mainly depends on plasma viscosity, red blood cell aggregation and mechanical properties of erythrocytes. As it was discussed previously, erythrocytes have reversible deformation capabilities during which cells' stiffness and rheology changes constantly depending on the amount of shear stress imposed on cells during circulation. It is also important to mention that aggregation of erythrocytes is an outcome of high shear stress especially during microcirculation. When there is high amounts of shear force like during microcirculation, the aggregates dismantle and individual erythrocytes deform in order to enter capillaries. In case of low shear stress, red blood cells attached to one another in the matter of seconds and form aggregates that have unique and size dependent mechanical characteristics. Hence it can be concluded that formation and size of aggregates has an inverse proportional relation to the amount of applied forces [7].

Apart from effects of aggregates on blood flow patterns, there are other factors that alter flow patterns of blood such as temperature, rheological properties of plasma and leucocytes and of course aging. For instance plasma as the suspension fluid for dispersed elements (erythrocytes, leukocytes, etc.), is a liquid with temperature dependent rheology. At 37 °C the viscosity of plasma varies between 1.10 and 1.35 cp and in case of an injury or disease these values are much higher [37].

Brain just like blood is another organ in human body that has been under extensive investigations so as to understand the rheological properties if its tissues. There are numerous studies regarding the biomechanical characteristics of tissues of central nervous system and brain, in particular [70]. The results of these studies revealed many inconsistencies for biomechanical properties of same tissues that are apparently tested under similar conditions. These inconsistencies may result from operator errors, age and gender of the subject tissue, tissue pretreatment protocols, etc. Unfortunately there are no accurate values for rheological properties of brain tissue which makes it difficult for scientists and engineers to apply obtained results in design and fabrication of biological scaffolds and material for diseases of CNS, but so far the database of biomechanical properties of brain and spinal cord of humans includes very enlightening information [12]. For instance three different regions of brain were investigated in term of biomechanical properties, in particular rheology. The selected sections of both white and grey matter were Corpus callosum (CC), Thalamus (Th) and Corona radiata (CR). The experiments were done by subjecting tissue specimens to repeated creep-recovery shear force. According to the results, all three tissues have both reversible and irreversible deformations happening on different sections. The former is due to the elastic nature of specific sections of the tissue and the latter occurs because of the viscous characteristic of those sections. Results also demonstrated that all three tissues have distinct biomechanical properties. Also it was observed that the highest stiffness value belonged to thalamus tissue and lowest stiffness belonged to corpus callosum [9].

Apart from these findings, biomechanical properties of brain tissue under the classification of white and grey matter were investigated. It is well known that human brain consists of mainly two layers named grey and white matter. The white matter which is the inner layer of the brain, consist of myelinated axons whereas the gray matter that is the outer layer, houses cell bodies, dendrites, and unmyelinated axons.

Studying biomechanical differences between grey and white matter can be useful in understanding many neurological advancement such as neurodevelopment and the environmental factors involved in morphological abnormalities of brain tissue such as brachycephaly and plagiocephaly, flat or asymmetric heads, and hydrocephalus, abnormal accumulation of cerebrospinal fluid in the brain [10].

By studying biomechanical characteristics of grey and white matter via indentation, it was revealed that white matter is approximately 39% stiffer than gray matter based on the data obtained for average modulus of both tissues to be at 1.895 and 1.389 kPa respectively. Viscosity test results also showed that white matter is more viscous than grey matter; hence, the response time of white matter to mechanical loading was much slower than grey matter [10].

Measuring the differences between mechanical properties of gray and white matter and understanding the reasons behind them as well as their impact on health of the brain can be of great importance in better understanding of neurological networks and developments and diagnosis of neurological disorders.

In 2016 Kofahl et al. modified MRI imaging machine in order to detect tumorous brain tissue based on the biomechanical differences between healthy and tumorous tissue. This combination of rheological characterization and novel imaging techniques is an excellent illustration of how results of extensive studies regarding rheological properties of different tissue types can be applied to modern detection and diagnostic techniques [40].

6 Rheological Properties of Scaffolds and Their Biomedical Applications

After addressing the concept, followed by clarifying the influence and importance of these rheological properties in cells and tissues, almost all of the discoveries made so far can be utilized in synthesis and fabrication of biomaterials and related structures such as 2D and 3D scaffolds as hydrogels or sponges. Just like cells and tissues, there has been extensive research focusing on creation of perfect artificial structures suitable for biomedical applications on both in vitro and in vivo uses, namely hydrogels, bone fillers and constructed heart valves.

Cardiovascular diseases are one of the main reasons of death in 21th century. Among the heart diseases, heart valve defects are one of the most common ones and unfortunately its rate seems to be increasing every year which arises the urgent need for artificial tissue engineered heart valves. Despite all efforts spent so far on fabrication and modification of functional tissue engineered heart valves that match characteristics of native human heart valve in terms of biocompatibility and mechanical function and in general implantability, engineered valves are far from perfect [29, 40]. One of the main biomechanical characteristics necessary for a fully functional heart valve is its rheological properties of the soft tissue. Heart valves are constantly under mechanical force. Hence for the valves to have an acceptable performance, it is necessary to respond to wide range of mechanical stimuli instantaneously and precisely. Dynamic and static mechanical properties of tissue including tensile strength, viscoelastic and rheological properties are the main regulators of valve's response. A tissue engineered heart valve must be as similar to native tissue as possible in terms of fatigue, flexural, rheological and viscoelastic properties [32].

Tissue engineered heart valves are fabricated via cell seeding or cell encapsulation into 3D scaffolds made of biodegradable and biocompatible materials. The seeded scaffolds are then placed in bioreactors for culturing. When the cells reach a certain level of viability, the scaffolds can be implanted. The aim of using bioreactor is to induce tissue formation rather than 2D cell layer formation [31]. As the tissue engineered scaffolds are implanted, they can obtain mechanical and structural properties of the native tissue as they adapt to in vivo conditions by remodeling, repair and tissue growth [32]. Since any implanted heart valve is subjected to a great amount of physical force upon implantation, it is imperative for the tissue to provide adequate mechanical strength and flawless function. This can be achieved provided that all mechanical properties of native tissue are known and precisely calculated and integrated into the fabricated artificial valve in a stable manner [20].

Despite the fact that data obtained for biomechanical properties of human heart valve are not very thorough due to the limited availability of fresh human heart valves and ideal measurement methods, a useful set of data are available regarding biomechanical properties of human heart valves [32, 39]. For instance transvalvular pressure of human mitral valve, aortic valve, tricuspid valve and pulmonary valve are reported at 120.0, 80.0, 25.0 and 10.0 mmHg respectively whereas values of the same pressure for bovine heart valves are found at 144.0, 92.4, 27.2 and 119.0 mmHg for bovine mitral valve, aortic valve, tricuspid valve and pulmonary valve respectively [32].

Although by knowing these and similar other values of mechanical and geometric characteristics of the native human and animal valves along with methods of manipulation and adjustment of mechanical properties of xenografts and polymeric scaffolds to desired values, fabrication of suitable tissue engineered heart valves have become possible, extensive and thorough investigation are still required to fabricate an exact replica of native human heart valve [32].

Bone diseases are another major problem of modern medicine and hot topic of interest for biomedical researchers. In human body bones have many responsibilities, namely providing a structural stability and a means for flawless motion along with acting as a physiological reservoir of hemocytoblasts and mesenchymal cells.

There are many of reasons for bone diseases such as trauma, tumors, degenerative diseases, osteoporosis, etc., that cause bone defects. Bone filler also known as bone cement is one of the popular methods for treating these bone defects by repairing damaged tissue via imparting desired biomechanical and physiological healing factors for the bones. Bone filler is so far the best alternative for allografts and xenografts which have problems such as availability, infection hematoma and rejection in case of allografts and reduced osteogenic capacity, reduced revascularization post implantation, etc. in case of xenografts [52].

Synthetic materials used in treatment of bone defects have problematic issues as well such as functional defects, difficulties in handling during implantation, unbalanced or insufficient local density, low adhesion problem, and mechanical instability with time. Despite these issues, among all treatment methods, bone fillers seem to be more efficient if and only if they are properly designed in terms of chemical compositions, biodegradability, biocompatibility, stiffness, flow properties, tensile strength and many other mechanical, chemical and biological properties [66].

Hydroxyapatite, tricalcium phosphate, hydroxyapatite cement, bioglass, methylmethacrylate polymer and porous polyethylene polymers are the most commonly used materials in the novel medicine as bone substitutes. There are other materials in form of composites which are used as bone cements such as zinc phosphate cement, zinc polycarboxylate cement and glass polyalkenoate cement which are less used nowadays. Bone cements are usually viscoelastic materials which look like toothpastes which turn into solid once injected or manually placed [23, 61]. Solidification occurs by an external cross linker. Each material has its own unique cross linker. For instance, PMMA, widely used in orthopedic surgery, is composed of two main components: PMMA powder and methyl methacrylate liquid. Once these two are mixed, a homogenous dough-like white material is obtained. Solidification of this dough after placement into the intended site happens via polymerization of methylmetacrylate. There is benzoyl peroxide present in the powder which starts producing free radicals as it reacts with N, N-dimethyl-p-toluidine present in the liquid [23].

There are many aspects that must be carefully studied and calculated in designing new bone filler materials such as chemical composition, biocompatibility and of course mechanical properties. Depending on the site where the bone cement needs to be injected, the flow properties of the filler is of vital importance along with all factors that can affect the quality and duration of polymerization such as pH, temperature, chemical compositions of plasma, etc. On the other hand the implanted cement must be completely biocompatible with similar mechanical strength as similar to desired value for the intended function as possible [23, 61].

7 Conclusion

Rheology is one of the characteristics of materials, which is categorized under mechanical properties, is a representative of flow properties of materials. Human body both in large scale (organs) and in small scale (cells) consists of mostly fluid. In material science fluids mostly show viscous behaviors. However, in case of fluids of human body, viscous behavioral patterns are useless due to the solid small components within cell and tissues, ECM, phenotype differences of cells and tissues along with compositional differences between different body fluids which lead to diverse structural and metabolic differences. Fortunately combination of concepts of elasticity and viscosity is able to define biomechanical nature of cells, tissues and related structures.

By advancement of science and technology over that past century a great number of methods have been developed for measuring mechanical properties of almost all types of materials in nano-scale and micron. In particular, investigating rheological properties of cells in micron and nano-scale has revealed incredibly comprehensive information. These invaluable data can help understand biomechanical and therefore physiological processes which take place in cells and their relation with cellular functions, diseases, mutations and in general, life.

There are many other studies regarding rheological studies of cells, tissues, organs, biomaterials, artificially engineered structure that have great applications in biomedical field. Despite the fact that almost all cell types have been biomechanically characterized along with almost all tissue types, and biomaterials such as polymers, ceramics, metals and composites fabricated in form of membranes, 3D scaffolds, hydrogels, fibrous structures, etc. there are still so many unsolved problems and unanswered questions concerning fabrication of ideal devices and implants that need to be solved and answered.

The number of researches conducted to this day regarding rheological properties of biomaterials is uncountable but unfortunately only a small portion of it was discussed in this chapter as in introductory to the world of rheology in living things. Wish of solving the riddle of life can only come true if all is revealed in subatomic level and that is the ultimate goal of today and the inevitable certainty of the future.

Acknowledgements The authors would like to thank Prof. Dr. Herbert J. Meiselman, Prof. Dr. Aysegül Temiz Artmann and Prof. Dr. Oguz K. Baskurt for their invaluable research in this field as the result of their lifetime of intelligence and hard work.

References

1. Alenghat, F. J., & Ingber, D. E. (2002). Mechanotransduction: All signals point to cytoskeleton, matrix, and integrins. *Science's STKE : Signal Transduction Knowledge Environment, 119,* pe6.
2. Amblard, F., Maggs, A. C., Yurke, B., Pargellis, A. N., & Leibler, S. (1996). Subdiffusion and anomalous local viscoelasticity in actin networks. *Physical Review Letters, 77*(21), 4470–4473.
3. Artmann, G. M. (1995 Sep–Oct). Microscopic stiffness and relaxation time of red blood in a flow chamber of stiffness and relaxation time of red blood cells in a flow chamber. *Biorheology, 32*(5), 553–570.
4. Artmann, G. M., Sung, K. L., Horn, T., Whittemore, D., Norwich, G., & Chien, S. (1997). Micropipette aspiration of human erythrocytes induces echinocytes via membrane phospholipid translocation. *Biophysics Journal, 72,* 1434.
5. Ayala, Y. A., Pontes, B., Ether, D. S., Pires, L. B., Araujo, G. R., Frases, S., et al. (2016). Rheological properties of cells measured by optical tweezers. *BMC Biophysics, 9,* 5.
6. Badalà, Federico, Nouri-mahdavi, Kouros, & Raoof, Duna A. (2008). NIH public access. *Computer, 144*(5), 724–732.
7. Baskurt, Oguz K., & Meiselman, Herbert J. (2003). Blood rheology and hemodynamics. *Seminars in Thrombosis and Hemostasis, 29*(5), 435–450.
8. Bausch, A. R., Ziemann, F., Boulbitch, A. A., Jacobson, K., & Sackmann, E. (1998). Local measurements of viscoelastic parameters of adherent cell surfaces by magnetic bead microrheometry. *Biophysical Journal, 75*(4), 2038–49. http://dx.doi.org/10.1016/S0006-3495(98)77646-5.
9. Boudjema, F., Lounis, M., Khelidj, B., & Bessai, N. (2015). Rheological regional properties of brain tissue studied under cyclic creep/recovery shear stresses. *Journal of Physics: Conference Series, 602*(1), 6.
10. Budday, S., Nay, R., de Rooij, R., Steinmann, P., Wyrobek, T., Ovaert, T. C., & Kuhl, E. (2015). Mechanical properties of gray and white matter brain tissue by indentation. *Journal of the Mechanical Behavior of Biomedical Materials, 46,* 318–330. http://dx.doi.org/10.1016/j.jmbbm.2015.02.024.
11. Chan, C. J., Whyte, G., Boyde, L., Salbreux, G., & Guck, J. (2014). Impact of heating on passive and active biomechanics of suspended cells. *Interface Focus, 4*(2), 20130069.
12. Cheng, S., Clarke, E. C., & Bilston, L. E. (2008). Rheological properties of the tissues of the central nervous system: A review. *Medical Engineering & Physics, 30*(10), 1318–1337.
13. Chien, S. (2007). Mechanotransduction and endothelial cell homeostasis: The wisdom of the cell. *American Journal of Physiology-Heart and Circulatory Physiology, 292*(3), H1209–H1224.
14. Chien, S., Sung, K. L., Skalak, R., Usami, S., Tözeren, A. (1978, November 24). Theoretical and experimental studies on viscoelastic properties of erythrocyte membrane. *Biophysical Journal, 24,* 463–487.
15. Chmiel, B., Karkoszka, H., Cierpka, L., & Wiecek, A. (2005). Rheological properties of red blood cells in kidney transplant recipients: The role of lipid profile and type of immunosuppresion. *Transplantation Proceedings, 37*(4), 1885–1888.
16. Crocker, J. C., & Hoffman, B. D. (2007). Multiple-particle tracking and two-point microrheology in cells. *Methods in Cell Biology, 83*(7), 141–178.

17. Darling, E. M., Topel, M., Zauscher, S., Vail, T. P., & Guilak, F. (2008). Viscoelastic properties of human mesenchymally-derived stem cells and primary osteoblasts, chondrocytes, and adipocytes. *Journal of Biomechanics, 41*(2), 454–464.
18. Di Terlizzi, R., & Platt, S. (2006). The function, composition and analysis of cerebrospinal fluid in companion animals: Part I—Function and composition. *Veterinary Journal, 172*(3), 422–431.
19. Doherty, G. J., & McMahon, H. T. (2008). Mediation, modulation, and consequences of membrane-cytoskeleton interactions. *Annual Review of Biophysics, 37*(1), 65–95.
20. Driessen, N. J. B., Mol, A., Bouten, C. V. C., & Baaijens, F. P. T. (2007). Modeling the mechanics of tissue-engineered human heart valve leaflets. *Journal of Biomechanics, 40*(2), 325–334.
21. Drury, J. L., & Dembo, M. (1999). Hydrodynamics of micropipette aspiration. *Biophysical Journal, 76*(1 Pt 1), 110–128.
22. Engelking, L. R. (2011). Chemical composition of living cells. *Textbook of veterinary physiological chemistry* (pp. 2–6). Academic Press. eBook ISBN: 9780123919106.
23. Farrar, D. F., & Rose, J. (2001). Rheological properties of PMMA bone cements during curing. *Biomaterials, 22*(22), 3005–3013.
24. Franck, C., Maskarinec, S. A., Tirrell, D. A., & Ravichandran, G. (2011). Three-dimensional traction force microscopy: A new tool for quantifying cell-matrix interactions. *PLoS ONE, 6*(3).
25. Goldmann, W. H. (2012). Mechanotransduction in cells. *Cell Biology International, 36*(6), 567–570.
26. Guck, J., Ananthakrishnan, R., Mahmood, H., Moon, T. J., Cunningham, C. C., & Käs, J. (2001). The optical stretcher: A novel laser tool to micromanipulate cells. *Biophysical Journal, 81*(2), 767–784.
27. Guilak, F., Tedrow, J. R., & Burgkart, R. (2000). Viscoelastic properties of the cell nucleus. *Biochemical and Biophysical Research Communications, 269*(3), 781–786.
28. Haase, K., & Pelling, A. E. (2015). Investigating cell mechanics with atomic force microscopy. *Journal of the Royal Society, Interface/the Royal Society, 12*(104), 20140970.
29. Hale, C. M., Sun, S. X., & Wirtz, D. (2009). Resolving the role of actoymyosin contractility in cell microrheology. *PLoS ONE, 4*(9).
30. Han, M. K. L., & de Rooij, J. (2016). Converging and unique mechanisms of mechanotransduction at adhesion sites. *Trends in Cell Biology, 26*(8), 612–623. https://doi.org/10.1016/j.tcb.2016.03.005.
31. Hasan, A., Memic, A., Annabi, N., Hossain, M., Paul, A., Dokmeci, M. R., et al. (2014). Electrospun scaffolds for tissue engineering of vascular grafts. *Acta Biomaterialia, 10*(1), 11–25.
32. Hasan, A., Ragaert, K., Swieszkowski, W., Selimović, Š., Paul, A., Camci-Unal, G., et al. (2014). Biomechanical properties of native and tissue engineered heart valve constructs. *Journal of Biomechanics, 47*(9), 1949–1963.
33. Hochmuth, R. M. (2000). Micropipette aspiration of living cells. *Journal of Biomechanics, 33*(1), 15–22.
34. Hoffman, B. D., & Crocker, J. C. (2009). Cell mechanics: Dissecting the physical responses of cells to force. *Annual Review of Biomedical Engineering, 11*, 259–288.
35. Humphrey, J. D., Dufresne, E. R., & Schwartz, M. A. (2014). Mechanotransduction and extracellular matrix homeostasis. *Nature Reviews Molecular Cell Biology, 15*(12), 802–812. http://dx.doi.org/10.1038/nrm3896.
36. Icard-Arcizet, Delphine, Cardoso, Olivier, Richert, Alain, & Hénon, Sylvie. (2008). Cell stiffening in response to external stress is correlated to actin recruitment. *Biophysical Journal, 94*(7), 2906–2913.
37. Irace, C., Carallo, C., Scavelli, F., Esposito, T., De Franceschi, M. S., Tripolino, C., & Gnasso, A. (2014). Influence of blood lipids on plasma and blood viscosity. *Clinical Hemorheology and Microcirculation, 57*(3), 283–290.

38. Jay, A. W., & Canham, P. B. (1977). Viscoelastic properties of the human red blood cell membrane. II. Area and volume of individual red cells entering a micropipette. *Biophysical Journal, 17*(2), 169–178. http://dx.doi.org/10.1016/S0006-3495(77)85634-8.
39. Kalejs, M., Stradins, P., Lacis, R., Ozolanta, I., Pavars, J., & Kasyanov, V. (2009). St Jude Epic Heart Valve bioprostheses versus native human and porcine aortic valves—Comparison of mechanical properties. *Interactive Cardiovascular and Thoracic Surgery, 8*(5), 553–556.
40. Kofahl, A. L., Theilenberg, S., Bindl, J., Ulucay, D., Wild, J., Napiletzki, S., et al. (2016). Combining rheology and MRI: Imaging healthy and tumorous brains based on mechanical properties. *Magnetic Resonance in Medicine, 0,* 1–11.
41. Kollmannsberger, P., & Fabry, B. (2007). High-force magnetic tweezers with force feedback for biological applications. *Review of Scientific Instruments, 78*(11), 1–6.
42. Kraning-Rush, C. M., Califano, J. P., & Reinhart-King, C. A. (2012). Cellular traction stresses increase with increasing metastatic potential. *PLoS ONE, 7*(2).
43. Laperrousaz, B., Drillon, G., Berguiga, L., Nicolini, F., Audit, B., Satta, V. M., et al. (2016). From elasticity to inelasticity in cancer cell mechanics: A loss of scale-invariance. *AIP Conference Proceedings* (p. 1760).
44. Last, J. A., Liliensiek, S. J., Nealey, P. F., & Murphy, C. J. (2009). Determining the mechanical properties of human corneal basement membranes with atomic force microscopy. *Journal of Structural Biology, 167*(1), 19–24.
45. Lomakina, E. B., Spillmann, C. M., King, M. R., & Waugh, R. E. (2004). Rheological analysis and measurement of neutrophil indentation. *Biophysical Journal, 87*(6), 4246–4258.
46. Lombardo, M., Lombardo, G., Carbone, G., De Santo, M. P., Barberi, R., & Serrao, S. (2012). Biomechanics of the anterior human corneal tissue investigated with atomic force microscopy. *Investigative Ophthalmology and Visual Science, 53*(2), 1050–1057.
47. Lu, Y.-B., Franze, K., Seifert, G., Steinhäuser, C., Kirchhoff, F., Wolburg, H., et al. (2006). Viscoelastic properties of individual glial cells and neurons in the CNS. *Proceedings of the National Academy of Sciences, 103*(47), 17759–17764.
48. Lulevich, V., Zink, T., Chen, H. Y., Liu, F. T., & Liu, G. Y. (2006). Cell mechanics using atomic force microscopy-based single-cell compression. *Langmuir, 22*(19), 8151–8155.
49. Massiera, G., Van Citters, K. M., Biancaniello, P. L., & Crocker, J. C. (2007). Mechanics of single cells: Rheology, time dependence, and fluctuations. *Biophysical Journal, 93*(10), 3703–3713. https://doi.org/10.1529/biophysj.107.111641.
50. Matthews, B. D., Overby, D. R., Mannix, R., & Ingber, D. E. (2006). Cellular adaptation to mechanical stress: Role of integrins, rho, cytoskeletal tension and mechanosensitive ion channels. *Journal of Cell Science, 119*(Pt 3), 508–518.
51. Moeendarbary, E., & Harris, A. R. (2014). Cell mechanics: Principles, practices, and prospects. *Wiley Interdisciplinary Reviews: Systems Biology and Medicine, 6*(5), 371–388.
52. Morra, M., Giavaresi, G., Sartori, M., Ferrari, A., Parrilli, A., Bollati, D., et al. (2015). Surface chemistry and effects on bone regeneration of a novel biomimetic synthetic bone filler. *Journal of Materials Science: Materials in Medicine, 26*(4). http://dx.doi.org/10.1007/s10856-015-5483-6.
53. Munevar, S. (2004). Regulation of mechanical interactions between fibroblasts and the substratum by stretch-activated Ca2+ entry. *Journal of Cell Science, 117*(1), 85–92.
54. Müller, U. (2004). Integrins and extracellular matrix in animal models. *Handbook of Experimental Pharmacology, 165,* 217–241.
55. Nagayama, K., Adachi, A., & Matsumoto, T. (2011). Heterogeneous response of traction force at focal adhesions of vascular smooth muscle cells subjected to macroscopic stretch on a micropillar substrate. *Journal of Biomechanics, 44*(15), 2699–2705.
56. Nawaz, S., Sánchez, P., Bodensiek, K., Li, S., Simons, M., & Schaap, I. A. (2012). Cell visco-elasticity measured with AFM and optical trapping at sub-micrometer deformations. *PLoS ONE, 7*(9).
57. Oh, M.-J., Kuhr, F., Byfield, F., & Levitan, I. (2012). Micropipette aspiration of substrate-attached cells to estimate cell stiffness. *Journal of Visualized Experiments, 67,* e3886.

58. Pajerowski, J. D., Dahl, K. N., Zhong, F. L., Sammak, P. J., & Discher, D. E. (2007). Physical plasticity of the nucleus in stem cell differentiation. *Proceedings of the National Academy of Sciences of the United States of America, 104*(40), 15619–15624.
59. Park, Y., Best, C. A., Badizadegan, K., Dasari, R. R., Feld, M. S., Kuriabova, T., et al. (2010). Measurement of red blood cell mechanics during morphological changes. *Proceedings of the National Academy of Sciences of the United States of America, 107*(15), 6731–6736.
60. Pierini, F., Zembrzycki, K., Nakielski, P., Pawłowska, S., & Kowalewski, T. A. (2016). Atomic force microscopy combined with optical tweezers (AFM/OT). *Measurement Science and Technology, 27*(2), 25904.
61. Pryor, L. S., Gage, E., Langevin, C. J., Herrera, F., Breithaupt, A. D., Gordon, C. R., et al. (2009). Review of bone substitutes. *Craniomaxillofacial Trauma & Reconstruction, 2*(3), 151–160.
62. Pullarkat, P. A., Fernández, P. A., & Ott, Albrecht. (2007). Rheological properties of the eukaryotic cell cytoskeleton. *Physics Reports, 449*(1–3), 29–53.
63. Quinto-Su, P. A., Kuss, C., Preiser, P. R., & Ohl, C.-D. (2011). Red blood cell rheology using single controlled laser-induced cavitation bubbles. *Lab on a Chip, 11*(4), 672–678.
64. Raucher, D., Stauffer, T., Chen, W., Shen, K., Guo, S., York, J. D., et al. (2000). Phosphatidylinositol 4,5-Bisphosphate functions as a second messenger that regulates cytoskeleton–plasma membrane adhesion. *Cell, 100*(2), 221–228.
65. Riveline, D., Zamir, E., Balaban, N. Q., Schwarz, U. S., Ishizaki, T., Narumiya, S., et al. (2001). Focal contacts as mechanosensors. *The Journal of Cell Biology, 153*(6), 1175–1186.
66. Roohani-Esfahani, S.-I., Newman, P., & Zreiqat, H. (2016). Design and fabrication of 3D printed scaffolds with a mechanical strength comparable to cortical bone to repair large bone defects. *Scientific Reports, 6*(February 2015), 19468.
67. Rother, J., Nöding, H., Mey, I., & Janshoff, A. (2014). Atomic force microscopy-based microrheology reveals significant differences in the viscoelastic response between malign and benign cell lines. *Open Biology, 4*(5), 140046.
68. Sasaki, N. (2012). Viscoelastic properties of biological materials. *Annals of the New York Academy of Sciences,* 99–122. http://www.ncbi.nlm.nih.gov/pubmed/7076685.
69. Schulze, C., Wetzel, F., Kueper, T., Malsen, A., Muhr, G., Jaspers, S., et al. (2012). Stiffening of human skin fibroblasts with age. *Clinics in Plastic Surgery, 39*(1), 9–20.
70. Shuck, L. Z., & Advani, S. H. (1972). Rheological response of human brain tissue in shear. *Journal of Basic Engineering Trans ASME, 94 Ser D*(4), 905–911.
71. Sleep, J., Wilson, D., Simmons, R., & Gratzer, W. (1999). Elasticity of the red cell membrane and its relation to hemolytic disorders: An optical tweezers study. *Biophysical Journal, 77*(6), 3085–3095.
72. Stamenović, D. (2008). Rheological behavior of mammalian cells. *Cellular and Molecular Life Sciences, 65*(22), 3592–3605.
73. Supriya, B., Jun, D., Paul, B. C., & Dahms, T. E. S. (2012). *Viscoelasticity in biological systems: A special focus on microbes*. ISBN 978-953-51-0841-2, Publisher: InTech.
74. Thoumine, O., & Ott, A. (1997). Time scale dependent viscoelastic and contractile regimes in fibroblasts probed by microplate manipulation. *Journal of Cell Science, 110*(Pt 1), 2109–2116.
75. Tozeren, A., Skalak, R., Sung, K. P., & Chien, S. H. U. (1982). Viscoelastic behavior of erythrocyte membrane. *Biophysical Society, 39*(1), 23–32.
76. Trepat, X., Grabulosa, M., Puig, F., Maksym, G. N., Navajas, D., & Farré, R. (2004). Viscoelasticity of human alveolar epithelial cells subjected to stretch. *American Journal of Physiology, 287*(5), L1025–L1034.
77. Tseng, Y., Kole, T. P., & Wirtz, D. (2002). Micromechanical mapping of live cells by multiple-particle-tracking microrheology. *Biophysical Journal, 83*(6), 3162–3176.
78. Van Citters, K. M., Hoffman, B. D., Massiera, G., & Crocker, J. C. (2006). The role of F-actin and myosin in epithelial cell rheology. *Biophysical Journal, 91*(10), 3946–3956.

79. Verdier, C., Etienne, J., Duperray, A., Preziosi, L., et al. (2009). Review: Rheological properties of biological materials. *Comptes Rendus Physique, 10*(8), 790–811. http://dx.doi.org/10.1016/j.crhy.2009.10.003.
80. Von Wichert, G., Jiang, G., Kostic, A., De Vos, K., Sap, J., & Sheetz, M. P. (2003). RPTP-α acts as a transducer of mechanical force on αv/β3-integrin–cytoskeleton linkages. *The Journal of Cell Biology, 161*(1), 143–153. http://doi.org/10.1083/jcb.200211061.
81. Vonna, Laurent, Wiedemann, Agnès, Aepfelbacher, Martin, & Sackmann, Erich. (2003). Local Force Induced Conical Protrusions of Phagocytic Cells. *Journal of Cell Science, 116*(Pt 5), 785–790.
82. Walters, K. (1980). History of rheology. *Rheology* I.
83. Wang, N., Tolic-Nørrelykke, I. M., Chen, J., Mijailovich, S. M., Butler, J. P., Fredberg, J. J., & Stamenovic, D. (2002). Cell prestress. I. Stiffness and prestress are closely associated in adherent contractile cells. *American Journal of Physiology. Cell Physiology, 282*(3), C606–C616.
84. Zhou, S., Sokolov, A., Lavrentovich, O. D., & Aranson, I. S. (2014). Living liquid crystals. *Proceedings of the National Academy of Sciences of the United States of America, 111*(4), 1265–1270. http://www.ncbi.nlm.nih.gov/pubmed/24474746.

Author Biography

Bora Garipcan (Turkey) Ph.D. He received his B.Sc. in 1999 from the Department of Chemistry, Hacettepe University, Ankara, and his Ph.D. in Bioengineering in 2008 from the same university. His main research areas are; Biomimetic and Bioinspired materials, Biosensors. He is a member of the Bogazici University, Institute of Biomedical Engineering since July 2011. Garipcan`s research focuses on surface engineering of several medical surfaces (such as Ti, Si, Au, biodegradable polymers, and elastomers) by changing surface properties (stiffness, topography, chemistry and biochemistry) for controlled and directed cell behavior (adhesion, proliferation, differentiation).

Histeresis Loops and Yield Stress in Cell Research

Yong-de Shi

1 Introduction

If there is a force to act on a system, the system will produce a reaction force. The former belongs to the acting force and the latter to the reacting one. Between both of the forces, which possess dependence and hysteresis of each other, there exist distinct differences in values and in the time they emerge. Through the analysis of these data, structural change of a system can be determined, depending on the history of applied and then withdrawn forces on the system. An example is that, if your hand takes a hammer to hit a rock, you will feel a painful shock in your hand, which tells you the rock is very hard; in the contrary, if you take the hammer to hit cotton material, you won't feel any pain in your hand, which tells you the cotton material is very soft. This example demonstrates that the analysis of acting and reacting forces on a system will tell us information about the materials' structure and their possible function. This analysis is called the hysteresis loop technique. It is not only used to observe the quantities of acting and reacting forces, but also to obtain information about the following relationships: (1) the time lag between the acting and reacting forces (generally called 'mechanical hysteresis'), in case a force field is applied and then withdrawn from a system; and (2) the reaction force development curve, when the system is exposed to an increasing force field from zero to a maximum and vice versa back to zero. Results of above-mentioned measurements can be plotted as mechanical hysteresis loop curves of stress versus time. According to the shape of the hysteresis loop curve, structural changes in the system, energy storage or dissipation, and possibly information about the biological function of the system can be obtained. Yield stress is easily determined in a hysteresis loop when the shear rate falls back to zero. The determination of

Y. Shi (✉)
Shanghai Medical College, Fudan University, 138 Yi Xue Yuan Road, Shanghai 200032, China
e-mail: ydshi@ssbiophysics.com.cn

G. M. Artmann et al. (eds.), *Biological, Physical and Technical Basics of Cell Engineering*, https://doi.org/10.1007/978-981-10-7904-7_6

hysteresis and yield stresses involve multi-dimensional digital record keeping and automation of controlling and sampling action forces, reaction forces, temperature, timing, etc. Today's high speed computer and information technologies are able to fulfill these functions. However, research in medicine and biology requires dealing with biological samples, which usually provide an additional level of difficulty in comparison to traditional engineering disciplines, due to biological samples being on a very small scale and the reaction forces from tested systems are also very small. Thus, biology experts neglect such techniques. This article will be a review of work done with hysteresis loop and shear stress techniques for cell based research.

2 Requirements for Determining Hysteresis Loops

To generate a series of triangular waves of shear rates, including time intervals and the total duration time (Fig. 1), it takes a rheometer, controlled by a servo motor, and suitable computer software. In a tested system, the automation has to generate specified shear rates from zero to a maximum in certain time intervals within a total duration time, and then reverse the shear rates from the maximum back to zero, using the same time intervals and total duration.

During such time courses of applied increasing and decreasing shear rates to a system, the system's responses in values of shear stress (reaction forces), temperature values, and interval times (in seconds) need to be automatically recorded with real time sampling. Instrumentation used for the experiments described in this article, were a HT Rheometer with an air-bearing design, created by Senior Engineer Jiao S. K., and a commercially available rheometer from USA Brookfield programmable DVII Rheometers.

Fig. 1 Tri-angular shear rate curve; shear rate (1/s) versus time (s)

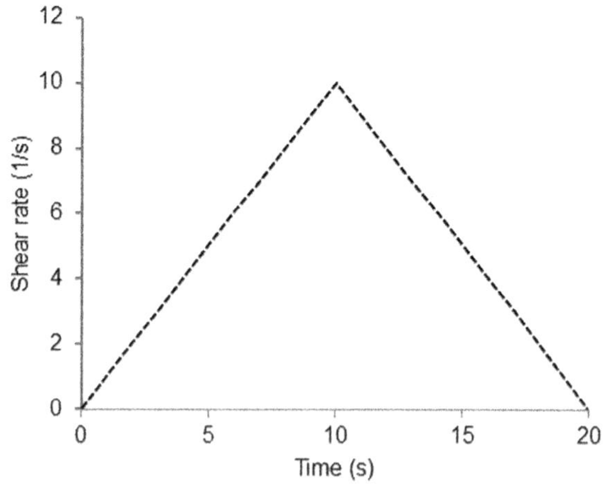

3 Classification of Hysteresis Loop Types

Results of experimental work in our laboratories showed that hysteresis loops can be divided into 5 classifications as shown in Table 1.

Table 1 Classifications of various hysteresis loops and their theoretical bases

Types	**Schemes** The vertical axis shows stress responses from the tested system; the horizontal axis shows shear rates applied	**Theoretical considerations** Prediction of structural changes inside the tested system will be discussed for varying shear forces applied
I	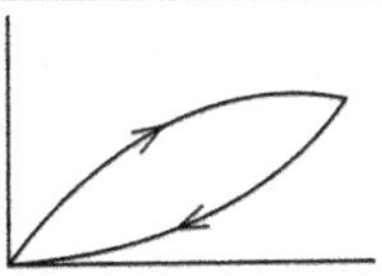	**(I) Mono clockwise loop** The tested system dissipates the energy supplied by the servo motor of the rheometer; the enclosed area shows the dissipated energy of one cycle (erg/cm^3/s, 10^{-1} kg m^{-1} s^{-3}). The structure in the system is changed from aggregation to disaggregation, so with to be softening or liquefying
II	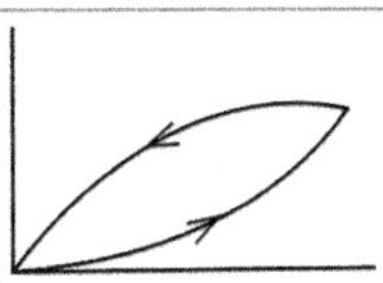	**(II) Mono counter-clockwise loop** The tested system absorbs the energy from the servo motor or from the system itself; the enclosed area is the summation of the absorbed energy (10^{-1} kg m^{-1} s^{-3}). The structure in the system is changed from disaggregation to aggregation, so with to be hardening or thickening
III		**(III) "8" double-loop shape** There is an intersection point between two loops; the loop becomes a counter-clockwise loop close to the origination point (as in type II), and a clockwise loop far away from the origination point (as in type I)
IV		**(IV)** Through theoretical deduction, **another "8" double-loop shape** could appear, again with an intersection point between two loops (the loop close to the origination point is clockwise, and the loop far away from the origination point is counter-clockwise). However, our lab did not find such type
V	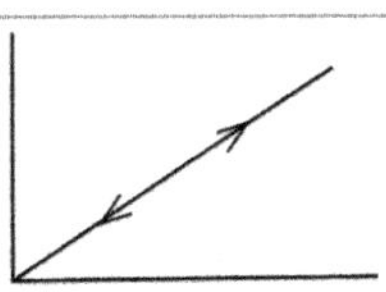	**(V) No loop (pure liquids)** An upward curve and a downward curve are overlapping as a straight line. For example in water, the system doesn't show any structural change when it is exposed to shear stresses or withdrawn from shear forces

4 Hysteresis Loop Types of Human Whole Blood Samples

Experimental results in our laboratory identified the shape of the hysteresis loop curve found in the human whole blood system to have a shape of the digit "8" type with an intersection point between two loops, therefore being classified as type III (Table 1).

Historically, already in the 1970s, Huang et al. performed experiments to determine whole blood hysteresis loops [1]. However, Huang et al. only found a clockwise loop, but did not identify a counter-clockwise loop in his findings. Huang's research team established a generalized rheological equation to describe the hysteresis loop curve of human whole blood [1]. In 1986, Chen et al. verified the clockwise loop far away from the origination point, but also uncovered a counter-clockwise loop close to the origination point with an intersection location at 1–2 s^{-1}. According to Chen's findings, he suggested that Huang's equation should be modified [2]. In the 1990s, investigations on a series of normal and ischemic stroke patients in our laboratories verified the findings as reported earlier by Chen and his team. In 1998, Fang et al. established a new constitutive equation to describe the "8" shape hysteresis loop in whole human blood [3]. In 2016, Shi W. et al. reported a digital solution using a continuous kinetics equation for non-linear hysteresis loop curves in human blood [4]. This article will further review continuous studies in this field performed in our laboratories.

In our recent experimental studies, 86 human blood samples (cases) were tested, of which 44 were normal subjects (30 males and 14 females) and 42 were ischemic stroke patients after a first stroke attack (25 males and 17 females) [5]. The instrument used in our tests was a HT Rheometer (Shanghai Astronaut Bureau), designed by Senior Engineer Jiao S. K. Experimental results are displayed in Fig. 2. Useful information from the "8" shape hysteresis loops were extracted for explanations in biology and clinical medicine as in the following: (A) *The intersection point*: It divides the "8" into two loop areas. On the one hand, the loop close to the origination point (zero) is a counter-clockwise loop and reflects that the whole blood system absorbs the energy from the servo motor to form red blood cell (RBC) rouleaux under low shear rate conditions. On the other hand, the loop far away from the origination point (at zero) is a clockwise loop and reflects that the whole blood system dissipates the energy from the servo motor to disperse RBC rouleaux into single RBCs or smaller rouleaux under high shear rate conditions. The intersection point of the "8" shape for normal subjects is located at 1 s^{-1}, resembling the velocity gradient in cm/s/cm (Fig. 2, dotted curve). The intersection point of the ischemic stroke patient group, however, is 1.4 s^{-1} (Fig. 2, solid line), showing an average increase of 40% in comparison to the normal subject group. (B) *The enclosed loop area*: The area enclosed by the two curves represents the energy density velocity in erg/cm^3/s. On the one hand, the counter-clockwise loop close to the origination or zero point encloses an area of 0.17 erg/cm^3/s for the male normal subject group, and 0.39 erg/cm^3/s for the male ischemic stroke patient group, revealing an increase of 130%. For the female normal subject group, the enclosed area is 0.09 and 0.19 erg/cm^3/s for the female ischemic stroke patient group, revealing an increase of 111%. The above results show that the RBC rouleaux formation in the human whole blood system of ischemic patients is stronger in comparison to normal subjects (>100%). On the other hand, the clockwise loop far away from the origination or zero point encloses an area of 0.15 erg/cm^3/s for the male normal subject group and 0.40 erg/cm^3/s for the male ischemic stroke patient group, revealing an increase of 167%. For the female normal subject group the enclosed are as 0.15 and 0.28 erg/cm^3/s for the female ischemic stroke patient group, revealing an increase of 87%.

The results show that the RBC rouleaux formation in the whole blood system of ischemic patients requires more energy to be dispersed into single RBC or smaller rouleaux than those of normal subjects. (C) *The yield stress values*: When the shear rate decreases from a maximum back to zero, the yield stress values can be identified by the interval point of the vertical axis with the downward loop curve. The yield stress values from both patient groups, either male or female, are higher than those from both normal groups, respectively. The above results can be interpreted that the ischemic stroke patients require a higher energy level in order to maintain their blood in a well-flowing state.

In some individual cases, the above mentioned parameter can be found to be several times higher than the normal group's measurements. In such cases, the patients should be treated urgently, since either an ischemic stroke has already taken place, or will happen with very high probability as a serious indication of the onset of an (micro-) vascular disease. In addition, if longitudinal comparisons of blood hysteresis loops continuously increase to a higher level with a trend to the left area in the coordinate system, this change may indicate the need of the blood circulatory system to dissipate in vivo a substantial higher amount of energy than normal subjects. The quantity of the necessary energy can be divided into two parts: one

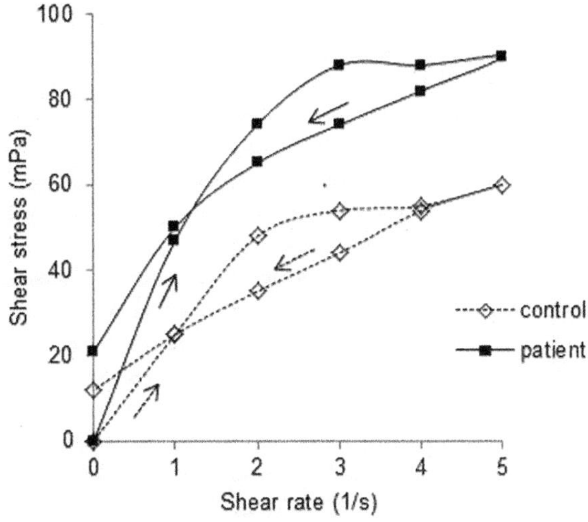

Fig. 2 Comparison between average values from a female ischemic stroke patient group (17 cases, solid line) —■— and a female normal subject group (14 cases, dotted line) —◆—, the yield stress value is 12 (SD = ±6) mPa for the normal subject group and 21 (SD = ±13) mPa for the ischemic stroke patient group

part serves the blood to stay in a liquid state, and the other part serves to overcome the forces of inner friction to keep the blood flowing. Differences in whole blood hysteresis loop morphology between males and females can be explained by females in reproductive age having a 10% lower hematocrit value than males of the same age group. This leads to localizations of hysteresis curves slightly lower in the coordinate system in females. However, the differences between males and females are not considered significant after females' loss of productive ability at a higher age, or no difference in whole blood hysteresis loop morphology can be seen in comparison to the male counter group. Additional detailed information about blood yield stress methodologies and its applications in medicine, especially in geriatrics, can be found in other publications from our laboratories [6, 7], but would go beyond the scope of this article.

5 Hysteresis Loop Type of Human Sperm

In recent reports, it had been discussed that the quality of human semen is declining [8]. The reasons for this development may be connected to two factors: changes in cellular development and environmental influences. Changes in cellular development of the human ejaculate may include biochemical, biophysical, and cytological in vivo conditions during the course of maturation and transportation of spermatozoa and seminal plasma. Examples of such conditions that influence the quality of human semen are sperm density and distribution, protein modification and rearrangement of the sperm surface, changes in the sperm surface charges, scavenging of reactive oxygen species, levels of hepatocyte growth factors, zinc, and other molecules, and the functional processing of fertility in epididymis. The mentioned biochemical factors may lead to abnormal gene expression and/or altered gene regulation. Environmental factors, such as different levels of electromagnetic radiation due to geographical regions and latitudes, seasonal changes, occupational risk factors, and other conditions may also lead to changes in the quality of the human ejaculation system.

The quality of human semen is routinely assessed by the ejaculate volume, sperm density, and observations of sperm morphology and motility. The sperm motility test is a very popular parameter used in the diagnosis of medical infertility conditions. From a bio-mechanical perspective, sperm motility should be determined on spermatozoa themselves, using the seminal plasma as medium. The bulk fluidity of human semen should also be considered. Some other reports mentioned ejaculate viscosity. However, viscosity values are inconstant and depending on the extent of the thinning process and the time passed since ejaculation. Preliminary thinning of the ejaculate takes about 5–10 min after ejaculation and the completion to total ejaculate thinning may take up to 20–40 min. The viscosity of the thinning semen can be estimated by the length (2–3 cm) of the spinning filaments, using a syringe or a glass rod, and the semen thinning time can be recorded using a simple timer on a viscometer filled with 0.5 ml of thinning semen. The accuracy of these methods, used in clinics, seems to be not very satisfactory or standardized. Therefore, it is necessary to study the relationship of fluid behavior and ejaculation time, in order to describe sperm motility more accurately and to characterize the mechanical properties of semen. This article is focused on the characteristics of yield stress and hysteresis loops for human semen during liquefaction, based on the results from experiments in our laboratories [8].

Immediately after ejaculation ejaculates were transferred into the test cup of the Brook Field Programmable DV-II Rheometer to determine the hysteresis loops and yield stresses on fresh human ejaculates. Data were measured every minute for a time period of 20 min [8]. Results suggested the following conclusions: (A) Yield stress values dropped from more than 3000 mPa to 60 mPa in about 5 min after ejaculation; (B) The shape of hysteresis loops of shear stress versus shear rates changed from a counter-clockwise loop, enclosing a large area, to a loop, enclosing a very small area after about 4–5 min. This can be interpreted that the ejaculated semen initially exhibited a semi-solid or viscoelastic body behavior in the first 5 min after ejaculation, and thereafter changed into a thinning fluid with very low viscosity.

Figure 3 shows 20 hysteresis loop curves, each obtained in 1-min intervals at 1–20 min time points after ejaculation. The first curve, obtained 1 min after ejaculation (series 1), shows a marching direction to become a counter-clockwise loop and to enclose a very large area, which cannot fully be displayed in Fig. 3 due to limits of the testing range of the instrument used. The rheometer's maximum value is 3000 mPa, however, the estimated intersection point with the vertical axis or the yield stress value is over 5000 mPa. The second hysteresis curve, obtained 2 min after ejaculation, also shows a large enclosed area (1389 erg/cm^3/s) and indicates the yield stress value being at 1900 mPa. The third minute curve continues to show a counter-clockwise directional loop with an enclosed area of 411.8 erg/cm^3/s and a yield stress value of 900 mPa. The fourth minute curve shows an enclosed area of 36.75 erg/cm^3/s and a yield stress value of 200 mPa. In the fifth minute hysteresis curve, the enclosed area has decreased to 8.5 erg/cm^3/s and the yield stress value dropped to 60 mPa. This trend continued in all curves obtained between the 6th and the 20th minute. At above sixth minute hysteresis curves appear crowded and overlapping (Fig. 3). Integral calculations showed that the enclosed areas in the hysteresis loops of the 6th to the 20th minute ranges from 60 erg/cm^3/s to near zero, and the yield stress ranges from 60 to 30 mPa.

Functional human ejaculation is expected to have a yield stress of more than 5000 mPa and is considered a specialized semi-solid or viscoelastic body. The stress value of a healthy human ejaculate is expected to be about 40,000 Pa. Before ejaculation, spermatozoa are stored and protected in the epididymal fluid in the epididymis. The semen ejaculated is composed of spermatozoa and seminal plasma. The latter includes secretions from the epididymis, ductus deferens, ampulla, seminal vesicles, and the prostate gland, among others. Seminal plasma is a chemical mixture, and from a physio-mechanical perspective to a complex viscoelastic body. In the epididymal fluid the sperm is quiescent, receiving nutrition during maturation, but with limited motility. At the event of ejaculation, motility is initiated for capacitation. Shortly after ejaculation, the semen changes its physical

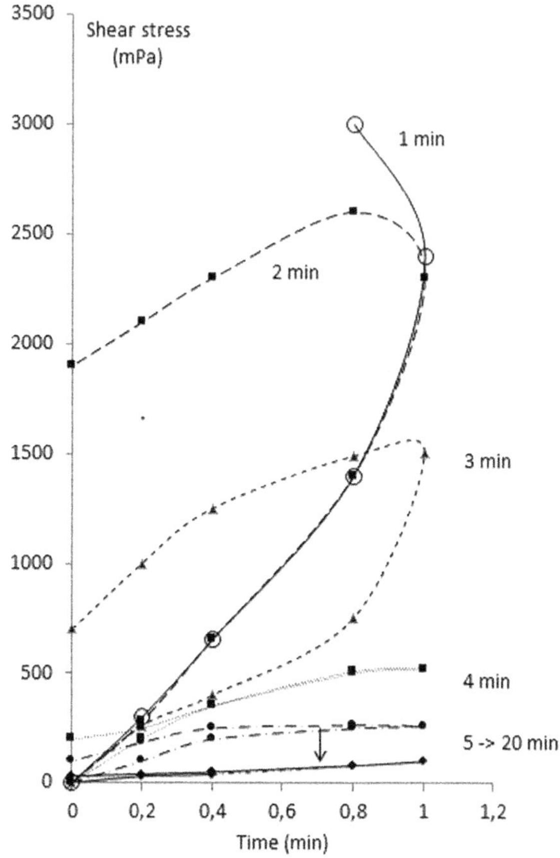

Fig. 3 Hysteresis loops of a human ejaculate: a hysteresis curve was obtained every minute for a 20 min time period. The measurement times are indicated in the figure. After the 5th minute, the hysteresis loops are overlapping, indicating that the human ejaculate changed from a viscoelastic body to a liquid state about 5 min past ejaculation [8]

properties from semi-solid or viscoelastic into a thinning fluid, loosing its high yield stress and its viscoelastic counter-clockwise hysteresis loop. After ejaculation this change is necessary for the sperm, in order to secure active motility and to penetrate the cervical mucus to reach the ovum for fertilization. From a bio-mechanical point of view, the parameter of the semen changes after ejaculation can be precisely established, such as standardized yield stress and enclosed area values of a counter-clockwise hysteresis loop. These standardized parameters reflect the semen thinning course from viscoelastic to semi-solid over a specified time period and can serve as a more precise, medical semen quality test in the future. Furthermore, according to above shown experimental results, human semen's yield stresses drop down from a high value of 5000 mPa to a low value of about 60–30 mPa in about 4.45 min. At the same time the counter-clockwise hysteresis loop is lost, indicating that the ejaculate has changed from a viscoelastic or semi-solid state into a fluid state. Since these two events take place at a natural coincidental time line, a logical explanation could be a possible configuration change of effected molecules of the semen. A model of this transformation may be incorporating a change of the tri-dimensional network structure of seminal plasma before its liquefaction into dissociation into mono-molecules after its liquefaction. This model is matching the findings of our experiments, relating a tri-dimensional structure with high yield stresses and counter-clockwise hysteresis loops, and mono-molecules in a liquid revealing very low yield stresses and loss of the counter-clockwise hysteresis loop [8].

6 Hysteresis Loop Comparison of Plant and Animal Cells

Already in a very popular children's book series in China, named the '100 thousand WHYs', the question of how a leave of a plant of *Mimosa pudica* can move is addressed. However, the children's book uses an explanation for plant movement based on increasing turgor pressure provided by osmosis on the pulvinus of the leave in comparison to a decreasing turgor pressure on the other side of the leave. Can this be applied to the movement of the plant *Mimosa pudica L.* as a reaction to very small forces applied to the petioles of the leaves? Taking a closer look at the mechanism why *Mimosa pudica L.* will move by a small force appears not that simple and is not clear.

On the other hand, in lectures for medical students in the Department of Biophysics, Shanghai Medical College, Fudan University, where the author was an instructor from 1960–1998, the movement of the plant *Mimosa pudica L.* leaves was used to explain the term 'thixotropy'. Adding the example of how *Mimosa pudica* moves it leaves as a reaction to a very small force initiated by touch and slowly recovers its original shape after the force of touch is removed, helped the students to understand the process of thixotropy more clearly. Biomaterials, such as the *Mimosa pudica* leaves, can change their flow properties when they are exposed to a force action, and can recover their original state when the force is withdrawn. This phenomenon is called thixotropy.

Following the question about how the plant *Mimosa pudica L.* can move its leaves in reaction to a very small force, our laboratories used hysteresis loops to possibly shed light on the underlying mechanisms. Following the idea from the children's book '100 thousand WHYs', petioles and leaves of the *Mimosa pudica L.* plant were used to make a homogenous suspension in water. Hysteresis loops on the plant suspensions were tested and recorded using the rheological instrument and computer software from BROOKFIELD ENGINEERING LABORATORIES, INC, USA [9]. The shear rates were adjusted to triangular waves of 0-1-0, 0-2-0, 0-4-0, 0-6-0, and 0-10-0 s^{-1}. Figure 4 shows the results of triangular wave measurements of the first experiment, using 0-1-0 s^{-1}. Follow-up experiments with above listed triangular wave ranges are not displayed in this article, since they were revealing similar results as established in the first experiment (see [9] for more details). Figure 4 shows the comparison of experimental results from three homogenate suspension samples of (a) plant *Mimosa pudica*, (b) pig skeleton muscles, and (c) another plant called *Thyllanthus urinaria*, which has nearly the same shape as *Mimosa pudica*, but lacks the movement ability. As visible in Fig. 4a, only the suspension from the *Mimosa pudica* plant revealed a significant counter-clockwise hysteresis loop and a certain yield stress value. The other two suspension samples (Fig. 4b, c) do not show such loops or yield stress values. These results indicate that only in the *Mimosa pudica* homogenate suspension, protein or cytoskeletal substances may be present that can change their structural state from granules into networks, from disorder into order, and from fluidity into viscoelasticity, when exposed to a very small shear stress field. The related protein or cytoskeletal substances had to have the ability to re-form their structures, while suspended in the *Mimosa pudica* solution. Such remolding behavior involves the generation of a certain energy level, in order to be able to couple molecules forming a structural change. Therefore, such changes would need energy in the suspension itself. In the contrary, the other two suspension samples (B- from pig skeleton muscles and C- from the plant *Thyllanthus urinaria*) do not exhibit such significant phenomena.

Figure 5 exhibits the results of experiments with varying yield stress values in the *Mimosa pudica L.* suspension only. The system reveals that applied triangular waves of increasing shear rates (horizontal axis) linearly increases the wave's peaks of shear rates (vertical axis). Figure 6 shows the comparison between the 3

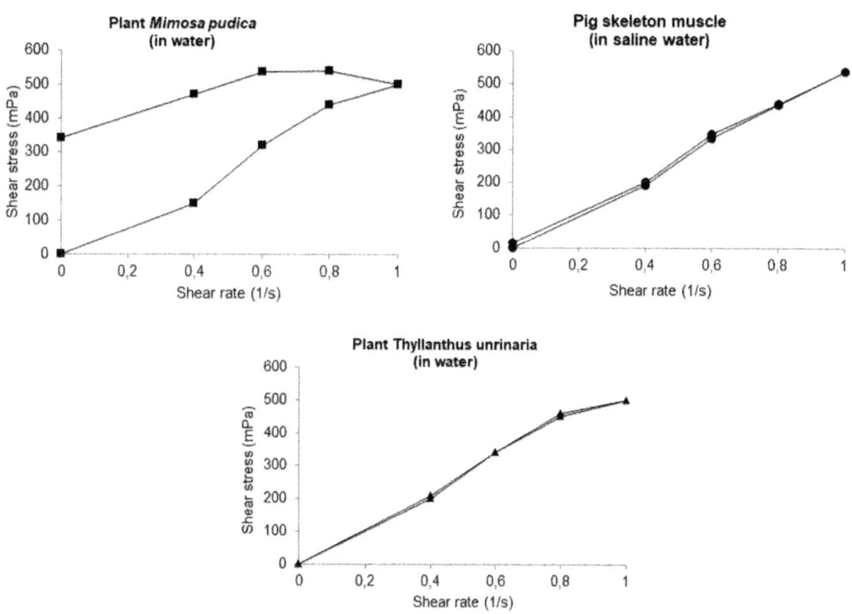

Fig. 4 Hysteresis loop comparison of three samples: Plant *Mimosa pudica*, Pig skeleton muscle, Plant *Thyllanthus unrinaria* (homogenized suspensions). Only the sample of Plant *Mimosa pudica* shows a significant counter-clockwise hysteresis loop and a significant yield stress value

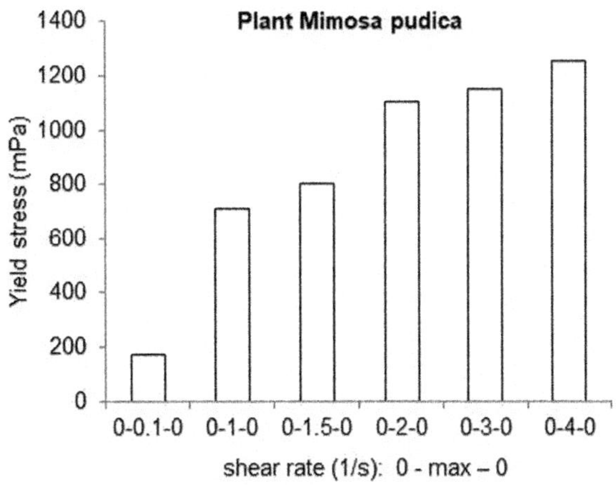

Fig. 5 Relationship between yield stresses in the *Mimosa pudica* system (vertical axis, unit in mPa) and shear rate triangular waves (horizontal axis, unit in s^{-1}). Hystersis loops are originated from zero to a maximum and returned to zero in a triangular fashion: 0—Max—0)

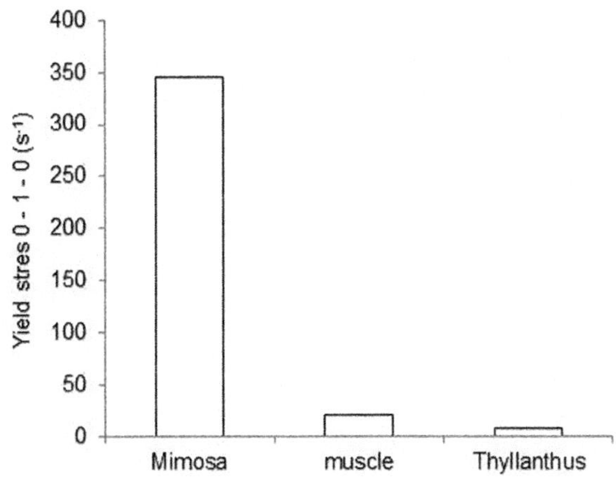

Fig. 6 Comparison of the yield stress values among 3 suspension systems of (1) *Mimosa pudica*, (2) Pig skeleton muscle, and (3) *Thyllanthus urinaria* plant. Suspensions are prepared in the same concentration of 15% fresh tissue precipitation. Triangular shear rate waves of 0-1-0 s^{-1} were applied to all three systems. Only the *Mimosa pudica* suspension shows a significant yield stress value

suspension systems of *Mimosa pudica*, pig skeleton muscle, and the plant *Thyllanthus urinaria* under an applied shear rate triangular wave of 0-1-0 s^{-1}. As discussed before, only the *Mimosa pudica* suspension system (first column) shows a significant yield stress value.

7 Summary

Hysteresis loop and yield stress techniques belong to the supplemental tools to show structural change, when biomaterials are exposed to a shear rate field and react with a mechanical shear action. Information gained from hysteresis loop analysis about the molecular and structural changes in biological systems can be coupled with measurement from molecular biology and biochemistry methods, in order to enhance techniques like isolation, purification, crystallization, and X-ray diffraction of proteins and cytoskeletons.

Acknowledgements We thank Prof. Zhi-jiu Qin, Prof. Zhen-sheng Tang, and Prof. Chuan-zhen Lv for their co-investigation on the research grant entitled "Studies on hemorheology of cerebral vascular diseases", received from the Central Health Ministry in 1990. For work on the instrument, we thank Senior Engineer Prof. Shu-kui Jiao, who designed the HT Rheometer for us to test the samples. The instrument belongs to the Shanghai 9th Hospital, administrated by Prof. Jing-wen Yang and Prof. Zhong-qing Le, who both helped us to execute the measurements. Blood samples were obtained of patients of the emergency Departments of the Huashan and Zhongshan Hospitals.

We thank Dr. Luan-feng Pan (Fudan University), Dr. Fei-kun Yang, and Dr. Si-qi Wang.

Last not least we thank Dr. Luan-feng Pan (Fudan University), and Dr. Fei-kun Yang for their cooperation.

References

1. Huang, C. R., & Fabisiak, W. (1978). A rheological equation characterizing both time dependent and the steady state viscosity of whole human blood. *AIChE Symposium Series, 74* (182), 19–21.
2. Chen S. Q., & Liang Z. J. (1986). The study of Hysteresis loop of whole human blood and Huang's equation characterizating thixotropy of blood system. In W. F. Chen L. W. Yuan & Y. Z. Xu (Eds.), *Advances in rheology* (pp. 492–499). Beijing, China: Academic Press. (in Chinese).
3. Fang, B., & Jiang, T. Q. (1998). A novel constitutive equation for viscoelastic-thixotropic fluids and its application in characterization of blood hysteresis loop. *Chinese Journal of Chemical Engineering, 6*(3), 264–270. (in Chinese).
4. Shi, W., Jiao, S. K., Yang, J. W., Le, Z. Q., & Shi, Y. D. (2016). Digital solution of continuous kinetics for blood non-linear hysteresis loop curves. *Chinese Journal of Hemorheology, 26*(4), 389–396. (in Chinese).
5. Shi Y. D. (1990). Thixotropy, viscoelasticity and hysteresis in blood system. In Y. D. Shi (Ed.), *Horizons in hemorheology*, Part I (pp. 67–81). Shanghai: Shanghai Scientific and Technological Literature Publishing House.
6. Shi, Y. D. (2013). Yield stress of blood. *Chinese Journal of Hemorheology, 23*(2), 199–202. (in Chinese).
7. Shi, Y. D. (2016). Blood yield stress and aging. *Chinese Journal of Hemorheology, 26*(2), 135–142. (in Chinese).
8. Shi, Y. D., Pan, L. F., Yang, F. K., & Wang, S. Q. (2004). A preliminary study on the rheological properties of human ejaculate and changes during liquefaction. *Asian Journal of Andrology, 6*(4), 299–304.
9. Shi, Y. D., Pan, L. F., Yang, F. K., & Artmann, G. M. (2003). Stress-shear hysteresis loops of homogenate suspension from leaves and petioles of *Mimosa pudica*. *Journal of Plant Physiology and Molecular Biology, 29*(4), 275–280. (in Chinese).

Author Biography

Professor Yong-de Shi (China), Ph.D. He is a dedicated biophysicist from Shanghai. He was born in 1938, in Zhejiang Province. He has experienced hard times at times. Later he held various academic appointments at Shanghai Medical University: from 1978 to 1984 as Biophysics Lecturer, from 1984 to 1990 Associate Professor in Biophysics; and from 1990 until his retirement as a full professor. He has been a member of the Shanghai Biophysical Society since 1998 and has been its Secretary General since 2012 until today. Between 1992 and 2001 he was again and again for weeks or months on invitations of the college and Professor Gerhard Artmann Visiting Professor at the Aachen University of Applied Sciences. Both have an original, shared scientific interest in the Blood Rheology. Despite his late international career, Professor Shi has published 211 peer reviewed papers to date, some in Chinese and some in international journals. In addition, he has authored a total of ten scientific books, as co-editor with Professors Shu Chien (USA) and Masahiro Sokabe (Japan). Professor Shi has traveled widely and has seen much of the world. He has also often welcomed international guests in Shanghai. He loves his job, his family, and the beauty of his country, the great Shanghai, but also the Yellowstone Park in the USA, and his memories of human encounters in Aachen and the wonderful nature of Germany.

Part II
Technology

Functional Toxicology and Pharmacology Test of Cell Induced Mechanical Tensile Stress in 2D and 3D Tissue Cultures

Gerhard M. Artmann, Jürgen Hescheler, Haritha Meruvu, Sefa Kizildag and Aysegül Artmann

1 Introduction

1.1 Mechanical Cell Forces and Tensile Stress

Mechanical forces/tensile stresses are critical determinants of cellular growth, differentiation and migration patterns in health and disease [13, 19, 27, 61, 43]. They co-regulate the mechanical balance between cells and the extracellular matrix in nearly all tissues outside and within organs of the body. In addition, they are associated with the regulation of mechanically driven biochemical and genetic cellular processes [1, 35]. Tensile stresses determine intercellular traction, as well as internal cytoskeleton mediated tensile stress transmittance to the extracellular matrix (ECM) through cell-matrix adhesion proteins [22, 30].

In loving memory of our former student Taylan Demirci a passionate young scientist who died in 2010.

G. M. Artmann (✉) · J. Hescheler
Institute of Neurophysiology, University of Cologne, Robert Koch Strasse 39, Cologne D-50931, Germany
e-mail: g.artmann@gmx.net

H. Meruvu
Department of Bioengineering, Gaziosmanpasa University, Tokat, Turkey

S. Kizildag · A. Artmann
Institute for Bioengineering, Medical and Molecular Biology, University of Applied Sciences Aachen, Juelich, Germany
e-mail: A.Artmann@fh-aachen.de

S. Kizildag
Department of Medical Biology, Faculty of Medicine, Dokuz Eylul University, Izmir, Turkey

G. M. Artmann et al. (eds.), *Biological, Physical and Technical Basics of Cell Engineering*, https://doi.org/10.1007/978-981-10-7904-7_7

Table 1 Mechanical force, rigidity and tensile stress [related SI units]

Tissue tensile stress	Cell force	Tissue rigidity	Tissue strain
$\sigma = \frac{F}{A}\left[\frac{\mathrm{N}}{\mathrm{m}^2}\right]$	$F[N]$	$\frac{stress}{strain} = \frac{\sigma}{\varepsilon}\left[\frac{\mathrm{N}}{\mathrm{m}^2}\right]$	$\varepsilon = \frac{\Delta l}{l_o}$

The terms “force” and “tensile stress” should not be mixed up. “Cell Force” is more intuitively accessible and is used in most biological papers, while the term “tensile stress” is more abstract and therefore less commonly used. The term “force” describes pulling forces in cells given in Newton “N”. “Tensile stress” instead, acts when a force pulls at a cross sectional area perpendicular to it. It is given in N/m^2 where “m^2” stands for the product of thickness × width of the (idealized) rectangular cross-section of a tissue sample [27, 62]. In some applications where the thickness of a cultured tissue sample is very much smaller than its width, for example when cells are cultured as approximately 2 μm thick monolayers on a CellDrum membrane of 16 mm in diameter, then the “tensile stress” is given in N/m. The latter implies a “force” pulling perpendicularly to an imagined cutting line “cut” perpendicular through the monolayer plane. The “line length” is given in “m” (Table 1). These are simple yet necessary statements when talking about “force” and “tensile stress” induced by tissues or cells, respectively.

Although all cytoskeletal proteins have an impact on the cell’s mechanical properties, actin is the basis of mechanically driven movement in all known eukaryotic cells. Phagocytosis, cytokinesis, cell crawling and muscle contraction all depend upon structures build from actin. Typically, actin represents about 5–10% of total amount of proteins within a cell and 20% of protein in muscle. Actin forms assemblies of filaments and plays a key role in force transmission from the ECM to the cell’s nucleus and vice versa [51, 59].

In fibroblasts the primary function of the protein actin is building up filaments (Fig. 1). Actin monomers, G actin, polymerize to filamentous F actin. A high actin concentration at polymerization onset allows for faster formation of filaments. Actin filaments exhibit both ends, a slow growing minus as well as a fast growing plus end [50].

Myofibroblasts are an intermediate form of smooth muscle cells and fibroblasts containing contractile actin and myosin fibers. These cells endogenously contribute high amounts of collagen to the ECM. Myofibroblasts contract via smooth muscle cell actin-myosin complexes, which contains more alpha-smooth muscle actin.

1.2 Focal Adhesions

Focal adhesions are large cellular protein assemblies (Fig. 1, red colored dots) which help regulate the “mechanical balance” in tissues. They are a prerequisite to regulate cell internal signals in response to the cellular mechanical environment. Via focal adhesions cells transmit forces to the extracellular matrix (ECM) and

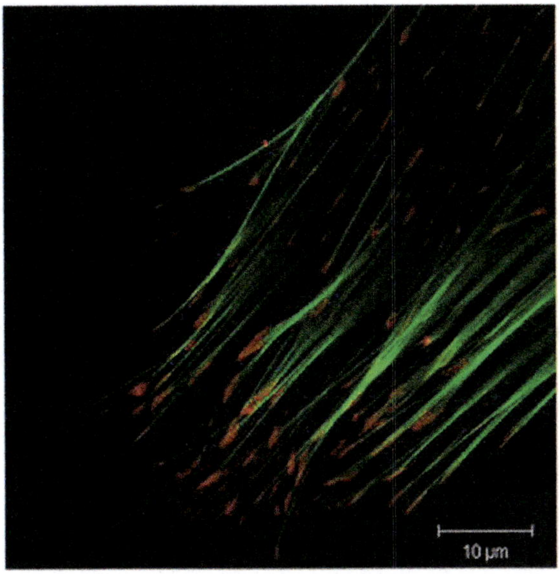

Fig. 1 Immunofluorescence image of actin filaments (fibers, green) and the focal adhesion protein vinculin (red dots) in a fibroblast. Focal adhesions are integrin-containing, multi-protein structures forming mechanical cell membrane spanning "bridges" between intracellular actin fibers and the extracellular matrix and/or a surface, respectively, in adhesion dependent cell types [80]

other interacting cells. In other words, they represent sub-cellular structures mediating regulatory effects (i.e., signaling events) of a cell in response to ECM adhesion [18]. The attachment of cells to its substratum which may be a basement membrane in vivo or the culture dish in vitro, thus, is mediated by focal contacts or focal adhesions [34]. These multi-domain proteins adhere to the surface of the tissue culture dish and expose specific sequences that are recognized by cell-surface receptors. The cell-surface receptors involved in focal-contact formation belong to the integrin superfamily. Integrins are non-covalently associated complexes of two distinct, high molecular weight polypeptides called α and β integrin acting as transmembrane linker in a variety of cells. They span the plasma membrane in the region of focal contacts. Its cytoplasmic domain binds through a chain of various acting proteins including talin, vinculin and α actinin to actin filaments of the cortical cytoskeleton. The ability of fibroblasts to perceive extracellular stress to and from transmittable contractile force to the ECM is crucial for regulating their activity during connective tissue remodeling [30, 71, 61].

1.3 Extracellular Matrix

Cell adhesion is defined as cellular binding to a surface, which can be another cell, a surface or an extracellular, organic matrix. Extrinsic structures determining a tissue's mechanical integrity are mainly formed by the extracellular matrix (ECM). The ECM is a molecular complex mainly consisting of fibers and an amorphous inter-fibrillary matrix composed by components like glycoproteins and proteoglycans. The ECM also contains molecules such as growth factors, cytokines, matrix-degrading enzymes and their inhibitors [78]. The idea of a dynamic reciprocity between the ECM on the one hand and the cytoskeleton and nuclear matrix on the other hand is widely accepted. In this model, ECM molecules interact with cell surface receptors, which then transmit signals across the cell membrane to molecules in the cytoplasm; these signals initiate a cascade of events through the cytoskeleton into the nucleus, resulting in the expression of specific genes, whose products, in turn affect the ECM in diverse ways [11].

Basically all cell-ECM interactions [36] participate directly in processes like cell migration, growth, differentiation, programmed cell death (known as apoptosis), cell adhesion, activating intracellular signaling and contractile response. Matrix components and the mechanical forces that cells experience markedly influence the maintenance of cellular phenotypes and effect cell shape, polarity and differentiated function. In most tissues, the ECM is constantly being remodeled, particularly in processes like wound repair or tumor cell invasion. Extracellular matrices are specialized for particular function, such as strength (tendon, connective tissue) or adhesion (basement membrane of epithelia) [60].

1.4 Cell and Tissue Tensile Stress Measurement Technology

An important physiological function of cells in terms of biomechanics is the tensile stress they are able to generate either alone as single cell, as 2D-cell monolayer, or embedded in a 3D cell-ECM culture, respectively. Quantifying these tiny tensile stresses is a real technological challenge. On top of all other requirements as there are sensitivity, reproducibility, electronic drift, signal-to-noise ratio, and temperature further preconditions MUST be considered. Tensile stress measurements require defined boundary conditions to be considered and results obtained with a certain experimental setup should not depend on specific technological designs. Although these are general requirements for measurements in mechanics, there are major problems for biological samples. The cells are alive, meaning their metabolism needs to be maintained. They also age, changing their mechanical properties, among other things. They must be kept as much as possible in their natural biomechanical and biochemical environment since, importantly, cells exist in a delicate balance with their mechanical environment. Cells in a soft matrix provide data other than cells in a stiff or very rigid matrix [37]. This is a factor that is by far not adequately taken into account in many cell experiments by biologists and medical scientists. Protein and gene expression of cells depend quantitatively and qualitatively on the mechanical properties of their environment, the extracellular matrix. For example, fibroblasts that suddenly find a foreign material with extremely high modulus of elasticity as a result of pacemaker implantation, rather than soft tissue as before, extremely change their extracellular matrix. Doctors even speak of encapsulation of the pacemaker [40].

Furthermore, it is often important to follow the time course of the mechanical measurements of cells over days and weeks. Then the selected boundary conditions of the measurement setup must be chosen so that the cells remain vital for one thing and for another degenerate in no way. This is the case for example in toxicology tests with cardiomyocytes [14, 39]. These prerequisites define the technological-biological measures to be taken in experimental setups and test systems aiming at cell tensile stress measurements at all scales from tissue via cells to subcellular structures.

Methods for force measurement on cells and subcellular structures use tools to mechanically deform cells in combination with transducers and/or imaging technologies, respectively [24]. These approaches include atomic force microscopy [28], magnetic cytometry, carbon fiber based systems etc. Qualitative measurements on individual cells have been made using atomic force microscopy to investigate cytoskeletal substructures [63]. The nano-indention technique is another tool to characterize mechanical properties of living cells to understand biomechanical and biophysical processes, such as disease progression and cell–material interactions [43]. At the same time, advancement in the field enabled high-resolution structural visualization of living materials. For instance the two-photon fluorescence microscopy may provide a more sensitive detection technique than classical histological methods. Multi-photon microscopy assesses for example the nano-mechanical contribution of elastin networks to overall tissue mechanics in multiscale models [57]. Mechanical investigations on a molecular scale are approached by Förster resonance energy transfer (FRET) [75], photo-quenching, loss of fluorescence or changes in fluorophore emission properties for correlating molecular tensile stresses to molecular, cellular or tissue functions [21, 66]. FRET-based Molecular Tension Microscopy uses a mechanical tension-sensitive FRET biosensor genetically encoded within a protein of interest and depicts spatiotemporal maps of molecular tensions detected [29]. These approaches can also be complemented by in vitro experiments using soft gel substrates (traction force microscopy), elastic micropillars [64] and gel matrices to quantify traction forces generated by cultured cells individually and collectively in 2D and 3D geometries [16]. Techniques based on birefringence can be used to quantify tissue scale stress but suffers setbacks like requiring flat, transparent samples and delicate calibration. Tissue-scale laser ablation measures tissue stress to viscosity ratio but requires sample and laser alignment and allows only a few experiments per sample [67]. Information on cellular processes in tissues/organs can be derived from the mechanical properties of living cells (see Chap. 8), cell layers and tissues. Particularly elasticity measurements offer a valuable insight. Before 1970, mechanical behavior of tissues and organs were interpreted from the whole organ behavior (like pressure-volume relations of the heart). Later on attention has been paid to the mechanics of non-adhesion dependent blood cells like red blood cells.

Micropipette aspiration techniques were applied to determine the viscoelastic behavior of individual cells and initially used for analyzing mechanical membrane properties of red blood cells (RBCs) and other blood cell types [20, 38]. However, this technique together with all further technologies developed for the mechanical analysis of individual cells have major drawbacks limiting their applications to a great extent, in particular if considered for routine use. They do not keep cells in a normal (mechanical) in vivo environment. Furthermore, the micropipette fabrication and cell sample preparation is quite skeptic, complicated and time consuming [69]. On the other hand (and with different scientific aims) micropipettes can be very useful in biophysical cell experiments. This technique lead to the discovery that hemoglobin molecules of different animal species as well as of humans must be seen not only as oxygen and carbon-dioxide transporting molecule but also as a molecular thermometer sensing a species' body temperature [8, 6].

Most other cells are adhesion dependent. They only survive when mechanically stretched [48, 61]. Cells exert mechanical tensile stress and transmit it to their environment (substrate, ECM). This may cause substrate/ECM deformation that can be visualized and investigated further for cellular tensile stress induction if substrate/ECM material properties and biomechanical laws are considered properly [27]. The silicon wrinkling technique was pioneering such approaches and proved to visualize the traction forces of individual cells on ultra-thin flexible membranes. It uses a highly viscous polymeric fluid polydimethyl siloxane crosslinked at the surface forming a thin elastic film over it. A cell on its surface creates wrinkles that are characteristic of the mechanical stresses generated by the cell (Fig. 2).

Improvements were made later on by include the "tuning" of the elastic compliance of the elastic film. However, deformation data can be analyzed only semi quantitatively because the buckling of thin polymer film is a nonlinear phenomenon according to the theory of elasticity [15, 33].

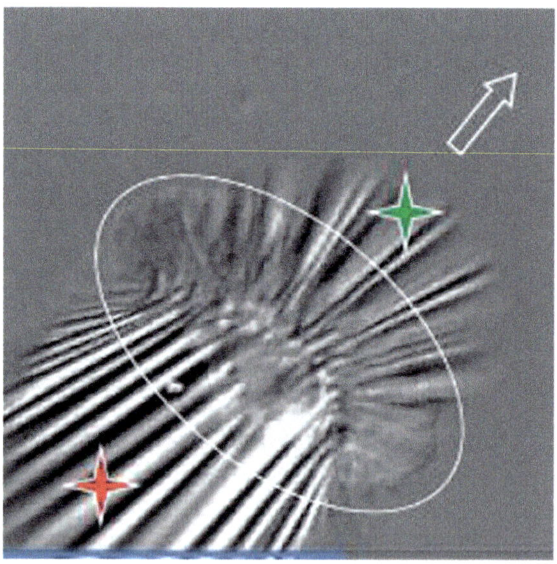

Fig. 2 Single fibroblast crawling spontaneously on a thin silicone sheet (<1 μm thick). The elliptically sheathed area covers the cell's body which is attached to the surface. Arrows indicate the cell's current moving direction. Stars mark the instantaneous wrinkling area in front of (green) and behind (red) the cells body (adopted from [33])

Other augmented experiments realized the impact of a 3D environment on individual cells and began analyzing mechanical properties of standardized cell seeded artificial tissues [25, 49]. Available contemporary sensor technology allowed analyzing the mechanical behavior of tissues or tissue cultures by stretching them in one (uniaxial) or two (biaxial) directions, respectively. The uniaxial tissue-stretch-technique (Fig. 3) is an optical-electromechanical system for measuring the force-deforming behavior of uniaxial or biaxial loaded slabs of soft tissue specimen [54]. Although the set-ups appear intuitively appropriate, look simple and easy to use, many circumstances and boundary conditions do not suggest using such devices. They suffer a major technical drawback caused by specimen deformation due to unreliable grip (slippage or tear-out near clamps). Further factors limiting the application of such technique are soft, gel-like mechanical consistency of cultured tissue strips, experiments carried out in a culture medium bath, and surface tensile stress problems. Furthermore, rectangular gels suffer from "necking" due to the stress variations imposed during uniaxial force measurements. The uniaxial loading also introduces a physiologically unknown parallel aligning in response to the applied mechanical tensile stress and there is augmented likelihood that the object to be investigated is actively modified by the experimental setup.

In summary and in a nutshell all devices and methods are able to monitor somehow mechanical properties of cultured tissue constructs or individual cells. However, they suffer common drawbacks like use of complicated setups, not accounting for the environmental influences, absence of well-defined and biologically relevant boundary conditions, and last not least not considering the specific needs of a sterile and cell culture compatible experimental setup. These drawbacks may cause misinterpretation of experimental results marring the investigator's experimental expertise. Hence there is a quintessential need for a technique with potential upscaling for pharmaceutically relevant sample throughput.

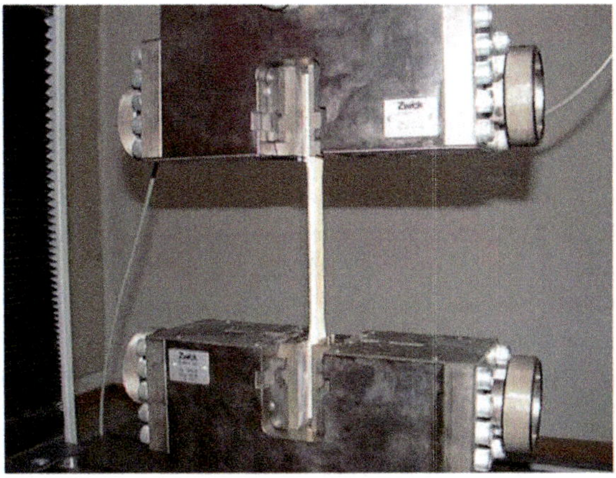

Fig. 3 Uniaxial stretching test setup. The white tissue testing strip (center) is mounted between two holders (above and below) and is being elongated by them. The force applied and the resulting elongation are monitored simultaneously and plotted as force versus elongation curves [15, 54]

2 The CellDrum Solution

The CellDrum principle was patented in 2000 [3]. This chapter introduces the terms and procedures involving applications of the CellDrum technology as an innovative routine testing solution. In this perspective we discuss about time independent or gradually changing mechanical tensile stress patterns, rhythmic changes of mechanical tensile stress and, the future perception of individualized high throughput screening (HTS) in pharmacology and toxicology.

2.1 Measurement Principle

All adhesion dependent cells exert mechanical tensile stress to the substrate they adhere upon [6, 47, 63]. There are many feedback mechanisms and signal transduction cascades involving regulation of the dynamics between mechanical tensile stresses, cell motility and cell growth. These facts gave a reason to develop the CellDrum technology. This, however, was just a technological expression of the wide field of Mechanobiology [27, 43]. The technology dedicates itself to analyze the effects of mechanical tensile stress of their environment (extracellular matrix, ECM, proteins, and biomaterial surfaces) to cells and vice versa. Cell matrix interactions normally are regulated and cell forces are transmitted through focal adhesion complexes. They transmit mechanical signals from outside the cell to the nucleus and from the nucleus to the cells environment, respectively. Thus, mechanical effects have a great impact on the fate of cells in a matrix [43, 61]. Normally, when cells adhere to a culture dish/surface, the (small) cellular mechanical tensile stress does not bend it, it remains flat. This is because the underlying substrate is too stiff to bend. However, when underlying substrates of any cell culture assembly approximately match the order of magnitude of cell layer forces then that tensile stress will be able to deform the substrate. This can be made visible and the extend of surface bending quantified. The CellDrum [3, 73] uses as substrate a thin (normally 4 μm), circular (16 mm) silicone membrane on which the cells are cultured (Fig. 4). A culture medium container (cylinder) is sealed watertight from below with an ultra-thin silicone membrane (pink colored, Fig. 4a). The ratio membrane thickness divided by the cylinder diameter of 16 mm is 0.00025! Cells are cultured from inside the chamber on top of the membrane which is functionalized for good cell attachment. The CellDrum is filled with cell culture medium. The culture medium weight bulges the membrane downwards. This indentation is measured. Membrane indentation is at its maximum when no cells are attached to the membrane. An additionally cultured cell monolayer on top of the membrane or a 3D tissue culture, respectively, will lift the membrane upwards. This is because cells generate intrinsic lateral mechanical tensile stress inside the cell

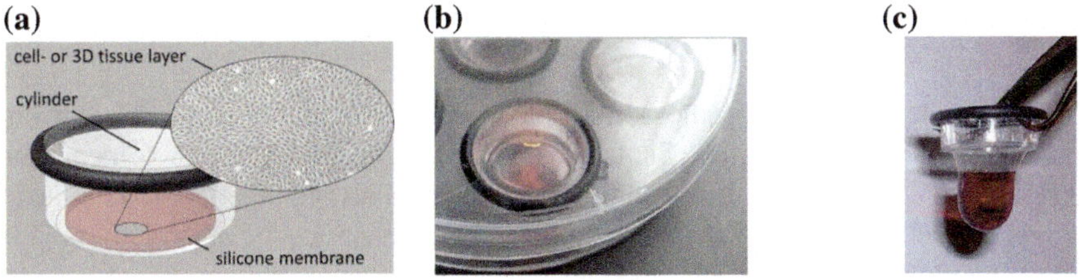

Fig. 4 CellDrum: **a** Scheme with enlarged cell monolayer, **b** CellDrum holder for culturing purposes, **c** A smaller than 2 μm thick membrane allows its complete bending under the weight of the drop. The drop is surrounded by the membrane drop-shaped which prevents it from tearing off. If we would culture beating cardiomyocytes on the inside of the membrane the drop would contract like a real heart with no further stimulation

layer and try to shorten it. Since cells are attached to the membrane surface, tensile stress is transmitted via cellular focal adhesions to the membrane and lifts it this way [3, 56, 72, 73]. When a cell layer contraction is induced by chemical compounds the membrane lifts up further. If we reverse this and wash out the extra tensile stress causing compound from the cell monolayer (if possible), then the tensile stress in the cell layer decreases—the membrane bends again downwards. Thus, the lateral cell monolayer tensile stress relaxes to the mechanical state it had before compound application. If we then detach all cells of the monolayer biochemically from the membrane, then their contribution to the total mechanical tensile stress of cell monolayer plus membrane bias is lost. Thereafter, the total stress, and thus the deflection of the membrane, should assume the state it had before the addition of cells. If the membrane deflection does not fully return downwards to the indentation of the "unused, pure" membrane the residual indentation difference must be the result of the ECM that has been secreted by the cells during the cell monolayer culture period. This value could become another important mechanical parameter of cell mechanics research that could not previously be measured. The CellDrum technology inventors would call this an "interesting side effect". It might reveal insight into the stiffness of the ECM for example depending on cell culture time.

In summary, with the CellDrum system only the membrane indentation is measured at each condition. Now, if simple laws of mechanics are applied, the mechanical stress caused by the cell layer can be calculated. Since the mechanical tensile stress of the membrane without cells is known for each CellDrum, it can be subtracted and the result is the membrane tensile stress alone, as if it was bulged without any supporting membrane underneath. For cell culturing purposes, multiple such CellDrums are placed in a CellDrum holder which than can be put in the incubator for cell culturing. Cell culturing procedures are the same as used regularly. Ongoing new technology development aims at automated culture medium change inside the incubator.

2.2 Finite Element Simulation

For finite element analysis the MSC Patranv2001R3 and NastranV707 software (MSC. Software Corporation, Santa Ana, USA) were used [73]. Two membrane tensile stress causing load application models were analyzed (Fig. 5), (1) a uniaxial stretched rectangular membrane resembling tissue constructs of other authors [13], and (2) a planar, concentrically biaxial stretched circular membrane (Fig. 6). As for the membrane material, a Young's modulus of 2.5 MPa (vs. 210 GPa for steel!) and a Poisson's ratio of 0.49 were assumed. For calculations, a linear, elastic approach and a freely scalable unit-force at the boundary nodes being either uniaxial or concentrically were applied. The resulting pseudo-color images indicate the levels of the von Mises stress. FEM analysis for uniaxial load of a rectangular strip (Fig. 5a) revealed that at constant load the relative von Mises stress in the strip varied from $8.0 * 10 - 5$ to $1.4 * 10 - 4$. The circular membrane (Fig. 5b) loaded circumferentially showed almost no stress variation all over its area. Slight variances along the circumferential border of the membrane model (light blue) result from theoretical force transmission points. They became visible because of the limited number of boundary points considered in the calculations. The stresses discussed here all are stresses acting in-plane with the tested samples.

We tested as well how two different hydrostatic loads would bend the same circular membrane (Fig. 5b). As a result, the former isotropic tensile stress distribution across the membrane area turns into a rotationally symmetric one with small radial stress variations (Fig. 6a), and (Fig. 6b) the tensile stress along the membrane's radius from center to the periphery differs significantly. In Fig. 6a, a small hydrostatic pressure of 100 Pa was modelled. In full accordance to our intuition, this tendency becomes much more pronounced at higher load (500 Pa) (Fig. 6b). Thus, there is not only one membrane tensile stress value anymore given which would characterize the whole membrane area. For experimental purposes with the CellDrum, however, up to a ratio of the membrane diameter/maximum membrane

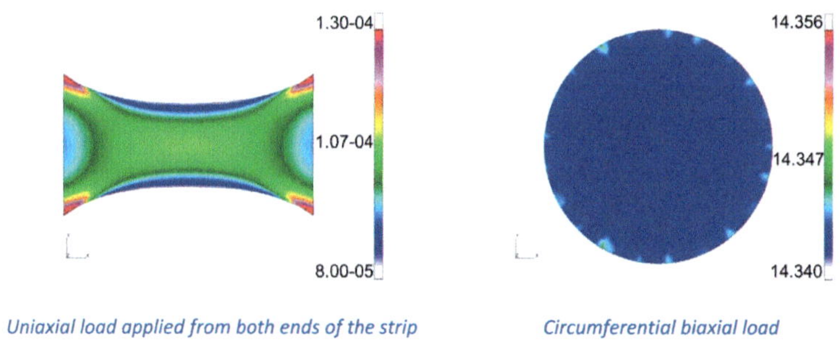

Fig. 5 Color encoded von Mises stress distribution in a **a** uniaxial stretched rectangular membrane, and **b** concentrically biaxial stretched circular membrane. The load was applied in-plane of the membranes. Light blue areas of the circular membrane (image **b**) result from the limited number of fix points considered in the model

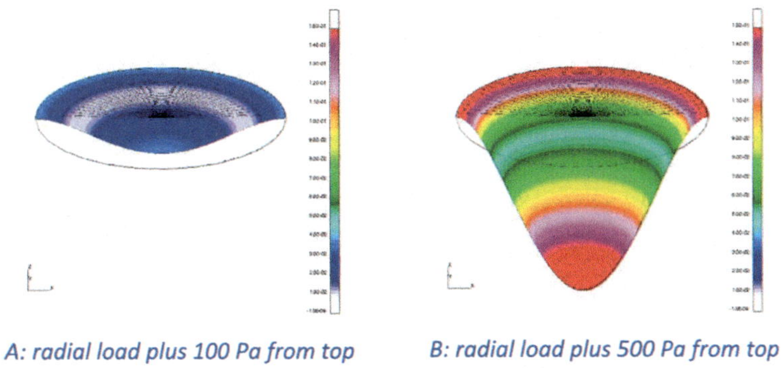

Fig. 6 Finite Element Method results of two concentrically pre-stretched circular silicone membranes. In addition to this, two different hydrostatic pressures act from above and cause significant membrane indentations. Model input parameters: membrane Ø = 16 mm, thickness = 100 µm, Young's modulus (silicone) = 2.5 MPa, Poisson's ratio = 0.49, boundary condition: edge clamped, shift = 0, twisting = 0, deformation scale 1:1 as compared to true size. For better visualization only the rear part of the membrane is shown. Colors represent the von Mises-Yield Criterion. Color scales for both images are identical

indentation of ~10 the stress and 50 Pa hydrostatic pressure variations across the membrane can still be neglected, in other words can be assumed to be constant.

3 "Load-Deflection-Mode"—Slowly Changing Cell Layer Tensile Stress

3.1 *Principle Setup*

Most cell types change their mechanical tensile stress which they transmit to each other or to their environment, respectively, slowly and not fast in milliseconds or rhythmically as cardiomyocytes do. The mode of CellDrum applications described was named "Load-Deflection-Mode". In this mode the cell monolayer/tissue construct tensile stress over pressure is measured. Figure 7 shows the CellDrum setup and Fig. 8 the principle function.

Fig. 7 CellDrum (rotationally symmetrical) scheme: CellDrum cylinder (dark grey), membrane with culture medium (typically 400 µL) on top (pink), cut-out—enlargement of a cell monolayer (red) when attached to the downwards bulged membrane

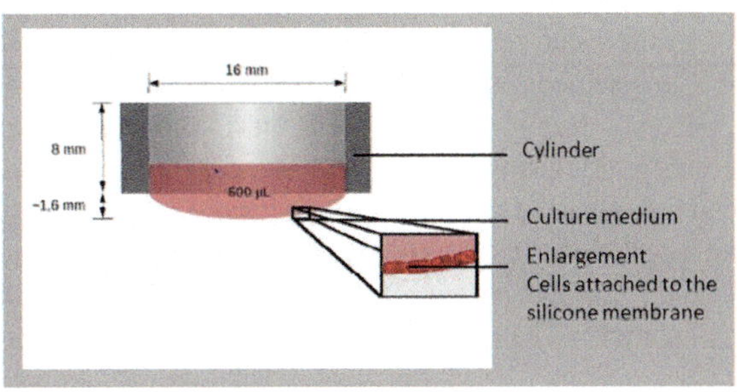

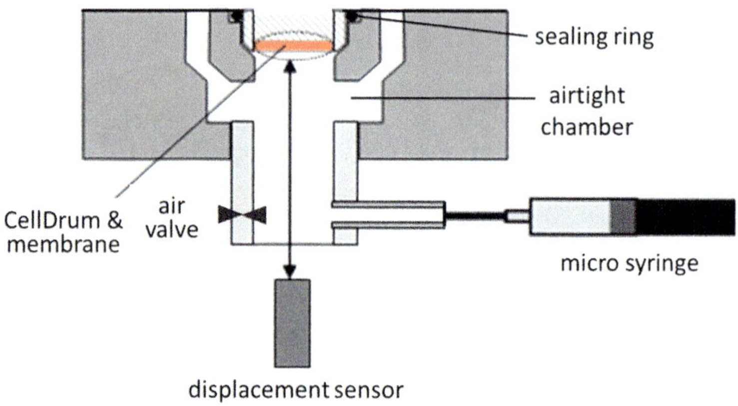

Fig. 8 Load-Deflection-Mode: Experimental setup for slowly changing mechanical tensile stress measurements. Pink colored strip: cell culture layer plus membrane when its plane is adjusted with the CellDrum cylinder's lower end-plane. Lower arc: initial membrane position right after CellDrum insertion into the measurement chamber. At this position the air valve is open for pressure equilibration to the environment. After closing, the air pressure inside the airtight chamber is increased via the micro syringe pump. With increasing pressure, the membrane continuously bulges upwards reaching the horizontal cylinder plane and bulges up further to the preset end of air pumping. It is usually stopped after the membrane's upper arc reached the same magnitude of the lower arc before air application

The CellDrum is partially filled with culture medium. From inside the cylinder and on top of the membrane cells are cultured as 2D (monolayer) or 3D cell-ECM-gel. For proper functioning of the CellDrum measurement principle a cell to membrane surface attachment via focal adhesions is necessarily required. In many cases fibronectin coating of the silicone surface alone may not be sufficient. Almost always a chemical or physical surface conditioning is required in such applications. Due to the culture medium weight, the membrane bulges out downwards. The CellDrum is placed into the measurement and seals it airtight. The chamber contains an air pressure sensor and an inlet for a micro syringe (Fig. 9). Before the measurement begins, the air pressure inside the chamber must be equilibrated to the atmospheric pressure via an air valve. The central indentation of the membrane is measured from below. There are several methods for membrane displacement measurements available. Here a laser sensor as described elsewhere is used [74].

In the Load-deflection mode (Fig. 9) the displacement 'Δh' at the beginning of the experiment is negative since it is bulged downwards below the CellDrum cylinder's bottom plane. When the pump presses air into the chamber, the membrane is lifted upwards. It passes the cylinder bottom plane and bulges up further, causing positive membrane deflection. Pressure increase 'Δp' and indentation 'Δh' are measured simultaneously (Fig. 9). From these data the mechanical tensile stress of the cell monolayer (2D, N/m) or the tissue construct (3D, N/m^2), respectively, is calculated. The development of a completely new technology is often

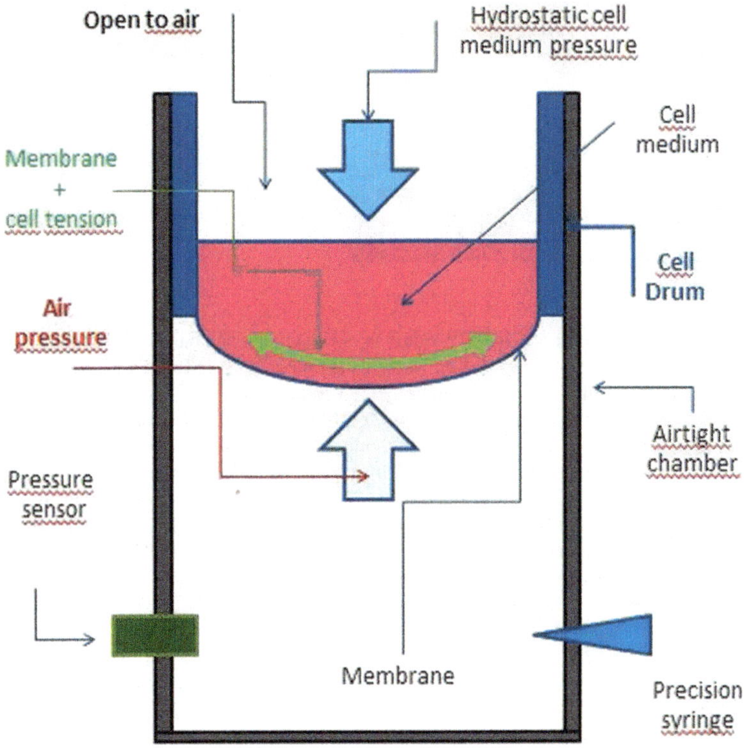

Fig. 9 Load-deflection-mode (extended scheme): cells are cultured on top of the silicone membrane (below the green bended arrow). The scheme shows the balance of forces in the setup: blue arrow from top; hydrostatic culture medium pressure, light blue arrow from below; counteracting air pressure as pumped into the air tight chamber following pressure equilibration with the outside atmosphere. Green arrow (alongside the membrane) shows the counteracting tensile stress of the membrane to the hydrostatic medium pressure. With cells tightly attached to the membrane an additional tensile stress alongside the silicone membrane is exerted which lifts the membrane further. From this indentation on, the system is "lifted" in addition to the membrane-plus-cell culture's intrinsic mechanical tensile stress upwards by the air pressure added to the airtight chamber by the precision syringe

technologically manifested in several stages of development, so also with the CellDrum technology. For example, Fig. 10 shows three generations of the Tissue Tension Analyzer for use in Load Deflection Mode.

3.2 *Mechanical Tensile Stress Calculation*

At small membrane indentations the bulged membrane can be seen as a spherical segment (Fig. 11). Thus, for calculating the membrane tension, σ, Laplace's Law

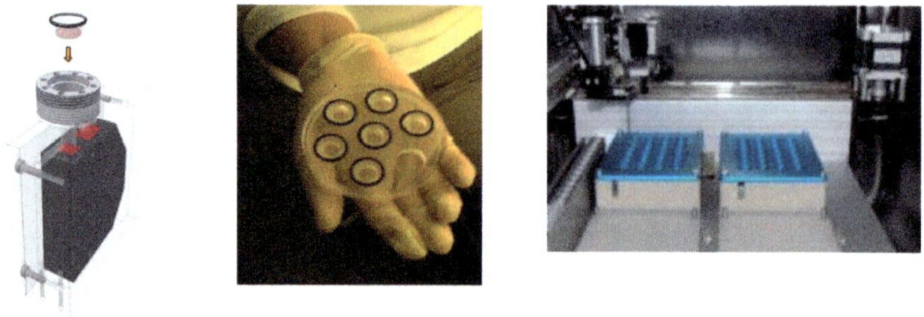

Fig. 10 Load-deflection-mode;Three generations of tissue tension analyzers; left image: **a** CTT I, single-well tissue tension analyzer (black: laser triangulator with red colored sensor fiel, above: ring shaped chamber which is closed airtight by the CellDrum inserted from the top; air valve and syringe pump are not shown); middle image: **b** CTT II, seven-well tissue tension analyzer; only its CellDrum holder is shown in photo; seven individual CellDrums are filled with medium, seeded with cells which are cultured further in an incubator; for measurement purposes, the holder with CellDrums is positioned on the CTT II instrument body forming seven independent measurement setups which can be individually analyzed; right image: **c** CTT MTS, 24-Well Tissue Tension Analyzer; the photo shows two CTT MTS 24-Well systems for Medium Through Screening in an incubator with 24 CellDrums

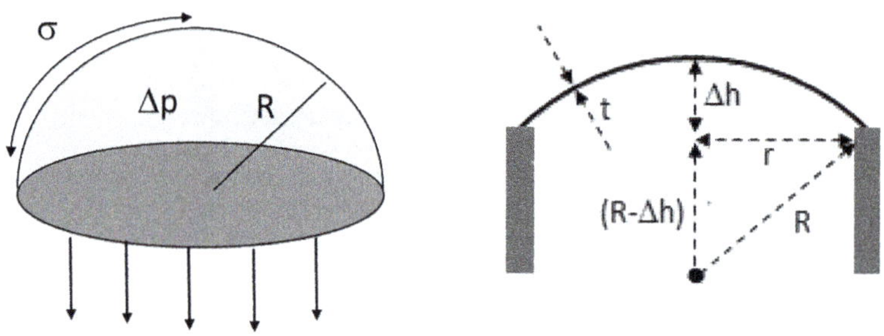

Fig. 11 Scheme for calculating the CellDrum's membrane tensile stress

can be applied although the membrane's circumferential edge is fixed at the cylinder. Hydrostatic pressure Δp of the cell culture medium bulges the membrane downwards forming a spherical cap at a curvature radius R.

The tensile stress σ of the membrane at small deflections Δh can be calculated using Laplace's formula for a thin-walled sphere with thickness t:

$$\sigma = \frac{\Delta p \cdot R}{2 \cdot t} \qquad (*)$$

Radius R is obtained from the deflection Δh with the Pythagorean Theorem:

$$R^2 = (R - \Delta h)^2 + r^2$$

where r is the radius of the CellDrum (r = 8 mm). This can be solved for:

$$R = \frac{r^2}{2 \cdot \Delta h}\left(1 + \frac{\Delta h^2}{r^2}\right)$$

With h ≪ r we neglect $\Delta h^2/r^2$ and simplify the formula for R to

$$R = \frac{r^2}{2 \cdot \Delta h} \qquad (**)$$

Combining equations (*) and (**), we obtain the membrane stress σ [Pa] as function of the differential pressure Δp [Pa] and the resulting deflection Δh of the membrane [μm], and the dimensions of the CellDrum [μm]:

$$\sigma = \frac{\Delta p}{\Delta h} \cdot \frac{r^2}{4 \cdot t} \qquad (***)$$

For calculations using Formula (***) the precise membrane thickness of each individual CellDrum, s, is needed. For this purpose a common interferometric procedure was adopted.

3.3 Data Acquisition in the Load-Deflection Mode

The experiments described in Sect. 3 were carried out with the Single Well Tissue Tension Analyzer, CTT I, in the so-called "Load-Deflection-Mode". Figure 12a shows the setup and Fig. 12b a typical load deflection curve. In order to obtain such curve experimentally, the CellDrum plus attached cell layer is placed into the airtight, temperature-controlled measurement chamber and equilibrated for 500 s. After this period, the strain of the membrane is adjusted to minus 0.3–0.5%. The corresponding pressure level is taken as zero pressure Δp in Formula (***). As for experimental data acquisition, additional air pressure Δp is slowly added lifting the membrane upwards. The pressure ΔP versus center deflection Δh is measured stepwise. The data points (Fig. 12b, blue dots) are than mathematically fitted. From this fit the slope Δh/Δp at zero membrane bending is calculated, inserted in Formula (***), and with geometrical membrane parameter the tensile stress σ is calculated. For absolute cell layer tensile stress measurements it is necessary to subtract the in-plan pre-stress of the membrane which must have been measured before cells were seeded on it. Data obtained with cells alone but without any tensile stress active compound are taken as controls for drug effect measurements.

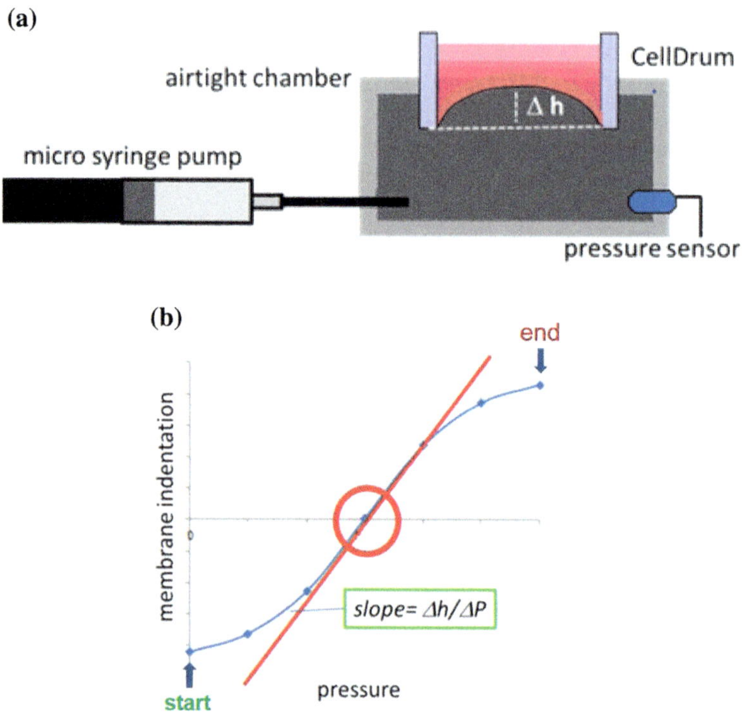

Fig. 12 **a** Chamber setup for data acquisition in the Load-Deflection-Mode: The membrane has been bulged upwards by the micro syringe pump to the a pre-set maximum membrane strain after a pressure deflection curve was measured. **b** Scheme of a load deflection data set. The membrane deflection at the "start" and the "end", respectively, of a data point sampling are indicated. The slope as take for tensile stress calculation in framed in green. At the dot circled in red marks the membrane is in-plane with the CellDrum's lower end plane

3.4 *Validity Experiments of the Load-Deflection-Mode*

The procedure above is repeated after the CellDrum with cells (control) was brought back to the lower membrane indentation by opening the air tight chamber (Fig. 9) to the atmosphere. It was regularly observed that the lower membrane deflection of controls returned back to its starting position, thus, was fully reversible. This indicates that load application to the extent chosen in the experiment did not affect cell-cell or cell-membrane attachments. Otherwise, the deflection would not be reversible.

For subsequent drug effect investigation based on control cell cultures the respective cell active compound is added for 5 min. This was carried out cumulatively for stepwise increased drug concentrations at each measurement cycle by replacing 20% volume of the culture medium with the same volume of culture medium containing the compound. A result was obtained as Fig. 13 shows.

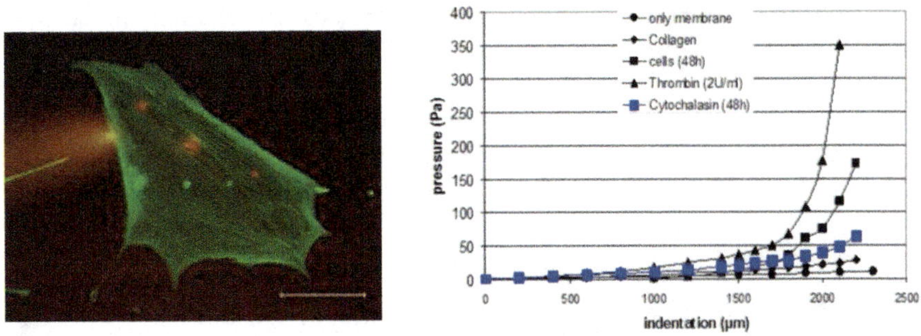

Fig. 13 Left: individual 3T3 fibroblast (bar 10 µm) with Phalloidin stained actin cytoskeleton. The cell is attached to the CellDum's silicon surface by focal adhesions. Right: Pressure indentation (deflection) curves of a 3D fibroblast 3D tissue construct cultured on a CellDrum membrane kept in-plane with the CellDrum bottom plane during culturing by supporting air pressure from below. After 48 h of culturing the load(pressure)-deflection curve was measured at various conditions as indicated in legend and text below. Culture condition: 3T3 Fibroblasts in Collagen I-Gel, 500 µm thickness, 10^6 cells/ml, FCS 10%; cytochalasin D was dissolved in dimethylsulfoxide (DMSO) and stored at −20 °C. Before use the final concentration was adjusted with DMEM. Experiments with fetal calf serum (FCS) were carried out using one and the same aliquot of FCS throughout experiments (Trzewik 2007)

By way of example, the pressure versus indentation curves from Fig. 13 will be commented on here. The membrane itself without cell culture was extremely soft and deformed fast at low ΔP. The cell containing collagen gel right after seeding stiffened the gel-membrane compound only weakly. At this point it should be mentioned, that the silicon membrane alone is not only highly but fast deformable. In other words, after applying a step wise increased ΔP the deformation needs a very short time to reach its new deformation stage. However, when gels and/or cells are cultured on the surface the silicone-gel- compound exhibits a higher viscosity, thus longer "relaxation times" are required to reach the new equilibrium deformation stage. Practical consequences are, that any change of pressure (membrane deformation) should be carried out slowly not to damage the cell-ECM integrity. This is only a hint. It has not been studied systematically.

Forty eight hours of fibroblast culturing in collagen stiffened the membrane-gel-compound drastically. The 3T3 fibroblasts not only proliferated during this period and more total cell generated tensile stress was developed. It is as well because cells formed contacts to each other and to the gel's proteins, thus, stiffened the membrane-cell monolayer compound "membrane". Thrombin known to cause fibroblast contraction increased the tissue construct's stiffness drastically. Depolymerization of the actin cytoskeleton is manipulated by actin-binding Cytochalasin disrupting the normal actin polymerization kinetics. The effects of Cytochalasins are specific to actin filaments, where they bind to the plus end of actin filaments resulting in an indirect depolarization of the actin fibers. In other words, due the Cytochalasin D action, used here, the mechanical tensile stress transmission pathway via the cellular cytoskeletons is interrupted.

As can be seen, the pressure deflection curve does not return to the curve of collagen gel alone as seen before. This is interpreted as a consequence of a residual stiffness caused by the extracellular matrix that has developed during culturing (compare our remarks about the remaining membrane stiffening after cell monolayer culture in Sect. 2.1). The latter effect is only an observation and should be studied in more detail.

3.5 *Efficacy of Cosmetic Ingredients and of Natural Remedies*

Animal experiments are not permitted any longer to be used in cosmetic research. The CellDrum technology and its unique features will become a requisite technological replacement for testing newly developed active ingredients (compounds, blends) on appropriate human 2D/3D tissue constructs. Moreover, with the CellDrum system applied to human tissue constructs much more information can be obtained than in animal experiments. Another imperative advantage implies that CellDrum results obtained with human cells will become more valuable for drug research, as results of animal tests can only be partially transmitted to humans.

Thus the CellDrum technology in its Load-Deflection-Mode may become a testing tool for researchers in the cosmetic industry incisive for mechanically active formulations of their ingredients, their prerequisite being "Does our particular ingredient show a positive effect on cell tensile stress and to what extent?" A "translation" of such scientifically sounding question is "Does our ingredient have the potential to smoothen skin wrinkles in vivo?" A very typical and normal question of the real industry world out there! These most legitimate applied research tasks involve sometimes big business and are often confident matters. Moreover, in many cases pure substances are questioned about their efficacy. Often active ingredients (of a skin cream for example) contain mixtures of compounds whose formulation will never be disclosed to the public, likewise are questions about the efficacy of "Natural Remedies" used for example against hypertension. In such cases the CellDrum may act as a cell assay revealing valuable information on the efficacy of Natural remedies and mixtures of multiple compounds. An in vitro standard testing approach is routinely establishing negative/positive controls as well as a more novel "damaging model" method. Such models reflect only parts of complex physiological situation in cells. Yet when chosen carefully they still might be a practical tool gaining first insights into an "ingredients efficacy" in certain questions.

Mechanical tensile stress in living tissues and tissue cultures is involved in multiple cellular processes. The project as partially presented in Fig. 14 uses primary human dermal fibroblasts, (NHDF), (Promo Cell GmbH, Heidelberg, Germany) derived from young (18–20 years old) and elderly (61–65 years old) donors, respectively. The aim of the project was to quantify cosmetic ingredient

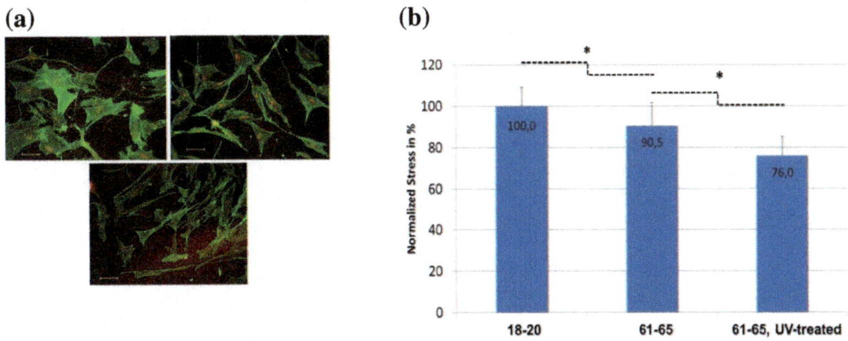

Fig. 14 **a** Micro photographs of actin stained primary human skin cell (NHDF) monolayers. Upper row, left photo: donor age 18–20 years; right: donor age 61–65 years. Lower photo: donor age 61–65 years and UV treated. **b** Normalized NHDF cell monolayer tensile stress of 18–20, 61–65 years old donor cells, and of 61–65 years old donor cell that have been exposed to UV irradiation (N = 12, +1 SD, $*p < 0.05$)

effects (blends of unknown substance compositions) on NHDF monolayer tensile stress containing primary human skin cells donated by young and elderly primary cell donors. In this applicable demonstration NHDF are used to explore the efficacy of a cosmetic company's active ingredients of skin cream.

In the demonstration presented here, a particular important question was about the establishment of a "damaging" or a "comparative" model, respectively. In this example the "comparative model" is represented by NHDF of elder donors. The "damaging model" is instead the application of UV light to the cells of elder donors. The question posed by the ingredient producer in this case is, whether their ingredient can improve the abnormal mechanical tensile stress of elder donor NHDF cells as compared to cells of young donors. Microscopic images show, that NHDF of elder donors have a somehow "eroded" cytoskeleton (Fig. 14a). This results in a significant lowered mechanical tensile stress of elder donors' NHDF cell monolayer versus those of young donors by ~10% (Fig. 14b). The same figure shows that an additional extrinsically caused cell damaging due to UV radiation (following a standardized UV light irradiation protocol) reduces the cell tensile stress exhibited by cells of elder donors further, by in total ~24%! Cell exposure to UV light enhances the expression of matrix metallo-proteinases (MMP) followed by an ECM degradation. The phenomenon is called skin photo aging [41].

As application the effects of cosmetic ingredients (blends) to NHDF cell monolayer tensile stress was measured (Fig. 15). NHDF cells of one 59 year old donor (grey bar) are taken as control (100% tensile stress). TGF β known to induce NHDF contraction (stiffening of the cell monolayer) elevated the cell tensile stress significantly by about 15% versus controls (green bar). Four different blends were added to cells of elder donors. Blend II and blend IV enhanced the mechanical tensile stress to about the extent of TGF β (shadowed in pink). Blend I and III showed only a tendency towards TGF β results but no significant increase. The

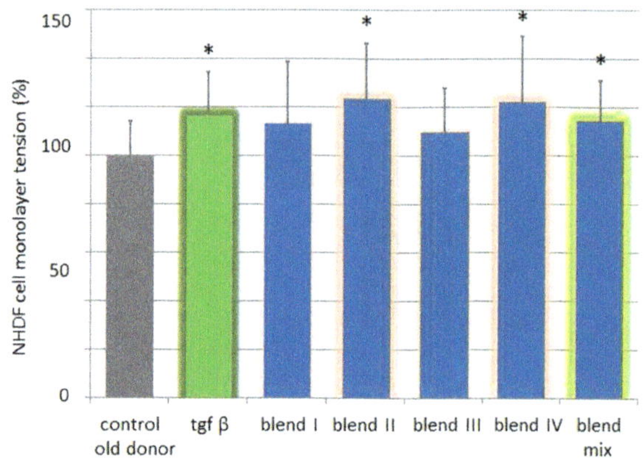

Fig. 15 Effects of cosmetic ingredients (blends) to NHDF cell monolayer tensile stress of one 59 year old donor (grey). As positive tensile stress inducing control 20 ng/mL TGF-β at an incubation time of 24 h was used (green). Than blends as indicated in the figure were added to control cells (N = 26 repeats per group, averaged and normalized, error bars +1 SD, Wilcoxon Signed Rank Test, $*p < 0.05$)

effect of a mix of blend I–IV (I: 2.5%, II: 24.5%, 48.9%, and 24.1%) was significant as well (shadowed in green), but less pronounced than blend II and IV, respectively, alone.

This is only one result out of many that have been obtained. However, it illustrates that the CellDrum technology in its Load-Deflection-Mode may serve as a sensitive tool for such type of questions and applications. The results underline a high potential for many more applications. However, there arises a new technological demand: The need for a High-through-put CellDrum system in order to screen many combinations at once and at low costs, to optimize a mix of pharmacological compound and/or to exclude drug interactions.

In conclusion, the CellDrum technology gives the opportunity to make comparative estimations of pharmaceuticals/substances, to find out about their potency to induce or reduce respectively cellular contractile forces. There exist many so far not sufficiently studied natural remedies which might be worth studied further with the technology described here. Another yet most important application of the CellDrum in both of its modes, however, will be pharmacology and toxicology testing. Despite many other such tests for screening new substances and to test new active principles the CellDrum technology represents a method that answers the question, whether a compound exhibits a wanted or an unwanted, respectively, cell mechanical tensile stress induction effect on (human) cell cultures in vitro. The relative simplicity of the CellDrum technology will allow routine testing with large sample numbers in the very near future.

4 "Self-exciting Mode"—Rhythmically Contracting Cardiomyocytes

4.1 *Brief Introduction on Cardiomyocytes in the Heart*

Cardiomyocytes represent the active contracting cells of the heart comprising atrial as well as ventricular cells in the respective functional parts of the heart. Their functions are critical to the proper form and regulation of the beating human heart. Human Induced Pluripotent Stem Cells (hIPS) from healthy as well as diseased patients have the potential to give rise to all phenotypes of cardiomyocytes, making them a good model system to study human cardiac diseases that affect different parts of the heart. There are three types namely, atrial myocytes, ventricular myocytes and pacemaker myocytes. Adult atrial myocytes and ventricular myocytes are examples of non-pacemaker action potentials in the heart. Because their action potentials undergo rapid depolarization (upstroke), these are sometimes referred to as "fast response" action potentials. Atrial cardiomyocytes make an important contribution to the refilling of ventricles with blood, which enhances the subsequent ejection of blood from the heart. The dependence of cardiac function on the contribution of atria becomes increasingly important with age and exercise [12]. It has been reported that 95% of the cardiomyocytes from differentiating human Embryonic Stem Cells after treatment with Noggin and retinoic acid had atrial-like action potentials; treatment with retinoic acid increases atrial-specific gene expression and percentage of atrial myocytes but decreases expression of ventricular-specific genes [23, 32]. Treatment of differentiating hESCs with Noggin and retinoic acid inhibitors increased the expression of ventricular-specific genes and 83% of the obtained cardiomyocytes had ventricular-like action potentials. Weng et al. reported a protocol that generates 93–100% ventricular myocytes from different hESC and hiPSC lines based on electrophysiological properties of the cardiomyocyte population and resulting ventricular-like cardiomyocytes exhibited typical ventricular ionic currents [76].

Cardiac pacemaker myocytes carry impulses that are responsible for the beating of the heart, they can spontaneously generate, send out electrical impulses and also electrical impulses from cell to cell. Kapoor et al. [46] demonstrated that de novo cardiac pacemaker myocytes can be obtained by direct reprogramming of rodent ventricular myocytes via viral delivery of the transcription factor required for embryonic pacemaker tissue development, the translational potential of this work has been suggested by preclinical studies carried out with pigs demonstrating that this strategy can be translated to large animals. Ionta et al. found that overexpression of Shox2, a transcription factor critical for embryonic pacemaker tissue development, in differentiating mouse ESCs resulted in upregulation of pacemaker-specific genes. The obtained pacemaker-like cells exhibited enhanced automaticity and are able to pace adult rat hearts [32, 42].

4.2 The CellDrum a Complementary In Vitro Tool to the Langendorff Heart

For more than hundred years explanted animal hearts were used for pharmacology and toxicology testing of inotropic drugs. Since the fundamental work by Langendorff the isolated perfused heart (Fig. 16) has been the ultimate experimental model in cardiovascular research [53].

However, from the moment of organ harvesting the heart sets on deteriorating; unphysiologic formations of edemas occur, and the rates of deterioration are affected together with many other factors. This includes the skill of the operator avoiding contusion injury, the species, the composition of the perfusion fluid, the presence or absence of various drugs, animal age, heart rate/cardiac output and the temperature at which the studies are carried out. This rate of deterioration can be crucial in the design and interpretation of studies. This is in particular true when it is necessary to use time-matched controls with corrections for baseline deterioration when comparing groups like papillary or trabeculae preparations obviously have to cope with the same limitations [68]. Some more objections are obvious: One animal needs to be killed for only one experiment, physiological data of animals cannot be simply transferred to humans, data obtained with just one organ explant cannot be generalized for a whole animal population, and last not least, high-through-put testing is ethically inacceptable. On the other hand, no in vitro test can fully replace animal organ testing. The Langendorff heart is a naturally grown organ and, thus, represents still best the endogenous regulation of the heart.

However, taking into account the broad experimental potential of the CellDrum Technology it may become the ultimate in vitro method of choice for pharmacology and toxicology testing as well as for applied and basic heart research. It may also serve well in basic medical research. One such feature for example, so far not yet experimentally approached with the CellDrum technology, could be studying the molecular and cellular mechanisms underlying the Frank-Starling effect [65]. They

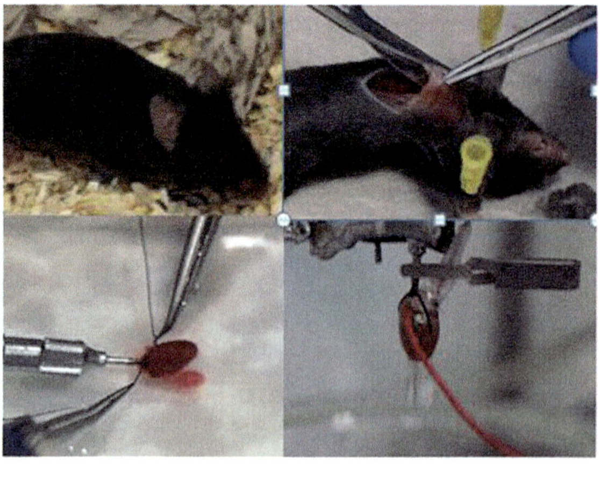

Fig. 16 Preparation of a Langendorff heart from a healthy mouse. The authors thank the very talented scientist Dr. med. rer.nat habil. Petra Kleinbongard from the University of Duisburg-Essen for providing us with the photos to the figure

remain still elusive. The question raised is why an increase in the heart's ventricular filling, causing elongation of cardiomyocyte sarcomere length, induces an increase in the ventricular mechanical work caused by strain-stimulated cardiomyocyte's ability to perform more physical work at increasing strain levels (mechanical tensile stress).

4.3 "Self-exciting-Mode"—Proof of Principle with Cardiomyocytes

Whereas the frequency of cardiomyocyte cultures can be analyzed routinely with various other methods, the CM induced mechanical tensile stress cannot be measured on a routine basis as demanded in pharmacology and toxicology applications. In 2010, the CellDrum technology was first successfully applied to autonomously beating cardiomyocytes (CMs). An important proof of principle test was carried out by that time with neonatal CMs from Wistar rats using the CellDrum technology in its self-exciting mode (Figs. 17 and 18).

Experiments are aimed at deriving rhythmical cardiomyocyte induced pressure signals corresponding to cardiomyocyte induced mechanical tensile stress. Neonatal cardiomyocytes from Wistar rats are cultured on a CellDrum membrane and a pressure versus time signal is derived. For tensile stress measurements the CellDrum is placed into the airtight measurement chamber. Normally 300,000 cardiomyocytes at a time are used per CellDrum. Cells adhere, form contacts and start beating autonomously and simultaneously. No further electrical excitement is needed. Afterwards the syringe raises the chamber pressure inducing an offset pressure in the chamber. It bulges membrane and cell layer upwards.

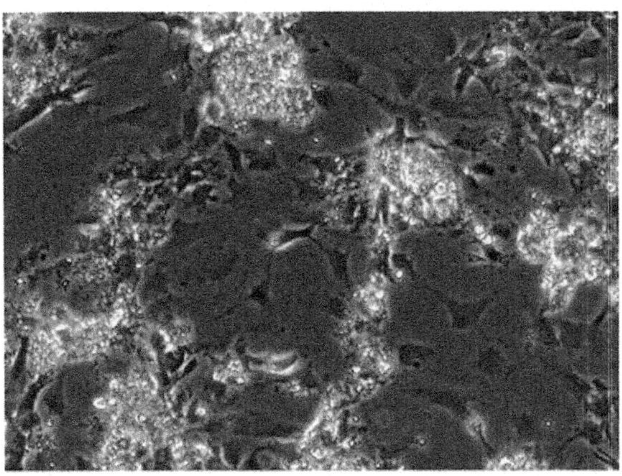

Fig. 17 Micro photograph of a cardiomyocyte monolayer of neonatal Wistar rats on a CellDrum membrane

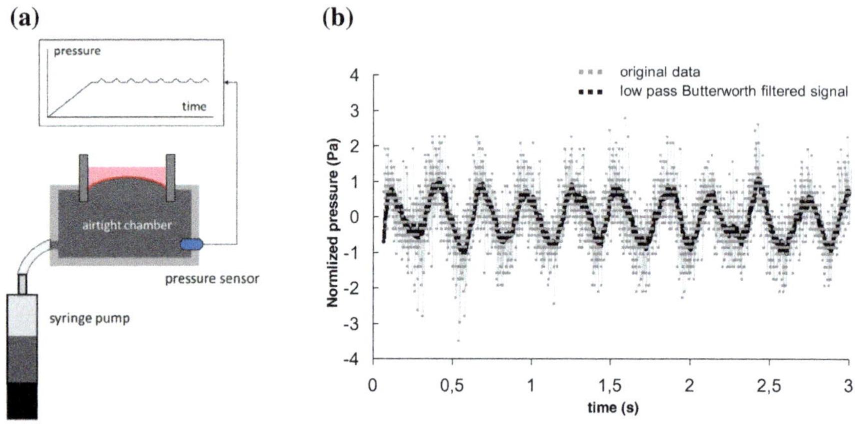

Fig. 18 "Self-Exciting-Mode" for quantifying cardiomyocyte mechanical tensile stress. **a** Scheme of the experimental setup: CellDrum cylinder (dark grey), membrane and cell monolayer (red), culture medium (pink), plus syringe pump and pressure sensor with access to the airtight chamber. **b** About 10–15 contraction cycles are measured. Although rhythmic beats are somehow visible, the original signal is very noisy. Low pass Epanechnikov filtering eliminates noise and reveals smooth individual mechanical beats. The application of filters needs careful considerations in order not to "filter out" important signal content

Cardiomyocytes continue beating autonomously. Their rhythmical mechanical tensile stress modulates an additional sinusoidal pressure to the offset pressure (Fig. 18a). Figure 18b shows an original plot containing information on the frequency and beat tensile stress amplitude. In these experiments a monolayer of cells was sandwiched in-between two ultrathin collagen layers. These tests were performed by Trzewik [73].

This pressure signal is measured over time at high resolution. The original signal is noisy and needs software filtering. An Epanechnikov filter is used to smoothen the input curve (Fig. 18b). A peak-to-peak data analysis over a sufficient number of subsequent filtered beats gives the average frequency and the beat amplitude in N/m (cardiomyocyte monolayer, 2D) or N/m^2 (cultured 3D cardiomyocyte tissue construct), respectively.

In another experiment, another monolayer of neonatal Wistar rat cardiomyocytes embedded in-between two ultrathin collagen layers was tested. The aim was showing that the rhythmicity observed in the signal was caused by self-exciting beating cardiomyocytes. Figure 19a shows a corresponding data recording. Clearly rhythmicity can be observed even without signal filtering. It disappeared fully after killing the cells with a 10% Triton X-100 solution (Fig. 19b). Respective signals observed after noise reduction are shown in Fig. 19c (living cells) and Fig. 19d (dead cells). After applying Fourier transform software there appeared a strong 3.2 Hz frequency peak from curves taken with alive cells before Triton X-100 application (Fig. 19e). Such peak was NOT anymore observed after Triton treatment (Fig. 19f). Consequently, the sinusoidal rhythmicity in Fig. 19, E is due to

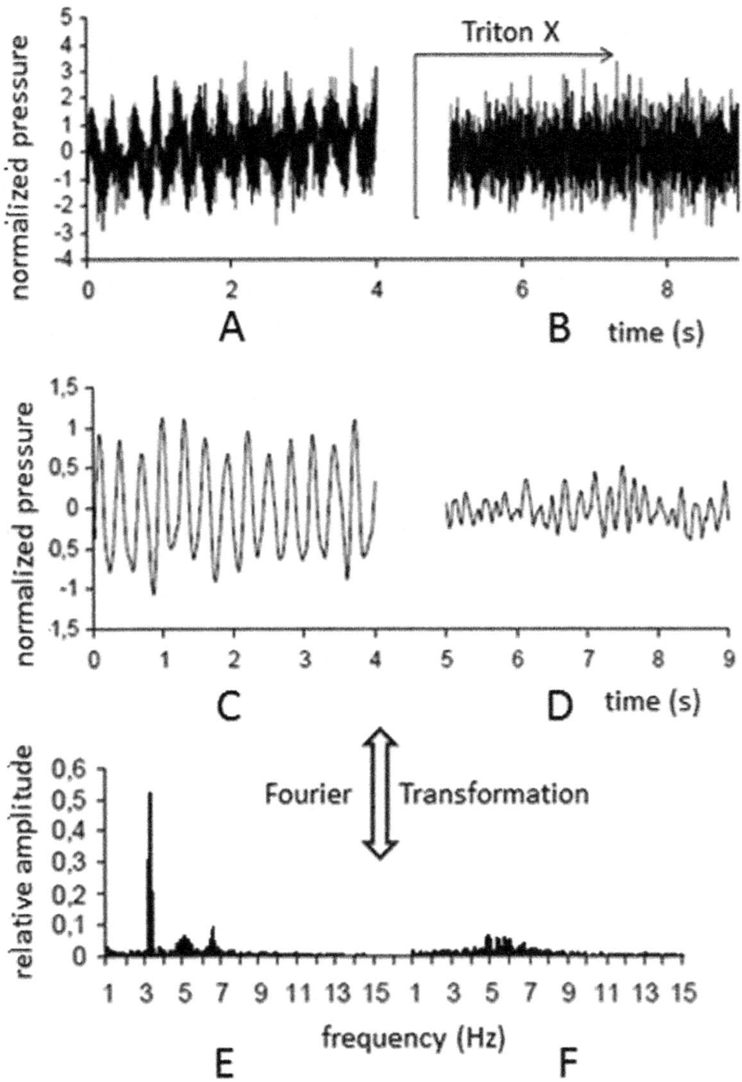

Fig. 19 Life cell–dead cell proof: the graphs A, C, and E, were obtained with an alive cell culture. However, B, D, and F represent signals after the application of Triton X-100 killing the cells (original data after the offset pressure subtraction). **Above**: Original pressure signal, **a** before, and **b** after Triton X application. **Middle**: **c** and **d** Low pass filtered pressure signals. **Below**: **e** The Fourier transformed signal reveals a frequency peak at 3.2 Hz (192 beats/min) of the alive cell culture and **f** no such peak was seen after Triton X-100 treatment

cellular rhythmical beating of alive cells. In this test a cell induced tensile stress of $1.28 \pm 0.13\ \mu N/mm^2$ at a frequency of 3.2 Hz (192 beats/min) was measured.

The experiments above reveal that contracting cardiomyocytes induce a measurable rhythmical signal. The road for developing the CellDrum technology towards measuring routinely cardiomyocyte tensile stress and inotropic drug effects was opened.

4.4 Pharmacological Experiments with the "Self-exciting" Mode

In an early experiment the effect of noradrenalin and a β blocker [17, 70] on the contraction frequency and the beating strength of primary Wistar rat cardiomyocytes was examined using the CellDrum in its "self-exciting-mode (Fig. 20).

Noradrenalin acts on the heart in a positive chronotropic, dromotropic, inotropic, bathmotropic und lusitropic manner. In the experiment show in Fig. 20, initially, cardiomyocytes are beating at 4.4 Hz. After ~6 min, 10–4 M noradrenalin is added. An immediate increase of the contraction frequency is observed reaching a peak at 11% above the initial level after 4 min. Within the next 5.5 min the frequency drops to 6% which is followed by a linear decrease at 7.3 MHz/min. Thirty five minutes after experiment begin, 10–6 M β blocker were added to the noradrenalin pre-treated cell culture. The rate of drug induced frequency change was −15.5 MHz/min. After ~60 min the frequency reached a plateau at minus 4.5%, thus below the initial frequency. Such experiments might easily be repeated and carried out with cells grown in culture dishes. However, results might be different.

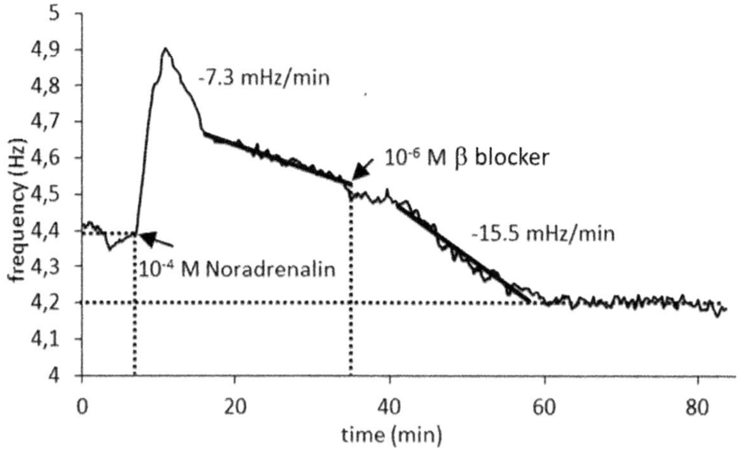

Fig. 20 Effect of Noradrenalin and a β blocker on the frequency of cardiomyocytes of primary neonatal Wistar rats [6]

We would like to underline at this point another special feature of the CellDrum technology: Cells are seeded on a soft and deformable substrate—the ultrathin silicone membrane. Cardiomyocytes are used to a permanent mechanical and rhythmical biaxial elongating and contracting environment. Such mechanical boundary condition is greatly mimicked by the CellDrum; however it is not by a stiff culture dish surface. This may affect biochemical and physical responses of cells to their mechanical environment [52]. Cardiomyocytes in the CellDrum do not only exhibit rhythmic beating pattern; they perform real physical work. Cells use glucose from the cell culture medium producing energy allowing them to work against the culture medium weight without any other intervention (no electrostimulation). Due to cell contractions the culture medium is lifted upwards against gravity. This work is partially represented in the beat amplitude of the CellDrum signal.

Summarizing so far, using the CellDrum technology, the frequency and simultaneously the mechanical beating strength of primary cardiomyocytes from neonatal Wistar rats can be measured and principle drug action can be shown. In other words, the new element of these data presented here is that the mechanical cell tensile stress (beating strength) of cell layers can be measured **in addition** to the frequency. Thus, in the future new testing strategies can be designed in order to understand drug action better than with experiments lacking mechanical beating strength information. The CellDrum opens up a new perspective performing long-term drug testing experiments over weeks. Moreover, much less Langendorff heart experiments may be needed in the future to verify CellDrum data by whole organ studies [10, 53].

4.5 Inotropic Drug Effects on Human IPS Derived Cardiomyocytes

In March 2016, the first article on contraction forces of cultured tissue containing **human** induced pluripotent stem cells (hIPS CMs) was published [31]. The results of this study are presented below.

Before doing so, however, let us briefly turn to a similar study, which was reported a few months later, in May 2016. The authors reported on force measurements on cultured tissue containing IPS-derived human CMs Although a different force transducer principle has been used in the modified method [58], forces of uniaxial strained cultured tissues with all the disadvantages associated therewith continue to be measured. Such setup is obviously not suitable for sample quantities in the medium or high throughput range. In contrast, with the CellDrum technique, mechanical stresses with physiological, two-dimensional clamping are measured with high precision. Furthermore, the CellDrum technology is basically reasonably applicable for medium throughput set-ups up to 96 CellDrum units. The data shown were obtained using the CTT I setup (Fig. 10).

Two different drugs targeting cardiac Ca^{2+} channels, S-Bay K8644 and Verapamil, respectively, are applied to hiPS-derived cardiomyocytes. Representative recordings of S-Bay K8644, a Ca^{2+} channel agonist, and the antagonist verapamil are shown in Fig. 21. S-Bay K8644 enhances the beating strength amplitude of the CR-cycle by 60% above control reaching a plateau at 90 nM. Concurrently, the beating frequency reduced between 90 and 160 nM significantly. Verapamil, a Ca^{2+} channel antagonist, reduced the beating strength amplitude (Fig. 22) to 25% at 300 nM. The frequency was increased linearly to 160% at 300 nM. There is no saturation observed [31].

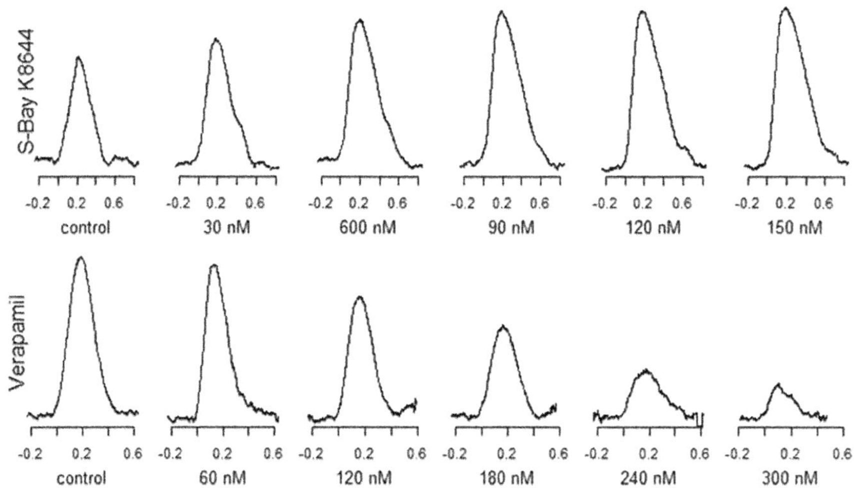

Fig. 21 Representative recordings of S-Bay K8644 (Ca^{2+} channel agonist, above) and Verapamil (antagonist, below) action on hIPS derived cardiomyocytes

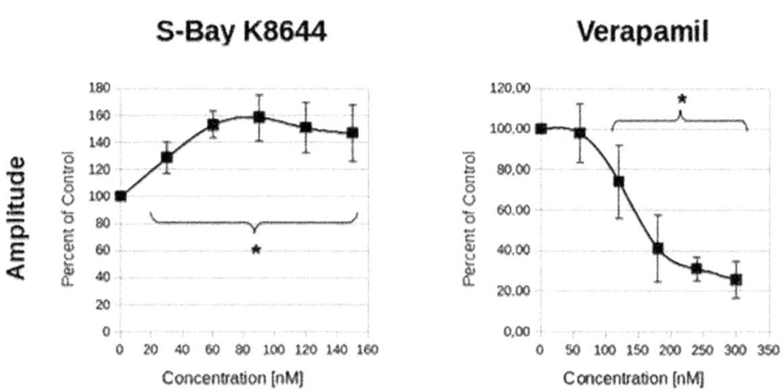

Fig. 22 Dose dependency of S-Bay K8644 and Verapamil action on hIPS derived cardiomyocytes

4.6 *Medium Throughput (MTS) CellDrum Based Test System*

The CellDrum technology offers a unique technological solution towards a functional MTS toxicology and pharmacology test of cell induced mechanical tensile stress in 2D and 3D cultured tissue constructs, in particular hIPS derived cardiomyocyte cultures. Figure 23 shows a set-up for a 24-well MTS toxicology and pharmacology solution. Multi-well solution development is in progress. New multi-well developments based on CellDrum technology are right now in the development stage. Those are meant for routine use and large numbers of experiments performable in a decent lab time. At the same time smaller CellDrum volumes in multi-well setups reduce the amount of test substances needed for experiments easily by factor one hundred. Multi-well systems enable comparative estimations of pharmaceuticals with equal indications, to find out the individual potency of pharmaceuticals and/or substances, to do toxicological tests (adverse reactions), to do screening of new substances and to test new active compounds. For smaller test numbers and research applications a set of smaller and more affordable instruments are available.

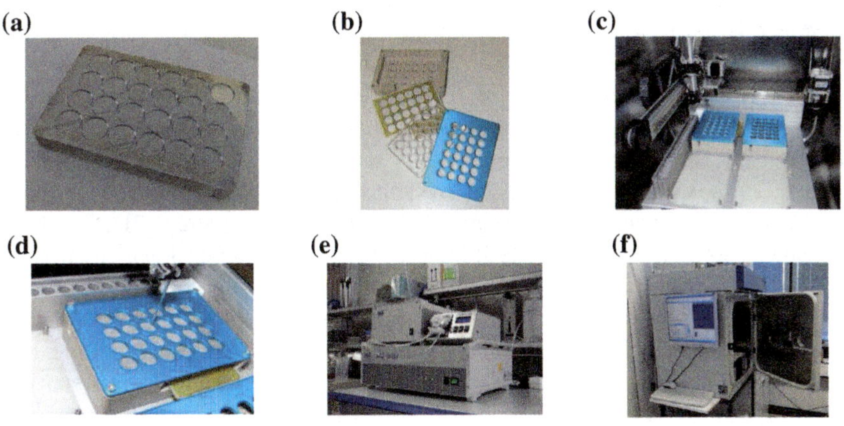

Fig. 23 **a** 24-well CellDrum array as seen from below. The membrane spans the drilling in the array block watertight. **b** 24-well array, components: From top: base with pressure chambers and air inlets, circuit board & electrodes, chamber block, top-lid (blue). **c** Four 24-well arrays can be positioned simultaneously into the incubator for automated measurements. The incubator contains a pipet robot controlled from outside the incubator. **d** The pipette is used for media exchange during cell culturing. Once cells are ready for experiments, pipettes help mix substance solutions from stock solutions and place them into each one of the 24 CellDrum-wells for measurements setting on right after. **e** External hardware, pumps, valves, steering units etc. **f** Incubator opened for experiment preparation with touchscreen. Exposure times, dosages, kinds of substances are controlled via touch screen. A PC steers the whole process and analyses data

5 Conclusions

The Cell Drum Technology represents an ideal high precision experimental setup to test cultured cells for their ability to generate mechanical tensile stress at various biochemical and -physical conditions. It enables precise measurements of this parameter using various cell types and at different conditions. It has the potential to simulate elevated blood pressure, oxygen partial pressure, and drug supply as well as to be upgraded to at least medium high-throughput testing. At standardized conditions the Cell Drum can simulate to certain extend conditions including ischemia/reperfusion by changing the oxygen partial pressure. Human IPS derived cardiomyocytes represent a neonatal signature which is different from adult heart cells and, thus, their response to certain drugs. Efforts need to be made towards maturing those neonatal cells in so far unknown ways.

We conclude that the Cell Drum technology is an accurate high precision method to quantify mechanical properties of cardiac myocyte cell layers in vitro. It combines a high degree of handling simplicity and allows for precise conclusions on cellular mechanics. The major advantage of this approach is its potential scalability towards medium-throughput-screening systems. The application of Laplace's law on circular tissue constructs enables biaxial measurement of cellular tensile stress. Hence, cellular alignment is not necessary for the acquisition of the exerted stress. As a result, this setup eliminates common deviations caused by improper control of the geometry of the specimen. As a research tool with well-defined biomechanical boundary conditions and its potential for upgrading the technology to high throughput tests the Cell Drum satisfies many needs for drug testing of cardiomyocytes as well as the mechanical characterization of other cells and tissues. The experimental setup offers the possibility to simulate acute and chronic cardiovascular hypertension by adjustment of the membrane deflection through application of pneumatic pressure. Adopting the current 2D Cell Drum technology to a 3D model may lead it closer to the three-dimensional environment in ventricular tissue. In addition, the Cell Drum benefits from the fact that cell are not grown on a stiff substrate but on highly flexible membranes mimicking to some extend the natural compliance of soft tissue. Yet, the highly anisotropic three dimensional environment and several other factors still rely on full organ tests suffering from the drawback of a non-scientific experimental setup—the heart of an animal. However, the parallel mathematical modeling and simulation allows in principle a transfer to different types of cardiac myocytes as well as to the anisotropic behavior of oriented tissues.

At present, human iPS stem cells promise to bring about a turning point affecting numerous applications from pharmacology to toxicology. Besides ethical advantages over animal models and the prevention of concerns regarding the translatability of the pharmacological results, hiPS derived cardiomyocytes can be produced in an infinite amount from individuals with known medical and clinical history. Yet, the promises of the new technology bear a certain number of unanswered questions to date. In particular, iPS-derived cardiomyocytes are suspected to not reflect the maturity of adult somatic cells, especially concerning the functionality of ion channels and expression of specific receptors and signaling cascades. Despite these issues, we have shown consistent and physiological responses of hiPS-derived cardiomyocytes on inotropic compounds. Hence, the findings of this study support the hypothesis that iPS-derived cardiomyocytes can provide an adequate model for pharmacological and toxicological studies. Future studies with hiPS-derived electro-competent cardiac fibroblasts and biomimetic three-dimensional connective tissue scaffolds may improve the physiological conditions towards a more concise cardiac tissue model. The parallel mathematical modeling and numerical simulation may be further developed to understand and predict maturation of iPS-derived cells and predict the transfer to adult cells.

The CellDrum technology is basically reasonably applicable for medium throughput set-ups up to 96 CellDrum units. There is currently no technological solution available for the economic and routine high-throughput measurements required for tens of thousands of wells per day in the future. In the future, functional (mechanical) high-throughput test principles will have to be developed. They will help avoid many organ specific, circulatory, metabolic and stress effects as present in whole organ studies. On the other hand whole organ studies cannot be replaced by the CellDrum whenever the whole organ physiology matters.

Acknowledgements Some of our teachers remain a role model for our entire life. Gerhard Artmann's role models remain the professors Y. C. Fung, Shu Chien (both UCSD, San Diego, USA), Georg Büldt (Research Center Juelich, Germany) and Peter Paufler, physicist and crystallographer (University of Dresden, Germany). For providing stem cell expertise and free disposal of hIPS derived, highly purified and well characterized hIPS derived cardiomyocytes as well as for his excellent scientific advice and kindness we owe special thanks to Professor Jürgen Hescheler.

All technological developments and scientific tasks related to the CellDrum technology were carried out in the laboratories and under the scientific direction of Professors Gerhard M. as well as Aysegül Artmann. Significant contributions to the work were made by Professor J. Trzewik our former Ph.D. student and major CellDrum co-inventor. Working with him was scientifically and technologically very productive. Thank you, Juergen. Professors Artmann thank their doctoral candidate M. Gossmann for his lab work. As his teachers, however, they are thoughtfully looking back at the time with him. Another but extremely talented doctoral candidate was P. Linder. He worked for more than a decade and a half intensively with us. We thank the young Prof. Dr. Ilya Digel for the excellent cooperation in the Institute of Bioengineering. In particular, we thank him for many stimulating scientific discussions and his ongoing loyalty.

Quite a number of CellDrum projects were financed by numerous grants to Professors Artmann received by the State of North Rhine Westphalia (NRW) as well as by funding institutions of the Government of the Federal Republic of Germany (BMBF, BMWI).

References

1. Anfinsen, C. B. (1973, July 20). Principles that govern the folding of protein chains. *Science, 181*(4096), 223–230.
2. Artmann, G. M., Digel, I., Zerlin, K. F., Maggakis-Kelemen, C., Linder, P., Porst, D., et al. (2009). Hemoglobin senses body temperature. *European Biophysics Journal, 38*(5), 589–600.
3. Artmann, G. M. (2000, August). Device and method for the measurement of forces from living materials. Patents, Ref. No: US020040033482A1, 08/2000, AU000007638401A, 08/2000, EP000001311850B1, 08/2000, CA 2420141, 07/2002.
4. Artmann, G. M. (2014). (https://www.researchgate.net/publication/301608217_EEE_Laborherz.
5. Artmann, G. M. (2015, November). 6. Deutscher Querdenker Kongreß, Award Winner: Category—"Pioneering".
6. Artmann, G. M., & Chien, S. (2008). *Bioengineering in cell and tissue research*. Berlin, Heidelberg: Springer.
7. Artmann, G. M., Burns, L., Canaves, J. M., Temiz-Artmann, A., Schmid-Schönbein, G. W., Chien, S., et al. (2004). Circular dichroism spectra of human hemoglobin reveal a reversible structural transition at body temperature. *European Biophysics Journal, 33*(6), 490–496.
8. Artmann, G. M., Kelemen, C., Porst, D., Büldt, G., & Chien, S. (1998). Temperature transitions of protein properties in human red blood cells. *Biophysical Journal, 75*(6), 3179–3183.
9. Artmann, G. M. https://de.wikipedia.org/wiki/Gerhard_Artmann.
10. Bell, R. M., Mocanu, M. M., & Yellon, D. M. (2011, June). Retrograde heart perfusion: The Langendorff technique of isolated heart perfusion. *Journal of Molecular and Cellular Cardiology, 50*(6), 940–950.
11. Bissell, M. J., Hall, H. G., & Parry, G. (1982, November 7). How does the extracellular matrix direct gene expression? *Journal of Theoretical Biology, 1*(7), 31–68.
12. Bootman, M. D., Higazi, D. R., Coombes, S., & Roderick, H. L. (2006). Calcium signalling during excitation-contraction coupling in mammalian atrial myocytes. *Journal of Cell Science, 119,* 3915–3925.
13. Brown, R. A., Prajapati, R., McGrouther, D. A., Yannas, I. V., & Eastwood, M. (1998, June). Tensional homeostasis in dermal fibroblasts: Mechanical responses to mechanical loading in three-dimensional substrates. *Journal of Cellular Physiology,* 323–332.
14. Burridge, P. W., Li, Y. F., Matsa, E., Wu, H., Ong, S., Sharma, A., et al. (2016). Human induced pluripotent stem cell-derived cardiomyocytes recapitulate the predilection of breast cancer patients to doxorubicin-induced cardiotoxicity. *Nature Medicine, 22*(5), 547–556.
15. Burton, K., & Taylor, D. L. (1997, January 30). Traction forces of cytokinesis measured with optically modified elastic substrata. *Nature, 385,* 450–454.
16. Campàs, O., Mammoto, T., Hasso, S., Sperling, R. A., O'Connell, D., Bischof, A. G., et al. (2014). Quantifying cell-generated mechanical forces within living embryonic tissues. *Nature Methods, 11,* 183–189.
17. Cerbai, E., Crucitti, A., Sartiani, L., De Paoli, P., Pino, R., Rodriguez, M. L., et al. (2000, January). Long-term treatment of spontaneously hypertensive rats with Losartan and electrophysiological remodeling of cardiomyocytes. *Cardiovascular Research, 45*(2), 388–396.
18. Chen, C. S., Alonso, J. L., Ostuni, E., Whitesides, G. M., & Ingber, D. E. (2003, July 25). Cell shape provides global control of focal adhesion assembly. *Biochemical and Biophysical Research Communications, 307*(2), 355–361.
19. Chien, S. (2007, March). Mechanotransduction and endothelial cell homeostasis: The wisdom of the cell. *American Journal of Physiology—Heart and Circulatory Physiology, 292*(3), H1209–H1224.

20. Chien, S., Sung, K. L., Skalak, S., Usami, A., & Tözeren, A. (1978, November). Theoretical and experimental studies on viscoelastic properties of erythrocyte membrane. *Biophysical Journal, 2,* 463–487.
21. Cost, A. L., Ringer, P., Grashoff, A. C., & Grashoff, C. (2015). How to measure molecular forces in cells: A guide to evaluating genetically-encoded FRET-based tension sensors. *Cellular and Molecular Bioengineering, 8*(1), 96–105.
22. David, B. (2000, May 4). A surprising simplicity to protein folding. *Nature, 405,* 39–42.
23. Devalla, H. D., Schwach, V., Ford, J. W., Milnes, J. T., Haou, S., Jackson, C., et al. (2015). Atrial-like cardiomyocytes from human pluripotent stem cells are a robust preclinical model for assessing atrial-selective pharmacology. *EMBO Molecular Medicine, 7,* 394–410.
24. Dumont, S., & Prakash, M. (2014, November 5). Emergent mechanics of biological structures. *Molecular Biology of the Cell, 25*(22), 3461–3465.
25. Eastwood, M., McGrouther, D., & Brown, R. A. (1994, November 11). A culture force monitor for measurement of contraction forces generated in human dermal fibroblast cultures: Evidence for cell-matrix mechanical signalling. *Biochimica et Biophysica Acta, 1201*(2), 186–192.
26. Eder, A., Vollert, I., Hansen, A., & Eschenhagen, T. (2016, January 15). Human engineered heart tissue as a model system for drug testing. *Advanced Drug Delivery Reviews, 96,* 214–224.
27. Fung, Y. C. (1993). *Biomechanics: Mechanical properties of living tissuses.* New York: Springer.
28. Gawain, T., Burnham, N. A., Camesano, T. A., & Wen, Q. (2013). Measuring the mechanical properties of living cells using atomic force microscopy. *Journal of Visualized Experiments, 76,* 50497.
29. Gayrard, C., & Borghi, N. (2016). FRET-based molecular tension microscopy. *Methods, 94,* 33–42.
30. Giancotti, F. G., & Ruoslahti, E. (1999, August 13). Integrin signaling. *Science, 285*(5430), 1028–1033.
31. Goßmann, M., Frotscher, R., Linder, P., Neumann, S., Bayer, R., Epple, M., et al. (2016). Mechano-pharmacological characterization of cardiomyocytes derived from human induced pluripotent stem cells. *Cellular Physiology and Biochemistry, 38,* 1182–1198.
32. Hamel, V., Cheng, K., Liao, S., Lu, A., Zheng, Y., Chen, Y., et al. (2017, February 20). De novo human cardiomyocytes for medical research: Promises and challenges. *Stem Cells International, 2017,* 4528941. https://doi.org/10.1155/2017/4528941.
33. Harris, A. K., Wild, P., & Stopak, D. (1980, April 11). Silicone rubber substrata: A new wrinkle in the study of cell locomotion. *Science, 208*(4440), 177–179.
34. Heath, J. P., & Dunn, G. A. (1978). Cell to substratum contacts of chick fibroblasts and their relation to the microfilament system. A correlated interference-reflexion and high-voltage electron-microscope study. *Journal of Cell Science,* 197–212.
35. Heisenberg, C. P. (2017). Cell biology: Stretched divisions. *Nature, 543*(7643), 4443–4444.
36. Helenius, J., Heisenberg, C. P., Gaub, H. E., & Muller, D. J. (2008). Single-cell force spectroscopy. *Journal of Cell Science, 121,* 1785–1791.
37. Heras-Bautista, C. O., Katsen-Globa, A., Schloerer, N. E., Dieluweit, S., Abd El Aziz, O. M., Peinkofer, G., et al. (2014). The influence of physiological matrix conditions on permanent culture of induced pluripotent stem cell-derived cardiomyocytes. *Biomaterials, 35*(26), 7374–7385.
38. Hochmuth, R. M. (2000). Micropipette aspiration of living cells. *Journal of Biomechanics, 33*(15), 15–22.
39. Holmgren, G., Synnergren, J., Bogestål, Y., Améen, C., Åkesson, K., Holmgren, S., et al. (2015). Identification of novel biomarkers for doxorubicin-induced toxicity in human cardiomyocytes derived from pluripotent stem cells. *Toxicology, 328,* 102–111.
40. Horn, M. A., Graham, H. K., Richards, M. A., Clarke, J. D., Greensmith, D. J., Briston, S. J., et al. (2012, July). Age-related divergent remodeling of the cardiac extracellular matrix in

heart failure: Collagen accumulation in the young and loss in the aged. *Journal of Molecular and Cellular Cardiology, 53*(1), 82–90.
41. Hwang, E., Park, S.-Y., Sun, Z.-W., Shin, H.-S., Lee, D.-G., & Yi, T. H. (2014). The protective effects of fucosterol against skin damage in UVB-irradiated human dermal fibroblasts. *Marine Biotechnology (NY), 16*(3), 361–370.
42. Ionta, V., Liang, W., Kim, E. H., Rafie, R., Giacomello, A., Marbán, E., et al. (2015, January). SHOX2 overexpression favors differentiation of embryonic stem cells into cardiac pacemaker cells, improving biological pacing ability. *Stem Cell Reports, 4*(1), 129–142.
43. Iskratsch, T., Wolfenson, H., & Sheetz, M. P. (2014). Appreciating force and shape—The rise of mechanotransduction in cell biology. *Nature Reviews Molecular Cell Biology, 15,* 825–833. https://doi.org/10.1038/nrm3903.
44. Jin, T., Li, L., Siow, R. C. M., & Liu, K.-K. (2015). A novel collagen gel-based measurement technique for quantitation of cell contraction force. *Journal of The Royal Society Interface, 12(106),* 20141365–20141365.
45. Joung, Y.-H. (2013). Development of implantable medical devices: From an engineering perspective. *International Neurourology Journal 17*(3), 98.
46. Kapoor, N., Liang, W., Marbán, E., & Cho, H. C. (2013). Direct conversion of quiescent cardiomyocytes to pacemaker cells by expression of Tbx18. *Nature Biotechnology, 31,* 54–62.
47. Khalili, A. A., & Ahmad, M. R. (2015). A review of cell adhesion studies for biomedical and biological applications. *International Journal of Molecular Sciences, 16*(8), 18149–18184.
48. Kirmse, M., Otto, H., & Ludwig, T. (2011). Interdependency of cell adhesion, force generation and extracellular proteolysis in matrix remodeling. *Journal of Cell Science, 124,* 1857–1866.
49. Kolodney, M. S., & Wysolmerski, R. B. (1992, April 1). Isometric contraction by fibroblasts and endothelial cells in tissue culture: A quantitative study. *The Journal of Cell Biology, 117,* 73–82.
50. Korn, E. D., Carlier, M. F., & Pantaloni, D. (1987, October 30). Actin polymerization and ATP hydrolysis. *Science,* 238(4827), 638–644.
51. Kumar, V., Abul, A. K., & Aster, J. C. (2014). *Robbins & Cotran pathologic basis of disease* (9th Ausg.). Philadelphia: Elsevier.
52. Kurazumi, H., Kubo, M., Ohshima, M., Yamamoto, Y., Takemoto, Y., Suzuki, R., et al. (2011). The effects of mechanical stress on the growth, differentiation, and paracrine factor production of cardiac stem cells. *PLoS One, 6*(12).
53. Langendorff, O. (1895). Untersuchungen am überlebenden Säugetierherzen. *Pflugers Arch, 61,* 291 ff.
54. Lanir, Y., & Fung, Y. C. (1974). Two-dimensional mechanical properties of rabbit skin. I. Experimental system. *Journal of Biomechanics, 7*(1), 29–34.
55. Lehto, H., Talo, A., Tirri, R., & Vornanen, M. (1983, August). Membrane potential oscillations in enzymatically isolated rat myocardial cells. *Acta Physiologica, 118*(4), 385–391.
56. Linder, P., Trzewik, J., Rüffer, M., & Artmann, G. M. (2010, January). Contractile tension and beating rates of self-exciting monolayers and 3D-tissue constructs of neonatal rat cardiomyocytes. *Medical & Biological Engineering & Computing, 48,* 59–65.
57. Liu, K.-K., & Oyen, M. L. (2014). Nanobiomechanics of living materials. *Interface Focus, 4*(2), 20140001–20140001.
58. Mannhardt, I., Breckwoldt, K., Letuffe-Brenière, D., Schaaf, S., Schulz, H., Neuber, C., et al. (2016, July 12). Human engineered heart tissue: Analysis of contractile force. *Stem Cell Reports, 7*(1), 29–42.
59. Pollard, T. D. (1986). Assembly and dynamics of the actin filament system in nonmuscle cells. *Journal of Cellular Biochemistry, 31*(2), 87–95.
60. Ringer, P., Colo, G., Fässler, R., & Grashoff, C. (2017, April 4). Sensing the mechano-chemical properties of the extracellular matrix. *Matrix Biology* (in press).
61. Ruoslahti, E. (1997, May 30). Stretching is good for a cell. *Science, 276*(5317), 1345.

62. Schimid, S. G., Kosawada, T., Skalak, R., & Chien, S. (1995, May 01). Membrane model of endothelial cells and leukocytes. A proposal for the origin of a cortical stress. *Journal of Biomechanical Engineering, 117*(2), 171–178.
63. Shen, Y., Nakajima, M., Kojima, S., Homma, M., Kojima, M., & Fukuda, T. (2011, October 5). Single cell adhesion force measurement for cell viability identification using an AFM cantilever-based micro putter. *Measurement Science and Technology, 22*(11), 115802.
64. Soiné, J. R. D., Hersch, N., Dreissen, G., Hampe, N., Hoffmann, B., Merkel, R., & Schwarz, U. S. (2016). Measuring cellular traction forces on non-planar substrates. *Interface Focus, 6*(5), 20160024
65. Solaro, R. J. (2007, December). Mechanisms of the Frank-Starling Law of the heart: The beat goes on. *Biophysics Journal, 15, 93*(12), 4095–4096.
66. Stabley, D. R., Jurchenko, C., Marshall, S. S., & Salaita, K. S. (2012). Visualizing mechanical tension across membrane receptors with a fluorescent sensor. *Nature Methods, 9,* 64–67.
67. Sugimura, K., Lenne, P. F., & Graner, F. (2016). Measuring forces and stresses in situ in living tissues. *Development, 143,* 186–196.
68. Sutherland, F. J., Shattock, M. J., Baker, K. E., & Hearse, D. J. (2003, November). Mouse isolated perfused heart: Characteristics and cautions. *Clinical and Experimental Pharmacology and Physiology, 30*(11), 867–878.
69. Theret, D. P., Levesque, M. J., Sato, M., Nerem, R. M., & Wheeler, L. T. (1988, August 1). The application of a homogeneous half-space model in the analysis of endothelial cell micropipette measurements. *Journal of Biomechanical Engineering, 110*(3), 190–199.
70. Tirri, R., & Lehto, H. (1984). Alpha and beta adrenergic control of contraction force of perch heart (*Perca fluviatilis*) in vitro. *Comparative biochemistry and physiology C, 77*(2), 301–304.
71. Tomasek, J. J., Gabbiani, G., Hinz, B., Chaponnier, C., & Brown, R. A. (2002, May). Myofibroblasts and mechano-regulation of connective tissue remodelling. *Nature Reviews Molecular Cell Biology, 3*(5), 349–363.
72. Trzewik, J., Ates, M., & Artmann, G. A. (2002). A novel method to quantify mechanical tension in cell monolayers. *Biomedizinische Technik/Biomedical Engineering, 47*(s1a), 379–381.
73. Trzewik, J. (2008). http://citeseerx.ist.psu.edu/viewdoc/download?doi=10.1.1.632.9489&rep=rep1&type=pdf.
74. Trzewik, J., Linder, P., & Zerlin, K. F. (2008). How strong is the beating of cardiac myocytes?—The cell drum solution. In *Bioengineering in cell and tissue research*. Berlin, Heidelberg: Springer. ISBN 978-3-540-75408.
75. Verma, D., Bajpai, V. K., Ye, N., Maneshi, M. M., Jetta, D., Andreadis, S. T., et al. (2017). Flow induced adherens junction remodeling driven by cytoskeletal forces. *Experimental Cell Research, 359*(2), 327–336.
76. Weng, Z., Kong, C.-W., Lihuan, R., Ioannis, K., Lin, G., Jiaozi, H., et al. (2014, February 25). A simple, cost-effective but highly efficient system for deriving ventricular cardiomyocytes from human pluripotent stem Cells. *Stem Cells and Development, 23*(14), 1704–1716.
77. Wobus, A. M., Wallukat, G., & Hescheler, J. (1991). Pluripotent mouse embryonic stem cells are able to differentiate into cardiomyocytes expressing chronotropic responses to adrenergic and cholinergic agents and Ca^{2+} channel blockers. *Differentiation, 48*(3), 173–182.
78. Zigrino, P., Brinckmann, J., Niehoff, A., Lu, Y., Giebeler, N., Eckes, B., et al. (2016). Fibroblast-derived MMP-14 regulates collagen homeostasis in adult skin. *Journal of Investigative Dermatology, 136*(8), 1575–1583.
79. Zimmermann, W. H., Eschenhagen, T. (2005, December). Engineering myocardial tissue. *Circulation Research, 97*(12), 1220–1231.
80. https://en.wikipedia.org/wiki/Focal_adhesion.

Author Biography

Professor Gerhard M. Artmann (Germany) Dr. rer. nat. habil. He received his diploma in physics from the University of Dresden (1974, GDR), his doctorate in physics from the RWTH Aachen (1988, West Germany), and his habilitation from the Technical University of Ilmenau (1998, Germany). In 1983/84 Artmann was for six months a political prisoner in East Germany. He spent three months in strict solitary confinement and another three months in prison with criminal, Communist-educated East German citizens. He was detained for refusing to use the weapon for a totalitarian regime. For Artmann, freedom is the foundation of everything, including science. He is convinced that the commitment to freedom is worth every inch of our courage and deprivation. From 1989 to February 2017 he worked at the University of Applied Sciences Aachen, Germany, as Professor of Biophysics and Bioengineering. His research has focused on fundamental biophysical research as well as the development of advanced human cell measurement techniques. He is currently working with Professor Jürgen Hescheler [77] at the University of Cologne and is researching stem cells with him. G. M. Artmann conducted research in the nineties of the past century at UC San Diego, USA together with the professors Y. C. Fung [27] and Shu Chien [19, 20]. He discovered a temperature transition of hemoglobin of many species and also in humans [2]. Artmann is also a German writer and published numerous short stories, as well as two novels, a poetry book and a radio play. He edited three textbooks on bioengineering in cell research [9]. He is married to Professor Dr. Aysegül Artmann and is grandfather of five very active grandchildren: Mila, Norea, Levi, Janis, and Leleth.

Development in Cell Manipulation Techniques for the Study of Single Cardiomyocyte Mechanics

Gentaro Iribe

1 Background

The heart is an elastic balloon that consists of cardiomyocytes, firboblasts, etc. The mechanical properties of such elastic balloon can be described by its pressure and volume relation. Therefore, it is important to understand the cardiac left ventricular pressure-volume (PV) relationship in order to evaluate the cardiac performance. Cardiac mechanics study based on PV relationship using in situ or excised whole heart preparation has been established in the 1970–80s [31]. Because of its simplicity, this framework is useful in the cardiovascular research field, including clinical study, to evaluate the integrated cardiac performance [4, 8, 30].

Many studies on the cardiac mechanics had used multicellular preparations (whole heart, papillary muscle, etc.). However, there are potential sources of errors in investigating the intrinsic mechanical properties of the contractile component of cardiac muscle by using multicellular preparations. One of the problems to perform mechanics study in a multicellular preparation is the presence of the connective tissue and endothelium, which cause significant modification of the viscoelastic and contractile properties of the tissue [23, 36]. Another considerable complication is intrinsically heterogeneous activity of the cardiac muscle elements that causes non-uniform contractions of the cardiomyocytes [17, 19]. This means that some cells are contracting, while other cells are lengthened even during isovolumic contractions; therefore, it is difficult to evaluate the cell behaviors during the isometric contractions using multicellular preparations. Single cell preparations allow artifacts prevention and analysis of the intrinsic mechanical characteristics of the contractile component.

G. Iribe (✉)
Okayama University Graduate School of Medicine,
Cardiovascular Physiology, Dentistry and Pharmaceutical Sciences,
2-5-1, Shikata-Cho, Kita-Ku, Okayama 700-8558, Japan
e-mail: iribe@okayama-u.ac.jp

G. M. Artmann et al. (eds.), *Biological, Physical and Technical Basics of Cell Engineering*, https://doi.org/10.1007/978-981-10-7904-7_8

The externally homogeneous mechanical activity of the heart results from the integration of heterogeneous cell function [17]. In addition, mechanical activity of the cells influences their electrical activities (mechano-electric coupling) [26]. To study the heterogeneous mechanical properties combined with cellular mechano-electric coupling under physiological loading conditions, researchers need to handle the single cell preparations for measurement of the force-length (FL) relationship, which is the analog of PV relationship in the whole organ.

In this chapter, we briefly review the concept of cardiac mechanics from the viewpoint of the PV relationship, and introduce various methods and recent developments in the cell manipulation technique for studying the dynamic single cardiomyocyte mechanics.

2 Left Ventricular Pressure-Volume Relationship

For the PV relationship under the physiological cardiac contraction, the time course of the ventricular pressure and volume through linkage with its anatomy is important. Figure 1a shows a schematic illustration of the left heart system, which includes the left atrium (LA), the left ventricle (LV), and the aorta. Additionally, two valves, the mitral valve, placed between the LA and the LV, and the aortic valve, placed between the LV and the aorta, are included in this system. These valves ensure one-way flow through the pump by preventing the backward blood flow during the LV cycle of contraction and relaxation.

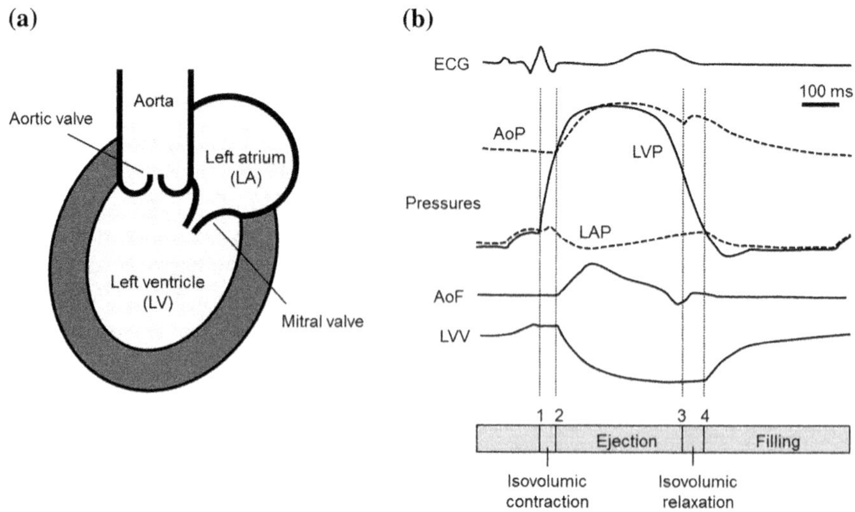

Fig. 1 Schematic diagram of the left heart system (**a**) and time course of the heart contraction (**b**). Modified from [11]

Figure 1b shows a schematic diagram of the time course of changes in the aortic pressure (AoP), aortic blood flow (AoF), LV pressure (LVP), LV volume (LVV), LA pressure (LAP), and electrocardiogram (ECG) during one cardiac beat. Time point 1 in Fig. 1b represents the end-diastolic point that corresponds to the beginning of the LV mechanical contraction. After the LA contraction, the LVP rises sharply as a consequence of the LV active contraction. The LVV remains constant until the aortic valve opens because the LVP is below AoP; thus the blood cannot open the closed aortic valve. Concomitantly, although the LVP is higher than the LAP, the mitral valve closure does not allow backward blood flow into the LA. Therefore, this period is called the isovolumic contraction phase. Time point 2 corresponds to the opening of the aortic valve. The aortic valve opens when the LVP exceeds the AoP. The evolving pressure gradient accelerates AoF until it reaches a maximum. Subsequently, the LVP gradually declines, initially at a similar rate as the AoP. AoF decelerates until the aortic valve closes. This marks the end of the period called the ejection phase. Time point 3 represents the end-systolic point that corresponds to the beginning of the LV relaxation and the closure of the aortic valve. After the aortic valve closure, the LVP decreases rapidly, while the LVV remains constant, until the mitral valve opens. This period is referred to as the isovolumic relaxation. Time point 4 corresponds to the opening of the mitral valve. The mitral valve opens when the LVP decreases below the LAP. The blood flow from the LA to the LV is driven by this pressure gradient. This period is called the filling phase.

The PV relationship offers an important characterization of the mechanical properties of the beating heart. When one plots the changes in the LVP as a function of the LVV, a loop-like shape can be traced, describing the PV relationship (Fig. 2a). The numbers on the 'corners' of this PV-loop correspond to the time labels in Fig. 1b. The PV relation curve shows more clearly the isovolumic phase (1–2) ejection (2–3), the isovolumic relaxation (3–4), and filling (4–1). The width of the PV-loop corresponds to the stroke volume, which is the amount of blood ejected from the heart during one beat.

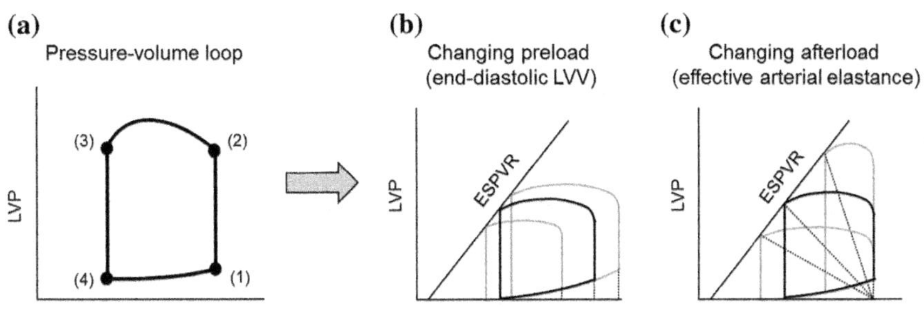

Fig. 2 Pressure-volume loop under physiological loading condition (**a**) and load-dependent changes in loop shape (**b** and **c**). Note that end-systolic pressure-volume relation (ESPVR) is load-independent. LVP, left ventricular pressure; LVV, left ventricular volume

3 Preload and Afterload

Two major components of the mechanical load on the cardiac muscle are the preload and afterload. The preload can be defined as the initial length of the cardiac muscle prior to the contraction. The afterload can be described as the load (i.e. the arterial pressure or the ventricular wall stress) that the cardiac muscle must contract against.

In the late 19th century to the early 20th century, Otto Frank and Ernest H. Starling discovered the "law of the heart", which illustrates the preload-dependency of the cardiac pump function, one of the most fundamental mechanical property of the cardiac muscle. This law states that any parameter that characterize the cardiac performance (cardiac output, systolic pressure, twitch force of the muscle, etc.) depends on the preceding parameters that characterize the preload (end-diastolic LVP and/or LVV, muscle length, etc.). In the PV relation curve, the law of the heart is represented as an increase of the loop area (cardiac external work during one beat) in a similar shape along with the end-diastolic LV volume [33] (Fig. 2b). Although this property of the cardiac muscle is so fundamental and well-known, the subcellular mechanisms that underlie the Frank-Starling law of the heart are still under the debate [29].

The definition of the term "afterload" is confusing. Either the arterial pressure or the arterial resistance (impedance) can be an index of the afterload. Sunagawa et al. showed that the impedance of the arterial system can be approximated to the slope of the relationship between the stroke volume and the ventricular end-systolic pressure, which is called the effective arterial elastance (Fig. 2c).

Suga and Sagawa investigated the cardiac LV pump behavior by further analysis of the PV diagram [27, 32, 33]. Suga et al. reported that in the absence of changes in contractility, the end-systolic PV relationship (ESPVR) is independent of the pre- and afterload variations in excised cross-circulated canine hearts: ESPVR follows a single, linear relationship (Fig. 2b, c).

4 Frog Cardiomyocyte

In the single cardiomyocyte mechanics study, cellular FL relation is an important analog of the PV relationship. Amphibian cardiomyocytes are about 5–10 μm in width and 250–400 μm in length. They are relatively compliant compared to the mammalian cardiomyocytes; thus, their cell length can be easily changed by stretch.

The essential techniques to record the FL relationships for single cardiac cell mechanics study are the cell ends holding in order to apply the stretch and twitch force measurements. Tarr et al. held both cell ends by wrapping each cell end around a glass beam coated with a cell adhesive, poly-L-lysine (Fig. 3a). One of the two glass beams was stiff in order to firmly fix the position of one cell end. Moving the position of the stiff glass beam allowed changes in cell length. The other glass beam was thin and compliant enough in order to use its displacement to measure the contractile force, which was expressed in nano-Newtons. The compliance of the glass beam was calibrated by a force transducer [34, 35]. This technique allows not only mechanics measurements, but also combining the electrophysiological patch clamp measurements by attaching a glass electrode to the cell membrane on the outside of the stiff glass beam [35]. This area is theoretically stabilized and the patch detachment due to the cell movement induced by stretch or contraction can be avoided (Fig. 3a).

Further single cell mechanics studies on isolated frog myocytes had been performed by Tung et al. using different cell holding and force measurement techniques [36] (Fig. 3b). Their originally developed force transducer consisted of a light source with a compliant glass optical fiber force probe; the free end of the probe faced another fixed glass optical fiber that was connected to a photodiode. An isolated single frog ventricular myocyte was held by the suction of each cell end into a tip of glass pipette, and the cell was mounted over the free end of the compliant optical fiber force probe in an inverted U-shaped configuration. The cell contraction bended the optical fiber force probe, causing misalignment of the cross sections of the two optical fibers that resulted in a change of the transmitted light

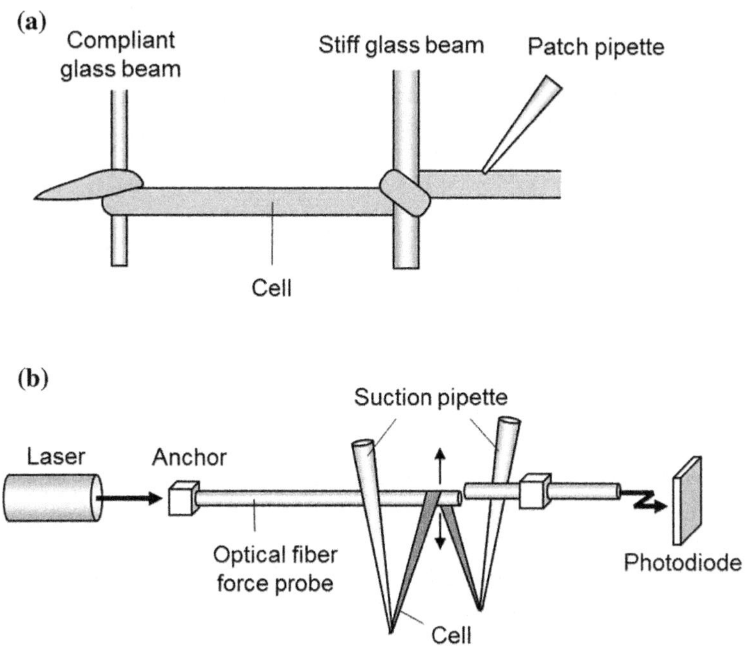

Fig. 3 Examples of cell end holding technique and force measurement in frog cardiomyocytes **a** Cell ends are held by wrapping around poly-L-lysine coated stiff and compliant glass beams and the twitch force is measured based on the displacement of the compliant glass beam. Redrawn from [35]; **b** cell ends are held by suction pipettes and the twitch force is calculated based on the change in transmitted light intensity caused by optic fiber displacement. Redrawn from [36]

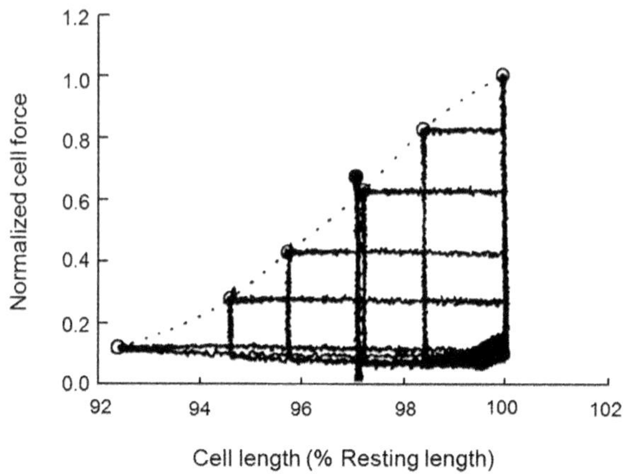

Fig. 4 Physiological work loop style contractions in various loading conditions in a frog cardiomyocyte. Redrawn from [24]

intensity which was proportional to the generated force and was detected by the photodiode. The cell length (preload) could be altered by moving the force probe upward (lengthen) or downward (shorten). To achieve isometric/isotonic contractions, either the cell length or the force signals were clamped using high-gain negative feedback servo-control loops. By electronically switching these modes in sequence, physiological loop style contractions could be mimicked [24] (Fig. 4). This technique revealed that the end-systolic FL relationship (ESFLR) is independent of the type of load in frog cardiomyocytes, contradicting the relationships observed in frog whole heart PV [9].

5 Mammalian Cardiomyocyte

5.1 Cell End Holding with Suction Pipettes

Holding the cell ends in mammalian cardiomyocytes is rather difficult because of their cell dimension. Their cardiomyocytes are shorter (100–150 μm in length), thicker (20 μm in width and about 7 μm in thickness) and, thus, stiffer compared to frog cardiomyocytes, meaning that the cell ends holding technique like wrapping (Fig. 3a) [34, 35] is not applicable. There were several attempts to hold the mammalian cell ends by using suction glass pipettes; however, the success rate of cell stretch was extremely low because of the cell damage due to the high stress on the sarcolemma at the tips of the micropipettes [2]. To reduce such stress at the cell ends, Brady et al. conceived a different holding technique with concentric double-barreled suction pipettes. The inner pipette was slightly recessed to allow the outer pipette to cover 3–5 sarcomeres around the cell ends. Therefore, the tip of the inner pipette firmly attached to the intercalated disk where the tension transfer physiologically occurs when the cell is stretched [3].

5.2 Cell End Holding with Adhesive Substances

Early experiments of cell end attachment with adhesive substances used fibrin [7], poly-L-lysine [28] or silicone adhesive [13]. For the attachment with fibrin glue, a glass pipette loaded with human fibrinogen was brought close to the cell end. The cell was placed in a perfusion chamber that contained bovine thrombin. Subsequently, a small amount of fibrinogen was pressure-ejected from the pipette to form fibrin that connected the glass pipette and the cell end. Usually, fibrin covered approximately 20–30% of the cell surface. In the attachment method using poly-L-lysine, an isolated cell was put on a pair of beveled glass rods coated with poly-L-lysine. To assure the cell attachment with poly-L-lysine electrostatic charge, 65% of the cell surface was covered with the two beveled glass rods. For the attachment of cell ends using silicone adhesive, 45 min were necessary to cure the silicone adhesive before the measurements. All of these methods are currently rarely used.

Very recently, a newly developed biocompatible adhesive (MyoTak™; Ionoptix, Milton, MA, USA) became available for cell attachment [12, 25]. MyoTak is a composition of extracellular matrix proteins (laminin, entactin, etc.). Before coating the glue, the tips of glass micro-rods are pre-coated with an aluminum silicate suspension and dried. Subsequently the tips are dipped in MyoTak. After the glue drying in the air for several minutes, the glue-coated rod shall to be kept wet, so that it can hold cell ends for up to several hours. In the most of the glue-based cell attachment methods except MyoTak, the glue-coated pipettes could be used only once, because of the cell debris stuck to the tool after the measurements. Adhesive power of MyoTak is not "too strong." Therefore, lifting the glue-coated rod out of the chamber solution can remove the stuck cell debris by the surface tension at passage through the liquid-air interface, allowing another cell measurement.

5.3 Cell End Holding with Carbon Fibers

Although various theoretically feasible cell holding techniques have been developed in early studies, the success rate of the measurements has been low in mammalian cardiomyocytes until carbon fiber technique was introduced. In the late 1980s, Le Guennec et al. developed a technique that used carbon fibers attached to the opposite cell ends to manipulate the preload. The carbon fiber-based stretching gave rise to a relatively homogeneous increase in the sarcomere length, and allowed the calculation of the passive and active forces from optically monitored changes in the carbon fiber bending [20].

In this technique, thin carbon fibers (10–12 μm in diameter), obtained by high temperature treatment of polyacrilonitryl, were used for cell ends holding. Usually, either a pair of the double short fibers [5, 10, 37, 38] or a different type of a thick and rigid fiber [21, 22, 39, 40] was attached to one cell end to fix the attachment point. The other cell end was attached with a single carbon fiber, which was compliant, and therefore, could be used for force calculation from its bending (Fig. 5a, b). Iribe et al. developed a bi-directional carbon fiber-based FL control system, which used cells that are suspended from two compliant carbon fibers. In this setting, the cell center remained stationary relative to the optical axis, thereby avoiding sarcomere blurring, which might happen during the "single-sided displacements," and improved resolution of the optical detail (Fig. 5c) [14].

Carbon fibers can be attached simply by touching the cell surface. Although the actual mechanism of the attachment is not clear, the electrostatic forces between the carbon fiber surface and the myocyte surfaces are most likely implied [37]. However, in practice, this initial attachment is not firm enough to stretch the mammalian cardiomyocytes. To reinforce this attachment, the carbon fiber needs to be gently pressed several times, then left unstimulated for several minutes [37], or electrically stimulated at 2–4 Hz for 3 min under gentle press [14].

Reusability is one of the advantages of carbon fiber technique. Similar to the MyoTak-coated glass micro-rod, cell debris that sticks to the carbon fibers during measurements can be removed by lifting the fiber out of the chamber solution.

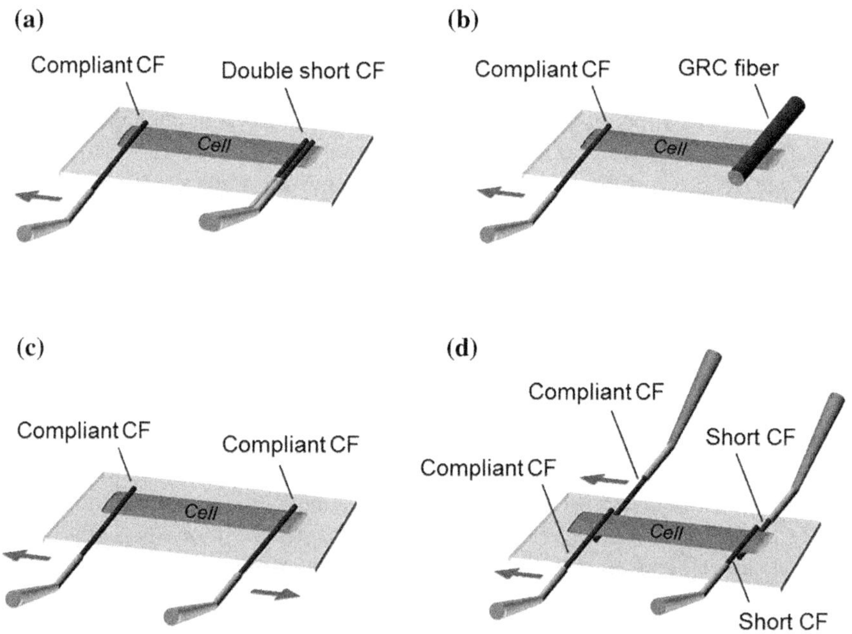

Fig. 5 Examples of cell end holding using carbon fiber (CF) technique **a** double short fibers are attached to fix one cell end, and the other cell end is attached with a compliant single CF for force measurements; **b** fixed cell end is held by a Graphite Reinforced by Carbon (GRC) fiber, which attaches to the cell surface more firmly compared to the conventional CF; **c** both cell ends are held by compliant CFs for bi-directional force-length control; **d** each cell end is held by two CFs attached to both upper and lower cell surfaces for more attachment force

When the debris is difficult to remove by this method, dipping the fiber into the NaOH solution for several hours can clean the fiber without affecting its adhesive ability. Therefore, the carbon fiber normally can be reused until it is broken.

Some further developments of the carbon fiber technique have been made to further improve the cell attachment. White et al. reported a higher probability of successful attachment by placing the fibers in front of a TV and discharging the screen before each attempt of attachment in order to enhance the electrostatic adhesion [37]. Yasuda et al. used a newly developed Graphite Reinforced by Carbon (GRC) fiber that contained many fine graphite crystal granules to increase its effective surface area and surface charges. They found that GRC fiber attached to the cell surface much more easily and firmly compared to the conventional carbon fiber. In these methods, the cells were held by a pair of carbon fibers attached to the upper surface. However, cell end holding force in these conventional methods is not enough to apply a stable stretch beyond 2.0 μm in end-diastolic sarcomere length. Even in case of successful cell attachment, extensive axial stretch in these methods may cause inhomogeneous cell stretch (longer sarcomeres near upper cell surface). To overcome this limitation, Iribe et al. attached two carbon fibers to both upper and lower cell surfaces to improve the cell holding and apply more stretching force (Fig. 5d) [16]. This method allows single cell mechanics study under a wider range of mechanical loading conditions than was previously possible with the conventional carbon fiber technique.

Stiffness of the fiber and the amount of cell shortening (amount of the fiber bending) determine the force applied to the fiber. The fiber bending is calculated from the position of the tip of the carbon fiber, which is detected from the optically obtained cell image, and the position of the carbon fiber holder, which is usually obtained from the position signal of a piezo translator to which the carbon fiber holder is mounted. The difference between the position of the fiber tip and the mount position yields fiber bending (Fig. 6). Stiffness (K) of the carbon fiber is determined by the length (L) and the diameter (D) of the carbon fiber, and its modulus of elasticity ($E = 32.784 \times 10^9$ N/m^2), according to the following equation [38].

$$K = 3\pi ED^4 \times \left(64L^3\right)^{-1}$$

The diameter of the fiber is supposed to be constant; therefore, the appropriate length of the fiber is the only variable to yield a fiber with optimal stiffness to measure twitch force of single cardiomyocytes. However, the fibers are not usually manufactured with a precisely controlled diameter along the entire length; therefore, the actual fiber stiffness calibrated by a force transducer with known stiffness may differ from the theoretical value calculated from the equation above.

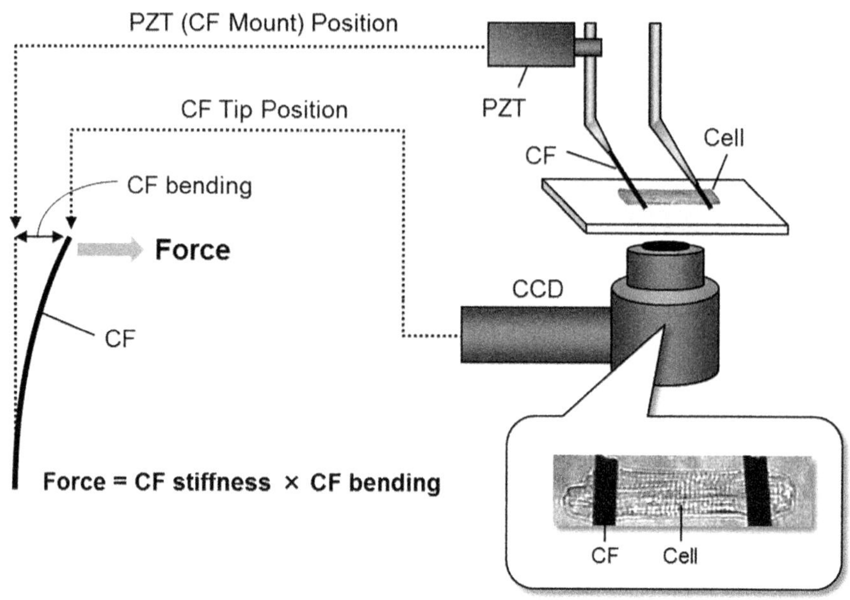

Fig. 6 Force measurement using carbon fiber technique. Force applied to the carbon fiber (CF) is determined by CF bending, which is equal to the distance between CF tip and mount position, and CF stiffness. PZT, piezo translator; CCD, charge-coupled device

5.4 *Afterload Control in Carbon Fiber Technique*

Early studies using carbon fiber technique did not control position of the carbon fibers during the contraction. Hence, as the cell length shortened, the carbon fiber bent more during the contraction, meaning that the contractile force increased. This setting has been successfully used for the characterization of the cardiomyocyte mechanical properties during the auxotonic contractions [1, 5, 6, 10, 15, 18, 37, 38]. In the auxotonic contraction setting, the afterload (the slope of the relation of cell shortening and end-systolic force, which is an analog of the effective arterial elastance in the whole heart setting) is identical to the stiffness of the carbon fiber, and therefore it cannot vary.

Although the carbon fiber technique enabled various preloads in single cell mechanics study, the afterload control in the mammalian cardiomyocytes using the carbon fiber technique has been challenging. In carbon fiber technique, the optically obtained signal of the position of the tip of the carbon fiber is essential for calculating both cell length and force signals. In order to change the afterload (i.e. the effective stiffness of the carbon fiber), the carbon fiber position (i.e. the distance between the carbon fibers, or cell length) needs to be controlled using real-time feedback control loops. If the negative feedback gain is large enough to compensate the cell shortening, the cell contracts in an isometric manner and the carbon fiber is virtually stiffened. If the positive feedback gain is large enough to compensate carbon fiber bending, the cell contracts in an isotonic manner and the carbon fiber is virtually softened.

Although the real-time feedback control is theoretically feasible, it is difficult in practice because of the noise in the signal of the carbon fiber position. Vibration of the carbon fiber due to the amplified noise signal may kill the cell. In frog myocytes, it is feasible to use high-gain feedback control loops because long twitch duration (1000–1500 ms) in frog myocytes is advantageous for the signal-to-noise ratio in length signal [36] compared to the mammalians (rodents in many cases), in which the twitch duration is approximately 100–200 ms. In spite of the difficulties, Yasuda et al. reported a successful afterload control with the real-time analog feedback control system for isometric and isotonic contractions in rat myocytes [39, 40]. However, they confirmed that the real-time feedback control was very difficult in their further attempt to reproduce the physiological work-loop style contractions, in which the isometric mode and isotonic mode should be switched with feedback control during one beat. Instead, they used an adaptive feedback control on a per beat basis. In this method, the feedback command for the next beat is calculated from the difference between either force or length curve of the current beat and that of the aimed mode (isotonic or isometric) with an empirically determined gain value to reach the convergence within three or four contractions [21]. This method enabled afterload control in order to reproduce the physiological work-loop style contraction. The resultant loop shape still shows unclear borders between the isometric and isotonic phases (corresponding to the point 2 and 3 in Fig. 2a), indicating the difficulty in afterload control with feedback using carbon fiber technique (Fig. 7a).

Another successful attempt to reproduce the work-loop style contraction in mammalian cardiomyocytes used an adaptive feed-forward control approach [14]. In this method, the length signal was recorded for 10 beats and averaged to use as the source signal for piezo command signal design to achieve the aimed loading states. For isometric contractions, the negated source signal was applied to induce

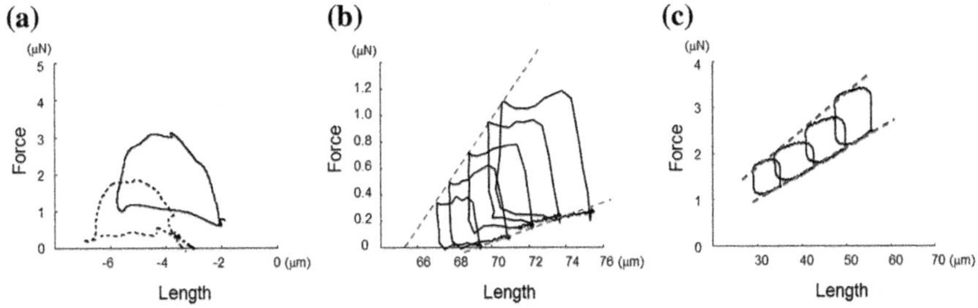

Fig. 7 Physiological work loop style contractions in various loading conditions with different afterload control methods in mammalian cardiomyocytes. **a** Adaptive feedback control with carbon fiber technique in a rat myocyte. Redrawn from [21]; **b** feed-forward control with carbon fiber technique in a guinea pig myocyte. Redrawn from [14]; **c** real-time feedback control with a laser interferometry force transducer in a rat myocyte. Redrawn from [12]

an outwardly directed counter-movement of the piezo translator, and the gain was adjusted until the carbon fiber tip position remained near-constant. For isotonic contractions, the source signal was applied to induce an inwardly directed movement of the piezo translator, and the gain was adjusted to minimize the changes in carbon fiber bending. The piezo command for the physiological work-loop style contraction consisted of four commands of outward/inward movements in order to achieve four phases of the isometric/isotonic contraction/relaxation. Each command was a part of the inversely or normally gained source signal. These four gains were manually adjusted until the desired work-loop conditions were obtained. Adaptive feed-forward approach is advantageous from the viewpoint of noise resistance, and therefore, the stable control of the afterload, allowing clearly reproduction of the four phases in the work-loop (Fig. 7b). Using this technique, single cell mechanics during physiological work-loop style contractions in isolated guinea pig cardiomyocytes were investigated in detail [14]. Although feed-forward approach gives stable afterload control, the manual gain adjusting process requires a fair amount of experience for prompt measurements.

5.5 Development in Force Measurement

The force measurements using carbon fiber bending is an established technique that is currently most used in this field because of its simplicity and reliability. However, the accuracy of the measured force value itself depends on the accuracy of the force calibration by the force transducer. Conventional commercially available force transducers are usually optimized for the measurement of forces in the scale of mN, while single myocyte twitch force ranges up to several μN. Therefore, carbon fiber force calibration using these force transducers suffers from a large noise level in this force range. Obviously, the direct twitch force measurements using these force transducers suffers from same problem.

Recently, Helmes et al. developed a new optical force transducer, which measures the bending of a cantilever as an indicator of force using laser interferometry like in an atomic force microscopy system [12]. This force transducer system allowed forces measurement with up to 0.5–10 μN with 5–10 nN accuracy, which depended on the stiffness of cantilevers. For instance, with a 20 N/m stiffness cantilever, the measurable force had a force resolution of 20 nN on a 2 μN range with a noise band of 15 nN. The cantilever was a small rectangular sheet of gold-coated glass with a high resonance frequency ($\sim$7 kHz). This extremely low noise level and high resonance frequency compared to the carbon fiber technique allows the afterload control in real-time feedback loop (Fig. 7c). Contrary to the feed-forward control used in the carbon fiber technique, this system enables instantaneous control of different pre- and afterload conditions.

6 Summary

Although the majority of research focused at load-dependent characterization of cardiac mechanics has been conducted in whole heart preparations, isolated cardiomyocytes are an advantageous preparation in this field because sub-cellular mechanisms can be assessed in greater detail in single cardiomyocytes, compared to intact tissue preparations. Since carbon fiber technique was introduced, the technique has been improved and mammalian single cell mechanics study is now practically feasible. A recently developed optical force transducer system and cell-end holding technique using MyoTak enable instantaneous force/length feedback control.

Cardiac myocytes are required to sequentially contract and relax through their existence. In this context, beat-by-beat adaptation to dynamically changing mechanical loads is one of the core physiological requirements for cardiomyocytes. The ability to apply such a physiological loading condition to single cell preparation provides a further method for better understanding of "real" cardiac physiology.

Acknowledgements The author thanks to Dr. Yohei Yamaguchi, Dr. Toshiyuki Kaneko and Dr. Anastasia Khokhlova for sharing their wide experience in single cell mechanics experiments.

References

1. Bollensdorff, C., Lookin, O., & Kohl, P. (2011). Assessment of contractility in intact ventricular cardiomyocytes using the dimensionless "Frank-Starling Gain" index. *Pflugers Archiv: European Journal of Physiology, 462,* 39–48.
2. Brady, A. J. (1991). Mechanical properties of isolated cardiac myocytes. *Physiological Reviews, 71,* 413–428.
3. Brady, A. J., & Farnsworth, S. P. (1986). Cardiac myocyte stiffness following extraction with detergent and high salt solutions. *American Journal of Physiology, 250,* H932–H943.
4. Burkhoff, D. (2013). Pressure-volume loops in clinical research. *Journal of the American College of Cardiology, 62,* 1173–1176.
5. Cazorla, O., Le Guennec, J. Y., & White, E. (2000). Length–tension relationships of sub-epicardial and sub-endocardial single ventricular myocytes from rat and ferret hearts. *Journal of Molecular and Cellular Cardiology, 32,* 735–744.
6. Chuan, P., Sivaramakrishnan, S., Ashley, E. A., & Spudich, J. A. (2012). Cell-intrinsic functional effects of the α-cardiac myosin Arg-403-Gln mutation in familial hypertrophic cardiomyopathy. *Biophysical Journal, 102,* 2782–2790.
7. Copelas, L., Briggs, M., Grossman, W., & Morgan, J. P. (1987). A method for recording isometric tension development by isolated cardiac myocytes: Transducer attachment with fibrin glue. *Pflugers Archiv: European Journal of Physiology, 408,* 315–317.
8. Delnoy, P. P. H. M., Ottervanger, J. P., Luttikhuis, H. O., Vos, D. H. S., Elvan, A., Ramdat Misier, A. R., et al. (2009). Pressure-volume loop analysis during implantation of biventricular pacemaker/cardiac resynchronization therapy device to optimize right and left ventricular pacing sites. *European Heart Journal, 30,* 797–804.
9. Frank, O. (1990). The basic shape of the arterial pulse. First treatise: Mathematical analysis. *Journal of Molecular and Cellular Cardiology, 22,* 255–277.

10. Gannier, F., White, E., Garnier, D., & Le Guennec, J. Y. (1996). A possible mechanism for large stretch-induced increase in $[Ca^{2+}]_i$ in isolated guinea-pig ventricular myocytes. *Cardiovascular Research, 32,* 158–167.
11. Guyton, A. C., & Hall, J. E. (2006). *Textbook of medical physiology* (11th ed.). Philadelphia: Elsevier Saunders.
12. Helmes, M., Najafi, A., Palmer, B. M., Breel, E., Rijnveld, N., Iannuzzi, D., et al. (2016). Mimicking the cardiac cycle in intact cardiomyocytes using diastolic and systolic force clamps; measuring power output. *Cardiovascular Research, 111,* 66–73.
13. Hofmann, P. A., Hartzell, H. C., & Moss, R. L. (1991). Alterations in Ca^{2+} sensitive tension due to partial extraction of C-protein from rat skinned cardiac myocytes and rabbit skeletal muscle fibers. *Journal of General Physiology, 97,* 1141–1163.
14. Iribe, G., Helmes, M., & Kohl, P. (2007). Force-length relations in isolated intact cardiomyocytes subjected to dynamic changes in mechanical load. *American Journal of Physiology Heart and Circulatory Physiology, 292,* H1487–H1497.
15. Iribe, G., Kaihara, K., Ito, H., & Naruse, K. (2013). Effect of azelnidipine and amlodipine on single cell mechanics in mouse cardiomyocytes. *European Journal of Pharmacology, 715,* 142–146.
16. Iribe, G., Kaneko, T., Yamaguchi, Y., & Naruse, K. (2014). Load dependency in force-length relations in isolated single cardiomyocytes. *Progress in Biophysics and Molecular Biology, 115,* 103–114.
17. Katz, A. M., & Katz, P. B. (1989). Homogeneity out of heterogeneity. *Circulation, 79,* 712–717.
18. King, N. M. P., Methawasin, M., Nedrud, J., Harrell, N., Chung, C. S., Helmes, M., et al. (2011). Mouse intact cardiac myocyte mechanics: cross-bridge and titin-based stress in unactivated cells. *Journal of General Physiology, 137,* 81–91.
19. Krueger, J. W., & Pollack, G. H. (1975). Myocardial sarcomere dynamics during isometric contraction. *Journal of Physiology, 251,* 627–643.
20. Le Guennec, J. Y., Peineau, N., Argibay, J. A., Mongo, K. G., & Garnier, D. (1990). A new method of attachment of isolated mammalian ventricular myocytes for tension recording: length dependence of passive and active tension. *Journal of Molecular and Cellular Cardiology, 22,* 1083–1093.
21. Nishimura, S., Yasuda, S., Katoh, M., Yamada, K. P., Yamashita, H., Saeki, Y., et al. (2004). Single cell mechanics of rat cardiomyocytes under isometric, unloaded, and physiologically loaded conditions. *American Journal of Physiology Heart and Circulatory Physiology, 287,* H196–H202.
22. Nishimura, S., Kawai, Y., Nakajima, T., Hosoya, Y., Fujita, H., Katoh, M., et al. (2006). Membrane potential of rat ventricular myocytes responds to axial stretch in phase, amplitude and speed-dependent manners. *Cardiovascular Research, 72,* 403–411.
23. Ohayon, J., & Chadwick, R. S. (1988). Effects of collagen microstructure on the mechanics of the left ventricle. *Biophysical Journal, 54,* 1077–1088.
24. Parikh, S. S., Zou, S. Z., & Tung, L. (1993). Contraction and relaxation of isolated cardiac myocytes of the frog under varying mechanical loads. *Circulation Research, 72,* 297–311.
25. Prosser, B. L., Ward, C. W., & Lederer, W. J. (2011). X-ROS signaling: rapid mechano-chemo transduction in heart. *Science, 333,* 1440–1445.
26. Quinn, T. A., Kohl, P., & Ravens, U. (2014). Cardiac mechano-electric coupling research: Fifty years of progress and scientific innovation. *Progress in Biophysics and Molecular Biology, 115,* 71–75.
27. Sagawa, K. (1981). The end-systolic pressure-volume relation of the ventricle: definition, modifications and clinical use. *Circulation, 63,* 1223–1227.
28. Shepherd, N., Vornanen, M., & Isenberg, G. (1990). Force measurements from voltage-clamped guinea pig ventricular myocytes. *American Journal of Physiology, 258,* H452–H459.
29. Shiels, H. A., & White, E. (2008). The Frank-Starling mechanism in vertebrate cardiac myocytes. *Journal of Experimental Biology, 211,* 2005–2013.

30. Steendijk, P., Tulner, S. A., Bax, J. J., Oemrawsingh, P. V., Bleeker, G. B., van Erven, L., et al. (2006). Hemodynamic Effects of Long-Term Cardiac Resynchronization Therapy: Analysis by Pressure-Volume Loops. *Circulation, 113,* 1295–1304.
31. Suga, H. (1990). Ventricular energetics. *Physiological Reviews, 70,* 247–277.
32. Suga, H., & Sagawa, K. (1974). Instantaneous pressure-volume relationships and their ratio in the excised, supported canine left ventricle. *Circulation Research, 35,* 117–126.
33. Suga, H., Sagawa, K., & Shoukas, A. A. (1973). Load independence of the instantaneous pressure-volume ratio of the canine left ventricle and effects of epinephrine and heart rate on the ratio. *Circulation Research, 32,* 314–322.
34. Tarr, M., Trank, J. W., Leiffer, P., & Shepherd, N. (1979). Sarcomere length-resting tension relation in single frog atrial cardiac cells. *Circulation Research, 45,* 554–559.
35. Tarr, M., Trank, J. W., & Goertz, K. K. (1986). Voltage-tension relations in single frog atrial cardiac cells. *Circulation Research, 59,* 447–455.
36. Tung, L., & Morad, M. (1988). Contractile force of single heart cells compared with muscle strips of frog ventricle. *American Journal of Physiology, 255,* H111–H120.
37. White, E., Le Guennec, J. Y., Nigretto, J. M., Gannier, F., Argibay, J. A., & Garnier, D. (1993). The effects of increasing cell length on auxotonic contractions; membrane potential and intracellular calcium transients in single guinea-pig ventricular myocytes. *Experimental Physiology, 78,* 65–78.
38. White, E., Boyett, M. R., & Orchard, C. H. (1995). The effects of mechanical loading and changes of length on single guinea-pig ventricular myocytes. *Journal of Physiology, 482,* 93–107.
39. Yasuda, S. I., Sugiura, S., Kobayakawa, N., Fujita, H., Yamashita, H., Katoh, K., et al. (2001). A novel method to study contraction characteristics of a single cardiac myocyte using carbon fibers. *American Journal of Physiology Heart and Circulatory Physiology, 281,* H1442–H1446.
40. Yasuda, S., Sugiura, S., Yamashita, H., Nishimura, S., Saeki, Y., Momomura, S., et al. (2003). Unloaded shortening increases peak of Ca2+ transients but accelerates their decay in rat single cardiac myocytes. *American Journal of Physiology Heart and Circulatory Physiology, 285,* H470–H475.

Author Biography

Professor Gentaro Iribe, M.D., Ph.D. (Japan) After an 8-year career as an anesthesiologist, Dr. Gentaro Iribe studied whole-heart mechano-energetics with Dr. H. Suga at Okayama University, Japan. Subsequently he moved to the UK to study single cardiac cell mechanics in combination with computer modeling and received his Ph.D. in medicine from the University of Oxford, working with Dr. P. Kohl. He then returned to Okayama University, where he examined cardiac mechano-electric coupling, working with Dr. K. Naruse. He is an associate professor at Okayama University since 2016.

Growth Modelling Promoting Mechanical Stimulation of Smooth Muscle Cells of Porcine Tubular Organs in a Fibrin-PVDF Scaffold

Minh Tuấn Dương, Volker Seifarth, Ayşegül Artmann, Gerhard M. Artmann and Manfred Staat

1 Introduction

In modern industrial countries, live expectancy has increased due to improved medical standards, hence, a whole pallete of organ abnormalities, disorders or trauma can be cured [1, 2]. However, organ maldevelopment and injury of the genitourinary tract and especially the ureters and urethras can result in severe complications that may badly influence quality of life. Reconstructive surgery and tissue replacements like tubularized intestine segments for ureter reconstruction often bring about for example stricture formation, fistula, stenosis, obstruction, metaplasia, hydro-(neoureter) nephrosis, mucous production and/or chronic inflammatory reactions [3, 4].

Furthermore, there is a demand for developing alternative reconstruction methods in the field of urology. Novel promising techniques in reconstructive surgery deal with Tissue Engineering (TE) procedures on which for example autologous cells, taken from bladder biopsies, are seeded on tubular scaffolds to create bio-artificial tissue constructs [5–7]. In addition, a bioreactor may be a promising technology to generate also an almost functional tissue substitute in

M. T. Dương (✉) · V. Seifarth · A. Artmann · G. M. Artmann · M. Staat
Institute for Bioengineering, University of Applied Sciences Aachen, Heinrich-Mußmann-Str. 1, 52428 Jülich, Germany
e-mail: minhtuan.duong9@gmail.com

G. M. Artmann
e-mail: g.artmann@gmx.net

M. T. Dương
University of Erlangen-Nuremberg, Immerwahrstraße 1, 91058 Erlangen, Germany

M. T. Dương
Hanoi University of Science and Technology, No. 1, DaiCoViet Road, Hanoi, Vietnam

G. M. Artmann et al. (eds.), *Biological, Physical and Technical Basics of Cell Engineering*, https://doi.org/10.1007/978-981-10-7904-7_9

mimicking the physiological surrounding of the tissue of interest [8] in which the bioreactor gives a stimulus to the cells to copy the nature. For urological TE purposes, Qiang Fu et al. demonstrated that mechanical extension stimulation can improve the viability of adipose derived stem cells (ADSCs) induced to differentiate into SMCs and form a muscular tube of urethral TE [9]. However, a cellular alignment in both circular and longitudinal directions could not yet be achieved.

Moreover, computational growth modelling is often adopted to predict the tendency of cell growing or atrophying under loading conditions. For soft tissue growth, density is mostly considered to be constant, meanwhile changes take place in volume. The decomposition of the total growth deformation into a growth part and an elastic part, applied to a material characterized by a 3D (three-dimensional) simple Fung-type potential [10]. Moreover, the multiplicative decomposition of the deformation gradient into elastic deformation and growth part for isotropic, transversely isotropic and orthotropic biomaterials were presented in the framework of finite deformation [11]. The idea for isotropic growth for an isotropic material was then realized by [12] with their algorithmic realisation in a finite element setting in which the isotropic stretch ratio is introduced as internal variable at the integration point level. Subsequently, this idea was applied to analyse the isotropic growth for arterial walls and in-stent restenosis related to atherosclerosis [13]. However, the material describing the arterial wall is the neo-Hookean which is not realistic because arteries are anisotropic [14]. In this paper, we first investigate the smooth muscle cells in a tubular fibrin scaffold by mechanical stimulation and then computational growth modelling is analysed with two growth models, which are isotropic growth and transversely isotropic growth for isotropic muscle cells in order to accurately simulate and observe behaviour of cells.

2 Mechanical Stimulation

2.1 Materials and Methods

2.1.1 Cell Isolation

Smooth muscle cells were isolated from porcine bladders that were obtained from a local abattoir, dissected and sliced into pieces of about 1 mm^3 as in [15]. Tissue samples (20) were placed in T75 flasks (Nunc, Germany) and incubated with a standard cell culture medium (provided by Invitrogen and Sigma Aldrich, Germany). Medium was changed regularly after three to four days and cells were passaged up to four times, before they were applied in fibrin scaffolds.

2.1.2 Fibrin Gel Preparation

Fibrinogen from human plasma (Merck Millipore, Germany) was prepared as described in [16]. Fibrinogen was dissolved in double distilled water (ddH_2O) at 25 mg/mL, before it was dialyzed over night against Trizma buffered saline (TBS), using a dialysis membrane with a cut off of 6000–8000 MW (Carl Roth, Germany). After sterile filtration, the fibrinogen concentration was determined with a spectrophotometer (Jasco, Germany) at 280 nm and an extinction coefficient of 1.55 [17].

Fibrin gels (n = 14) were prepared with 3 mL fibrinogen (12.5 mg/mL), 2.1 mL TBS that include 3×10^7 cells, 450 μL thrombin (40 IU/mL) and 450 μL calcium chloride (50 mM). In the end, a fibrinogen concentration of 6.25 mg/mL was achieved with a cell count of 5×10^6 cells/mL. Biohybrids were cultured with the standard cell culture medium supplemented with 1 mM L-ascorbic acid 2-phosphate (Sigma Aldrich, Germany) to enhance the tissue and collagen secretion and with 0.5 mg/mL tranexamic acid (Pfizer, Germany) to prevent fibrinolysis [18, 19].

2.1.3 Scaffold Preparation and Cell Seeding

Figure 1 shows a casting mould for the preparation of a tubular composite scaffold made up of fibrinogen and polyvinylidene fluoride (PVDF). The PVDF mesh (RWTH Aachen, Germany) formed the inner layer of the tubular structure. This mould was produced out of Teflon and polyoxymethylene (POM) and was manufactured at Aachen University of Applied Sciences, Germany.

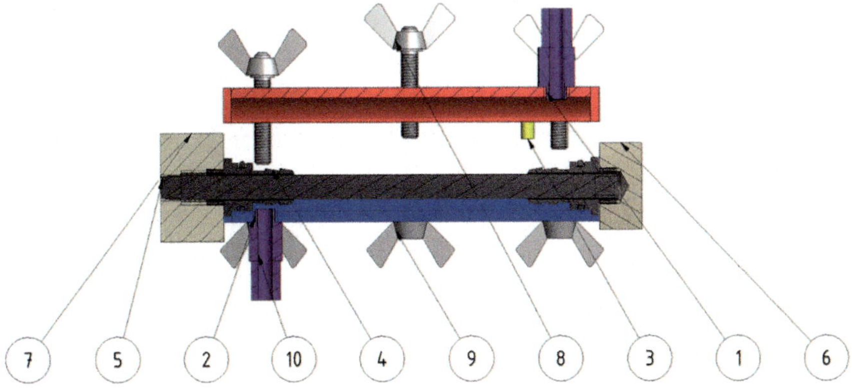

Fig. 1 Casting mould for composite biohybrids consisting of: top (*1*), bottom (*2*), pins for positioning (*3*), fittings (*4*), central rod (*5*), front cap (*6*), end cap (*7*), wing bolts/nuts (*8*)/(*9*) and Luer adapters (*10*) [21]

The fibrinogen solution and the cell suspension were injected into the mould by a two-chamber application system (Medmix, Switzerland). The polymerization inside the mould starts within several seconds and results in a 2 mm thick and 75 mm long tubular structure with an inner diameter of 6 mm. The biohybrid was transferred into the bioreactor after 45 min of polymerization at room temperature.

2.1.4 Mechanotransduction

A kyphoplasty catheter (Joline, Germany) was inserted into the lumen of the biohybrid in order to mimic a urine drop. The catheter was inflated by a syringe pump and was pulled through the tube by a linear actuator. The mechanical experimental stimulator system was partially manufactured by the HITEC Zang GmbH (Herzogenrath, Germany). Subsequently, the catheter was deflated and pushed forward, again. This mechanical stimulation was performed with a frequency of 0.015 Hz. The inflation of the catheter followed a sigmoidal waveform as it is depicted in Fig. 2 with a maximal strain of 20% [20, 21].

2.1.5 Histological Analysis

Tissue fragments were fixed in Carnoy's fluid for at least 8 h, before they were dehydrated and embedded in paraffin. Dehydration, embedding, slicing in 5 μm sections and haematoxylin and eosin (H&E) staining were performed.

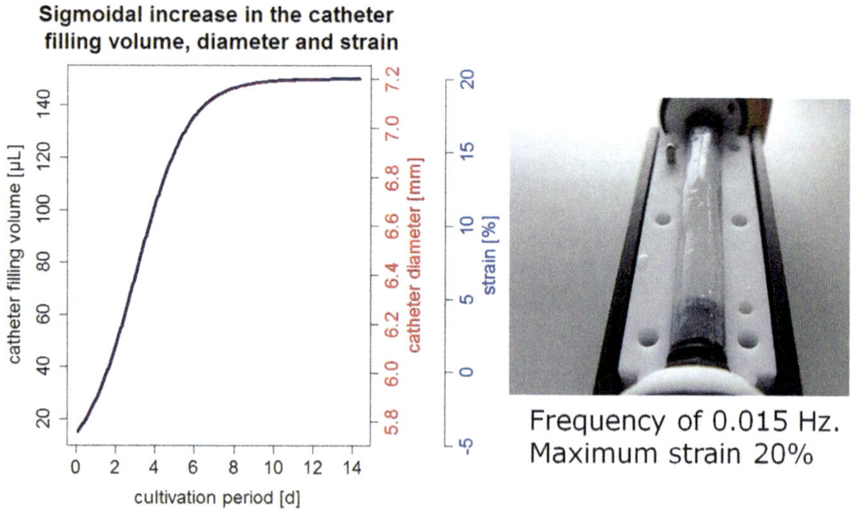

Fig. 2 A sigmoidal waveform and a casting of scaffold [21]

The analysis of the alignment of cells in histological cross sections and scanning electron microscopic pictures from the outer surface of the tubes was performed with LabVIEW (National Instruments, USA). Cellular and structural orientations were analysed according to the size of the interquartile ranges.

2.1.6 Biochemical Analysis

Diffusion experiments were performed as previously described in [15]. Samples of Krebs buffer with a volume of 1 mL were taken after 0.5, 1, 1.5, 2, 3, 4, 5, 6, 7, 8 h. The sample volume was replaced by fresh buffer and the concentration of urea and creatinine were then determined (Uniklinik RWTH Aachen, Germany).

2.1.7 Mechanical Properties

Mechanical properties were analysed by the rupture pressure of the tissue fragments. A chamber was constructed for inflation test at bursting pressure (Aachen University of Applied Sciences, Germany).

Five to ten cm long tubular structures were sliced longitudinally and were arranged in the pressure chamber in a way that an increasing pressure was applied onto the extraluminal surface of the biohybrids. A syringe pump was controlled and the data acquisition were performed with an interface (HiTec Zang GmbH, Germany).

2.2 *Results and Discussion*

Figure 2 shows homogeneous, bubble free 75 mm long fibrin-VDF composite prosthesis that polymerized for 45 min in the casting mould at room temperature [22]. For mechanical stability, the rupture pressure revealed that there are no significant differences between mechanically stimulated and unstimulated prosthesis. Table 1 indicates that there was a tendency to increasing mechanical stability for stimulated prostheses. A similar tendency was observed for extended incubation periods. The rupture pressure increased after two weeks of incubation and with mechanical stimulation. On cellular orientation, histological cross sections stained with HE show an orientation of cells for stimulated prosthesis in comparison to untrained prosthesis. This can be observed due to elongated cell morphologies, see Fig. 3. Scanning electron microscopy from the outer tubular surface exposed that mechanically stimulated tissues show a preferred orientation of cells along the tubular structure of the prosthesis [8]. Obviously, the intensity of mechanical stimulation is dependent on various parameters and different models: the applied cell type, cell source, cell passage as well as in the experimental setup. Nevertheless, it is seen to be stress level and cell type specific.

Table 1 Bursting pressure[21]

Incubation period	$P_{\text{untrained}}$ [mbar]	P_{trained} [mbar]
One week	1210 ± 310	1530 ± 50
Two weeks	1450 ± 530	2120 ± 200

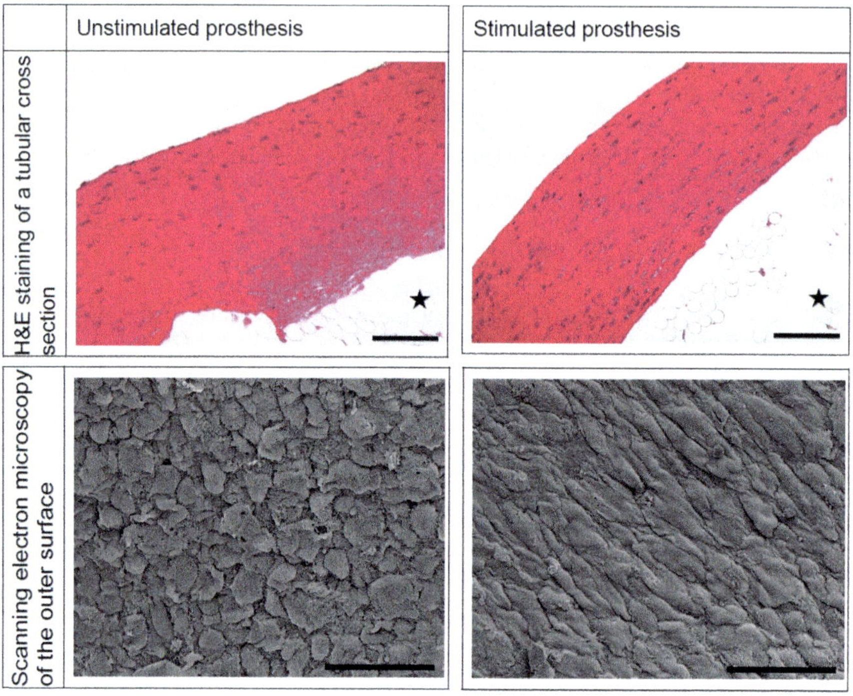

Fig. 3 Morphological appearance of mechanically unstimulated (left) and stimulated (right); H&E staining [21]

Tissue engineering as multidisciplinary field in science might lead to new technologies for the replacement of defect genitourinary tract organs. For that reason, different cultivation methods as well as scaffold materials are continuously tested and developed. The application of mechanical forces plays a crucial role for the tissue generation and orientation of cells. Cells react on mechanical stresses and control the remodelling of the tissues composition and mechanical properties [23–26]. Smooth and continuous tubular composite biohybrids can be prepared reliable and are useful for the application in the bioreactor. Even when the PVDF mesh material is used as luminal structure that is not integrated in the tubular wall as seen by others [27–29]. This was done in order to stabilize the tubular structure and to prevent a damage of the gel by the kyphoplasty catheter. Up to date, bioreactors for urological applications can be found nearly exclusively for bladder wall substitutes of planar scaffold materials. Studies on bladder wall substitutes show benefits with increased cellular proliferation rates for dynamic cultivation conditions, mimicking a physiological intravesical pressure situation [30, 31]. For that reason, planar scaffold materials are deflected frequently. The presented study shows a novel stimulation procedure mimicking a urine bolus in a seamless, tubular scaffold material. Therefore, a kyphoplasty balloon catheter was used to simulate a physiologic movement of a urine bolus. To our knowledge, no comparable bioreactor system can be found in literature.

To date, tubular biohybrids are cultivated and trained in bioreactors, where the cell culture medium is transferred by peristaltic pumps leading to increased intraluminal pressures and diameters [26, 32]. Another method is to stretch ring shaped constructs with hooks [33]. Both methods lead to the same result with circularly arranged cells and improved mechanical properties, which is comparable to the situation in this study.

3 Computational Growth Modelling

The growth of cells in the fibrin tubular structure can be modelled and simulated by using growth theory for soft tissue in which the density is mostly unchanged. In general, growth can be described in different ways. For instance, growth can be an added mass, which can occur through cell division (hyperplasia), cell enlargement (hypertrophy), secretion of extracellular matrix, or accretion at external or internal surfaces. Atrophy (negative growth) can occur through cell death, cell shrinkage, or resorption. In most cases, hyperplasia and hypertrophy are mutually exclusive processes [34]. Depending on the age of the organism and the type of tissue, one of these two growth processes dominates. Furthermore, to fully investigate the tissue growth, remodelling should be considered, especially for medical application such as implants. The remodelling involves changes in material properties. These changes, which often are adaptive, may be brought about by alterations in elastic modulus, internal structure, strength, or mass density. For example, bone and heart muscle may change their internal structures through reorientation of trabeculae and muscle fibres, respectively. In addition, in a weightless environment, bone may lose both stiffness and mass density. Remodelling of biological soft tissue consisting of fibres embedded in a soft matrix (e.g. ureters, arteries and intestines) arises from changes in mechanical properties due to variations in collagen content, reorientation, type of fibres and fibre thickness as well as variations in muscle dimension. However, the model of growth in this paper does not take into account the remodelling for simplification. Herein, the growth is modelled for account the change in shape and mass, the material properties are considered to be only slightly changed and are therefore ignored in the simulation.

In this section, the kinematics and balance laws of finite growth are described. The isotropic growth and transversely isotropic growth are also described and adopted to solve a cube in triaxial tests and a cylinder with an internal pressure. The simulation results show that the growth models can be applied to simulate the porcine tubular organs in a fibrin-PVDF scaffold. In our approach, the staggered approach in the finite element setting was used.

3.1 Applied Approach

Let Ω_0 be the reference configuration of the body (organ), which is assumed to be stress free. Let us define the deformation map ψ, mapping the reference material point X of a material particle in the configuration Ω_0 to its current point $\boldsymbol{x}$ in the spatial configuration Ω at time t. Then $\mathbf{F} = \text{Grad}\psi(\boldsymbol{X}, \mathrm{t})$ is a material deformation gradient that relates an infinitesimal reference element d$\boldsymbol{X}$ from Ω_0 to d$\boldsymbol{x}$ in the deformed configuration Ω_g at time t. In Fig. 4, the intermediate (growth) configuration Ω_g is obtained by instantaneous elastic de-stressing of the current configuration Ω to zero stress [11]. The mass of an infinitesimal volume element in Ω_0 is dm_0. The mass of the corresponding elements in Ω_g and Ω is dm.

The multiplicative decomposition of the deformation gradient into growth $\mathbf{F}_g$ and elastic deformation $\mathbf{F}_e$ [11] is

$$\mathbf{F} = \mathbf{F}_e \cdot \mathbf{F}_g, \qquad \det(\mathbf{F}_g) \neq 1, \det(\mathbf{F}_e) = 1. \tag{1}$$

In (1), $\det(\mathbf{F}_g) \neq 1$ since growth does not require volume preservation [35].

Different from an unchanged mass in the classical continuum mechanics, the mass for a growth object is no longer a conserved quantity. Therefore, transformations of the important quantities of the different configurations must be performed. Herein, the scalar value ρ_0 signifies the reference density of a mass

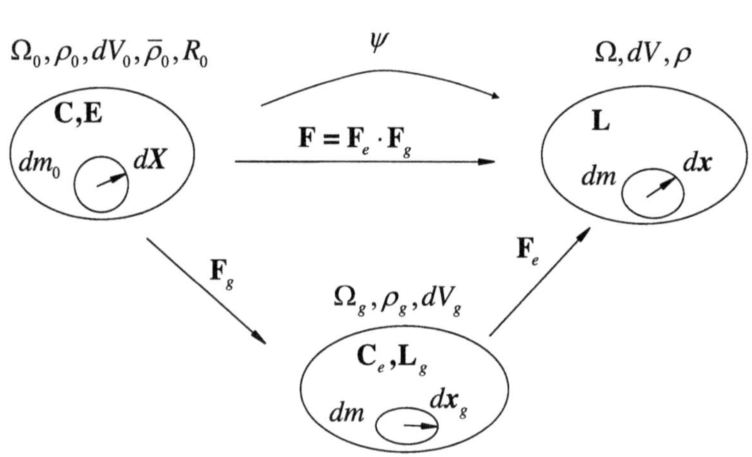

Fig. 4 Multiplicative decomposition of deformation gradient $\mathbf{F}$ into its growth $\mathbf{F}_g$ and elastic $\mathbf{F}_e$ parts

element. Its counterpart in the current and in the growth configurations are denoted by ρ and ρ_g, respectively [12].

A volume element in the reference configuration is $\mathrm{d}V_0$, its counterparts in the current and the growth configurations are $\mathrm{d}V$, and $\mathrm{d}V_g$, respectively. Besides, the Jacobian $J = \det(\mathbf{F}) = \frac{\mathrm{d}V}{\mathrm{d}V_0}$ of the total deformation gradient, the Jacobians J_g and J_e of the growth and the elastic deformation gradients, respectively, are introduced as $J_g = \det(\mathbf{F}_g) = \frac{\mathrm{d}V_g}{\mathrm{d}V_0}$ and $J_e = \det(\mathbf{F}_e) = \frac{\mathrm{d}V}{\mathrm{d}V_g} = \frac{J}{J_g} = \frac{\rho_g}{\rho}$. To model the growth of tissue the governing equations for balance of mass, balance of momentum and density transformation need to be solved.

3.1.1 Balance of Mass

The balance of mass and the balance of momentum for an open system in thermodynamics are universal equations [36]. Thus, both the balance of mass and the balance of momentum must be solved simultaneously or sequentially in growth simulations [37], leading to a requirement of modifying the stiffness matrices. The balance of mass for an open system is

$$\mathrm{Div}\boldsymbol{R} + R_0 = \frac{\mathrm{d}\rho_0}{\mathrm{d}t}, \tag{2}$$

where $\boldsymbol{R}$ is the mass flux vector, R_0 is the mass source per unit volume in the reference configuration. Cell migration can be incorporated phenomenologically through the mass flux $\boldsymbol{R}$. On the other hand, the mass source R_0 can incorporate phenomena such as increased cell growth, cell division or cell enlargement [13]. The balance of mass can be referred to as the 'biological equilibrium' or the 'homeostatic equilibrium' [37]. Without loss of generality, a mass flux through the surface of the considered mass element is neglected in the following discussion [13].

3.1.2 Balance of Momentum

The differential equations of motion are derived from the first Euler's law of motion of the continuum with the usual form as in conventional continuum mechanics with the material divergence [11].

$$\mathrm{Div}(\mathbf{F} \cdot \mathbf{S}) + \rho_0 \boldsymbol{b}_0 = \mathrm{Div}\mathbf{P} + \rho_0 \boldsymbol{b}_0 = \rho_g^0 \frac{\mathrm{dv}}{\mathrm{dt}}, \tag{3}$$

where $\mathbf{S}$ denotes the second Piola-Kirchhoff stress and the nominal stress (first Piola-Kirchhoff stress) $\mathbf{P} = \mathbf{F} \cdot \mathbf{S}$ is related to the nominal traction $\boldsymbol{T}$ by $\boldsymbol{T} = \boldsymbol{N} \cdot \mathbf{P}$, where $\boldsymbol{N}$ is the unit normal vector to the surface Γ_0 bounding the reference volume

Ω_0 and ρ_g^0 is also called the time rate of mass growth per unit reference volume. The nominal traction $\boldsymbol{T}$ is composed of two terms as

$$\boldsymbol{T} = \boldsymbol{T}^{\text{closed}} + \boldsymbol{T}^{\text{open}} \tag{4}$$

in which apart from the classical term in continuum mechanics $\boldsymbol{T}^{\text{closed}}$, the additional growth dependent contribution $\boldsymbol{T}^{\text{open}}$ is taken into account [37]. Similarly, the body force is expressed as

$$\boldsymbol{b}_0 = \boldsymbol{b}_0^{\text{closed}} + \boldsymbol{b}_0^{\text{open}}. \tag{5}$$

The balance of momentum is typically referred to as the 'mechanical equilibrium' [37]. This equation of the open thermodynamic system implies that the growth tensor $\mathbf{F}_g$ (or its determinant J_g) leads to volume changes and in addition causes mass changes. The balance of mass signifies that mass changes result from changes in density represented by the mass flux $\boldsymbol{R}$ and the source R_0 [37].

3.1.3 Density Transformation

The grown mass element dm [12] consists of the reference mass element dm_0 and the mass produced by the mass source R_0 (mass flux $\boldsymbol{R}$ is ignored) during the growth time $[t_0, t]$ as

$$dm = dm_0 + \left(\int_{t_0}^{t} R_0 \text{dt}\right)\text{dV}_0 = \rho_0 \text{dV}_0 + \left(\int_{t_0}^{t} R_0 \text{dt}\right)\text{dV}_0 \tag{6}$$

$\bar{\rho}_0$ is the density in the reference configuration too $\bar{\rho}_0 = \rho_0 + \int_{t_0}^{t} R_0 \mathrm{d}t$.

3.1.4 Essential Balance Equations

The material time derivative of $\bar{\rho}_0$ results in the well-known local balance of mass (continuity equation) in the reference configuration $\dot{\bar{\rho}}_0 = R_0 = r_g^0$, where r_g^0 is also called the time rate of mass growth per unit reference volume [11].

The current form of the continuity equation

$$\frac{\mathrm{d}\rho}{\mathrm{d}t}(\mathrm{d}V) + \rho\frac{\mathrm{d}}{\mathrm{d}t}(\mathrm{d}V) = r_g \mathrm{d}V, \tag{7}$$

where r_g is the time rate of the mass growth per unit current volume [11] and $r_g^0 = J r_g$. For $r_g > 0$ mass growth occurs, and for $r_g < 0$ mass resorption (atrophy or negative growth) takes place.

3.1.5 Density Preservation

The assumption of density preservation from the reference state to the intermediate configuration, i.e. $\rho_0 = \rho_g = \text{const.}$, implies that the volume of the mass element has to change in order to generate a mass change [12]. This effect of a volume change is denoted as growth if the volume increases or as atrophy if the volume decreases [34]. When using the density preservation, the mass source is rewritten as

$$R_0 = J_g \rho_g \mathrm{tr}\mathbf{L}_g = J_g \rho_0 \mathrm{tr}\mathbf{L}_g. \tag{8}$$

where $\mathbf{L}_g$ is the velocity gradient $\mathbf{L}_g = \dot{\mathbf{F}}_g \cdot \mathbf{F}_g^{-1}$ with the material time derivative. Obviously, the mass source can directly be determined if the growth deformation gradient $\mathbf{F}_g$ and its material time rate $\dot{\mathbf{F}}_g$ are given. The density preservation is thus suitable for soft tissues [13].

3.1.6 Constitutive Equations

A mass specific strain energy function $\Phi = \Phi_{\mathrm{iso}} + \Phi_{\mathrm{trans}}$ in terms of the volume specific strain energy functions (W_0, W_g and W) is defined as [38].

$$\Phi = \frac{W_0}{\bar{\rho}_0} = \frac{W_g}{\rho_g} = \frac{W}{\rho} \tag{9}$$

with $W_0\left(\mathbf{F}_e, \mathbf{A}_0^i\right) = J_g W_g\left(\mathbf{F}_g, \mathbf{A}_g^i\right) = JW\left(\mathbf{F}, \mathbf{A}^i\right)$, where W_0, W_g and W are volume specific strain energy functions in the reference, the intermediate and the current configurations, respectively. The elastic strain energy per unit current mass is then given by a function of the elastic Green-Lagrange strain $\mathbf{E}_e$ (or its invariants) and the structure tensors $\mathbf{A}_0^i = \boldsymbol{a}_0^i \otimes \boldsymbol{a}_0^i$, $i = 4, 6$ for the first fibre family and the second fibre family, respectively i.e. $\Phi\left(\mathbf{E}_e, \mathbf{A}_0^i, \bar{\rho}_0\right)$. Generally, the strain energy density functions are usually formulated in terms of the overall deformation gradient $\mathbf{F}$, the growth tensor $\mathbf{F}_g$, the structure tensor $\mathbf{A}_0^i$ and the density $\bar{\rho}_0$ such as $\Phi = W_0\left(\mathbf{F}, \mathbf{F}_g, \mathbf{A}_0^i, \bar{\rho}_0\right)$ [11, 13, 38]. Herein, it is more convenient to manage the strain energy functions which are described in terms of the invariants of the elastic right Cauchy Green tensor $I_1 = \mathrm{tr}(\mathbf{C}_e)$, $I_3 = \det(\mathbf{C}_e)$, where $\mathbf{C}_e = \mathbf{F}_e^T \cdot \mathbf{F}_e$. In this work, the compressible neo-Hookean model was used with the density strain-energy function written as

$$\bar{\rho}_0 \Phi_{\mathrm{iso}}(I_1) = \frac{\mu}{2}(I_1 - 3) - \mu \ln(J_e) + \frac{\lambda}{2} \ln^2(J_e), \tag{10}$$

where $\mu = \frac{E}{2(1+\nu)}$ (with the Poisson's ratio ν) is equivalent to the small strain shear modulus, $\lambda = \frac{E\upsilon}{(1+\nu)(1-2\nu)}$ is equivalent to the first small strain Lamé's parameter.

$\Phi_{\text{trans}}(I_4)$ is the transversely isotropic strain energy in terms of invariants and can be used to simulate transversely isotropic growth $\left(I_4 = \boldsymbol{a}_0^4 \cdot \mathbf{C}\boldsymbol{a}_0^4\right)$. That means the growth along the fibres dominates others in the rest directions. It is worth noting that a compressible model must be utilized in such a way that the material can grow. Therefore, the volumetric energy, which enforces the incompressibility condition, is ignored.

$$\Phi_{\text{trans}}(I_4) = \frac{k_1}{2k_2}\left\{\exp\left(k_2[I_4 - 1]^2\right) - 1\right\}. \tag{11}$$

The second Piola-Kirchhoff stress tensor in the material configuration is written as

$$\mathbf{S} = 2\frac{\partial W_0}{\partial \mathbf{C}} = 2\bar{\rho}_0\frac{\partial \Phi}{\partial \mathbf{C}}. \tag{12}$$

Using the push forward operation to the growth configuration on $\mathbf{S}_e$ yields

$$\mathbf{S}_e = \mathbf{F}_g \cdot \mathbf{S} \cdot \mathbf{F}_g^T = 2\bar{\rho}_0 \mathbf{F}_g \cdot \frac{\partial \Phi}{\partial \mathbf{C}} \cdot \mathbf{F}_g^T = 2\bar{\rho}_0 \frac{\partial \Phi}{\partial \mathbf{C}_e} = \mathbf{S}_{e,\text{iso}} + \mathbf{S}_{e,\text{trans}}. \tag{13}$$

The second Piola-Kirchhoff stress for isotropic part is calculated as

$$\mathbf{S}_{e,iso} = 2\bar{\rho}_0\frac{\partial \Phi(I_1)}{\partial \mathbf{C}_e} = \mu\left(I_1 - \mathbf{C}^{-1}\right) + \lambda \ln(J_e)\mathbf{C}^{-1} \tag{14}$$

and the second Piola-Kirchhoff stress for transversely isotropic part (only one fibre family) is evaluated as

$$\mathbf{S}_{e,\text{trans}} = 2\bar{\rho}_0\frac{\partial (I_4)}{\partial \mathbf{C}_e} = 2k_1(I_4 - 1)\exp\left(k_2[I_4 - 1]^2\right)\boldsymbol{a}_0^4 \otimes \boldsymbol{a}_0^4. \tag{15}$$

The constitutive tensor $\mathbb{C}$ is then defined as

$$\mathbb{C} = 2\frac{\partial \mathbf{S}_e}{\partial \mathbf{C}_e} = 2\frac{\partial\left(\mathbf{S}_{e,\text{iso}} + \mathbf{S}_{e,\text{trans}}\right)}{\partial \mathbf{C}_e}. \tag{16}$$

The constitutive tensor of the isotropic part is written as

$$\begin{aligned}\mathbb{C}_{iso} = 2\frac{\partial \mathbf{S}_{e,iso}}{\partial \mathbf{C}_e} &= 2\left[-\frac{\partial \mathbf{C}_e^{-1}}{\partial \mathbf{C}_e}\mu + \lambda\left(\frac{\partial \ln J_e}{\partial \mathbf{C}_e} \otimes \mathbf{C}_e^{-1} + \ln J_e \frac{\partial \mathbf{C}_e^{-1}}{\partial \mathbf{C}_e}\right)\right] \\ &= \lambda \mathbf{C}_e^{-1} \otimes \mathbf{C}_e^{-1} + (\mu - \lambda \ln J_e)\mathbb{I}\end{aligned} \tag{17}$$

where $\mathbb{I} = \frac{1}{2}\left(C_{e,IK}^{-1}C_{e,JL}^{-1} + C_{e,IL}^{-1}C_{e,JK}^{-1}\right)$. Similarly, the constitutive tensor of the transversely isotropic part is computed as

$$\mathbb{C}_{trans} = 2\frac{\partial \mathbf{S}_{e,\text{trans}}}{\partial \mathbf{C}_e} = 4k_1\{\exp\left(k_2[I_4-1]^2\right) + 2k_2(I_4-1)^2\exp\left(k_2[I_4-1]^2\right)(\boldsymbol{a}_0^4 \otimes \boldsymbol{a}_0^4) \otimes (\boldsymbol{a}_0^4 \otimes \boldsymbol{a}_0^4)\}. \tag{18}$$

3.2 Computational Growth Models

3.2.1 Isotropic Growth

When considering the smooth muscle cells of porcine tubular organs in a fibrin-PVDF scaffold is isotropic, the isotropic growth model applied in this case is as

$$\mathbf{F}_g = \vartheta \mathbf{I}. \tag{19}$$

Evolution equations for stretch ratios ϑ are given in terms of isotropic scalar functions of the second Piola-Kirchhoff stress tensor $\mathbf{S}_e$ and the right Cauchy-Green strain tensor $\mathbf{C}_e$, assumed to be $\dot{\vartheta} = k_\vartheta(\vartheta)\text{tr}(\mathbf{S}_e \cdot \mathbf{C}_e)$.

To prevent an unlimited growth at an arbitrary nonzero state of stress, it is proposed that during the mass growth [11].

$$k_\vartheta(\vartheta) = k_{\vartheta 0}^{+}\left(\frac{\vartheta^{+} - \vartheta}{\vartheta^{+} - 1}\right)^{m_\vartheta^{+}}, \tag{20}$$

where $\vartheta^{+} > 1$ is the limiting value of the growth stretch ratio. In the case of the mass resorption, the corresponding expression is

$$k_\vartheta(\vartheta) = k_{\vartheta 0}^{-}\left(\frac{\vartheta - \vartheta^{-}}{1 - \vartheta^{-}}\right)^{m_\vartheta^{-}}. \tag{21}$$

In contrast to [11], we adopt the driving force, which is the Mandel stresses $\mathbf{M}_e = \mathbf{S}_e \cdot \mathbf{C}_e$ [12, 13].

3.2.2 Transversely Isotropic Growth

A transversely isotropic mass growth law is reasonably capable of characterizing a transversely isotropic material, e.g. muscle with a bundle of longitudinal fibres. Therefore, this model can be further employed for the smooth muscle cells of porcine tubular organs in a fibrin-PVDF scaffold due to the experimental data mentioned above showing the orientation of cells. The unit vector specifying the fibre orientation is denoted by $\boldsymbol{a}_0^4$ in the reference configuration Ω_0. The intermediate configuration Ω_g is defined with the same fiber orientation with respect to the fixed frame of reference. The isotropic growth deformation gradient is defined by assuming that the fibres are embedded in the material [11], this is ensured by defining $\mathbf{F}_g$ such that $\boldsymbol{a}_0^4$ is one of its eigendirections

$$\mathbf{F}_g \cdot \boldsymbol{a}_0^4 = \eta_g \boldsymbol{a}_0^4. \tag{22}$$

An infinitesimal fibre segment in the configuration is parallel to the vector $\boldsymbol{a}^4 = \mathbf{F}_g \cdot \boldsymbol{a}_0^4$, obtained from $\boldsymbol{a}_0^4$ by elastic stretching and rotation [11].

The expression for the growth part of the deformation gradient due to transversely isotropic mass growth [11] is

$$\mathbf{F}_g = \upsilon \mathbf{I} + (\xi - \upsilon)\boldsymbol{a}_0^4 \otimes \boldsymbol{a}_0^4. \tag{23}$$

The Eq. (23) is similar to the transversely isotropic mass growth for orthotropic material, except the direction of $\boldsymbol{a}_0^4$. The difference in growth direction leads to different formulae for the growth stretch ratios. Specifically, $\mathbf{F}_g \cdot \boldsymbol{a}_0^4 = \eta_g \boldsymbol{a}_0^4$ so that the growth stretch ratio in the fibre direction is ξ, while υ is the growth stretch ratio in any orthogonal direction. The inverse of the growth deformation tensor is

$$\mathbf{F}_g^{-1} = \frac{1}{\upsilon}\mathbf{I} + \left(\frac{1}{\xi} - \frac{1}{\upsilon}\right)\boldsymbol{a}_0^4 \otimes \boldsymbol{a}_0^4 \tag{24}$$

$\mathbf{F}_e = \mathbf{F} \cdot \mathbf{F}_g^{-1}$ gives an explicit expression for the elastic part of the deformation gradient

$$\mathbf{F}_e = \frac{1}{\upsilon}\mathbf{F} + \left(1 - \frac{\xi}{\upsilon}\right)\boldsymbol{a}_0^4 \otimes \boldsymbol{a}_0^4. \tag{25}$$

3.2.3 Evolution Equations for Stretch Ratios

The constitutive formulation is completed by specifying appropriate evolution equations for the growth stretch ratios υ and ξ. The evolution equations for the stretch ratios and must be given in terms of isotropic scalar functions of the stress tensor $\mathbf{S}_e$ and the structural tensor $\mathbf{A}_0^4 = \boldsymbol{a}_0^4 \otimes \boldsymbol{a}_0^4$ [11]. The evolution equations of the stretch ratios are assumed to be obtained by including in the list of arguments their current values as

$$\dot{\upsilon} = f_{\upsilon}\left[\xi, \upsilon, S^e_{a_0^4}, \frac{1}{2}\left(\mathrm{tr}\mathbf{S}_e - S^e_{a_0^4}\right)\right], \tag{26}$$

$$\dot{\xi} = f_{\xi}\left[\xi, \upsilon, S^e_{a_0^4}, \frac{1}{2}\left(\mathrm{tr}\mathbf{S}_e - S^e_{a_0^4}\right)\right]. \tag{27}$$

The normal component of the second Piola-Kirchhoff stress $\mathbf{S}_e$ in the direction of the fibre is $S^e_{a_0^4} = \boldsymbol{a}_0^4 \cdot \mathbf{S}_e \cdot \boldsymbol{a}_0^4$, while $\frac{1}{2}\left(\mathrm{tr}\mathbf{S}_e - S^e_{a_0^4}\right)$ represents the average normal stress in the plane perpendicular to the fibre. These two stresses are believed to have a dominant mechanical effect on the transversely isotropic mass growth [11]. Note that the normal stress in the direction of the fibre can be expressed in terms of the Cauchy stress as

$$\boldsymbol{a}_0^4 \cdot \mathbf{S}_e \cdot \boldsymbol{a}_0^4 = J_e\left(\boldsymbol{a}_0^4 \cdot \mathbf{F}_e^{-1}\right) \cdot \boldsymbol{\sigma} \cdot \left(\boldsymbol{a}_0^4 \cdot \mathbf{F}_e^{-1}\right).$$

In the simplest case, the rates of the stretch ratios depend linearly on the stress components, such that

$$\dot{\upsilon} = k_{\upsilon}(\upsilon)\left[S^e_{a_0^4} - \nu_{\upsilon}\left(\mathrm{tr}\mathbf{S}_e - S^e_{a_0^4}\right)\right] \tag{28}$$

$$\dot{\xi} = k_{\xi}(\xi)\left[S^e_{a_0^4} - \nu_{\xi}\left(\mathrm{tr}\mathbf{S}_e - S^e_{a_0^4}\right)\right] \tag{29}$$

where ν_{ξ} and ν_{υ} are constants, or functions of ξ and υ. For the sake of simplicity, these constants are chosen as $\nu_{\upsilon} = \nu_{\xi} = \frac{1}{2}$.

To prevent an unlimited growth at an arbitrary non-zero state of stress, it is proposed that during mass growth

$$k_{\xi}(\xi) = k^{+}_{\xi 0}\left(\frac{\xi^{+} - \xi}{\xi^{+} - 1}\right)^{m^{+}_{\xi}} \quad \text{for } S^e_{a_0^4} > \frac{\nu_{\xi}}{1+\nu_{\xi}}\mathrm{tr}\mathbf{S}_e \tag{30}$$

$$k_{\upsilon}(\upsilon) = k^{-}_{\upsilon 0}\left(\frac{\upsilon - \upsilon^{-}}{1 - \upsilon^{-}}\right)^{m^{-}_{\upsilon}} \quad \text{for } S^e_{a_0^4} < \frac{\nu_{\upsilon}}{1+\nu_{\upsilon}}\mathrm{tr}\mathbf{S}_e. \tag{31}$$

3.2.4 Numerical Implementation

In isotropic growth case, the constitutive tensor is calculated as

$$\mathbb{C}_e = 2\frac{\mathrm{d}\mathbf{S}_e}{\mathrm{d}\mathbf{C}_e} = 2\left(\frac{\partial \mathbf{S}_e}{\partial \mathbf{C}_e} + \frac{\partial \mathbf{S}_e}{\partial \upsilon} \otimes \frac{\partial \upsilon}{\partial \mathbf{C}_e}\right) = \mathbb{C} + \mathbb{C}_{\upsilon}. \tag{32}$$

In addition, the constitutive tensor of the transversely isotropic growth model is written as

$$\mathbb{C}_e = 2\frac{\mathbf{dS}_e}{\mathbf{dC}_e} = 2\left(\frac{\partial \mathbf{S}_e}{\partial \mathbf{C}_e} + \frac{\partial \mathbf{S}_e}{\partial \xi} \otimes \frac{\partial \xi}{\partial \mathbf{C}_e} + \frac{\partial \mathbf{S}_e}{\partial \upsilon} \otimes \frac{\partial \upsilon}{\partial \mathbf{C}_e}\right) = \mathbb{C} + \mathbb{C}_\xi + \mathbb{C}_\upsilon. \quad (33)$$

Note that the elasticity-growth tensor comprises two main parts, the elasticity tensor $\mathbb{C}$ as in the passive response and the active response and the growth tensors $(\mathbb{C}_\xi + \mathbb{C}_\upsilon)$ that relate to remodelling of the material. Furthermore, $(\mathbb{C}_\xi + \mathbb{C}_\upsilon)$ tensors are not symmetric because they are generated from the tensor products of different tensors. Consequently, asymmetric solvers are required for this computation.

The incremental updates for the stretch ratios in the cases of isotropic growth and transversely isotropic growth are presented without loss of generality. Obviously, the strains, the stresses, and the elasticity tensor in (15,17,18) depend on the stretch ratios at any time step. Since (ξ, υ) are introduced as internal variables at integration points, the implicit Euler backward scheme is used to compute the stretch ratios at a time step [37], hence the stretches are formulated as

$$\xi_{n+1} = \xi_n + \dot{\xi}_{n+1}\Delta t \quad (34)$$

$$\upsilon_{n+1} = \upsilon_n + \dot{\upsilon}_{n+1}\Delta t \quad (35)$$

and their derivatives computed as

$$\frac{\partial \xi}{\partial \mathbf{C}_e} = \left(\frac{\partial \dot{\xi}}{\partial \mathbf{C}_e} + \frac{\partial \dot{\xi}}{\partial \xi}\frac{\partial \xi}{\partial \mathbf{C}_e}\right)\Delta t \quad (36)$$

$$\frac{\partial \upsilon}{\partial \mathbf{C}_e} = \left(\frac{\partial \dot{\upsilon}}{\partial \mathbf{C}_e} + \frac{\partial \dot{\upsilon}}{\partial \upsilon}\frac{\partial \upsilon}{\partial \mathbf{C}_e}\right)\Delta t. \quad (37)$$

3.3 Numerical Simulation of Growth

To exhibit the capability of modelling and simulating the growth of cells, a simple test case was investigated with different material laws, such as the neo-Hookean strain energy function and the model in [39]. Their parameters are tabulated in Table 2.

Table 2 Growth parameters for isotropic growth

Mode	$k_{\vartheta 0}^{+}$	$k_{\vartheta 0}^{-}$	m_{ϑ}^{+}	m_{ϑ}^{-}	ϑ^{+}	ϑ^{-}	Δt
Unlimited	1.5×10^{-3}	0.8×10^{-3}	2.5	3	2	0.5	1
Limited	6.5×10^{-3}	0.8×10^{-3}	1.5	3	1.24	0.5	1

It is well known that soft tissues are considered as nearly incompressible materials. However, the tissues grow under stress induced changes, and hence, the incompressibility condition need not strictly be imposed. Thus, the Poisson's ratio is chosen as $v = 0.4$.

3.3.1 Triaxial Tension and Compression Tests of a Cube

The cube is subjected to monotonic loading by prescribed displacements as in the training described above for the tubular fibrin structure. Loading can be simultaneously applied (e.g. in a step function with 5 times increments) along three orthogonal directions of one eighth of the cube since the intermediate configuration (growth configuration) is incompatible, see Fig. 5. The cube is simulated by finite elements using the hexahedral element (FEM-H8). The time step Δt (its unit depends on the unit of the measurements of the growth process or time counted for tissue enlargement, e.g. when this time is measured in hour or day, then Δt is measured in the same unit) found for each load increment are solved at the biological equilibrium state (all stresses are zero). Figure 5 clearly shows the numerical results of employing the neo-Hookean model with the shear modulus $\mu = 1.5$ kPa. The larger stresses $\boldsymbol{\sigma}$ induce the smaller number of time steps, i.e. the tissue grows faster. Correspondingly, the stresses in each load increment are decreasing slowly, see Fig. 5. At the final state, the increased volume of the cube is 3.375 times the initial volume. In this case, there is no growth limit since the load (prescribed displacement) is 1.4 times smaller than the limit stretch ratio $\vartheta^+ = 2.0$.

In contrast with the tension test, in the simulations of the compression test, there are reductions of the size of the cube that were compensated by the negative growth (Fig. 6) and the larger strain leading to a faster atrophy. The anisotropic material

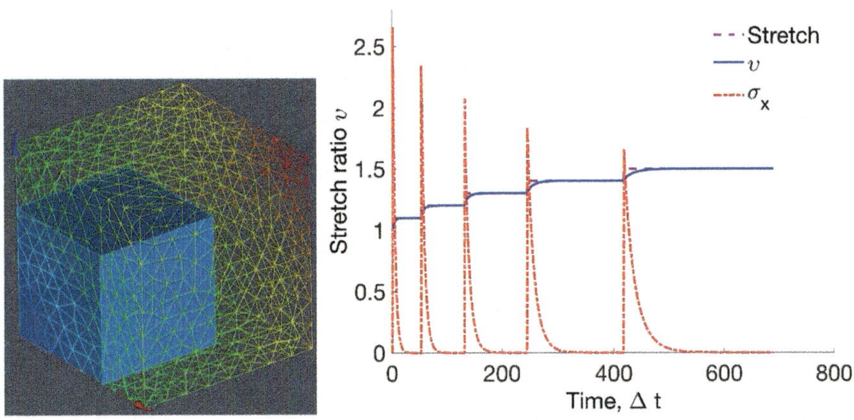

Fig. 5 Stretch and growth stretch ratio for a cube in tension test [40]

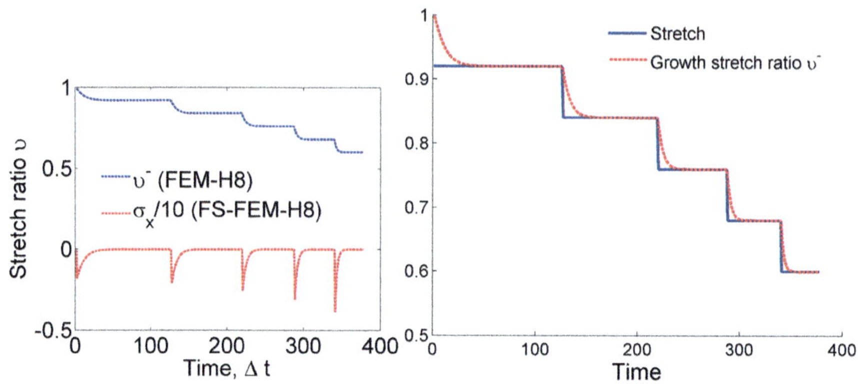

Fig. 6 Stretch and stretch ratio for cube of isotropic tissue in compression test [40]

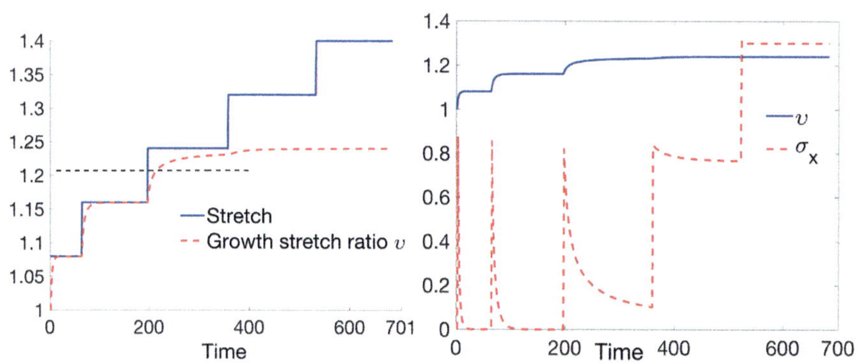

Fig. 7 Stretch and growth stretch ratio for a cube of anisotropic tissue in tension test with limit constant [40]

described in [39] and the growth constants in Table 2 adopted for limit growth simulation, result in residual stresses which cannot be compensated by growth in Fig. 7.

3.3.2 Tube Tension Test

For the first step of mimicking approximately urine dropping inside the ureter, an extension along axial direction of a tube was performed to investigate the growth of the tube, see Fig. 8 (left). The isotropic growth parameters are the same as in the case of the triaxial test, but the material model is transversely isotropic (11), combined with the neo-Hookean material with constants as $\mu = 1.28$ kPa, $k_1 = 1.00$ kPa, $k_2 = 1.91$, and the vector $\boldsymbol{a}_0^4 = [100]^{\mathrm{T}}$. The FEM boundary conditions

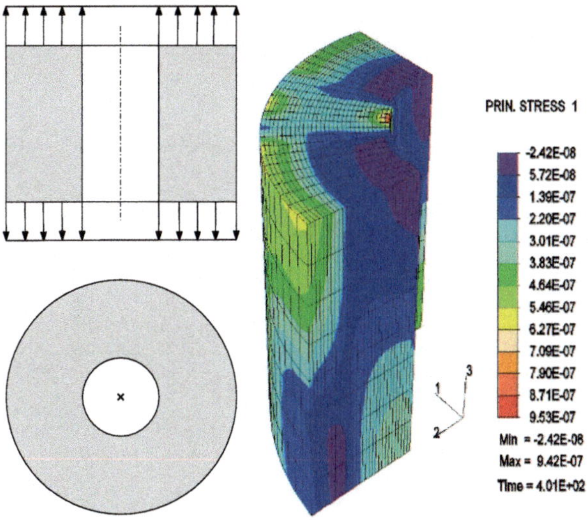

Fig. 8 Extension of a tube: boundary conditions and major principal stress [40]

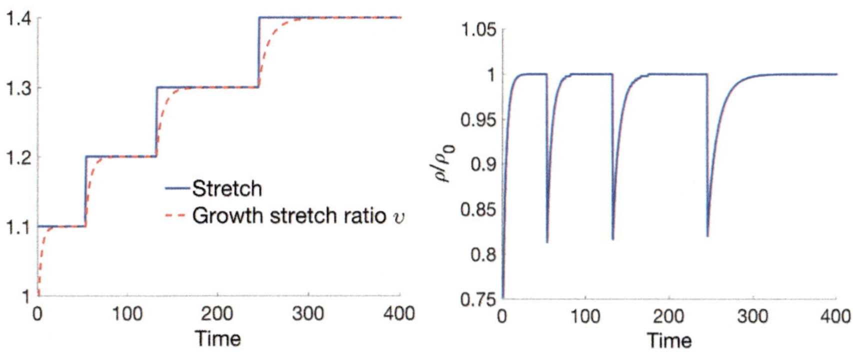

Fig. 9 FEM result of growth of extension of a tube [40]

and the major principal stress approaching zero are depicted in Fig. 8 in which we prescribe both ends of the tube with a specified stretch. For each increasing stretch imposed, the tissue needs to grow in order to compensate the imposed stretch to reach the zero-stress state. Specifically, Fig. 9 shows that after reaching the stretch of 1.4 with four loading increments, the first principal stress becomes zero due to compensation by growth with the density ratio approaching unity for the density preservation case. At the beginning of each stretch prescribed, the density curve drops suddenly and gradually increases to the magnitude of one. Consequently, the tube enlarges both in the radial and the longitudinal directions with a maximal stretch ratio of 1.4.

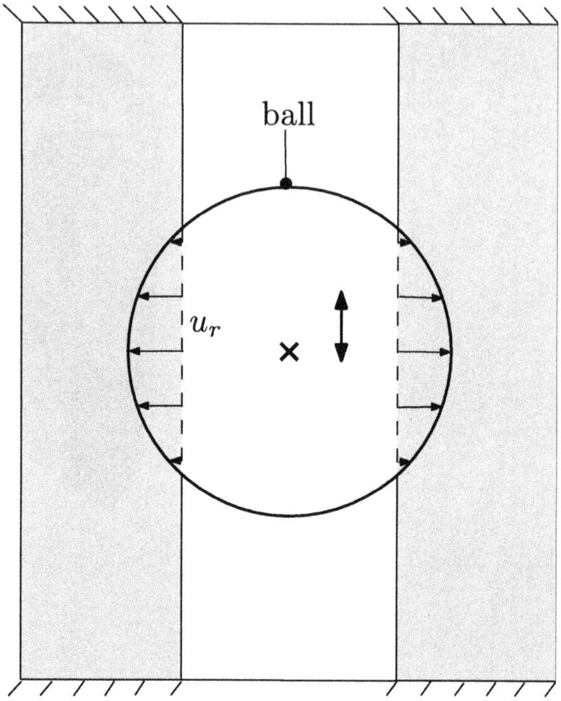

Fig. 10 Balloon travelling inside a tube mimicking urine drop in a fibrin-PVDF scaffold

Figure 10 describes the mimicking of urine with a ball or a hyperelastic kyphoplasty balloon catheter travelling inside the muscle in fibrin-PVDF scaffold. This is future work of the study applied to transversely isotropic growth and transversely isotropic material. This case is a dynamic simulation since the ball inducing displacement u_r, travels along the axial direction of the tube modelled as transversely isotropic. Hence, the growth of tissue will be solved and can be compared with experimental data such as the mechanical properties, shape, size and fibre orientation of the smooth muscles in a fibrin-PVDF scaffold.

4 Conclusions

Our work presents a novel stimulation procedure mimicking a urine bolus, applied to smooth muscle cells of porcine tubular organs in a fibrin-PVDF scaffold. The mechanical stimulation has a tendency to increase mechanical stability for stimulated prostheses. A similar tendency was observed for extended incubation periods. The rupture pressure is also increased after two weeks of incubation with mechanical stimulation. Furthermore, the stimulation results in a preferred orientation of cells along the tubular structure of the prosthesis. However, the intensity of mechanical stimulation is dependent on various parameters and different models: the applied cell type, cell source, cell passage as well as in the experimental setup. Nevertheless, it is seen to be stress level and cell type specific.

In this study, we focus on the change in shape and mechanical properties of cells with stimulated prostheses so that the growth theory can be applied to simulate the experiments. Although, growth is generally a very complex process and can account for the increase of tissue volume of inner organs due to daily activities and gravity this process can be characterised by simulation. With the mechanical stimulation presented for the experiments on smooth muscle cells of porcine tubular organs in a fibrin-PVDF scaffold, the isotropic and transversely isotropic growth models are certainly adequate for simulating the growth of smooth cells. Consequently, a standard protocol can be applied to conduct experiments and growth simulations are adopted to validate simulations with experimental data. That can be a novel challenge for the future work.

Acknowledgements The focal cyclic intraluminal mechanical stimulation is based on suggestions by G. M. Artmann. This project was supported in part by the German Federal Ministry of Education and Research of Germany through the FHprofUnt project "BINGO" (03FH073PX2) and by the German Federal Ministry of Economics and Technology through the project "UREPLACE" (KF0634101SB8).

Conflict of Interest

The authors declare that they have no conflict of interest.

References

1. Taulbut, M., Walsh, D., McCartney, G., Parcell, S., Hartmann, A., Poirier, G., et al. (2014). Spatial inequalities in life expectancy within postindustrial regions of Europe: A cross-sectional observational study. *BMJ Open, 4*(6), 1–9 (e004711).
2. Wohland, P., Bees, P., Gillies, C., Alvanides, S., Matthews, F. E., O'neil, V., et al. (2014). Drivers of inequality in disability-free expectancy at birth and age 85 across space and time in Great Britain. *Journal of Epidemiology and Community Health, 68,* 826–833.
3. Shalhav, A. L., Elbahnasy, A. M., Bercowsky, E., Kovacs, G., Brewer, A., Maxwell, K. L., et al. (1999). Laparoscopic replacement of urinary tract segments using biodegradable materials in a large-animal model. *Journal of Endourology, 13,* 241–244.
4. El-Assmy, A., El-Sherbiny, A. T., El-Hamid, M. A., Mohsen, T., Nour, E. M., Bazeed, M., et al. (2004). Use of single layer small intestinal submucosa for long segment ureteral replacement: A pilot study. *The Journal of Urology, 171,* 1939–1942.
5. Seifarth, V., Grosse, J. O., Gossmann, M., Janke, H. P., Arndt, P., Koch, S., et al. (2017). Mechanical induction of bi-directional orientation of primary porcine bladder smooth muscle cells in tubular fibrin-poly(vinylidene fluoride) scaffolds for ureteral and urethral repair using cyclic and focal balloon catheter stimulation. *Journal of Biomaterials Applications, 32*(3), 321–330.

6. Filippo, R. E. D., Yoo, J. J., & Atala, A. (2002). Urethral replacement using cell seeded tubularized collagen matrices. *The Journal of Urology, 168,* 1789–1793.
7. Orabi, H., AbouShwareb, T., Zhang, Y., Yoo, J. J., & Atala, A. (2013). Cell-seeded tubularized scaffolds for reconstruction of long urethral defects: A preclinical study. *European Urology, 63,* 531–538.
8. Huang, A. H., & Niklason, L. E. (2011). Engineering biological-based vascular grafts using a pulsatile bioreactor. *Journal of visualized experiments: JoVE, 52,* 1–14.
9. Fu, Q., Deng, C. L., Zhao, R. Y., Wang, Y., & Cao, Y. (2014). The effect of mechanical extension stimulation combined with epithelial cell sorting on outcomes of implanted tissue-engineered muscular urethras. *Biomaterials, 35,* 105–112.
10. Rodriguez, E. K., Hoger, A., & McCulloch, A. D. (1994). Stress-dependent finite growth in soft elastic tissues. *Journal of Applied Biomechanics, 27*(4), 455–467.
11. Lubarda, V. A., & Hoger, A. (2002). On the mechanics of solids with a growing mass. *International Journal of Solids and Structures, 39,* 4627–4664.
12. Himpel, G., Kuhl, E., Menzel, A., & Steinmann, P. (2005). Computational modelling of isotropic multiplicative growth. *Computer Modeling in Engineering & Sciences, 8,* 119–134.
13. Kuhl, E., Mass, R., Himpel, G., & Menzel, A. (2006). Computational modeling of arterial wall growth. *Biomechanics and Modeling in Mechanobiology, 6,* 321–331.
14. Holzapfel, G. A. (2000). *Nonlinear solid mechanics, a continuum approach for engineering.* Chichester: Wiley.
15. Montzka, K., Läufer, T., Becker, C., Grosse, J., & Heidenreich, A. (2011). Microstructure and cytocompatibility of collagen matrices for urological tissue engineering. *BJU International, 107,* 1974–1981.
16. Cornelissen, C. G., Dietrich, M., Krüger, S., Spillner, J., Schmitz-Rode, T., & Jockenhoevel, S. (2012). Fibrin gel as alternative scaffold for respiratory tissue engineering. *Annual Review of Biomedical Engineering, 40,* 679–687.
17. Dellenback, R. J., & Chien, S. (1970). The extinction coefficient of fibrinogen from man, dog, elephant, sheep, and goat at 280 mμ. *Proceedings of the Society for Experimental Biology and Medicine, 134,* 353–355.
18. Cholewinski, E., Dietrich, M., Flanagan, T. C., Schmitz-Rode, T., & Jockenhoevel, S. (2009). Tranexamic acid—An alternative to aprotinin in fibrin—Based cardiovascular tissue engineering. *Tissue Engineering Part A, 15,* 3645–3653.
19. Eyrich, D., Brandl, F., Appel, B., Wiese, H., Maier, G., & Wenzel, M. (2007). Long-term stable fibrin gels for cartilage engineering. *Biomaterials, 28,* 55–65.
20. Dorothea, L., Stollenwerk, K., Seifarth, V., Zraik, I. M., Vogt, M., Srinivasan, P. K., et al. (2016). Two differentially structured collagen scaffolds for potential urinary bladder augmentation: Proof of concept study in a Göttingen minipig model. *Journal of Translational Medicine, 15,* 1–16.
21. Seifarth, V. (2015). *Ureteral tissue engineering: Development of a bioreactor system and subsequent characterization of the generated biohybrids* (PhD Dissertation). University Duisburg-Essen, Germany.
22. Ternifi, R., Gennisson, J. L., Tanter, M., & Beillas, P. (2013). Effects of storage temperature on the mechanical properties of porcine kidney estimated using Shear Wave Elastography. *Journal of the Mechanical Behavior of Biomedical Material, 28,* 86–93.
23. Janmey, P. A., & McCulloch, C. A. (2007). Cell mechanics: Integrating cell responses to mechanical stimuli. *Annual Review of Biomedical Engineering, 9,* 1–34.
24. Murphy, S. V., & Atala, A. (2013). Organ engineering—Combining stem cells, biomaterials, and bioreactors to produce bioengineered organs for transplantation. *BioEssays, 35,* 163–172.
25. Nishi, M., Matsumoto, R., Dong, J., & Uemura, T. (2013). Engineered bone tissue associated with vascularization utilizing a rotating wall vessel bioreactor. *Journal of Biomedical Materials Research—Part A, 101,* 421–427.

26. Seliktar, D., Black, R. A., Vito, R. P., & Nerem, R. M. (2000). Dynamic mechanical conditioning of collagen-gel blood vessel constructs induces remodeling in vitro. *Annual Review of Biomedical Engineering, 28,* 351–362.
27. Tschoeke, B., Flanagan, T. C., Cornelissen, A., Koch, S., Roehl, A., Sriharwoko, M., et al. (2008). Development of a composite degradable/nondegradable tissue-engineered vascular graft. *Artificial Organs, 32,* 800–809.
28. Jockenhoevel, S., & Flanagan, T. C. (2011). In D. Eberli (Ed.), *Tissue engineering for tissue and organ regeneration*. InTech.
29. Geutjes, P., Roelofs, L., Hoogenkamp, H., Walraven, M., Kortmann, B., de Gier, R., et al. (2012). Tissue engineered tubular construct for urinary diversion in a preclinical porcine model. *The Journal of Urology, 188,* 653–660.
30. Wei, X., Li, D. B., Xu, F., Wang, Y., Zhu, Y. C., Li, H., et al. (2011). A novel bioreactor to simulate urinary bladder mechanical properties and compliance for bladder functional tissue engineering. *Chinese Medical Journal (Engl), 124,* 568–573.
31. Davis, N. F., Mooney, R., Piterina, A. V., Callanan, A., McGuire, B. B., Flood, H. D., et al. (2011). Construction and evaluation of urinary bladder bioreactor for urologic tissue-engineering purposes. *Urology, 78,* 954–960.
32. Wang, C., Cen, L., Yin, S., Liu, Q., Liu, W., Cao, Y., et al. (2010). A small diameter elastic blood vessel wall prepared under pulsatile conditions from polyglycolic acid mesh and smooth muscle cells differentiated from adipose-derived stem cells. *Biomaterials, 31,* 621–630.
33. Kanda, K., Matsuda, T., & Oka, T. (1993). Mechanical stress induced cellular orientation and phenotypic modulation of 3-D cultured smooth muscle cells. *ASAIO Journal, 39*(3), M668–M690.
34. Taber, L. A. (1998). Biomechanical growth laws for muscle tissue. *Journal of Theoretical Biology, 193,* 201–213.
35. Humphrey, J. D. (2002). *Cardiovascular solid mechanics-cells, tissues, and organs*. New York: Springer.
36. Kuhl, E. (2004). *Theory and numerics of open system continuum thermodynamics-Spatial and material settings* (Habilitation thesis). Technical University of Kaiserslautern, Germany.
37. Kuhl, E., Menzel, A., & Steinmann, P. (2003). Computational modeling of growth: A critical review, a classification of concepts and two new consistent approaches. *Computational Mechanics, 32,* 71–88.
38. Menzel, A. (2007). A fibre reorientation model for orthotropic multiplicative growth: Configurational driving stresses, kinematics-based reorientation, and algorithmic aspects. *Biomechanics and Modeling in Mechanobiology, 6,* 303–320.
39. Duong, M. T., & Staat, M. (2014). A face-based smoothed finite element method for hyperelastic models and tissue growth. In E. Oñate, J. Oliver & A. Huerta (Eds.), *Proceedings 11th World Congress on Computational Mechanics (WCCM XI), 5th European Conference on Computational Mechanics (ECCM V), 6th European Conference on Computational Fluid Dynamics (ECFD VI)* (pp. 2657–2668), Barcelona, Spain.
40. Duong, M. T. (2014). *Hyperelastic modeling and soft-tissue growth integrated with the smoothed finite element method—SFEM* (PhD Dissertation). RWTH Aachen University, Germany.

Author Biography

Minh Tuấn Dương (Vietnam), Dr. Ing. He is a lecturer of Hanoi University of Science and Technology. He was a researcher and postdoc at the Biomechanics Laboratory of the Institute for Bioengineering at Aachen University of Applied Sciences, focusing on biomechanical modelling and simulation of hyper-elastic materials using the finite element method (FEM) and Smoothed FEM (SFEM). He received his doctoral degree in mechanical engineering in 2014 at RWTH Aachen University, Germany. Currently, he is a postdoctoral researcher at the University of Erlangen-Nuremberg with a topic on modelling and simulation of a heart support system.

Towards Patient-Specific Computational Modeling of hiPS-Derived Cardiomyocyte Function and Drug Action

Ralf Frotscher and Manfred Staat

1 Why Modeling hiPS-CM Function?

Human-induced pluripotent stem cell-derived cardiomyocytes (hiPS-CM) have been used in diverse experiments to model cardiac function in vitro. They serve as a valuable basis for the investigation of normal electromechanical cardiac function, of cardiac medication and of mutations while avoiding animal experiments or risking human life. This kind of cells can be produced nearly ethically uncritically and leads the way to an increasing amount of cardiac experiments on cellular, cluster and tissue level. The consideration of patient-specific investigation of cardiomyocytes or hiPS-CM, however, e.g. with respect to normal function, diseases or drug action leads to a vast amount of experiments to conduct. This approach seems to be impracticable as it is very costly and time-consuming. A possible remedy is to support experimental investigation by parameterizable simulations. The variety of today's cellular models and electrophysiological and mechanical data of cardiac cells is substantial with respect to species (mouse, rat, guinea pig, frog, human and many more) and cell type (sinoatrial, atrial, ventricular cells and fibroblasts including spatial heterogeneity). Those models differ in complexity and parameterization and they are often based on a mosaic of experimental data. This chapter largely focuses on the fact that, although they are supposed to be similar, the results that have been obtained by different groups in similar experiments, vary a lot due to multiple reasons, like mixture of cell types, in vitro environment, maturity, cell production batch, temperature and donor. Of course, the differences in experimental results challenge the direct and unmodified applicability of the models in a particular application. In the following sections we discuss some early results that we

R. Frotscher · M. Staat (✉)
Biomechanics Laboratory, Institute of Bioengineering,
University of Applied Sciences Aachen, Heinrich-Mußmann-Straße 1,
52428 Jülich, Germany
e-mail: Staat@FH-Aachen.de

G. M. Artmann et al. (eds.), *Biological, Physical and Technical Basics of Cell Engineering*, https://doi.org/10.1007/978-981-10-7904-7_10

received and that motivated us to tackle adaptive cardiac modeling going towards computational investigation of patient-specific cardiomyocytes or of individualized hiPS-CM with respect to differentiation stages, maturation and drug treatment.

In order to answer the question that this section asks in its title, we here present a list of reasons as a motivation to model hiPS-CM *in silico*. Each of the items is elaborated in this chapter in more detail:

- *Number of experiments:* In all disciplines computational models are employed to virtually enhance the number of experimental results. In many fields the scientific world suffers from the inability of performing experiments on real cells and tissue in a sufficiently large number due to time, cost and manpower.
- *HiPS-CM with realistic behavior:* The production of hiPS-CMs with similar behavior like real cells requires much effort and is expensive. In fact it is well known that today hiPS-CM in many cases do not express a realistic phenotypical ion channel behavior. Even more so if malfunctions and mutations need to be represented by these real cell models. Based on computational modeling and simulation it is comparably easy to modify the cells and to investigate them.
- *Costs:* Computational models are built on paper and can be executed with a computer. On the other hand real experiments are expensive and require manual effort for each test case. The computation of single cell models is even very inexpensive. However, facing the need of simulating cell clusters, tissues and even the whole organ, it has to be admitted that the models are introducing controlled simplifications in order to remain computable.
- *Variation:* Bearing the drawbacks in mind, computational models of hiPS-CM and CM enable a number of different simulation executions employing different parameter sets. This way it is easily possible to e.g. investigate the response of several configurations of cells to an applied drug. The very controlled way of representing biological variability is a double-edged sword as the ability arises but on the other hand it is currently impossible to computationally imitate the variability on the global scale when billions of cells are interacting.
- *Feedback:* We are presenting some results that produced feedback on how to perform and to interpret the in vitro experiments based on computational results.
- *Computational experiments:* In a computational environment it is easy to perform a number of experiments that might require a much larger effort in the lab, by simple modification of model parameters. The number of experiments can thus be greatly increased and only the relevant results might become reproduced in the lab. This leads to a considerable decrease in lab efforts.
- *Tissue constitution:* The mixture of cell types in hiPS-CM-tissue is somewhat quantifiable but less controllable. In simulations it is possible to investigate any kind of heterogeneous and anisotropic tissue constitution. Based on selected experimental results it is thus possible to extrapolate the results to different tissue properties.

Of course the question could arise: why not directly modeling the original cell types instead of a partially inaccurate copy? The reason is that many experiments can only be performed on hiPS-CM in a sufficiently large number. In consequence that means that for the aforementioned reasons there must exist computational models that imitate those cells in order to achieve comparability with hiPS-CM experimental results. In the end however, cardiomyocytes and cardiac tissue are in the focus of interest. This leads to the necessity to computationally map experimental results of hiPS-CM to normal cardiomyocytes which is one of the major open challenges we will address in the end of this chapter again. Using such a mapping it would be possible to perform experiments on hiPS-CM and to interpret the results with respect to real cardiomyocytes. Recent overviews about the differences between hiPS-CM and native CM and about the challenges in using the former as a model for the latter are given in Denning et al. [7], Hartman et al. [17].

2 Challenges in the Modeling of hiPS-CM Function and Drug Action

After praising simulations of hiPS-CM it should clearly be indicated that there are challenges in the modeling of hiPS-CM that require a highly interdisciplinary work. The combined knowledge of cardiologists, cellular biologists, engineers, mathematicians and IT specialists is required to solve all the problems that arise in theory and practice. First of all hiPS-CM possess unclear cellular properties that cannot entirely be quantified. Computational models however need quantified data in order to be predictive. The second challenge is the multi-scale modeling. One can perform experiments either at the macroscopic or the microscopic level but not on both scales at the same time. HiPS-CM modeling of course takes place at the microscopic level. But how to transfer the results to the macroscopic level? In other words: bridging the scales from isolated cells to tissue to anisotropic tissue to the organ is one key challenge that needs to be addressed by modelers. Each and every model available today either models an isolated cell very accurately with respect to mechanics and electrophysiology or is able to very accurately represent macroscopic effects like mechanical heart contraction or electrophysiology.

One example is shown in Fig. 1 where the class-IV antiarrhythmic agent verapamil has been applied to an isolated computational Purkinje fiber cell and to a real hiPS-CM monolayer tissue (60% ventricular) in increasing concentrations [23]. The observable effect in the beating frequency is opposing which can even be supported by experimental results [15, 16, 24, 27, 31] that show that the reaction of cardiomyocytes with respect to verapamil indeed differs between isolated cells and cells as part of a tissue. It is thus undeniable that multi-scale modeling is required for appropriate modeling of cardiac function despite its indisputable complexity. The chronotropic effect of verapamil is modeled in the simulations by the half maximal effective concentration EC_{50} which represents the concentration of a drug where 50% of its maximal effect is observed [11, 10]. This is not a deterministic measure of a single cell but a stochastic measure of a cell population because EC_{50}

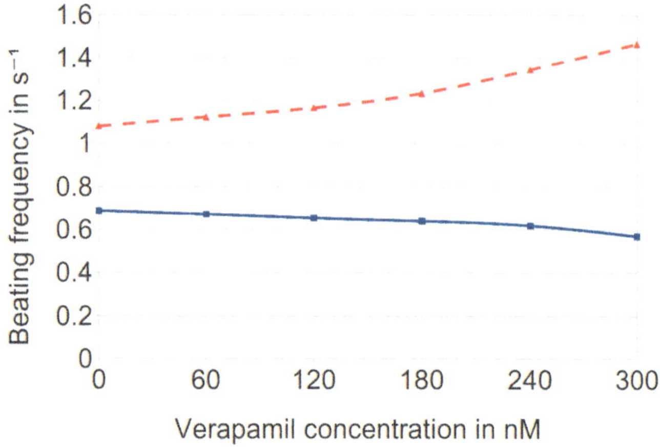

Fig. 1 Change of beating frequency of a modeled isolated cardiomyocyte (blue, solid line) and of a hiPS-CM monolayer (red, dashed line) at different verapamil concentrations

represents the concentration of a drug where 50% of the population exhibit a response after a specified exposure duration.

As the final and definitely most challenging influence we address the biological variability. The cellular behavior not only depends on species, cell type, day time, pH, mutations, phenotypical expressions and uncountable other external and internal influences. It also varies at so many levels that not all effects can be modeled or examined. Regarding the required model complexity that could reflect this number of effects one might only find stochastic processes as a sufficiently flexible modeling approach.

3 Isolated Cell Models

Most commonly, the available microscopic models[1] of cardiac cells are built to represent isolated cells although they are parameterized based on various experimental data that are differently obtained under varying conditions and even for mixed species depending on the cellular compartment. The cell models are specifically designed for the representation of a certain cell type. Within this chapter we indicate the several differently regarded cell types with an asterisk that herein can take the values n (nodal), s (sinoatrial), a (atrial), v (ventricular) and f (fibroblast).

No matter which kind of cellular model is regarded, a cellular membrane potential can be computed by the ordinary differential equation

[1]Implementations of relevant models are collected in OpenCell (www.cellml.org/tools/opencell).

$$\frac{\partial V_m^*}{\partial t} = \frac{1}{C_m^*}\left(I_{stim}^* - \sum_{i=1}^{n} I_i^*\left(g_{x_1}^*, g_{x_1}^*, \ldots\right)\right) \tag{1}$$

where $t, V_m^*, g_x^*, C_m^*, I_{stim}^*$ and I_i^* denote the time, the membrane action potential, the ion channel gates, the cell membrane capacitance, a potential electrical stimulus current and the individual ion channel currents, respectively. As it is obvious that the asterisked quantities are cell-type specific, the asterisk is omitted in the following due to brevity. Depending on the model complexity the number of regarded ionic currents n varies between two and approximately 20 and potentially more. Additionally the models often, but not always, define the relation to the mechanical excitation in terms of the tensile membrane stress resulting from the so-called excitation-contraction coupling as e.g. formulated in Niederer et al. [25], Niederer and Smith [26]

$$C_{a_b} = f_3(C_{a_i}, T, C_{a_b}) \tag{2}$$

$$z = f_2(z, \mathbf{B}, C_{a_b}), \tag{3}$$

$$T = f_1(\mathbf{B}, z), \tag{4}$$

where the ODE (2) determines the ion concentration C_{a_b} of calcium which binds to troponin C based on the free inner calcium ion concentration C_{a_i}. C_{a_b} determines the cellular activation measured by the activation parameter z in ODE (3). Equation (4) finally determines the active stress. The cellular excitation model determines the variation of C_{a_i} in time. Depending on the model complexity and on the experimental data that was available when creating the model, additional ordinary differential equations have been introduced in the various models for the potassium, sodium or cell-internal calcium kinetics.

Extensive overviews of models of isolated cardiac cells have been created [8] but due to the world-wide research effort invested in this field, it is impossible to provide a complete up-to-date overview. A more recent cell force model for instance has been published recently [20]. Herein we will focus on some few models that have been influential or directly have been used within our research.

3.1 Phenomenological Models

The work of [13] is very interesting as therein macrocopic modeling of the human heart directly incorporates microscopic modeling. The phenomenological FitzHugh-Nagumo model has been employed to describe the cellular cardiac processes. This model is phenomenological thus it is not applicable to our research as we are focussing on ion channel and electrophysiology modification that can only be performed on models that provide the required details. Phenomenological models however are very useful for obtaining insights into whole heart electrophysiology. A review of alternative deterministic and Markov models discusses many aspects of modelling cardiac physiology from single cell level to tissue level [6]. The review also considers publications on how to handle intra- and inter-individual variation or how to account for distinct subcategories of patients such as those with genetic channelopathies or underlying comorbidities.

3.2 Hodgkin-Huxley Models

Many of the currently applied models in single cell, tissue or organ simulations base on the concept of Hodgkin and Huxley [18], among which the models that we have been employing so far [23, 28, 29]. Hodgkin-Huxley models represent the phenotypical function of each ion channel by one or more gate differential equations which leads to a system of ordinary differential equations in the form of

$$\frac{\partial g_x^*(V_m)}{\partial t} = \alpha_x^+ \left(V_m^*\right)\left(1 - g_x^*\right) + \alpha_x^- \left(V_m^*\right) g_x^* \tag{5}$$

where t, V_m, g_x^*, α_x^+ and α_x^- denote the time, the membrane action potential, the ion channel gates, the activation and inactivation constants of the respective channel, respectively. Therein, x indicates the numerous different gates to describe the ion channel kinetics within a cell.

3.3 Markov Models

The quite recent approach of using Markov processes models the ion channel expression rather than its phenotype. In Clancy and Rudy [3] a Markov model of the fast sodium current I_{Na} that contributes to the depolarization phase of the membrane potential has been created. It replaces the standard Hodgkin-Huxley representation of this current in the Luo-Rudy model [22].

Figure 2 illustrates the Luo-Rudy model of the fast sodium channel. Six different states have been modeled, namely three different closed states $C1$–$C3$, fast and slow inactivation states IF and IS and an open state O_{Na}. The notation

$$C1 \underset{k_{OC1}}{\overset{k_{C1O}}{\rightleftarrows}} O_{Na} \tag{6}$$

expresses that the closed state $C1$ opens with the transition rate k_{C1O} and the open state O_{Na} closes with the transition rate k_{OC1}. Let $c1$ and o_{Na} be the probability for an ion channel to be in one of the states $C1$ and O_{Na}, respectively. In equilibrium $k_{C1O}c1 = k_{OC1}o_{Na}$. Similar equations characterize all the other transitions. The

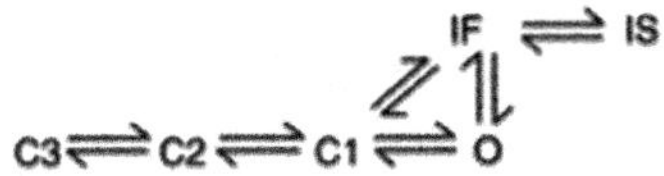

Fig. 2 The fast sodium channel of Luo-Rudy, 1994

probabilities of being in a certain state are computed by solving the associated system of ordinary differential equations (master equation). In the end there will be a certain probability o_{Na} of a fast sodium channel being in the open state O_{Na} and the corresponding ion current is computed as usual by

$$I_{Na} = G_{Na} o_{Na} (V_m - E_{Na}) \tag{7}$$

with maximal channel conductance G_{Na}, membrane potential V_m and reversal potential E_{Na}. In Hodgkin-Huxley models o_{Na} is replaced by a product of gate variables which govern opening and closure. The most important observations with respect to the Markov modeling approach are:

1. Replacing parts of a Hodgkin-Huxley type model by Markov processes does not change the general mathematical problem. The system to solve on the cellular level still is a potentially stiff system of ordinary differential equations.
2. The opening, closure, activation and inactivation kinetics of an ion channel can be modeled in very detail by introducing more states. Separation of slow and fast inactivation, for instance, is rather simple.
3. The heterogeneity of ion channels within one cell is directly reflected by the states in which they can be with a certain probability. This way the ion channel expression is modeled rather than its phenotype.

The main reasons to represent at least some ion channels of hiPS-CM using a Markov model are

- to model complex channel kinetics,
- to compute ion channel phenotypes, rather than prescribing them,
- the resulting heterogeneity of ion channels,
- to model known changes in protein function less phenomenologically,
- to individualize the models,
- to model a disease or mutants,
- to model drug effects or even design theoretical drugs, and
- to adapt them to mammalian cardiomyocytes for which significant experimental data can be found in the literature for model validation.

In the focus of our interest is that the model accurately reproduces the action potential of the hiPS-CM and that we can show its ability to show appropriate reactions to drug treatment. Adapting the model to mammalian cells can be done based on the extensive amount of available literature data and is most important for validation purposes. In a next step, given that some important differences between hiPS-CM and human CM are known and can be quantified, the model can serve as a basis to transfer the experimental results that we get for hiPS-CM to native human cells. Another point of interest is the modeling of the effect of age on the cell function. Although this task is much more challenging, future models like this might be applicable to the simulation of mutations and the changes they make to the cellular processes. Moreover, diseases and their treatment could be simulated such that the interpretation of clinical experimental results can be supported.

4 Cell Cluster and Tissue Models

So far, most of the considerations presented in here are with respect to single cell modeling and simulation. Hodgkin-Huxley based models of low complexity already have been used in tissue and whole organ simulations by applying different models in different regions of the heart, e.g. recently in Augustin et al. [1]. Given the fact that the contractile behavior of the tissue has to be computed at each node of the discretization, it is a matter of computation time, how detailed the models of ion channel, ion, troponin C, crossbridge or actin-myosin kinetics can be. The Markov approach discussed in Sect. 3.3 would introduce even more differential equations for each modeled ion channel which leads to a detailed model of the respective cell but in turn it is questionable how well these models can be applied in large scale simulations due to the resulting computational costs.

It has already been exemplarily indicated that the functional relation between the scales needs to be taken into account when attempting to simulate an accurate cellular and macroscopic response. If, due to limitation of computational resources, the microscopic level is not modeled in detail, macroscopic models usually become adopted phenomenologically. However, this approach is not effective when for instance drug effects are regarded. Ideally, the microscopic models influence and dictate the macroscopic tissue response as e.g. actions like drug application are effective at the cellular level. Moreover it has been illustrated already that isolated cells behave differently compared to the same cells in an environment (cluster, tissue, organ, anisotropic). This requires an accurate modeling of interactions between cells with respect to their containing cell cluster, tissue or organ.

One typically distinguishes between mono- and bidomain models. The former are models that homogenize the contributions of the different tissue constituents, mechanically as well as electrophysiologically. The latter modeling type respects the differences between at least two different tissue constituents, e.g. cell and extra-cellular matrix.

In our model of cardiac tissue we follow the monodomain approach and model the cellular interaction mechanically and electrophysiologically in terms of homogenization as formulated by Keip et al. [19]. The tissue we are investigating consists of four different cell types, namely ventricular (v), atrial (a) and nodal (n) CM and fibroblasts (f). Assuming that all tissues and cells experience the same strain, we arithmetically average the Cauchy stresses of the cell compound (c) as

$$\boldsymbol{\sigma}^c = \theta_v \boldsymbol{\sigma}^v + \theta_a \boldsymbol{\sigma}^a + \theta_n \boldsymbol{\sigma}^n + \theta_f \boldsymbol{\sigma}^f \tag{8}$$

with volume fractions θ_* that in total sum up to 1. Mechanically the homogenized cellular stresses then read as

$$\boldsymbol{\sigma} = \theta_t \left(2J^{-1}\mathbf{B}\partial\psi^m_{nH} \frac{(\mathbf{B}; C^m_{10})}{\partial\mathbf{B}} + \sum_{*=v,n,a,f} \theta_* \left(\frac{1}{4}\omega^*_0 J^{-1}\mathbf{B}\frac{\partial\psi^*_{nH}}{\partial B} + \frac{1}{2}T^*_0(\mathbf{B},t)\mathbf{I} \right) \right) - p\mathbf{I} \tag{9}$$

for a cardiac tissue. Therein, θ_t and θ_* refer to the portions of the tissue and of the cell types, respectively, J being the Jacobian determinant, $\mathbf{B}$ the left Cauchy-Green tensor, ψ^m_{nH} and ψ^*_{nH} the strain energies of the membrane and the cell types, respectively, ω^*_0 being weights, T^*_0 the cellular tension and finally p being the hydrostatic pressure. Details can be found in Frotscher and Staat [12]. A nice overview of the modeling of whole heart electrophysiology can be found in Clayton et al. [5].

5 Adaptivity of Cell Models

Quantitatively accurate computational modeling of any cell with respect to drug action, mutations or individualization often fails due to numerous influences on the model parameters, like cell type, mixture of cell types, maturity, cell production batch, the in vitro environment, temperature or donor. Moreover, individualization with respect to a specific patient and modification with respect to diseases of the widely used Hodgkin-Huxley models requires such an amount of experiments that cannot be performed in time-critical patient-specific treatment. It is already known that a possible remedy is the combination with Markov processes that allow the modeling of an ion channel genotype and determine the ion channel expression rather than modeling the ion channel phenomenologically. Changing the model with respect to known changes in protein function that result from a disease, a drug or a different cell type and species, experimental results for easily available hiPS-CM can directly be used to study those effects on the target cell type. In the following we describe how to combine electrophysiological data, mechanical experiments and the modeling of ion channel expression to simulate drug action, diseases and mutations on patient-specific human cardiomyocytes.

Although the experiments are first explained in Sect. 6.1 at this point we want to illustrate the adaptivity of cell models using some of the simulation results in order to demonstrate the necessity of model adaptation.

As an example, Fig. 3 shows the concentration-dependent change in contraction force represented through membrane deflection due to application of lidocaine in both, experiment and simulation. From this and other results it can be seen that the current model is able to qualitatively predict the inotropic response of the cardiac tissue with respect to drugs. However, the initial change in deflection, i.e. the change in deflection of untreated tissue is already different in experiment and simulation. This quite large difference can be observed although the model has been parameterized based on many pressure-deflection curves obtained from CellDrum experiments. Obviously the range in experimental results, i.e. in inotropy and chronotropy of the cells, is wide thus one way to produce quantitatively more correct results could be a phenomenological adjustment of the model. Including the discussed one, one can make multiple observations that all lead to reconsider the way of modeling the cell electrophysiology: The variation and standard deviation of results in electrophysiological and mechanical experiments is big and depends on multiple factors like cell type, mixture of cell types, in vitro environment, maturity, cell production batch, temperature and donor. We observed that both, the beating frequency and the cellular contraction are significantly different between different cell cultures that are supposed to show comparable behavior [9]. Our current model is not sufficiently flexible to reproduce the behavior of the heterogeneous cells in all the cases. One example is shown in Fig. 4 where the model predicts an opposite chronotropic effect of verapamil than observed in the experiment. One possible explanation is that the model has been developed for a different cell type (human Purkinje fiber cells) and therefore does not capture the actual behavior of the mixture of nodal and ventricular cells that are present in the tissue. In addition to

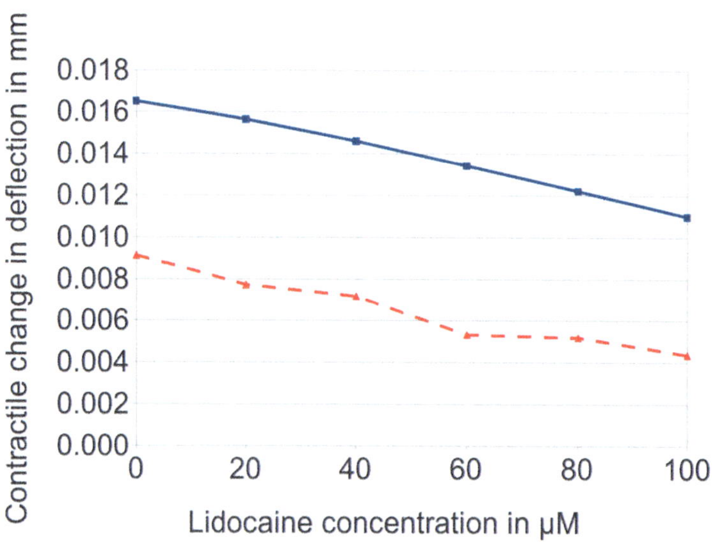

Fig. 3 Change in deflection of the CellDrum with respect to different lidocaine concentrations. The dashed red and solid blue curves are experimental and simulation results, respectively

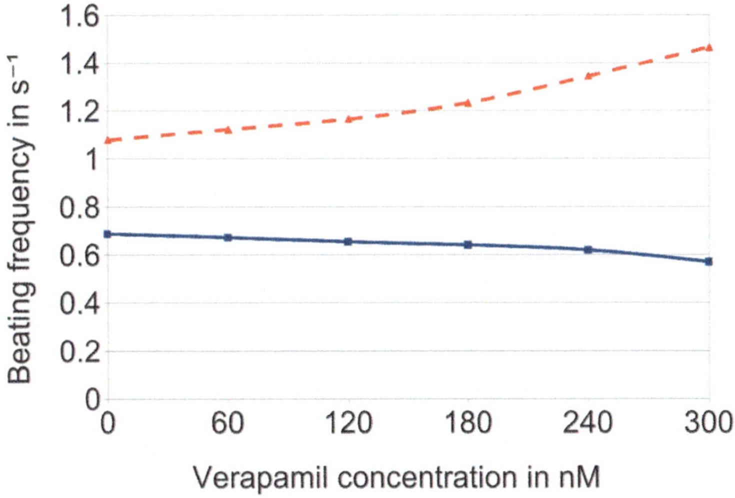

Fig. 4 Change in beating frequency of the cardiomyocytes with respect to different verapamil concentrations. The dashed red and solid blue curves are experimental and simulation results, respectively

that it has already been reported by other groups, that experimental results obtained for a cell monolayer even qualitatively differ from those obtained for single cells, cell clusters or whole organs [16].

A phenomenological adjustment of the simulations depending on the experiments is not simple and hardly possible because beating force and frequency are not decoupled. The so-called staircase effect shows that an increased beating frequency leads to a larger inner calcium concentration and vice versa which can hardly be represented phenomenologically. Thus, a simple adjustment of beating force and frequency would not reflect the proper changes in the inner workings of the cell. These observations all lead to a modeling that allows for individualization thus shows a certain flexibility. Related to that is the individualization of the Markov ion channel model. One way of individualization is the adjustment of the transition rates between the states. For instance, Bondarenko et al. succeeded in reparameterizing the Markov model of the fast sodium channel in Bondarenko et al. [2] to match mouse data. Another way of individualization is the introduction of additional states as for instance in Clancy and Rudy [3, 4]. In the former one the model shown in Fig. 2 has been supplemented by four additional states to model the changes in protein function arising from the KQP mutation in the SCN5A gene that causes long-QT syndrome. In the latter one the authors clearly showed the benefit of Markov models being able to reflect heterogeneity. Therein the I_{Na} model has been amended again by four states and some transition rates have been adjusted to account for the 1795insD mutation in SCN5A. This mutation causes both, a long-QT syndrome and a Brugada syndrome which result in an enhanced and a reduced function of the fast sodium channel at the same time. Being able to model those opposing effects is important to distinct both from each other and to get the appropriate cellular response. Although mutation is not in the focus of our work, these investigations show the modeling flexibility when using Markov models. The solutions of the system of partial differential equations obtained from Markov models of the cells including the drug effects can be understood as functions of the parameters in the drug models. Properties of drugs can be presented by free parameters in the modeling differential equations. The parameters can be chosen so that a best solution is achieved. Tveito and Lines [30] think this procedure one step further asking what parameters should a potential (not yet existing theoretical) drug have so that the solution best assumes the solution of a healthy model, in other words so that the drug best compensates disease or mutants.

6 Example for Model Adaptivity: Medication

In this section we want to present the ability of state-of-the-art models to reflect sample- and patient-specificity. Patient specificity requires patient-specific data that, at the level of detail that is required for cellular models, today is simply not obtainable. Reality-fitting cellular models have to be very well parameterized and a large number of experiments under varying conditions has to be performed for an accurate model. The closest we could come is sample-specificity, i.e. we adopted isolated cell models to macroscopic observations which we made with the CellDrum shown in Fig. 5.

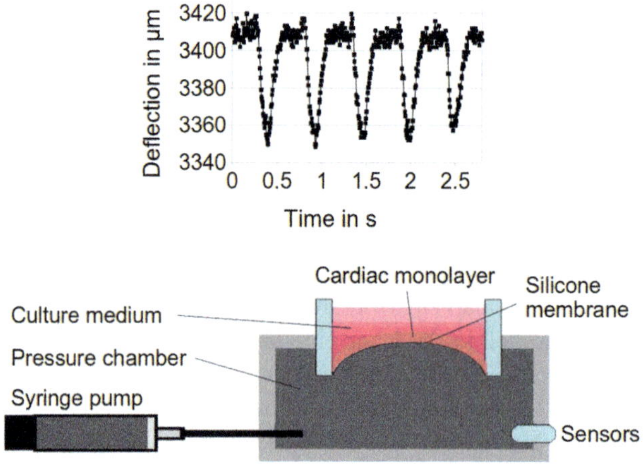

Fig. 5 CellDrum™ setup and deflection over time due to autonomous contraction

6.1 CellDrum and Drug Application

The experimental setup to compare our models has been developed in the Laboratory of Medical and Cell Biology at Aachen University of Applied Sciences, and is called CellDrum. Its intention is to measure cell contraction, frequency and forces. The CellDrum itself is a 3–4 μm thin, fully clamped, circular silicone membrane with a diameter of 16 mm. On top of the membrane, a monolayer consisting of hiPS-CM (Core.4U cell line with approximately 60% ventricular, 22% nodal and 18% atrial cardiomyocytes, Axiogenesis, Cologne, Germany) embedded in collagen can be chemically fixed to the silicone membrane. After some days of cultivation the sample can be put into the so-called Tissue Tension Analyzer that applies a pressure to the membrane and measures its resulting deflection. Moreover, within this setup it is possible to apply medication to the tissue and to investigate the changes in beating force, beating frequency, contraction duration and others. Details of the CellDrum can be found within this book and in Linder et al. [21].

6.2 Comparison Between Experimental and Computational Results

As a motivation for phenomenological model adjustments we started from Fig. 6. It shows the normalized contractile change in deflection as a measure of the beating force of the composite in the experiment and in two different simulations. It can be seen that if the ventricular model, which is parameterized for a 1 Hz stimulus, is paced at 1 Hz, the simulation can qualitatively capture the experimental result. The cells in the experiment, however show a beating frequency of 0.5 Hz in the control group that even decreases when lidocaine is applied. Consequently, for a quantitative agreement we paced the model at experimentally determined beating frequencies and observe that the so-called force-frequency relationship (FFR) overly inhibits the beating force. It is obvious that, although the FFR actually occurs in cardiomyocytes, the model overreacts to it with respect to the pacing. From this we conclude that the model needs to be modified by means of reparameterization with respect to the externally applied stimulus. Therefore, we run literature-supported parameter studies in order to determine the main influential ion currents on the autonomous beating frequency of a sinoatrial model. Figure 7 shows simulation results for drug application on the sinoatrial model that is altered by some modifications of different ion channel conductances g_x. It can be seen that some modifications lead to a model beating frequency that is much closer to the experimental one of approximately 0.5 Hz under control conditions. The effect of the drug application though varies a lot depending on the chosen model modification. More experimental and simulation results including results when applying other drugs are discussed in Frotscher et al. [10, 11], Goßmann et al. [14].

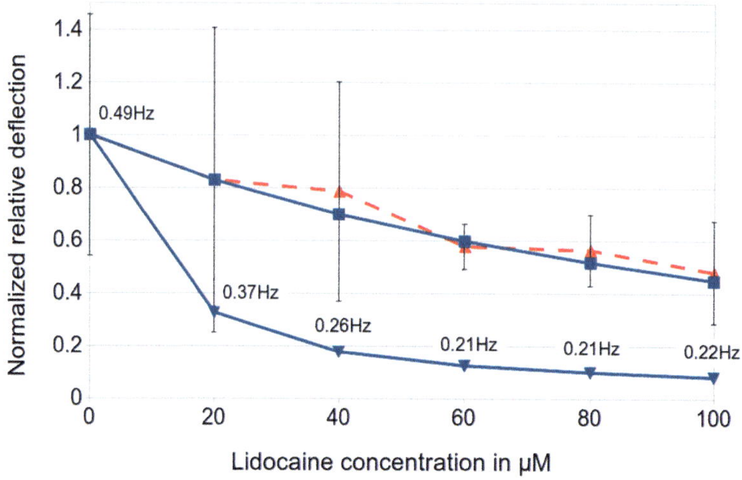

Fig. 6 Relative deflection of CellDrum tissue with respect to lidocaine concentration. The red curve, the blue curve with square and the one with triangular markers show the experimental, the simulated results at 0.5 Hz and at 1 Hz pacing, respectively

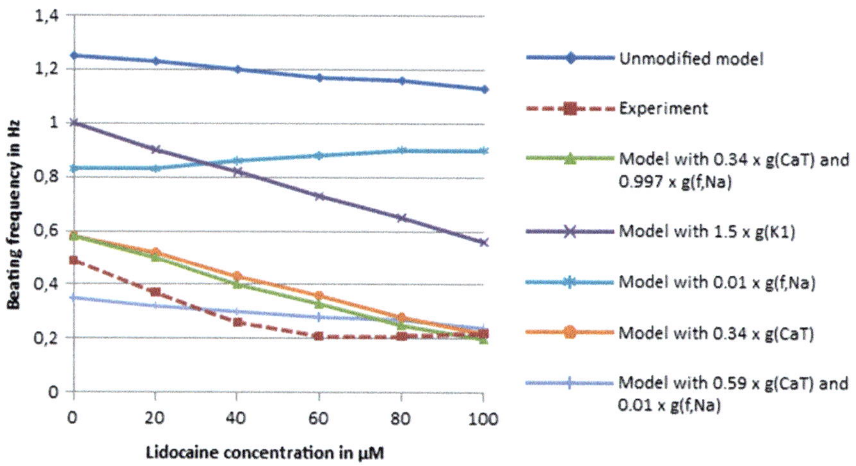

Fig. 7 The change in beating frequency of a sinoatrial cell model with respect to different changes in gate conductance when lidocaine is applied with increasing concentrations. The experimentally derived curve represents a heterogeneous monolayer

7 Open Challenges for Patient-Specific Modeling

Within a certain range of model modifications we succeeded in adapting the beating forces and frequencies of hiPS-CM tissues even though the force and frequency have been measured macroscopically and the adoptions have been made at the microscopic level. It has to be concluded that a sample-specific model adoption indeed is possible if the differences in phenotypical expression of cellular ion channels within the range of hiPS-CM can be determined. One open issue in the modeling is that the real differences in the cellular electrophysiology between the hiPS-CM tissue and the model are not known. In order to represent different behavior in the models anyway one currently has to intelligently guess the details and to modify the models accordingly. This model adoption thus is a challenging interdisciplinary work. Purely numerical thus phenomenological adaptations of the models are not reasonable because the model modifications need to remain realistic and physiologically correct. The derivation of an approach to perform physiologically correct sample-specific model adoption remains open.

However, the proper modeling of hiPS-CM based experimental results is not even sufficient because hiPS-CM themselves are only a model representing real cardiomyocytes. Today it is not yet possible to properly reflect the complex properties of mature human CM using hiPS-CM although they provide a great basis for performing experiments on human-like CMs. As long as hiPS-CM deficiently represent standard human CM, models might help in transferring hiPS-derived experimental results to healthy mature human cells. In the end this brought us to the necessity to map experimental results of hiPS-CM to normal cardiomyocytes which is the major open challenge.

As indicated in the introduction of this chapter, the objective is to model and to simulate mutation, medication and illness of cardiac tissue. This, however is impossible due to numerous external influences, due to biological variability, due to aging and due to the complexity of the involved bio-electrophysiological-chemical processes. Upcoming numerical models of hiPS-CM or CM shall contain or allow the possibility to be somewhat flexible in order to reflect the aforementioned variability of cells. On the basis of realistic models of healthy and mutant CM a vision of the future would be to adapt the models of drug effects so that disease can be healed in silico. This would guide towards the desired properties of new drugs.

Patient-specific simulations of cellular and cardiac function are currently far from being state-of-the-art although they are highly required. This is especially true regarding the fact that more and more experiments are performed on representative tissue like hiPS-CM-based tissue. Allowing computational models to be patient-specific will provide a valuable basis for experiments. However, this requires a close interaction between computational engineers, cardiologists and cell biologists.

References

1. Augustin, C. M., Neic, A., Liebmann, M., Prassl, A. J., Niederer, S. A., Haase, G., et al. (2016). Anatomically accurate high resolution modeling of human whole heart electromechanics: A strongly scalable algebraic multigrid solver method for nonlinear deformation. *Journal of Computational Physics, 305,* 622–646. https://doi.org/10.1016/j.jcp.2015.10.045.

2. Bondarenko, V. E., Szigeti, G. P., Bett, G. C. L., Kim, S.-J., & Rasmusson, R. L. (2004). Computer model of action potential of mouse ventricular myocytes. *American Journal of Physiology Heart and Circulatory Physiology, 287*(3), H1378–H1403. https://doi.org/10.1152/ajpheart.00185.2003.
3. Clancy, C. E., & Rudy, Y. (1999). Linking a genetic defect to its cellular phenotype in a cardiac arrhythmia. *Nature, 400*(6744), 566–569. https://doi.org/10.1038/23034.
4. Clancy, C. E., & Rudy, Y. (2002). Na^+ channel mutation that causes both Brugada and long-QT syndrome phenotypes: A simulation study of mechanism. *Circulation, 105*(10), 1208–1213. https://doi.org/10.1161/hc1002.105183.
5. Clayton, R. H., Bernus, O., Cherry, E. M., Dierckx, H., Fenton, F. H., Mirabella, L., … Zhang, H. (2011). Models of cardiac tissue electrophysiology: Progress, challenges and open questions. *Progress in Biophysics and Molecular Biology, 104*(1–3), 22–48. https://doi.org/10.1016/j.pbiomolbio.2010.05.008.
6. Davies, M. R., Wang, K., Mirams, G. R., Caruso, A., Noble, D., Walz, A., …, Polonchuk, L. (2016). Recent developments in using mechanistic cardiac modelling for drug safety evaluation. *Drug Discovery Today, 21*(6), 924–938. https://doi.org/10.1016/j.drudis.2016.02.003.
7. Denning, C., Borgdorff, V., Crutchley, J., Firth, K. S. A., George, V., Kalra, S., … Young, L. E. (2016). Cardiomyocytes from human pluripotent stem cells: From laboratory curiosity to industrial biomedical platform. *Biochimica et Biophysica Acta (BBA)—Molecular Cell Research, 1863*(7), 1728–1748. https://doi.org/10.1016/j.bbamcr.2015.10.014.
8. Fenton, F., & Cherry, E. (2008). Models of cardiac cell. https://doi.org/10.4249/scholarpedia.1868.
9. Frotscher, R., Koch, J.-P., Raatschen, H.-J., & Staat, M. (2014). Evaluation of a computational model for drug action on cardiac tissue. In E. Oñate J. Oliver & A. Huerta (Eds.), *Proceedings 11th World Congress on Computational Mechanics (WCCM XI), 5th European Conference on Computational Mechanics (ECCM V), 6th European Conference on Computational Fluid Dynamics (ECFD VI)* (pp. 1425–1436) Barcelona, Spain. https://doi.org/10.21269/6507.
10. Frotscher, R., Koch, J.-P., & Staat, M. (2015). Computational investigation of drug action on human-induced stem cell-derived cardiomyocytes. *Journal of Biomechanical Engineering, 137*(7), 71002. https://doi.org/10.1115/1.4030173.
11. Frotscher, R., Muanghong, D., Dursun, G., Goßmann, M., Temiz-Artmann, A., & Staat, M. (2016). Sample-specific adaption of an improved electro-mechanical model of in vitro cardiac tissue. *Journal of Biomechanics, 49*(12), 2428–2435. https://doi.org/10.1016/j.jbiomech.2016.01.039.
12. Frotscher, R., & Staat, M. (2015). Homogenization of a cardiac tissue construct. In P. Nithiarasu & E. Budyn (Eds.), *Proceedings of the 4th International Conference on Computational and Mathematical Biomedical Engineering—CMBE2015*, 29 June–1 July 2015, Cachan (Paris), France. Zeta Computational Resources Ltd., Swansea, UK (pp. 645–648). https://dx.doi.org/10.21269/7263.
13. Göktepe, S., & Kuhl, E. (2010). Electromechanics of the heart: A unified approach to the strongly coupled excitation-contraction problem. *Computational Mechanics, 45*(2–3), 227–243. https://doi.org/10.1007/s00466-009-0434-z.
14. Goßmann, M., Frotscher, R., Linder, P., Neumann, S., Bayer, R., Epple, M., … Artmann, G. M. (2016). Mechano-pharmacological characterization of cardiomyocytes derived from human induced pluripotent stem cells. *Cellular Physiology and Biochemistry, 38*(3), 1182–1198. https://doi.org/10.1159/000443124.
15. Guo, L., Abrams, R. M. C., Babiarz, J. E., Cohen, J. D., Kameoka, S., Sanders, M. J., … Kolaja, K. L. (2011). Estimating the risk of drug-induced proarrhythmia using human induced pluripotent stem cell-derived cardiomyocytes. *Toxicological Sciences : An Official Journal of the Society of Toxicology, 123*(1), 281–289. https://doi.org/10.1093/toxsci/kfr158.

16. Harris, K., Aylott, M., Cui, Y., Louttit, J. B., McMahon, N. C., & Sridhar, A. (2013). Comparison of electrophysiological data from human-induced pluripotent stem cell-derived cardiomyocytes to functional preclinical safety assays. *Toxicological Sciences, 134*(2), 412–426. https://doi.org/10.1093/toxsci/kft113.
17. Hartman, M. E., Dai, D.-F., & Laflamme, M. A. (2016). Human pluripotent stem cells: Prospects and challenges as a source of cardiomyocytes for in vitro modeling and cell-based cardiac repair. *Advanced Drug Delivery Reviews, 96,* 3–17. https://doi.org/10.1016/j.addr.2015.05.004.
18. Hodgkin, A. L., & Huxley, A. F. (1952). A quantitative description of membrane current and its application to conduction and excitation in nerve. *The Journal of Physiology, 117*(4), 500–544. https://doi.org/10.1113/jphysiol.1952.sp004764.
19. Keip, M.-A., Steinmann, P., & Schröder, J. (2014). Two-scale computational homogenization of electro-elasticity at finite strains. *Computer Methods in Applied Mechanics and Engineering, 278,* 62–79. https://doi.org/10.1016/j.cma.2014.04.020.
20. Land, S., Park-Holohan, S.-J., Smith, N. P., dos Remedios, C. G., Kentish, J. C., & Niederer, S. A. (2017). A model of cardiac contraction based on novel measurements of tension development in human cardiomyocytes. *Journal of Molecular and Cellular Cardiology, 106,* 68–83. https://doi.org/10.1016/j.yjmcc.2017.03.008.
21. Linder, P., Trzewik, J., Rüffer, M., Artmann, G. M., Digel, I., Kurz, R., … Temiz Artmann, A. (2010). Contractile tension and beating rates of self-exciting monolayers and 3D-tissue constructs of neonatal rat cardiomyocytes. *Medical & Biological Engineering & Computing, 48*(1), 59–65. https://doi.org/10.1007/s11517-009-0552-y.
22. Luo, C. H., & Rudy, Y. (1994). A dynamic model of the cardiac ventricular action potential. I. Simulations of ionic currents and concentration changes. *Circulation Research, 74*(6), 1071–1096. https://doi.org/10.1161/01.RES.74.6.1071.
23. McAllister, R. E., Noble, D., & Tsien, R. W. (1975). Reconstruction of the electrical activity of cardiac Purkinje fibres. *The Journal of Physiology, 251*(1), 1–59. https://doi.org/10.1113/jphysiol.1975.sp011080.
24. Mehta, A., Chung, Y. Y., Ng, A., Iskandar, F., Atan, S., Wei, H., … Shim, W. (2011). Pharmacological response of human cardiomyocytes derived from virus-free induced pluripotent stem cells. *Cardiovascular Research, 91*(4), 577–586. https://doi.org/10.1093/cvr/cvr132.
25. Niederer, S. A., Hunter, P. J., & Smith, N. P. (2006). A quantitative analysis of cardiac myocyte relaxation: A simulation study. *Biophysical Journal, 90*(5), 1697–1722. https://doi.org/10.1529/biophysj.105.069534.
26. Niederer, S. A., & Smith, N. P. (2008). An improved numerical method for strong coupling of excitation and contraction models in the heart. *Progress in Biophysics and Molecular Biology, 96*(1–3), 90–111. https://doi.org/10.1016/j.pbiomolbio.2007.08.001.
27. Sirenko, O., Crittenden, C., Callamaras, N., Hesley, J., Chen, Y.-W., Funes, C., … Cromwell, E. F. (2013). Multiparameter in vitro assessment of compound effects on cardiomyocyte physiology using iPSC cells. *Journal of Biomolecular Screening, 18*(1), 39–53. https://doi.org/10.1177/1087057112457590.
28. Stewart, P., Aslanidi, O. V., Noble, D., Noble, P. J., Boyett, M. R., & Zhang, H. (2009). Mathematical models of the electrical action potential of Purkinje fibre cells. *Philosophical Transactions. Series A, Mathematical, Physical, and Engineering Sciences, 367*(1896), 2225–2255. https://doi.org/10.1098/rsta.2008.0283.
29. ten Tusscher, K. H. W. J., & Panfilov, A. V. (2006). Alternans and spiral breakup in a human ventricular tissue model. *American Journal of Physiology Heart and Circulatory Physiology, 291*(3), H1088–H1100. https://doi.org/10.1152/ajpheart.00109.2006.

30. Tveito, A., & Lines, G. T. (2016). *Computing characterizations of drugs for ion channels and receptors using Markov models* (Lecture Notes in Computational Science and Engineering Vol. 111). Cham: Springer International Publishing. https://doi.org/10.1007/978-3-319-30030-6.
31. Yokoo, N., Baba, S., Kaichi, S., Niwa, A., Mima, T., Doi, H., … Heike, T. (2009). The effects of cardioactive drugs on cardiomyocytes derived from human induced pluripotent stem cells. *Biochemical and Biophysical Research Communications, 387*(3), 482–488. https://doi.org/10.1016/j.bbrc.2009.07.052.

Author Biography

Ralf Frotscher (Germany), Dr. Ing. He studied Applied Mathematics and Information Technologies at Aachen University of Applied Sciences. He was a researcher at the Biomechanics Lab at the same University focusing on electromechanical modeling and simulation of cardiac tissue and of surgical mesh implants in the female pelvic floor. He received his Dr.-degree in computational biomechanics in 2016 at the University Duisburg-Essen. Currently he is employed at TWT GmbH Science & Innovation, Stuttgart, as Research Manager where he is heading the department of Model-based Systems Engineering.

Mechanics of Soft Tissue Reactions to Textile Mesh Implants

Aroj Bhattarai and Manfred Staat

1 Introduction

Prosthetic or surgical meshes are permanently implanted inside the body to repair hernia and urogynecological disorders such as pelvic organ prolapse (POP) and incontinence that occurs due to weakened/damaged supporting tissues. After the first ever design of the polypropylene surgical mesh to repair abdominal hernia in the late 1950s [83], a novel use of biological materials convincingly reduced the rate of recurrence (less than 10%) compared to the conventional suture technique with 50% recurrence rate [14, 76]. Eventually, in the 1970s, surgeons began inserting surgical meshes to correct POP and stress urinary incontinence also. Advanced research in the last 50 years on the repair of abdominal wall hernia and pelvic floor dysfunctions has achieved significant improvements in the reconstruction techniques and a rapid growth in varieties of prosthetic meshes. Today, it is estimated that more than 200 different textile constructs are available with variations in polymer and pore structure [46] and more than 20 million meshes are implanted per year world-wide [9, 72].

In initial days of hernia repair, the prosthetic meshes constructed from stronger material such as polypropylene were popularly used due to their excellent biocompatibility, lowest failure rates, improved tolerance to bacterial infection and good cost/benefits of the polymer [1]. Such early mesh constructions were heavy weight with small pore size and larger mesh area. Over the years, despite clinical success, numerous post-operation complications were reported later with such implantations in the form of strong foreign body reaction such as inflammatory

A. Bhattarai (✉) · M. Staat
Biomechanics Laboratory, Institute for Bioengineering, University of Applied Sciences Aachen, Heinrich-Mußmann-Str. 1, 52428 Jülich, Germany
e-mail: Bhattarai@fh-aachen.de

G. M. Artmann et al. (eds.), *Biological, Physical and Technical Basics of Cell Engineering*, https://doi.org/10.1007/978-981-10-7904-7_11

reaction [13, 62], larger mesh shrinkage [3] and adhesion formation to adjacent viscera [54, 84] that ultimately affected the healing process [17, 18]. However, symptoms varied for several reasons in character and severity from patient to patient. Native tissue compliance loss with excessively stiff, dense, heavy and small pore mesh were also found [85]. Mesh shrinkage [3], scar tissue formation [85] and mechanical mismatch between native abdominal tissue and prosthesis [6, 24] were found to be the reason for recurrence, treatment failure that forces complete removal of the implant from the body. Early meshes with high mesh stiffness were also related to the likelihood of post-operative pain, unnatural or difficult organ movements, central mesh ruptures and biomechanical and biological degeneration of host tissue as erosion and exposure [3, 21, 26, 27, 52, 58].

Over the seven-year period between 1997 and 2005, influence of mesh weight and mesh porosity were identified to be decisive for biocompatibility and stabilization of the surgical treatment [2, 46, 50]. The concept have been widely accepted among surgeons and manufacturers that systematically updated the mesh design changing polymer type (polytetrafluroethylene (PTFE), polyethylene terephthalate (PET, PETE), expanded PTFE and polyvinylidene fluoride (PVDF)), pore size, pore geometry and biocompatibility of the mesh implants [20, 42]. Clinical trials showed the superiority of the light weight large porous textile constructions with significantly improved integration of the mesh into recipient tissues, decreased risk of bacterial colonization, reduced inflammation and fibrous reaction, and improved quality of life after repair [50, 51, 66, 68, 70]. Further in 2008, Mühl et al. [63] introduced the concept of effective porosity to characterize the textile porosity where large pore meshes preserves the effective porosity under deformation and avoids formation of scar bridges.

Mesh development is an ongoing process, a good mesh should optimally fulfill certain requirements, such as negligible foreign body reaction with no pathological fibrosis, flexibility with adaptable stiffness, adequate adhesion especially near the sutures, and mesh dynamics close to the anisotropic host tissue. Since, every prosthesis possess different mechanical properties, the response to the native tissues varies hugely making it difficult for surgeons to select the most appropriate mesh and its orientation for each type of patient with each type of disorder. To achieve a physiological behavior of the implanted mesh in the body, engineers and manufacturers should be aware of how the host tissue would respond with the implantation of such biomaterials. This chapter reviews some mesh related complications arising from the mechanical behavior of the mesh implants. Standard protocols for uniaxial tensile test in orthogonal symmetry directions are used to characterize the compressible dry meshes. The characterization should help predict how well a surgical mesh may contribute to the short-term and long-term success of the repair for hernia and urogynecological disorders.

2 Prosthetic Meshes

Biologically and mechanically competent tissue-mesh response is a key factor in repairing herniation and pelvic floor dysfunctions. An ideal synthetic mesh should always provide structural support to the organs, restore anatomical function and treat physiological disorders with post-operation complications as less as possible. In order to achieve idealistic functionality and biological compatibility of prosthetic implants, properties of (type, dimensions and mechanics) polymer and (size and deformation under force) pores should be carefully understood.

2.1 Textile Structures

Textile constructions can be monofilament or multifilament. Common monofilament meshes are very distensible, have thicker fibrils (100–150 μm) and large pore whereas, most multifilament meshes are softer, less distensible with thinner fibrils (20–30 μm) and smaller pores. Due to less inter-fiber space, multifilament meshes do not allow macrophages and neutrophils to enter through small pores. This provides comfortable environment for bacteria to survive within pores making such meshes at highest risk of infection.

2.2 Textile Porosity

Porosity is defined as the ratio of open space relative to the total area of the textile. It is an important parameter to predict the biocompatibility of the synthetic meshes. During the healing stage, granulomas are formed around mesh fibers as part of the foreign body reaction that become confluent with each other and encapsulate the entire mesh. A stiff scar plate is formed which reduces the mesh flexibility [38]. Mühl et al. introduced the concept of effective porosity that considers only the scar free or open pores [63]. Effective porosity excludes all pores, which are smaller in diameter than a critical diameter which depends on the mesh material, 1 mm for Polypropylene (PP) mesh and 0.6 mm for polyvinyldene fluoride (PVDF) [22, 63, 66]. Meshes with large pores preserve effective porosity and exhibit less inflammatory infiltrate, connective tissue, fistula formation, calcification and scar bridging, which allows increased soft tissue ingrowth.

2.3 An Ideal Polymer: Polyvinvlidene Versus Polypropylene?

Mechanically, an ideal synthetic mesh can be obtained from a mono filamentous large porous structure with anisotropic mechanical properties that mimics the mechanics of the host tissue. However, the biochemical compatibility of the meshes can be achieved through the choice of appropriate polymer material that affects the inflammatory response with granuloma formation. PP and PVDF are the mostly used polymers to construct surgical meshes due to their lesser amount of foreign body reaction compared to other available polymers. Though, PP shows inert response during infection, have excellent capacity for integration and are low cost meshes tested for years [55], they results in an intensified inflammatory reaction characterized by pronounced foreign body granuloma and are less stable. On the other hand, PVDF filaments have excellent biocompatibility and reduced adverse foreign body reaction, such as scar formation or pain claiming a higher biocompatibility and biostability than polypropylene [49]. A PVDF meshes has been successfully used as composite scaffold for cell cultures [25]. Some long-term advantages of the PVDF meshes against PP meshes are (a) no mesh curling under stress [66], (b) durability (lesser loss of tensile strength due to hydrolysis against PP: 7.5% vs. 46.6%) [53], (c) reduced risk of infection due to lower bacterial adherence, and (d) reduced mesh erosion [43].

3 Mechanics of Different Textiles

3.1 Specimen Preparation

Three different DynaMesh® meshes SIS direct soft (SD), CICAT and PRS soft (PRS) from FEG Textiltechnik mbH, Aachen, Germany have been tested. All of the selected meshes are monofilament made out of PVDF polymer but have a considerably different structure constructed for specific applications: SD is used for the repair of stress urinary incontinence, CICAT for abdominal hernia repair and PRS is used for uterine/vaginal vault prolapse correction. Numerous mechanical tests have been standardized to characterize the mechanical behavior of an anisotropic mesh materials. Among them, most common test are uniaxial tensile test, biaxial tensile test, shear test and inflation test.

For uniaxial tensile tests specimens (70 mm × 30 mm) were cut out of each mesh in longitudinal direction and transverse direction with respect to knitted pattern. The thickness and the pore size was measured using a digital microscope VHX-600 (Keyence, Japan). For inflation test, mesh specimen of 50 mm × 50 mm were cut and placed between two PVC plastic plates, see Fig. 1a. In [86] biaxial tests have been performed to characterize the passive behavior of the abdominal wall under three configurations: intact, after creating a defect simulating an

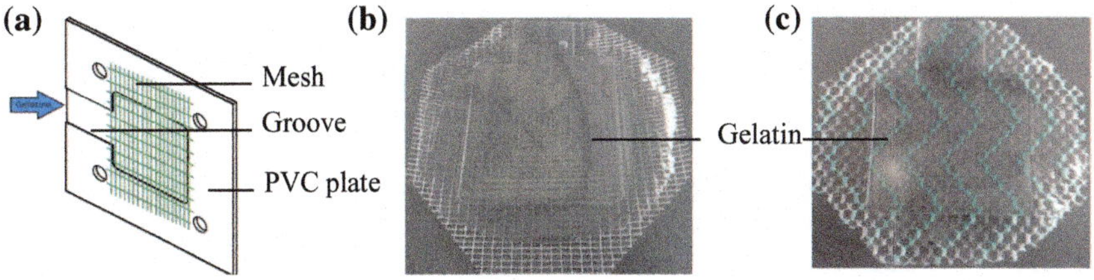

Fig. 1 Mesh specimen preparation for inflation (bulge) test: **a** schematic drawing of PVC plastic plate and mesh; **b** Dynamesh-PRS soft specimen with gelatin; **c** Dynamesh-CICAT specimen with transparent gelatin

incisional hernia, and after a repair with a mesh implanted intraperitonally. As alternative to animal testing ballistic gelatin type 3 of GELITA® (Eberbach, Germany), as a soft tissue surrogate was then poured inside the groove (1 mm deep), when cooled it forms a flexible gelatin-mesh complex (GMC), as shown in Fig. 1b, c. Gelatin has the practical advantage that its transparency and photoelastcity show detailed local behavior.

3.2 *Mechanical Test Set Up*

The uniaxial tensile tests are performed on a tension testing machine Z 010 (Zwick Roell AG, Ulm, Germany). Two uniaxial tests protocols are followed on dry mesh and explants (dry mesh and ballistic gelatin). The two ends of the testing meshes are clamped properly at a distance of 50 mm, see Fig. 2a. For strong grip and to protect the mesh against cutting by the fixation, paper or PVC foils are used. Digital image correlation (DIC) technique is used to measure the stretch during tests that is based on the optical measurement of the deformation of a speckle pattern made by grey graphite spray. The camera has taken pictures during the test with a frequency of 1 Hz for DIC and the deformation is analyzed with the software ISTRA 4D (Limess Messtechnik und Software GmbH, Krefeld, Germany) using a correlation algorithm.

Figure 2b shows second setup of biaxial test. GMC is clamped airtightly between two 2 mm thick stainless steel plates. Two cameras (uEye Modell UI-122xLE, WVGA (752 $\times$ 480); Fujinon lens HF25HA-1B) at two orthogonal observation planes measures the profile movement of the inflating GMC and captures the deformation of the anisotropic mesh. At the same time, air is blown to generate pressure inside the GMC that is measured by the pressure transducer. An automation program LabView (National Instruments, Austin, Texas) elaborates images with the predefined algorithm. The height of the inflated GMC was measured before gelatin layer bursts. Engineering parameters such as stress and strain are computed to evaluate the mechanical response of the GMC that can be compared with the in vivo response of the tissue-mesh complex.

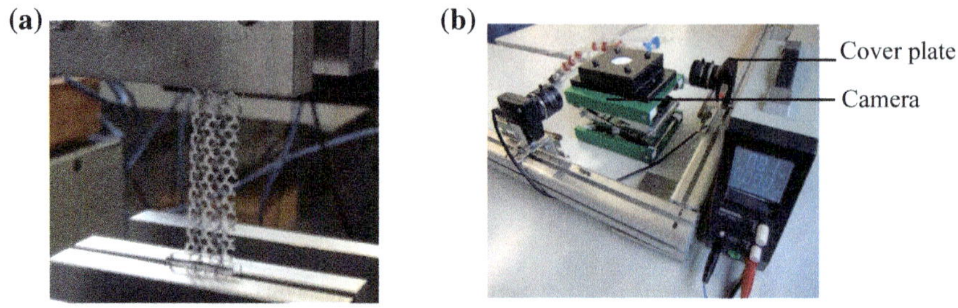

Fig. 2 Synthetic mesh clamped to **a** uniaxial tensile and **b** bulge test machines

3.3 Uniaxial Tensile Test on Rectangular Pore Dry Mesh

Uniaxial tensile tests on structurally similar dry PRS and SD mesh provided qualitatively similar results. As shown in Fig. 3b, c, the characteristic stress-stretch curves are nearly linear and orthotropic. Curves are plotted until complete failure of the implant. The SD mesh has larger toe stiffness and larger strain at inflection than the PRS mesh. Linear stiffness is measured that determines the strength and clinical effectiveness of the implantation. The initial Young's modulus in longitudinal direction of the PRS mesh ($E_L = 29.778$ MPa) is greater than for the SD mesh ($E_L = 26.73$ MPa), whereas, the initial transverse modulus is larger for the SD mesh ($E_T = 3.58$ MPa vs. $E_T = 1.919$ MPa). The meshes show nonlinear orthotropic behavior for which a polyconvex hyperelastic material model has been successfully identified in [36]. For simplicity we continue here to discuss the stiffness only linearly.

On the other hand, the CICAT mesh with hexagonal pore shape shows highly nonlinear mechanical behavior under uniaxial tensile test, see Fig. 3a. Compared to stretched SD and PRS meshes in longitudinal direction, CICAT mesh possessed the largest toe region and lower stiffness (E = 13.207 MPa) in the linear region.

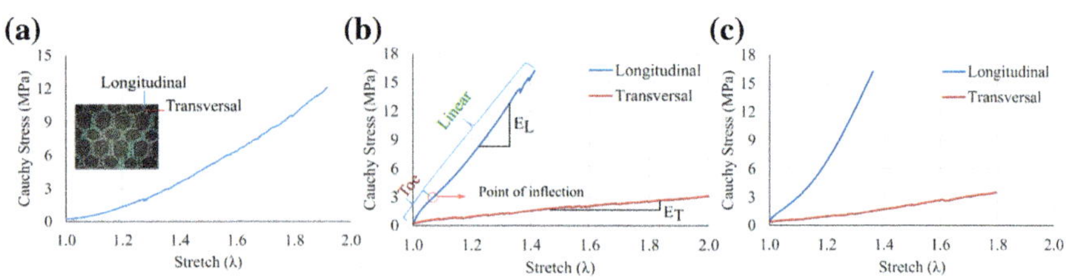

Fig. 3 Average stress-stretch curves of different dry meshes loaded in longitudinal and transverse direction for different meshes: **a** CICAT; **b** SD; **c** PRS

3.4 Optical Strain Measurement

The traditional extensometer technique provides an average strain over the specimen gauge length. For accurate strain measurement over the whole area between extensometer probes, especially at positions of necking, DIC tracks the positions of the same physical points during test. Lagrangian strains and stretch for large deformation are calculated and visualized by the color scaling using DIC analyses which are later converted into stretches for suitable presentation of results. In addition, for orthotropic prosthetic meshes, DIC technique benefits to measure the change of mesh porosity during tests.

3.4.1 SIS Direct Soft Mesh

In this part, the experiment performed using the DIC technique on the SIS direct soft mesh is presented. Two pairs of images at zero load and maximum load before failure of the mesh are compared when stretched in two orthogonal directions, see Fig. 4. Due to lesser number of woven fibers along transverse direction, axial elongation and transverse contraction of the specimen is significant. As shown in Fig. 4a, the textile construct shows a wide range of nearly linear stress-stretch curves when subjected to tension in longitudinal (E_L = 105.78 MPa) and transverse direction (E_T = 10.87 MPa). The dry meshes are found to be compressible with significant volume change during tensile test.

The stress-stretch behavior using uniaxial tensile test are qualitatively similar to that of DIC technique. However, computing Lagrangian strain and later converted into true strain and expressing them in terms of stretch reveals the maximum true stress is significantly large, see Figs. 3b and 4a. Under uniaxial tension load applied in longitudinal and transverse direction, the rectangular pore of the SD mesh becomes progressively bigger with increased fiber stretch. Stretching up to a stretch of λ = 1.4 in longitudinal direction, an increase in porosity of 2.4% was measured, see Fig. 4b. Despite of significant mesh contraction and higher stretchability near about 2, loading along transverse direction rather increases the mesh porosity by 3.93%.

3.4.2 Hernia Mesh

The uniaxial tensile tests of a PVDF hernia mesh shows a nonlinear stress-stretch behavior when loaded in longitudinal and transverse direction, see Fig. 5. Due to hexagonal pore geometry, pore deformation is extreme in low load range. In both loading directions, increased length and reduced width of the specimen is observed. Figure 6 shows deformed mesh when stretched in the longitudinal and the transversal direction of the mesh. Longitudinal (Fig. 6a) and transverse Lagrangian strains (Fig. 6b) for loading along both directions are shown in Fig. 6. Ciritsis et al.

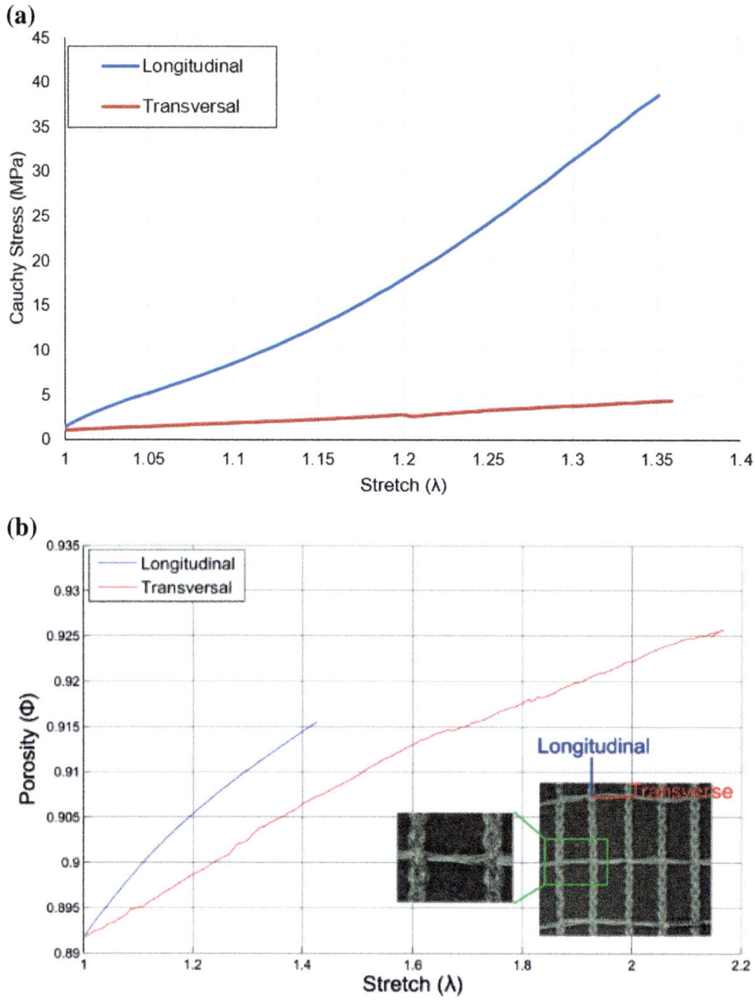

Fig. 4 Uniaxial tensile test on dry SD mesh using DIC technique; **a** the SD mesh elongates only by 40% if stretched to the thicker fiber (longitudinal) direction. In contrast, if rotated by 90°, an elongation of almost 100% occurs; **b** change of mesh porosity

[19] performed uniaxial tensile tests on PVDF constructed hernia mesh with hexagonal pore shape. As per the study, the porosity decrease was found to be 10.4 and 7.2% to a minimum in longitudinal and transverse directions, respectively.

In vivo hernia meshes loaded in biaxial tension show the increased porosity. In animal experiment, this led to an improved maturity of the scar indicated by a higher I/III collagen ratio which allowed the mesh implant to adapt to remain more flexible within the abdominal wall [19]. With respect to scar formation the concept of effective porosity has been suggested which could be only measured ex vivo with

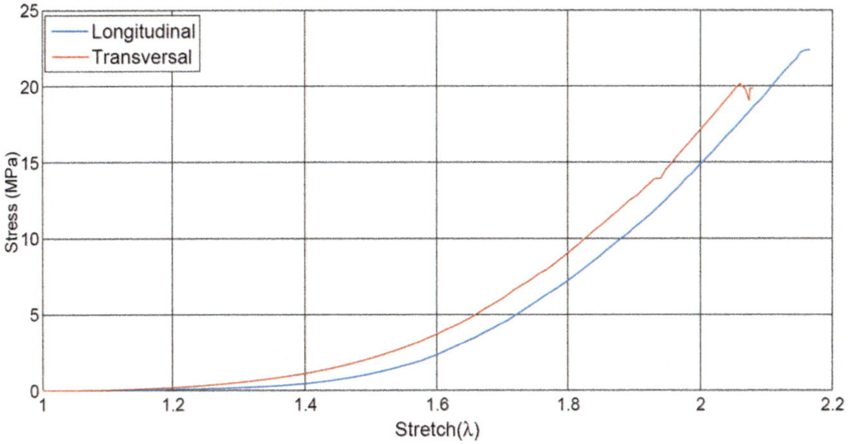

Fig. 5 Stress-stretch curves of anisotropic PVDF hernia mesh in uniaxial tensile test, see Fig. 2a

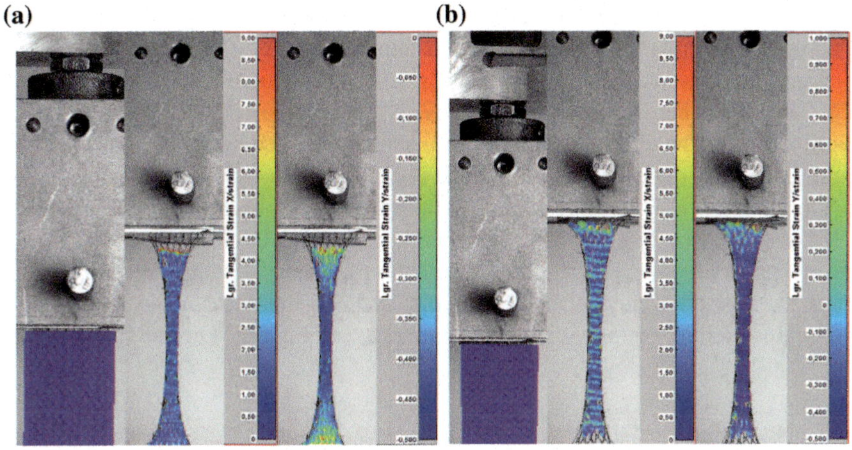

Fig. 6 **a** Lagrangian strains under loading in **b** longitudinal and **b** transverse fiber direction. Strain ε_x is extension in load direction and ε_y is contraction

a specially developed image processing. In [36] a new method has been presented which allows the calculation of effective porority from measurement of the global mesh deformation. Again the uniaxial tension was found to be critical because the pores collapse and effective porosity deceases to zero for the two tested meshes at stretches arround $\lambda = 1.4$.

3.5 Photoelastic Uniaxial Tensile Test on Tissue Surrogate-Mesh Specimen

The meshes are embedded in a transparent gelatin block that behaves similar to soft biological tissue but is transparent and photoelastic. Two ends of the mesh are clamped into a tension machine Z010 (Zwick Roell AG, Ulm, Germany). Applying tension to the gelatin-mesh complex (GMC) between two polarizing filters leads to isochromatic lines that represent lines of equal shear stresses. As shown in Fig. 7a, b, PRS mesh shows very little transverse contraction. However, due to axial force, the mesh surface is found to be increased by 18%. Since, there are no drastic reductions of the pore size, the mesh porosity and the effective porosity are satisfactorily preserved during uniaxial tensile test. Unlike PRS meshes, the CICAT mesh has different pore shapes in the loading direction and is completely unable to preserve its porosity; almost all pores are collapsed already under small tensile force, see Fig. 7c.

The failure modes of the tested meshes are observed from the photoelastic image of the uniaxial tension test that are quite distinct. The mesh shows a strong dependency of the GMC failure on the loaded direction. For the PRS soft mesh, loading along longitudinal direction sustains larger tensile force (124 N) that stresses the softer gelatin. The GMC specimen ruptures at lower stretch ($\lambda_{max} = 1.42$) whereas, along the transversal loading, the GMC is highly stretched ($\lambda_{max} = 2.05$). The zigzag (like steps of a staircase) orientation of transversal fiber families are unsupported at the boundaries and the mesh is pulled out at lower tensile load (22 N). This stresses the gelatin locally from where the rupture of the GMC begins, see Fig. 7b. On the other hand, the CICAT mesh with irregular hexagonal pores show large pore deformation along the loading direction. High shear stresses are generated in the gelatin at the boundaries of the mesh due to unequal stretching of the gelatin and the mesh, see Fig. 7c. Further stretching the

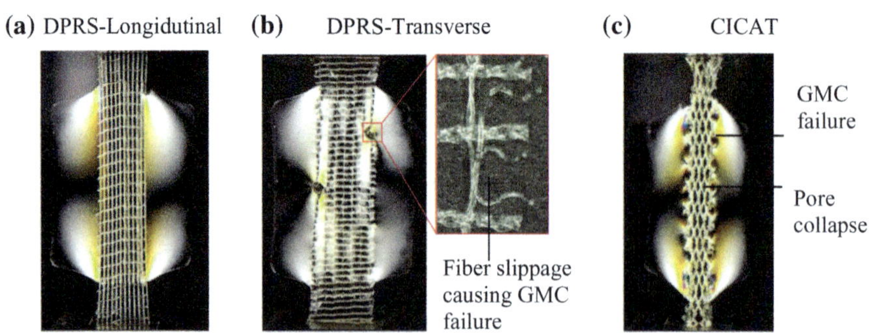

Fig. 7 Uniaxial tensile test on GMC specimen. DPRS mesh along **a** longitudinal direction (left: 80.89 N); **b** transverse direction (right: slippage of fiber (tearing out a seam) at less force, 15.89 N); **c** debonding of the CICAT mesh from the gelatin due to large transversal contraction and pore collapse at tensile force 60.404 N

GMC debonds the gelatin from the mesh along its edges and rupture of the GMC specimen occurs.

Such mesh edge failure have been observed after repair with the heavyweight small porous textiles [12] due to the heavy scar tissue formation, mesh shrinkage, mesh-size [50] and the mesh anisotropy [6]. However, pullout tensile tests and FE studies on monofilamentous and compliant Gynecare mesh (Ethicon, USA) shows a similar gradual debonding of the mesh from the tissue-surrogate gelatin in a zipper-like manner with huge pore deformation [31, 80]. Furthermore, large shear stresses at the lateral edges are generated when the mesh is highly stretched that is responsible for the failure of the compliant GC mesh. These reasons for similar mesh failure in two different classes of meshes are most likely due to the geometry, loading direction, compliance of the meshes and the generation of the normal and shear stresses.

3.6 *Tissue Surrogate Mesh Failure Under Biaxial Inflation Test*

Unlike, incontinence repair that supports mostly uniaxial physiological loads, abdominal hernia or pelvic organ prolapse repair uses a flat sheet of prosthetic meshes. They are designed to support the biaxial function of weak abdominal wall and pelvic organ and to restrain increased abdominal pressure on those surfaces developed with physical activities such as exercise, coughing, lifting, straining, and others. An inflation (bulge) test setup as shown in Fig. 2b on tissue surrogate gelatin-mesh complex theoretically mimics the physiological abdominal pressure and helps to determine the mechanical behavior after the implantation.

As shown in Fig. 8, the GMC specimen expands (inflates) with increased pressure from below. On the outer side of the GMC, small gelatin cups are formed across the mesh pores. Compliant CICAT-GMC specimen offers bigger cups due to (1.42 times in length and 2.41 times in width) larger pores than stiffer PRS-GMC specimen. The soft gelatin cups are highly stretched to burst at significantly lower pressure (97 kPa vs. 116 kPa) for CICAT-GMC specimen, thus failing at lower inflation height (6.89 mm vs. 11.44 mm), see Fig. 8a. For stiffer PRS mesh with small mesh pore the GMC bond is intact and bears larger stress. At high stress, the gelatin is scraped off from the edge of the cover plate to leak the pressure failing the GMC specimen. For inflation test, both meshes reveal materially linear response until failure and the structural differences of the meshes are found to be uninfluential. However, the maximum values of the principal stresses at the end of the experiment differ considerably in magnitude (483.79 MPa for CICAT and 1186.26 MPa for PRS). These differences are due to the pore geometry, pore size and the compliance of the two meshes.

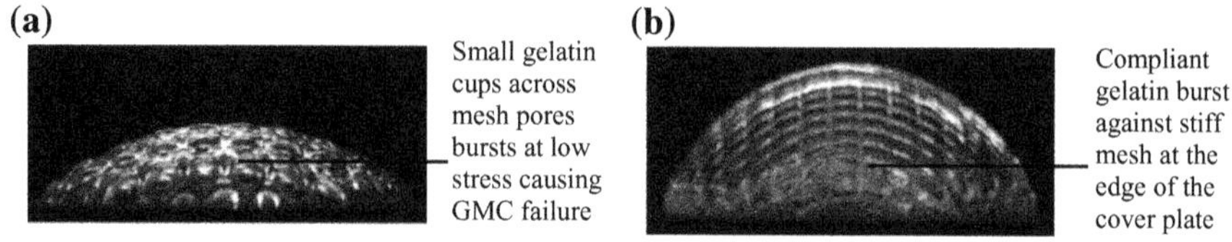

Fig. 8 Failure of tissue surrogate-mesh under biaxial inflation test: **a** Gelatin-CICAT mesh; **b** Gelatin-PRS mesh

3.7 Clinical Comparisons and Discussions

Immediately after implantation, biomaterials, such as prosthetic meshes are associated with the acute inflammation phase. Numerous blood and tissue proteins, such as cytokines or growth factors and leukocytes are deposited on the foreign body surface and injury site to form the provisional matrix and to clean the wound site [4, 40]. The types and concentration of such proteins are dependent on mesh surface properties and polymer hydrophilicity [87]. These proteins modulate the adhesion of monocytes/macrophages cells on the protein absorbed foreign object that are fused together to form larger foreign body giant cells (FBGCs) [16], also known as chronic inflammation. The chronic inflammation period for biocompatible materials should not be more than two weeks, otherwise may cause infection [5]. Adherent macrophages and FBGCs secrete fibroblast attractants, and the incoming fibroblasts from the surrounding tissue create an organized collagenous matrix around the biomaterial [78], called extracellular matrix (ECM) remodeling fibrosis. This vascular and collagenous capsule (usually 50–200 μm thick) isolates medical devices from the rest of the interstitial tissue at the end stage of wound healing process [86]. Later, abundant amount of fatty tissues and significantly less connective tissues penetrates completely into the pores of the mesh to form a layer of neo-tissue unit. But, why does the host tissue not heal normally around implants and what are the reasons for post-operation complications? These are very critical questions in surgical interventions for which thorough understanding of the foreign body response and the biocompatibility of the medical devices is important.

Biosafety (appropriate host response and in-growth of the patient's host tissue) and biofunctionality (effectiveness of biomaterial to further strengthen the defected area) are two elements to estimate the short-term and long-term biocompatibility and to predict post-operation complications and device failure/rejection [7]. In vivo assessment and cell culture examination are often performed to screen the tissue compatibility (toxicity) of implantable devices and to simulate the biochemical response of the host body [71, 73], the degree of which depends on the properties of the device, such as polymer composition, bacterial adhesion prior to surgery, smoothness and polymer surface area. However, biofunctionality (functional performance) of the mesh implants in vivo depends on the structural properties (effective porosity, pore size, pore deformation, filament type) and on the mechanical properties (tensile strength, isotropy, elasticity). A perfectly biocompatible mesh material should provide maximum safety, match the behavior of the anisotropic host tissue and sufficient structural stability to physiological loadings; otherwise. Adverse effects are often observed in the form of chronic pain, infection, mesh shrinkage, scar formation and mesh rejection [3, 13, 21, 27, 52, 54, 58, 59, 62, 84, 85].

3.7.1 Mesh Complications

Inflammation

Almost all modern biomaterials trigger a wide variety of adverse responses in vivo mainly due to inflammation right after implantation [47]. The biosafety of the mesh implant is characterized by the inflammation phase: acute and chronic, that are associated with the deposition of proteins and macrophages/monocytes. Longer acute inflammation directly enhances the development of a greater amount of fibrous tissue, whereas, longer chronic inflammation initiates infection [5]. An ideal, biocompatible material is assumed to deposit thin fibrous tissue around the mesh filament and to fill sufficiently larger pores mainly by fat tissues [50] that maintains the elasticity of the mesh even after the inflammation phase. These animal experiments with a low-weight, large pore size and monofilament made mesh reduces the inflammation, maintains the flexibility of the mesh and provides greater host tissue mobility during physiological loadings [46].

In contrast, heavy-weight, multifilament made mesh with small pores increases the surface areas in contact with the recipient host tissue that basically aggravate the acute and chronic inflammatory phase. The adverse effects are seen in the form of (a) huge macrophages and FBGCs deposition at the interface, (b) intense fibrosis forming thicker connective tissue around filament, (c) poor or no fat cell penetration to the pore, and (d) disordered collagen metabolism with much lower collagen (I/III) ratio, see Fig. 9. Changes in this ratio affect both tensile strength and mechanical stability and may increase the risk of recurrence. This consequently forms a thick continuous connective tissue, also known as scar plate with tangled bundles of collagen fibers around the filaments which grow to the adjacent filaments [40, 63], a phenomena called 'bridging' [15]. This bridging of newly formed inter-filament connective tissue (stiff scar plate) as shown in Fig. 9 (a) prevents the pores to deform freely, (b) reduces the mesh compliance, (c) restricts the mobility of the host tissue, (d) prevents further tissue ingrowth and (d) finally causes recurrence. As a result, the healing of the wound may be retarded.

The polymer and fiber surface greatly affect the inflammatory response. PVDF made meshes result in a significantly reduced foreign body granuloma size [49] with reduced diameters of the inner (inflammatory infiltrate) and the outer (connective tissue infiltrate) ring of granulomas [39] compared to conventional PP (deposits more collagen, longer inflammation and scar neo-tissue) [8], and polyester mesh (long-lasting chronic inflammatory response) [65].

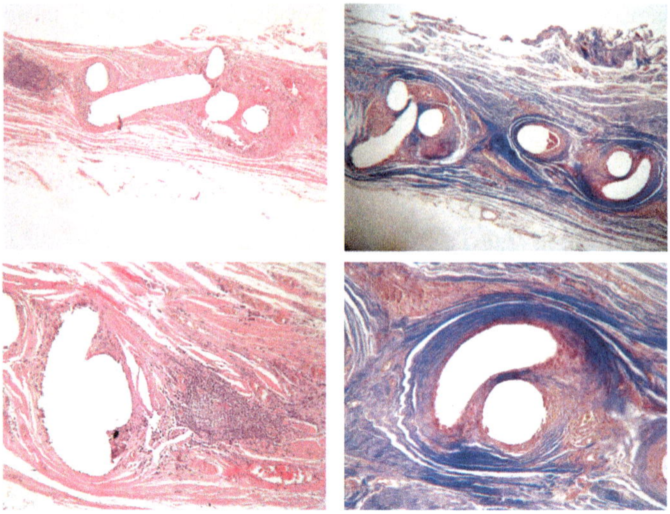

Fig. 9 Tissue response after suture repair. Histological analysis of explanted heavyweight polypropylene mesh stained with hematoxylin and eosin (H&E) or Masson's Trichrome. Histology reveals a large colony of leukocytes adjacent to the mesh filament and a thin layer of muscle surrounding the mesh filaments with an adjacent, dense layer of collagen surrounding the mesh filaments and is continuous between filaments 'bridging'. Images reprinted with permission courtesy of Melinda Harman, Ph.D. at Clemson University [15]

Seroma

Seroma is an ill-defined event, typically involving a collection of serous fluid in any potential fluid-occupying space that develops in the recipient body after surgery or any blunt injury. It usually appear under the surface of the skin as a swollen lump, like a large cyst when blood seeps out of the injured or ruptured blood vessels during operation. Seroma etiology remains unclear, its formation after incisional hernia repair with mesh is a most frequent problem due to local inflammatory response aggravated by the presence of the foreign body [10] and lasts for several weeks [81]. Chronic seromas are evidenced due to a long-term inflammatory reaction with heavyweight and small sized mesh monofilamentous PP and polyester meshes [46, 74].

Postoperative seroma formation anterior to mesh is very common up to 100% [35, 56, 81]; however, deep seroma behind the mesh develops an infection which ultimately requires surgical removal of the mesh [61, 77]. Further, a clear discharge of the serous fluid is non-harmful, however, bloody, colorful and odorous discharge indicates infection. Treatment options include observation for spontaneous resolution, percutaneous aspiration, closed suction drainage, abdominal binders, and sclerosant [57]. Though surgical drainage technique is used to prevent a seroma development and to reduce infection, seroma drainage and infection are inter-related. Drainage can introduce infection and in cases where the seroma causes discomfort or is infected then drainage is required [33], thus, conservative follow up and microbiological examination is required.

Adhesion

Postsurgical adhesions are serious and frequent complication resulting in a similar way that a scar tissue forms and vary from filmy to dense [39]. Such adhesions occur in almost all meshes as a result of inflammation [12] containing multiple foreign body granulomas that connect tissues or anatomic structures to the textiles. Studies shows 20–80% of the subjects develop adhesions after ePTFE and PP mesh implantation [28, 60]. Other than the polymer type, factors associated with adhesion formation include trauma, tight suturing, thermal injury, infection and foreign bodies. Traditional heavy PP meshes with small pores induce an intense fibrotic reaction that produce dense adhesion around 62% of the mesh area and provide a strong adherence to the abdominal [12]. An anti-adhesion film-like barrier around the mesh filaments have been an alternate to reduce adhesions [12], however, light weight, large porous PVDF constructs with PP coating seem to be superior with regard to the induced intensity of inflammation and filmy adhesion of 34.6% mesh area that could be lysed with traction [39].

Collagen Metabolism

Collagen metabolism has an intense influence on the wound healing after mesh repair [48]; decreased ratio of collagen I to III increases the risk of hernia recurrences [41]. The inflammatory process after surgical implantation is accompanied by a pronounced fibrosis to deposit the fibrillary collagen molecule that are stabilized by intracellular hydroxylation reaction forming hydroxylysine and hydroxyproline [80]. Lysyl oxidase enzymes mediates a cross-linking process to form a strong, stable collagen fibrils and fibers that provides strength, integrity and structure. Among 20 different collagen types, I and III are synthesized in huge amount. Collagen type I is stronger, mature and thick whereas type III is less cross-linked and immature that provides less tensile strength predominantly found in early wound healing [29]. Balanced collagen maturation and degradation by matrix metalloproteinases [64] to form type I collagen is a requirement for normal scar formation [29, 30]. Reduced ratio of collagen type I to III lead to thinner collagen fibrils, changed geometrical arrangement [37] with higher levels of non-polymeric soluble collagen [29] that contributes to a decreased tensile strength and instability of the connective tissue and induced scar tissue [11].

It has been proven that the fibroblastic ingrowth, chronic inflammation and scar formation is dependent on the weight and structure of the implanted mesh [12, 40, 46, 63, 70], but the quality and mechanics of such scar tissues are characterized by the extent of collagen type I/III [11, 67]. Studies on the ECM of the explanted hernia meshes constructed from PP, polyester and ePTFE shows lowered collagen type I/III and reduced tensile strength, a major reason for hernia recurrence [41]. Though, altered ratios of collagen can be seen within fibroblasts located at the edges of recurrent hernia [50], yet it is not clear if the type of mesh used has any effect [12].

Mesh Wrinkling

Wrinkling of the mesh is caused by failure or inadequate mesh fixation in defected place. In the sutureless technique to repair abdominal hernia, meshes are placed flat on the floor of the abdomen. With insufficient dissection to insert the meshes, during physiological movement, meshes have a tendency to wrinkle or curl increasing the potential for the formation of dead spaces where unintended mesh overlap occurs, see Fig. 10a. This process augments the mesh relative movement from its total implanted position leaving gaps at the edges for the protrusion of tissues, causing hernia recurrence, see Fig. 10b. Further, the localized fibroblastic infiltration into mesh pores are greatly altered that mechanically cause inhomogeneous inelastic mesh deformation between the implant (stiffer scar tissue around mesh at dead spaces) and the underlying soft tissue and biologically (a) leaves spaces for infection, (b) chronic postoperative pain, and mesh deterioration. A wrinkled mesh may also add to adhesion problems with viscera which can result in the uneven distribution of stress on the mesh and cause an undesirable distortion, premature weakening of the mesh, and hence predispose the wound to infection [82].

Other reasons of wrinkling have been identified due to (a) tearing edge fibers out of seam during mechanical loadings, (Fig. 7b) and (b) extreme pore deformation (Figs. 6a, b and 7c). Surgeons often cut the desired dimension of the mesh out of a large flat sheet to match the defect to be repaired; edge fibers loosely supported along the transverse woven direction get pulled out when the mesh is stretched. The fibers once out of the knitted mesh can no longer maintain their flat shape and start to curl around. In addition, pores of the compliant meshes can completely collapse under uniaxial mesh stretch. This process dramatically curls the mesh into an unclosed cylindrical geometry from its initial flat shape, see Fig. 6a, b. Repeated

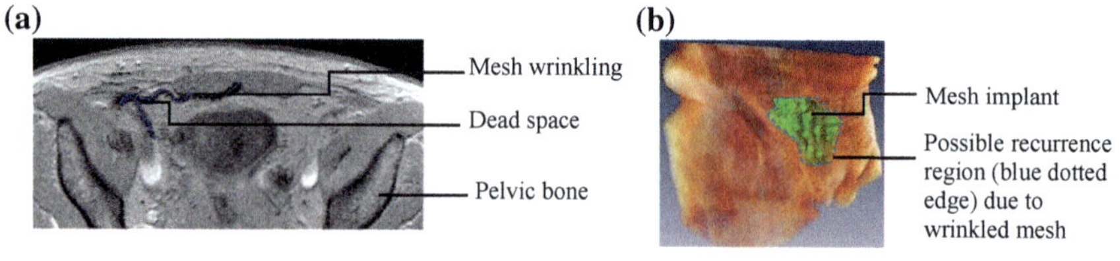

Fig. 10 *Segmentation with* Amira® (Zuse Institute Berlin, *FEI*™ *Bordeaux*) of *a*bdominal hernia repair using mesh implant [34]: **a** MRI image showing mesh wrinkled after placed in the abdomen: superior view (with permission of N. A. Krämer, Uniklinik RWTH Aachen, Germany); **b** representation of the position of the wrinkled mesh in abdomen (anterior-lateral view)

loading and unloading during rest and straining damages the ingrown tissue, impairing the mechanical biocompatibility of the implantation and finally recurring the damage. A physiologically relevant preconditioning prior to implantation might suppress such mesh wrinkling, but a hypothetic preconditioning procedure might make the mesh less compliant than initially expected.

Mesh Shrinkage

Shrinkage is most common in the surgical repair with the use of heavyweight small pore size PP meshes. Studies assume that it is not the mesh which undergoes shrinkage [50], which is supported by our strain measurement on mesh implants. Assuming the thickness of the mesh to remain constant during test, the mesh area increases on stretching, see Fig. 11. Mesh shrinkage results due to contraction of scar tissue around the mesh, starting with a constant water loss, followed by a surface area decrease [2]. Depending on the properties (pore size, weight and filament type) of the prosthesis, shrinkage of the mesh implants due to wound contraction are observed to be within a wide range from 5–62% [39, 50, 75]. Therefore, use of large implants to cover defects can justify the shrinkage effect, however, larger foreign body reactions, scarring, dyspareunia, shrinkage, mesh exposure and stiffening of the recipient host tissue can no longer be avoided [59]. Lightweight large porous PP meshes are specifically designed to reduce the bridging and scar formation [45], however, the use of highly inert PVDF material is found to significantly reduce the wound contraction, (9.3–19.9%) minimal loss of the original area [22, 39] provide long-term stability as a result of lower foreign body reaction and maintain original strength after years of implantation [32, 45, 49].

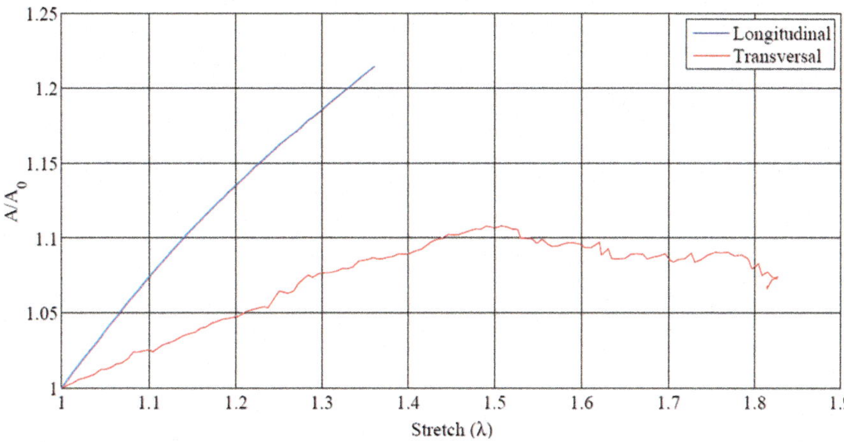

Fig. 11 Area change under uniaxial tensile test on SIS direct soft mesh. A_0 and A are the initial and current surface area of the mesh

Pain

Biomaterials, such as mesh implants are associated with a potential risk of chronic pain, as they are recognized as foreign body by the host tissue immediately after repair. The reasons for chronic pain are still unclear and vary from patient to patient. Studies suggest pain associated with mesh surgery can be (a) early pain immediately after surgery due to irritation from surgical material or nerve damage during surgery or (b) chronic pain existing for more than 3 months after surgery [50]. Repairs using small pore, heavyweight meshes with reduced effective porosity lead mainly to complaints about chronic pain that often requires the complete removal of the implant from the body [53, 69]. Post-retrieval studies on such explants obtained from similar patients with failed repair indicate irritation and destruction of nerve fiber and fascicles by the inflammation at the interface with the mesh [50]. Today's new generation meshes have been greatly improved using softer material, large pore and light weight meshes [44] which have proved to be a better mesh material with reduced level of inflammation and scar formation, which is directly related to pain [70].

Infection

Infection is another common FBR in repair using mesh implants, hindering the local clearance from bacteria [15]; Klinge and Klosterhalfen, 2002 that leads to a chronic inflammatory wound with marked scarring, loss of compliance, mesh contraction, migration, physiochemical changes, seroma, and in some cases, eventual mesh removal to resolve the problem [55]. The incidence of chronic mesh infection is highly related to the type of mesh material, type of filament, pore size, and porosity [23, 70]. Generally, microporous and multifilament meshes with increased surface of the mesh area and small pores are at higher risk of infection than large porous monofilament meshes (>75 μm). Macrophages and neutrophils are unable to enter small pores (<10 μm) but allows bacteria (<1 μm) to survive unchallenged within the pores. In contrast, there is some evidence that the persistence of the (a) colonies of Staphylococcus epidermis bacteria at the surface of the polymer fibers [50], and (b) acute/chronic inflammatory response beyond a 3 weeks period usually indicate higher risk of infection [5]. Chronic mesh infection following repair with small pore meshes or film like structures often requires removal of the infected mesh, which rarely results in hernia recurrence if sufficient fibrosis scarring remains [2]. Open pore meshes mostly can be treated conventionally.

3.7.2 Discussions

A physiological process of wound healing creates a combined reparative layer of implant and fat or connective tissue, which is assumed to be achieved from meshes with sufficient pore dimension and minimum surface area. Small pore sized mesh intensifies the fibrosis forming thicker scar tissue with poor or no fat cell penetration to the pore. Mechanical response of such heterogeneous neo-scar tissue hugely differs from the anisotropic recipient host tissue that may increase the risk of infection, chronic pain, complete removal of the implant from the body, and recurrence. Under extreme physiological movements and straining processes, higher tensile and shear stresses on the ingrown tissue develops, which increase the risk of postoperative complications such as mesh wrinkling and mesh erosion. Moreover, if mesh pores are heavily deformed and wrinkling of the mesh occurs (Fig. 5b, c), large stresses are developed at the edges of the interface (Fig. 6c). During this process, the implanted mesh may wrap the host tissue and comes in contact with the adjacent organs to irritate, degenerate and even lead to dysfunction of the organs.

Specimen, during surgery, constructed by cutting the desired dimension out of a large flat sheets to repair the defects often leaves free edges that gets seam out when it is stretched. The fibers once pulled out of the knitted mesh can no longer maintain their flat shape and start to stress the tissue locally. Successive tearing further debonds the ingrown tissue and the implanted mesh fails to perform the intended task in the body, see Fig. 6b. Repeated loading and unloading during rest and straining damages the ingrown tissue, impairing the mechanical biocompatibility of the implantation and finally recurring the damage. Thus, adequate structural stability as shown in Fig. 6a, which preserves the porosity under strain is essential for the proper function of such textile structures in a tensile environment in vivo.

The common trend in clinical practice for the reconstructive surgery has been the use of monofilamentous, lightweight large porous mesh in order to reduce the foreign body reaction and to minimize the risk of complications. Soft meshes with an adequate textile construction preserve their effective porosity and show structural stability under mechanical strain and lead to reduced scar formation. However, compliant meshes with additive softer material are often associated with intense pore deformation reducing the effective porosity [19, 36], therefore, a good compromise of the mesh stiffness (formstability), polymer type and pore size should be maintained. Rigorous clinical trials of such mesh should be done with respect to the deposition of physiologically regenerated tissue, for example, fat with the amount of stromal and fibroblasts, percentage of apoptotic cells, degree of foreign body response to compare the effectiveness of the available implants. This not only improves the behavior of existing meshes but insights to search for biomaterials with perfect biocompatibility in order to develop the ideal textile devices which can mimic the biomechanical properties of the tissues.

Calculation of the shear stress and maximum principal stresses at the tissue-implant interface and the estimation of change of the structural proteins after implantation could be quantities to compare the effectiveness of the meshes for such reconstructive surgeries. Depending on the geometry of the pore, mesh orientation and the loading direction, the physiological deformation of the implanted large pore mesh is beneficial if these stresses can be minimized and if adjacent tissue degeneration is successively avoided.

In conclusion, this study shows that meshes with appropriate pore geometry, adequate biaxial stiffness lead to a mechanically biocompatible structure during complex physiological conditions compared to conventional mesh geometry. Since the immune response and repair functions in the body are so complicated, it is not adequate to describe the biocompatibility of a single material in relation to a single tissue surrogate gelatin. However, these tests constitute an important step towards the laboratory animal testing and clinical trials that will determine the biocompatibility of the different meshes with variation in pore geometry, structure and material composition in a given application. Considering these shortcomings, the perfect mesh design should consider sufficient tensile strength, anisotropy/isotropy, elasticity, and provide the appropriate biomimetic environment to ensure cell survival and at the same time should preserve its porosity under strain.

Acknowledgements The first author has been partially funded by the German Federal Ministry of Education and Research through the FHprofUnt project BINGO (03FH073PX2). We would also like to thank our project partner FEG Textiltechnik mbH, Aachen, Germany for providing the prostheses, Nils Andreas Krämer, PD MD, Uniklinikum RWTH Aachen, Germany for providing MRI data, and Andreas Horbach, DI, and our students Christian Halbauer, Viola Gruben for their help with experiments and providing the results from their bachelor theses. The authors would also like to acknowledge Prof. Dr. Melinda Harman, Clemson University for providing permission to use the images reprinted in Fig. 9.

References

1. Alaedeen, D. I., Lipman, J., Medalie, D., & Rosen, M. J. (2007). The single staged approach to the surgical management of abdominal wall hernias in contaminated fields. *Hernia, 11*(1), 41–45.
2. Amid, P. K. (1997). Classification of biomaterials and their related complications in abdominal wall surgery. *Hernia, 1*(1), 15–21.
3. Amid, P. K. (2004). Shrinkage: fake or fact? In V. Schumpelick & L. M. Nyhus (Eds.), *Meshes: benefits and risks*. Berlin: Springer.
4. Anderson, J. M. (1988). Inflammatory response to implants. *ASAIO Transactions, 34*(2), 101–107.
5. Anderson, J. M., Rodriguez, A., & Chang, D. T. (2008). Foreign body reaction to biomaterials. *Seminars in Immunology, 20*(2), 86–100.
6. Anurov, M. V., Titkova, S. M., & Oettinger, A. P. (2012). Biomechanical compatibility of surgical mesh and fascia being reinforced: Dependence of experimental hernia defect repair results on anisotropic surgical mesh positioning. *Hernia, 16*(2), 199–210.
7. Arshady, R. (2003). Polymeric biomaterials: chemistry, concepts, criteria. In R. Arshady (Ed.), *Introduction to polymeric biomaterials: the polymeric biomaterials series* (pp. 1–62). London: Citus Books.
8. Baktir, A., Dogru, O., Girgin, M., Aygen, E., Kanat, B. H., Dabak, D. O., et al. (2013). The effects of different prosthetic materials on the formation of collagen types in incisional hernia. *Hernia, 17*(2), 249–253.
9. Bay-Nielsen, M., Kehlet, H., Strand, L., Malmstrøm. J., Andersen, F. H., Wara, P. et al. (2001). Quality assessment of 26,304 herniorrhaphies in Denmark: A prospective nationwide study. *Lancet, 358*(9288), 1124–1128.

10. Bendavid, R., & Kux, M. (2001). Seromas. In R. Bendavid, J. Abrahamson, M. E. Arregui, J. B. Flament, & E. H. Phillips (Eds.), *Abdominal wall hernias: Principles and management* (pp. 753–756). New York: Springer.
11. Birk, D. E., Fitch, J. M., Babiarz, J. P., Doane, K. J., & Linsenmayer, T. F. (1990). Collagen fibrillogenesis in vitro: interaction of types I and V collagen regulates fibril diameter. *Journal of Cell Science, 95*(Pt 4), 649–657.
12. Brown, C. N., & Finch, J. G. (2010). Which mesh for hernia repair? *Annals of the Royal College of Surgeons of England, 92*(4), 272–278.
13. Brown, G. L., Richardson, J. D., Malangoni, M. A., Tobin, G. R., Ackerman, D., & Polk, H. C. (1985). Comparison of prosthetic material for abdominal wall reconstruction in the presence of contamination and infection. *Annals of Surgery, 201*(6), 705–711.
14. Burger, J. W. A., Luijendijk, R. W., Hop, W. C. J., Halm, J. A., Verdaasdonk, E. G., & Jeekel, J. (2004). Long-term follow-up of a randomized controlled trial of suture versus mesh repair of incisional hernia. *Annals of Surgery, 240*(4), 578–585.
15. Casey, E. M. (2015). *Physical characterization of surgical mesh after function in hernia repair* (*Master Thesis*). Clemson University, South Carolina, USA. All Theses. Paper 2085.
16. Chen, E. H., Grote, E., Mohler, W., et al. (2007). Cell-cell fusion. *FEBS Letters, 581*(11), 2181–2193.
17. Choe, J. M., Kothandapani, R., James, L., & Bowling, D. (2001). Autologous, cadaveric, and synthetic materials used in sling surgery: comparative biomechanical analysis. *Urology, 58*(3), 482–486.
18. Chuback, J. A., Singh, R. S., Sills, C., & Dick, L. S. (2000). Small bowel obstruction resulting from mesh plug migration after open inguinal hernia repair. *Surgery, 127*(4), 475–476.
19. Ciritsis, A., Horbach, A., Staat, M., Kuhl, C. K., & Kraemer, N. A. (2018). Porosity and tissue integration of elastic mesh implants evaluated in vitro and in vivo. *Journal of Biomedical Materials Research Part B: Applied Biomaterials.* 106(2):827–833.
20. Cobb, W. S., Burns, J. M., Peindl, R. D., Carbonell, A. M., Matthews, B. D., Kercher, K. W., et al. (2006). Textile analysis of heavy-weight, mid-weight and light-weight polypropylene mesh in a porcine ventral hernia model. *Journal of Surgical Research, 136*(1), 1–7.
21. Cobb, W. S., Kercher, K. W., & Heniford, B. T. (2005). The argument for lightweight polypropylene mesh in hernia repair. *Surgical Innovation, 12*(1), 63–69.
22. Conze, J., Junge, K., Weiss, C., Anurov, M., Oettinger, A., Klinge, U., et al. (2008). New polymer for intra-abdominal meshes-PVDF copolymer. *Journal of Biomedical Materials Research. Part B, Applied Biomaterials, 87(2),* 321–328.
23. Deligiannidis, N., Papavasiliou, I., Sapalidis, K., Kesisoglou, I., Papavramidis, S., & Gamvros, O. (2002). The use of three different mesh materials in the treatment of abdominal wall defects. *Hernia, 6(2),* 51–55.
24. Dietz, H. P., Vancaillie, P., Svehla, M., Walsh, W., Steensma, A. B., & Vancaillie, T. G. (2003). Mechanical properties of urogynecologic implant materials. *International Urogynecology Journal and Pelvic Floor Dysfunction, 14*(4), 239–243.
25. Dương, M. T., Seifarth, V., Artmann, A.. T., Artmann, G. M., & Staat, M. (2018). Growth modelling promoting mechanical stimulation of smooth muscle cells of porcine tubular organs in a fibrin-PVDF scaffold. In G.M. Artmann, I.E. Digel, A. Zhubanova, A. Temiz Artmann (eds.), *Biological, Physical and Technical Basics of Cell Engineering* (pp. 211–234). Singapore: Springer Nature. 10.1007/978-981-10-7904-7_9.
26. Fenner, D. E. (2000). New surgical mesh. *Clinical Obstetrics and Gynecology, 43*(3), 650–658.
27. Feola, A., Barone, W., Moalli, P., & Abramowitch, S. (2013). Characterizing the ex vivo textile and structural properties of synthetic prolapse mesh products. *International Urogynecology Journal, 24*(4), 559–564.
28. Fischer, T., Ladurner, R., Gangkofer, A., Mussack, T., Reiser, M., & Lienemann, A. (2007). Functional cine MRI of the abdomen for the assessment of implanted synthetic mesh in patients after incisional hernia repair: initial results. *European Radiology, 17*(12), 3123–3129.

29. Fleischmajer, R., Perlish, J. S., Burgeson, R. E., Shaikh-Bahai, F., & Timpl, R. (1990). Type I and type III collagen interactions during fibrillogenesis. *Annals of the New York Academy of Sciences, 580,* 161–175.
30. Friedman, D. W., Boyd, C. D., Mackenzie, J. W., Norton, P., Olson, R. M., & Deak, S. B. (1993). Regulation of collagen gene expression in keloids and hypertrophic scars. *Journal of Surgical Research, 55*(2), 214–222.
31. Frotscher, R., & Staat, M. (2014). Stresses produced by different textile mesh implants in a tissue equivalent. *BioNanoMaterials, 15*(1–2), 25–30.
32. Göretzlehner, U., & Müllen, A. (2007). PVDF als Implantat-Werkstoff in der Urogynäkologie. *Biomaterialien, 8*(S1), 28–29.
33. de la Gutiérrez, P. C., Vargas Romero, J., & Diéguez García, J. A. (2001). The value of CT diagnosis of hernia recurrence after prosthetic repair of ventral incisional hernias. *European Radiology, 11*(7), 1161–1164.
34. Halbauer, C. (2014). Charakterisierung und Vergleich des mechanischen Verhaltens von in Gelatine gebetteten Netzimplantaten durch einen Aero-Bulgetest gegenüber einer FEM Simulation sowie die Entwicklung einer auf MRT-Scans basierenden 3D Visualisierungsmethode implantierter Netze. Unpublished bachelor thesis, Aachen University of Applied Sciences, Jülich.
35. Heniford, B. T., Park, A., Ramshaw, B. J., & Voeller, G. (2003). Laparoscopic repair of ventral hernias: Nine years' experience with 850 consecutive hernias. *Annals of Surgery, 238* (3), 391–399.
36. Horbach, A. J., Duong, M. T., & Staat, M. (2017). Modelling of compressible and othotropic mesh implants based on optical deformation measurement. *Journal of the Mechanical Behavior of Biomedical Materials, 74,* 400–410.
37. Hurme, T., Kalimo, H., Sandberg, M., Lehto, M., & Vuorio, E. (1991). Localization of type I and III collagen and fibronectin production in injured gastrocnemius muscle. *Laboratory Investigation, 64(1),* 76–84.
38. Jerabek, J., Novotny, T., Vesely, K., Cagas, J., Jedlicka, V., Vlcek, P., et al. (2014). Evaluation of three purely polypropylene meshes of different pore sizes in an onlay position in a New Zealand white rabbit model. *Hernia, 18*(6), 855–864.
39. Junge, K., Binnebösel, M., Rosch, R., Jansen, M., Kämmer, D., Otto, J., et al. (2009). Adhesion formation of a polyvinylidenfluoride/polypropylene mesh for intra-abdominal placement in a rodent animal model. *Surgical Endoscopy, 23(2),* 327–333.
40. Junge, K., Binnebösel, M., von Trotha, K. T., Rosch, R., Klinge, U., Neumann, U. P., et al. (2012). Mesh biocompatibility: Effects of cellular inflammation and tissue remodelling. *Langenbeck's Archives of Surgery, 397*(2), 255–270.
41. Junge, K., Klinge, U., Rosch, R., Mertens, P. R., Kirch, J., Klosterhalfen, B., et al. (2004). Decreased collagen type I/III ratio in patients with recurring hernia after implantation with alloplastic prostheses. *Langenbeck's Archives of Surgery, 389(1),* 17–22.
42. Klinge, U. (2007). Experimental comparison of monofile light and heavy polypropylene meshes: less weight does not mean less biological response. *World Journal of Surgery, 31*(4), 867–868.
43. Klinge, U., Binnebösel, M., Kuschel, S., & Schuessler, B. (2007). Demands and properties of alloplastic implants for the treatment of stress urinary incontinence. *Expert Review of Medical Devices, 4*(3), 349–359.
44. Klinge, U., & Klosterhalfen, B. (2012). Modified classification of surgical meshes for hernia repair based on the analysis of 1,000 explanted meshes. *Hernia, 16*(3), 251–258.
45. Klinge, U., Klosterhalfen, B., Müller, M., Ottinger, A. P., & Schumpelick V. (1998). Shrinking of polypropylene mesh in vivo: An experimental study in dogs. *European Journal of Surgery, 164*(12), 965–969.
46. Klinge, U., Klosterhalfen, B., Birkenhauer, V., Junge, K., Conze, J., & Schumpelick, V. (2002). Impact of polymer pore size on the interface scar formation in a rat model. *Journal of Surgical Research, 103*(2), 208–214.

47. Klinge, U., Park, J. K., & Klosterhalfen, B. (2013). The ideal mesh? *Pathobiology, 80,* 169–175.
48. Klinge, U., Si, Z. Y., Zheng, H., Schumpelick, V., Bhardwaj, R. S., & Klosterhalfen, B. (2000). Abnormal collagen I to III distribution in the skin of patients with incisional hernia. *European surgical Research, 32*(1), 43–48.
49. Klink, C. D., Junge, K., Binnebösel, M., et al. (2011). Comparison of long-term biocompatibility of PVDF and PP meshes. *Journal of Investigative Surgery, 24*(6), 292–299.
50. Klosterhalfen, B., Junge, K., & Klinge, U. (2005). The lightweight and large porous mesh concept for hernia repair. *Expert Review of Medical Devices, 2*(1), 103–117.
51. Klosterhalfen, B., Klinge, U., & Schumpelick, V. (1998). Functional and morphological evaluation of different polypropylene-mesh modifications for abdominal wall repair. *Biomaterials, 19*(24), 2235–2246.
52. Langer, C., Neufang, T., Kley, C., Liersch, T., & Becker, H. (2001). Central mesh recurrence after incisional hernia repair with Marlex are the meshes strong enough? *Hernia, 5*(3), 164–167.
53. Laroche, G., Marois, Y., Schwarz, E., Guidoin, R., King, M. W., Pâris, E., et al. (1995). Polyvinylidene fluoride monofilament sutures: Can they be used safely for long-term anastomoses in the thoracic aorta? *Artificial Organs, 19*(11), 1190–1199.
54. Law, N. W., & Ellis, H. (1988). Adhesion formation and peritoneal healing on prosthetic materials. *Clinical Materials, 3*(2), 95–101.
55. Leber, G. E., Garb, J. L., Alexander, A. I., & Reed, W. P. (1998). Long-term complications associated with prosthetic repair of incisional hernias. *Archives of Surgery, 133*(4), 378–382.
56. LeBlanc, K. A. (2001). The critical technical aspects of laparoscopic repair of ventral and incisional hernias. *American Surgeon, 67*(8), 809–812.
57. Lehr, S. C., & Schuricht, A. L. (2001). A minimally invasive approach for treating postoperative seromas after incisional hernia repair. *JSLS-Journal of the Society of Laparoendoscopic Surgeons, 5*(3), 267–271.
58. Liang, R., Abramowitch, S., Knight, K., Palcsey, S., Nolfi, A., Feola, A., et al. (2013). Vaginal degeneration following implantation of synthetic mesh with increased stiffness. *British Journal of Obstetrics and Gynaecology, 120*(2), 233–243.
59. Margulies, R. U., Lewicky-Gaupp, C., Fenner, D. E., McGuire, E. J., Clemens, J. Q., & Delancey, J. O. (2008). Complications requiring reoperation following vaginal mesh kit procedures for prolapse. *American Journal of Obstetrics & Gynecology, 199*(6), 678.e1–678.e4.
60. Matthews, B. D., Pratt, B. L., Pollinger, H. S., Backus, C. L., Kercher, K. W., Sing, R. F., et al. (2003). Assessment of adhesion formation to intra-abdominal polypropylene mesh and polytetrafluoroethylene mesh. *Journal of Surgical Research, 114*(2), 126–132.
61. Mohamed, M., Elmoghrabi, A., Shepard W. R., & McCann, M. (2016) Delayed onset seroma formation 'opting out' at 5 years after ventral incisional hernia repair. *BMJ Case Reports 2016*. https://doi.org/10.1136/bcr-2016-215034
62. Morris-Stiff, G. J., & Hughes, L. E. (1998). The outcomes of nonabsorbable mesh placed within the abdominal cavity: literature review and clinical experience. *Journal of the American College of Surgeons, 186*(3), 352–367.
63. Mühl, T., Binnebösel, M., Klinge, U., & Goedderz, T. (2008). New objective measurement to characterize the porosity of textile implants. *Journal of Biomedical Materials Research. Part B, Applied Biomaterials, 84*(1), 176–183.
64. Nagase, H., Visse, R., & Murphy, G. (2006). Structure and function of matrix metalloproteinases and TIMPs. *Cardiovascular Research, 69(3),* 562–573.
65. Orenstein, S. B., Saberski, E. R., Kreutzer, D. L., & Novitsky, Y. W. (2012). Comparative analysis of histopathologic effects of synthetic meshes based on material, weight, and pore size in mice. *Journal of Surgical Research, 176*(2), 423–429.

66. Otto, J., Kaldenhoff, E., Kirschner-Hermanns, R., Mühl, T., & Klinge, U. (2014). Elongation of textile pelvic floor implants under load is related to complete loss of effective porosity, thereby favoring incorporation in scar plates. *Journal of Biomedical Materials Research Part A, 102*(4), 1079–1084.
67. Pans, A., Albert, A., Lapière, C. M., & Nusgens, B. (2001). Biochemical study of collagen in adult groin hernias. *Journal of Surgical Research, 95(2),* 107–113.
68. Patel, H., Ostergard, D. R., & Sternschuss, G. (2012). Polypropylene mesh and the host response. *International Urogynecology Journal, 23*(6), 669–679.
69. Poobalan, A. S., Bruce, J., Smith, W. C., King, P. M., Krukowski, Z. H., & Chambers, W. A. (2003). A review of chronic pain after inguinal herniorrhaphy. *Clinical Journal of Pain, 19*(1), 48–54.
70. Post, S., Weiss, B., Willer, M., Neufang, T., & Lorenz, D. (2004). Randomized clinical trial of lightweight composite mesh for Lichtenstein inguinal hernia repair. *British Journal of Surgery, 91*(1), 44–48.
71. Ratner, B. D., Northup, S. J., & Anderson, J. M. (2004). Biological testing of biomaterials. In B. D. Ratner, F. J. Schoen, & J. E. Lemons (Eds.), *Biomaterials science: an introduction to materials in medicine* (2nd ed., pp. 355–360). San Diego: Elsevier.
72. Rutkow, I. M. (2003). Demographic and socioeconomic aspects of hernia repair in the United States in 2003. *Surgical Clinics of North America,* 83(5), 1045–1051, V–VI.
73. Saad, B., Abu-Hijleh, G., & Suter, U. W. (2003). Polymer biocompatibility assessment by cell culture techniques. In R. Arshady (Ed.), *Introduction to polymeric biomaterials: the polymeric biomaterials series* (pp. 263–299). London: Citus Books.
74. Salamone, G., Licari, L., Agrusa, A., Romano, G., Cocorullo, G., & Gulotta, G. (2015). Deep seroma after incisional hernia repair. Case reports and review of the literature. *Annali Italiani di Chirurgia,* 12:86 (ePub).
75. Scheidbach, H., Tamme, C., Tannapfel, A., Lippert, H., & Köckerling, F. (2004). In vivo studies comparing the biocompatibility of various polypropylene meshes and their handling properties during endoscopic total extraperitoneal (TEP) patchplasty: an experimental study in pigs. *Surgical Endoscopy, 18*(2), 211–220.
76. Schumpelick, V., Conze, J., & Klinge, U. (1996). Preperitoneal meshplasty in incisional hernia repair. A comparative retrospective study of 272 repaired incisional hernias. *Chirurg, 67*(10), 1028–1035.
77. Scott, P. D., Harold, K. L., Craft, R. O., & Roberts, C. C. (2008). Postoperative seroma deep to mesh after laparoscopic ventral hernia repair: computed tomography appearance and implications for treatment. *Radiology Case Reports, 3*(1), art. 10.128.
78. Shoshan, S. (1981). Wound healing. *International Review of Connective Tissue Research, 9,* 1–26.
79. Shoulders, M. D., & Raines, R. T. (2009). Collagen structure and stability. *Annual Review of Biochemistry, 78,* 929–958.
80. Staat, M., Trenz, E., Lohmann, P., Frotscher, R., Klinge, U., Tabaza, R., et al. (2012). New measurements to compare soft tissue anchoring systems in pelvic floor surgery. *Journal of Biomedical Materials Research. Part B, Applied Biomaterials, 100*(4), 924–933.
81. Susmallian, S., Gewurtz, G., Ezri, T., & Charuzi, I. (2001). Seroma after laparoscopic repair of hernia with PTFE patch: is it really a complication? *Hernia, 5(3),* 139–141.
82. Usher, F. C., & Gannon, J. P. (1959). Marlex mesh, a new plastic mesh for replacing tissue defects. *I. Experimental studies. AMA Archives of Surgery, 78*(1), 131–137.
83. Usher, F. C., Ochsner, J., & Jr, Tuttle L. L. (1958). Use of Marlex mesh in the repair of incisional hernias. *American Surgeon, 24*(12), 969–974.
84. van't Riet, M., de Vos van Steenwijk, P. J., Bonthuis, F., Marquet, R. L., Steyerberg, E. W., Jeekel, J., et al. (2003). Prevention of adhesion to prosthetic mesh: comparison of different barriers using an incisional hernia model. *Annals of Surgery*, 237(1), 123–128.

85. Welty, G., Klinge, U., Klosterhalfen, B., Kasperk, R., & Schumpelick, V. (2001). Functional impairment and complaints following incisional hernia repair with different polypropylene meshes. *Hernia, 5*(3), 142–147.
86. Williams, G. T., & Williams, W. J. (1983). Granulomatous inflammation-a review. *Journal of Clinical Pathology, 36*(7), 723–733.
87. Wilson, C. J., Clegg, R. E., Leavesley, D. I., & Pearcy, M. J. (2005). Mediation of biomaterial-cell interactions by adsorbed proteins: a review. *Tissue Engineering, 11*(1–2), 1–18.

Author Biography

Aroj Bhattarai (Nepal) He came from the land of nature. He received his M.Sc. in Computational engineering from the Technical University of Dresden (2010–2012). Since 2013 he joined the Biomechanics Laboratory at Aachen University of Applied Sciences as a Ph.D. candidate. His research (at the Institute of Bioengineering, Biomechanics Laboratory, Aachen University of Applied Sciences) mainly focuses on the finite element method to optimize the surgical repairs of the female pelvic floor dysfunctions using prosthetic mesh implants.

Extraction of Fusicoccin-like Compounds Using Nanocarbon Sorbents and Study of Their Biological and Cytotoxic Activity

Z. A. Mansurov, S. Azat and A. R. Kerimkulova

1 Introduction

1.1 Carbon Sorption Materials

The useful properties of activated carbon have been known since ancient times. This traces back to 1500 BC when Egyptians used charcoal as an adsorbent for medicinal purposes and a purifying agent. Around 420 BC it was observed that Hippocrates dusted wounds with powdered charcoal to remove their odor. Ancient Hindu societies purified their water by filtration through charcoal. In 1773, the Swedish chemist Karl Wilhelm Scheele was the first to observe adsorption of gases on charcoal. A few years later activated carbons began being used in the sugar industry as a decolorizing agent for syrup.

In the early 20th century the first plant to produce activated carbon industrially was built for use in sugar refining industry in Germany. Many other plants emerged in the early 1900s to produce activated carbons primarily for decolorization. During World War I activated carbon was used in gas masks for protection against hazardous gases and vapors.

Today, activated carbons are used to remove color from pharmaceutical and food products, as air pollution control devices for industrial and automobile exhaust, for chemical purification, and as electrodes in batteries. 500,000 tons per year of activated carbon are produced globally [1–7]. 80% of this is used for liquid phase applications, and 20% is used for solid phase applications.

Z. A. Mansurov (✉) · S. Azat · A. R. Kerimkulova
Department of Chemicalphysics and Materialscience,
Al-Farabi Kazakh National University, Almaty 050040, Kazakhstan
e-mail: zmansurov@kaznu.kz

Z. A. Mansurov · S. Azat · A. R. Kerimkulova
Institute of Combustion Problems, Almaty 050012, Kazakhstan

G. M. Artmann et al. (eds.), *Biological, Physical and Technical Basics of Cell Engineering*, https://doi.org/10.1007/978-981-10-7904-7_12

Activated carbon (AC) is a unique material with its immense capacity for adsorption from gas and liquid phases. It occupies a special place in terms of producing a clean environment involving water purification as well as separations and purification in the chemical and associated industries. In these roles, it exhibits a remarkable efficiency as the international production is a little more than a million tons per year, with perhaps 2 million tones being in continuous use. Broadly AC has been used for three main purposes:

1. AC is predominantly used for purposes of adsorption in liquid-phase. The wide-ranging scenarios for AC are:
   - drinking water availability, to improve taste, smell and color including removal of chlorinated compounds and other volatile organic carbons (VOCs);
   - purification of ground water purity from contaminants coming from disused sites of heavy industries;
   - treatment of both industrial and municipal wastewater;
   - mining operations require feed water treatment, metallic ion adsorption (gold and other metals), adsorption of excess flotation reagents and adsorption of natural organic matter (NOM);
   - pharmaceutical processes, including purification of process water, use with fermentation broths and purification of many products;
   - the food, beverage and oil industries for removal of color and unacceptable tastes;
   - the dry-cleaning industries require purification of solvents;
   - the electroplating industries require purification of wastewaters containing Pb, Cr, etc.;
   - household water purification, cleaning of aquaria and use in oven-extract hoods;
   - the sugar and sweetener industries need decolorization agents for production of white sugar, etc.
2. AC in different forms (granular, extruded, fiber or cloth) is used for production of pure gases in the chemical industry, to reduce pollutant gases to very low concentrations in a single stage, in protection against poisonous gases, in air conditioning, for removal of oil from compressed air, to separate gases in mixtures by sieving, etc.
3. Carbon materials have been used for some time in heterogeneous catalysis, acting as direct catalysts or as a catalyst support. Catalytic particles can be supported within the porosity to promote required catalytic conversions [8–10].

Activated carbon is a microcrystalline form of carbon with very high porosity and surface area. It may be visualized as foam solid that has a large surface within a rigid granule or particle structure of relatively small volume. Its chemical structure allows it to preferentially adsorb organic materials and other nonpolar compounds from gas or liquid streams.

Activated carbon has become one of the most technically important and most widely used adsorbents because of its high adsorptive capacity. Present technology demands a very large production of activated carbons with appropriate characteristics for each particular application. In general, activated carbon which is used in any of the most common applications must have adequate adsorptive capacity, chemical purity, mechanical strength, etc.

Furthermore, all these specifications should coexist with a low production cost. Activated carbon is obtained from a carefully controlled process of dehydration, carbonization and oxidation of organic substances. It can be prepared for research in the laboratory from a large number of materials. However, the most commonly used ones in commercial practice are peat, coal, lignite, wood and agricultural by-products such as coconut shell, almond shell, rice husk, etc.

Pyrolysis of the starting material with exclusion of air and without addition of chemical agent usually results in an inactive material with a specific surface area of the order of several m^2/g and low adsorption capacity. One can prepare carbon with a large adsorption capacity by activating the carbonized products with a reactive gas. The majority of activated carbon used throughout the world is produced by steam activation (physical activation). In this process, the carbonized product is reacted with steam over 900 °C.

Another procedure used in the production of activated carbon involves the use of chemical activating agents before the carbonization step. The most commonly used activating agents are phosphoric acid, zinc chloride and salts of sodium and magnesium etc. Many chemical substances act as dehydration agents and therefore may restrict the formation of tar during carbonization. Chemical activation is usually carried out at lower temperatures than the simple pyrolysis and the activation process with steam or carbon dioxide. The production at lower temperatures promotes the development of a porous structure because under these conditions elementary crystallites of smaller dimensions are formed [11, 12].

Most of the available surface area of activated carbon is nonpolar in nature. However, during production the interaction of surface with oxygen produces specific active sites giving the surface of slightly polar nature. As activated carbon is one of the most commonly used adsorbents in many industrial applications for its adsorptive capacity [13–15].

Most activated carbons contain pores of different sizes: micropores, transitional mesopores and macropores. According to the IUPAC definition, pores can be distinguished in three groups with respect to their dimensions.

- Macropores—Pores with larger than 50 nm;
- Mesopores—Pores with diameters between 2 and 50 nm;
- Micropores—Pores with diameters less than 2 nm.

Microporous carbons have a unique structure which essentially is a three dimensional network, or labyrinth of carbon atoms some in hexagonal arrangements and some as individual carbon atoms bonded together extremely close but not close packed. This bonding arrangement results in space existing between the atoms to create an interconnecting three dimensional passage way in which every space unit has a connection to all others within the carbon. The dimensions of the passage way (width) are those of molecules, about 0.5 nm, such that any molecule which enters into this space is subjected to intense dispersion forces from the carbon atoms (about eight per space) which make up the "surface" of the space.

These forces can "trap" the molecule for periods of time much longer than on an open surface and so the phenomenon of physical adsorption of gases is generated.

The site, place or space where a molecule can be trapped (adsorbed) is called an adsorption site. These adsorption sites can be modified in terms of their size (widened or narrowed) and in terms of their chemical composition [16]. The surface can be bonded to hydrogen, oxygen, chlorine, nitrogen, etc. to alter the polarity of the surface. Changing this polarity can enhance the adsorption process for polar molecules. The porosity can be widened by gasification with water vapor, oxygen or carbon dioxide. The parent materials may also be impregnated with zinc chloride, potassium hydroxide or phosphoric acid, these treatments improving adsorption characteristics, a process known as activation. The variables, within an adsorption system (dominantly an aqueous solution) controlling extents of adsorption include [17, 18]:

1. Volume of narrow micropores, <0.7 nm (VNM—cm^3/g), including the pore-size distributions.
2. Volume of wider micropores, 0.7–2.0 nm (VWM—cm^3/g), including the pore-size distributions.
3. Volume in mesopores, >2.0 and <50 nm (VME—cm^3/g), including the pore-size distributions.
4. The parameter "surface area" is a crude parameter and fails to quantify, adequately, the capacity of the all-important micropores and their size distributions.
5. The controlled additions of surface oxygen complexes from known sources, e.g. nitric acid, H_2O_2, etc., with controlled removal of the complexes, thermally or by reaction with hydrogen at temperature >750 °C.
6. The effects of additions of surface oxygen complexes upon pore volumes and pore size distributions (pore dimensions can be reduced).
7. The characterization of surface oxygen complexes.
8. The pH of the solution.
9. Ionic strength.
10. Temperature.

Nowadays in Kazakhstan the necessity of creation of carbon materials with high sorption properties has come. A pioneer in this field was "Laboratory of carbon nanomaterials" at the Institute of Combustion problems and at Al-Farabi KazNU created by Professor R. M. Mansurova. In this laboratory from waste rice grinding production by carbonation the first carbonized carbon biomaterial was obtained which has the ability to adsorb not only molecules, but even whole microbial cells.

In the work [19–21] on the carbonized carbon material from rice husk cells of the following microorganisms were sorbed. The choice of these microorganisms was due to the fact that they are widely used for detoxification of sewage. In this work we have obtained very interesting results. It is shown that microorganisms adsorbed on carbon materials remained viable in the presence of highly toxic ions of cadmium, lead and copper [22–24]. Sorption of microorganisms in the carbon sorbent from rice husk four times increased the viability of the microorganisms in the presence of toxic ions, in this case the total sorption capacity of these toxic ions increased several times. All these results suggest that unusual carbon material with sorbed living cells of microorganisms was created and it has the ability to high sorption of highly toxic ions. This new material can be widely used for industrial detoxification of polluted water contaminated with heavy metal ions.

In this laboratory headed by Professor R. M. Mansurova, carbon materials have been created by carbonization of the waste from local raw materials. Interesting work has been done on the sorption of toxic ions by obtained carbonized carbon materials derived from rice husk and wheat bran [25, 26]. It was found that activation of the obtained carbon materials with hydrogen peroxide and ammonia significantly increases the sorption capacity for heavy metal ions and radionuclides.

Thus, developed on the basis of vegetative raw materials of Kazakhstan carbonized carbon materials have a great potential for detoxification of water contaminated with ions of heavy metals and radionuclides. Moreover, they offer the prospect of using these materials for the enrichment of rare and expensive metals. Owing to excellent sorption quality of such materials it is necessary to expand and deepen the study on the creation of properties of new carbon materials derived from the vegetative raw materials of Kazakhstan [27].

The main trends of development in science and technology is currently focused on the study of nanoscale structural components of matter and functional materials. Progress in this area is of great theoretical and practical importance because such materials can be widely used in ecology, biology and medicine and just because in these areas of science the greatest social and economic effect from the introduction nanomaterials is predicted.

It should be noted that prospects for practical use of nanostructured sorbents are obtained by high temperature carbonization of cheap secondary vegetative raw materials. Carbon obtained by the carbonization of plant materials, keeps its original finely organized structure. By varying the conditions of carbonization, it is possible to obtain a complex of compositions of carbon from other materials. Some of these materials can find practical application. Availability and annual renewability of raw materials, low content of mineral impurities, developed porous structure, ecological production allow to obtain cheap, fast regenerable carbon sorbents [28–30].

1.2 The Properties of Fusicoccin and Its Application as an Anticancer Preparation

Fusicoccin was discovered in 1964 by the Italian scientist Alessandro Ballio as the phytotoxin of the phytopathogenic fungus "*Fusicoccum amygdali* Del". This phytotoxin was able to kill young almond trees as follows: this toxin opened stomata of leaves which did not close after. In consequence of the excessive transpiration and because of the weakness of the root system young trees were dying rapidly from drought, i.e. fusicoccin is a natural desiccant. In the laboratory of Professor A. Ballio the structure of this phytotoxin was determined as is shown in Fig. 1. Considering the fact that fusicoccin refers to natural terpenoids as well as the gibberellin Gronevald et al. developed a hypothesis about the kinship of gibberellin and fusicoccin. Indeed, they both are diterpenoids, metabolites of phytopathogenic fungi. Over time, the number of known gibberellins increased rapidly from few in 1960 to more than eighty by the end of the 80s. The same for fusicoccins: now more than 15 related compounds of this group are known (denoted by literal characters—A, B, C, etc.).

Preliminary data on the presence of substances of the fusicoccin family in higher plants were published first by Muromtsev and his co-wokers in 1980. Later for the detection of fusicoccin a gas chromatography, mass-spectrometry was used (GC/MS) with a preliminary fractionation of the material by the high performance liquid chromatography (HPLC). Initially, the authors have developed an identification

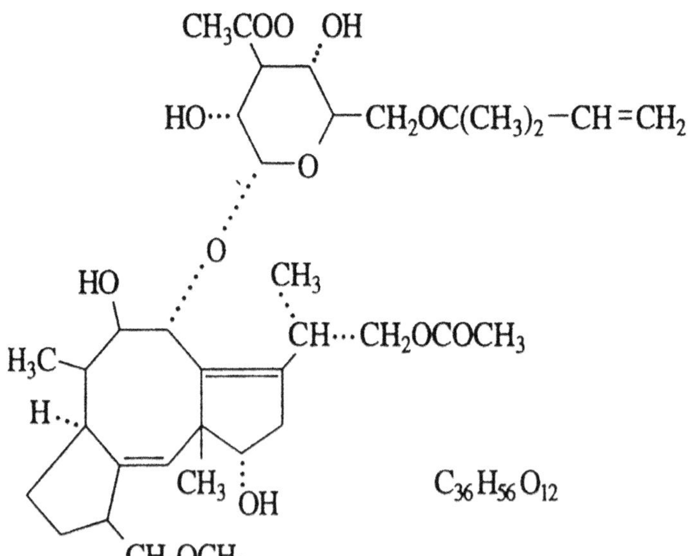

Fig. 1 Formula of fusicocin—phytotoxin of the phytopatogenic fungus *Fusicoccum amygdali* Del. [31]

method for fusicoccinal metabolites in the culture fluid of the fungus *F. amygdali* by the combined method of the gas chromatography and mass spectrometry of trisilil derivatives, when the detection is performed through several chosen characteristic ions. To obtain fusicoccin, the authors used corn cobs and cabbage leaves, which were homogenized in ethanol. The alcohol extracts were concentrated. The concentrate was extracted by chloroform. The chloroform extract was subjected to the liquid chromatography. The obtained peaks were subjected to the gas-liquid chromatography followed by mass spectrometry. The content of the endogenous fusicocin A in plants was 10^{-11}–10^{-12} M, which is 2–3 orders lower than that of the endogenous gibberellins. Hence it becomes clear that the problem of preparative quantities extraction of natural fusicoccin from higher plants is a very difficult experimental problem.

Then, the same authors tried another shot to extract natural fusicoccin from higher plants. At this time, as the object for the fusicoccin extraction the authors used a sterile culture of transformed horseradish roots distinguished by the intensive growth. This object was chosen for the following reasons. Firstly, it eliminates the possibility of contamination by microorganisms and fungi that may be sources of fusicoccin. Second, the culture of roots cells can be obtained in very large quantities due to their intensive growth.

From cultures of roots cells the authors obtained alcohol extracts. The extracts were evaporated in vacuo, the aqueous residues were extracted by chloroform, followed by determination of the presence of fusicoccin in aqueous and chloroform phases by immune methods. Fusicoccin was in chloroform, whereas in the aqueous phase there were found only trace amounts, that eliminates the possibility of effective interaction of such compounds as proteins and sugars with receptors and antibodies. Substances similar to fusicoccin were discovered at all stages of roots cultivation (since 14 days). It was found that the maximum content of fusicoccin in grown sterile roots is up to 150 nM per kilogram of roots.

Mass-spectrometric analysis of the obtained fractions showed mainly the presence of fusicoccin type A, also these analyses indicated the possible presence of another unidentified fusicoccin. As a result of undertaken studies Muromtsev and his co-workers made a conclusion that identification of endogenous fusicoccin is a very difficult task [32].

The first and main of the identified fusicoccins, fusicoccin A, is a glycoside of the carbotricyclic diterpene with a molecular weight of 680 Da gross and the formula $C_{36}H_{56}O_{12}$ (Fig. 1). The aglycon part of fusicoccin molecules is represented by the tricyclic system including one eight-membered and two five-membered rings. Besides the basic compound (fusicoccin A) the fungus produces some similar compounds (more than 15), which differ in the degree of acetylation, positions of acetyl groups, the degree of oxidation of the aglycone part of the molecule. Thus, ten fusicoccins represent monoacetates, diacetates and triacetates. Another series of fusicoccins has a more simple structure: their C_{20}-atom is not oxidized. The analysis of literature data shows that the fusicoccin molecule is not a unique structure and connections with the similar "skeleton" which are derivatives of the dicyclopentane and cyclooctane (the ring system 5-8-5), are widely distributed in the nature. They were found in fungi and algae, higher plants—liverworts, flowering plants, and even among animals (insects). If and when discovered and identified various representatives of this class of terpenoids the trivial names of organisms-producers are assigned to them. Thus, for example, the fusicoccin extracted from the liverwort *Plagiochila acanthophylla* is called fusicoplagin C, the fusicoccin extracted from the liverwort *Anastrepta orcadensin* is anadensin, the fusicoccin extracted from the giant kelp *Dictyota dichtoma* is epoxydictimen. It was proposed to combine compounds having dicyclopentane and cyclooctane "skeleton" into groups of terpenoids of the fusicoccane range. The basis of the range of such compounds was formed by the hypothetical trans-sin-trans C_{20} hydrocarbon that was named fusicoccine.

1. The further study of fusicoccin showed that they demonstrate various physiological and biochemical properties. This made it possible to class it among natural regulators of the plant growth.
2. The cells of many organs and tissues of higher plants respond to the treatment by fusicoccin with the increase of the volume, with which such effects can be associated as the opening of the stomata, cells growth and elongation. There are data on the possible role of fusicoccin as an endogenous regulator of seed germination and root formation.
3. Fusicoccin effectively activates seeds from the deep rest. The laboratory of Muromtsev first discovered that fusicoccin actively stimulates the rhizogenesis for number of cultures. Fusicoccin had a very interesting effect on germination of pea seeds. So it stimulated the growth of cotyledon cells and at the same time inhibited the growth of the embryonic axis, which indicates the polarity of the fusicoccin action. It was found that fusicoccin plays an important activating role at early stages of legumes contamination with microorganisms of the genus *Rhizobium* spp, i.e. fusicoccin activates the nodulation in legumes.
4. Fusicoccin activates absorption of carbon dioxide by columnar cells of the leaf. Also fusicoccin regulates the concentration of ascorbate in the apoplast of coleoptile cells *Vigna angularis*.
5. One of the most striking hormonal properties of fusicoccin is its anti-stress activity. Muromtsev and his co-workers showed that fusicoccin can increase the seed germination in conditions unfavorable for germination, for example: at elevated and low temperatures, excessive moisture, salination. The co-workers of the Institute of Plant Physiology showed that the seeds soaking (0.68 mg/l of fusicoccin) and also spraying of plants of winter wheat, rye, barley (0.34 mg/l of fusicoccin) leads to the increase of the frost hardiness of plants at booting and tillering phases. Moreover, the increase of the frost hardiness correlated well with the degree of development of the photosynthetic apparatus and sugar accumulation, as well as with the development strengthening of the endoplasmic reticulum in cells. It is shown that fusicoccin protects rice plants under salinity, increases the resistance of potato tubers to some diseases. Fusicoccin is undoubtedly one of the

most powerful anti-stress compounds for plants. The positive role of fusicoccin in adaptation to osmotic stress is shown in the work [33].

The available data on the study of the biochemical activity of fusicoccin are of particular interest for us.

For the first time [33] Ballio and his co-workers discovered fusicoccinal receptors on the plasma membrane isolated from the roots of maize seedlings. They also created proteoliposomes with the inclusion of isolated fusicoccinal receptors and studied properties of this artificial system. De Boer and his co-wokers developed a method for purification of FC-binding proteins on a special affine sorbent to which the biotinylated fusicoccin is ligated as the active group. A number of authors have also performed some studies on the extraction of fusicoccin-binding proteins [34].

For fusicoccin the alleged receptors are described—fusicoccin-binding proteins (FCBP) of two types. The first type has a high affinity for fusicoccin with a dissociation constant (Kd) about 10^{-10} M (high-affinity HA), and the second receptors type with a lower affinity, Kd about 10^{-7} M (low-affinity LA). They were first detected in the fraction containing plasma membranes of maize coleoptiles. It was shown in the publication that the ratio of the quantity of high-affinity receptors sites to low-affinity receptor sites on isolated membranes [HA]/[LA] is approximately one to two. The authors made a conclusion that only the high-affinity binding site is involved in action of fusicoccin. This point of view in the following years became dominant. Thus, the authors suggested that participation of the high-affinity site as a fusicoccin receptor is according to the reconstruction of the system in vitro consisting of fusicoccin-binding protein and H^+-ATPase A [35–37].

A small quantity or absence of low-affinity sites on most isolated membranes may be due to their inactivation during the isolation process. It is explained by a much greater lability of low-affinity binding sites in comparison with high-affinity ones. Currently, the issue of the functional role of the high- and low-affinity sites is the subject of research.

Membrane effects of fusicoccin belong to rapid answers: usually stimulation of protons emission begins immediately after adding fusicoccin and has no lag phase. In addition to the rapid membrane effects FC may have a prolonged, generalized action on plants that speaks in favor of its hormonal properties. Conspicuous is the fact that the current fusicoccin doses (10–20 mg/ha) are by two-three orders lower than that of other plant phytohormones and growth regulators (including those for gibberellin). In the publication of Babakov [38] the role of FC in the activation of protein kinase is discussed.

Claudio Olivari and his co-wokers [39] first stated that the activation process of H^+-ATPase by fusicoccin requires the presence of another protein. In his work Claudio Olivari found that the phenylarsine oxide was a specific inhibitor of activation process of H^+-ATPase by fusicoccin.

Very interesting is the action of fusicoccin on cells. First of all it concerns the acidification of the content of the cell under the influence of fusicoccin. The data that fusicoccin stimulates the protons emission in oat coleoptile cells were first specified in the work. The listed processes can be activated upon condition of the work stimulation of H^+-ATPase or reduction of the ions leakage through ion channels of a plasmalemma. Actually, according to the latest data, fusicoccin affects both the activity of H+-ATPase, and the conductivity of plasmalemma potassium channels [40].

1.3 Natural Small Molecules and Their Use as an Anticancer Drug

Some small molecules that are used in preclinical as well as clinical studies influence PPIs of the Bcl-2 protein family of anti-apoptotic proteins [41]. Several compounds have been identified that target the hydrophobic binding pocket, which forms the BH_3 binding site in Bcl-2, and prevent thereby the hetero-dimerization of Bcl-2 with pro-apoptotic members and thereby the induction of apoptosis. Another well known example is the nutlins (cis-imidazoline analogs) that target the p53/MDM2 interaction. These compounds bind selectively to MDM2 and prevent thereby the interaction with p53 and enhance the activity of the p53 pathway. It is interesting that healthy wild-type p53 cells retained their viability when treated with these nutlins, while wild-type p53 expressing cancer cells are killed. Besides these examples many other small molecules are known that prevent PPIs that are beneficial in the treatment of cancer. However, so far not many small molecules are known that affect cancer cells by stabilizing PPIs.

In recent years several reviews have been published about the isolation of (novel) natural compounds and active small molecules with an anticancer activity from various sources. Plants supplied several of these compounds while other anticancer drugs have been isolated from marine organisms and microorganisms like fungi. Although the working mechanism of many active small molecules is not yet (completely) clarified, most these compounds have been shown to belong to diverse structural classes, including polyketides, terpenoids, terpenes, alkaloids, and steroids and to influence PPIs. Furthermore, many drugs (>60%) that are currently used in cancer treatment are derived from natural occurring small molecules, either directly or indirectly. Paclitaxel (taxol) is, for instance, isolated from the bark of the Pacific yew tree Taxus brevifolia [42] and the vinca alkaloids vinblastine and vincristine are isolated from the Madagascar periwinkle Catharanthus roseus.

Fusicoccanes are a large group of active small molecules that refer to the terpenoids. They have a typical dicyclopenta[a,d] cyclooctane skeleton (5-8-5 core ring structure), to which various side chains are linked [43]. They are found throughout nature and are amongst others produced by various fungi, higher plants, liverworts, and even insects. A number of biological effects have been described for fusicoccanes on both plants and other organisms, and several of them are caused by the selective targeting of a particular protein or PPI. Some well-studied

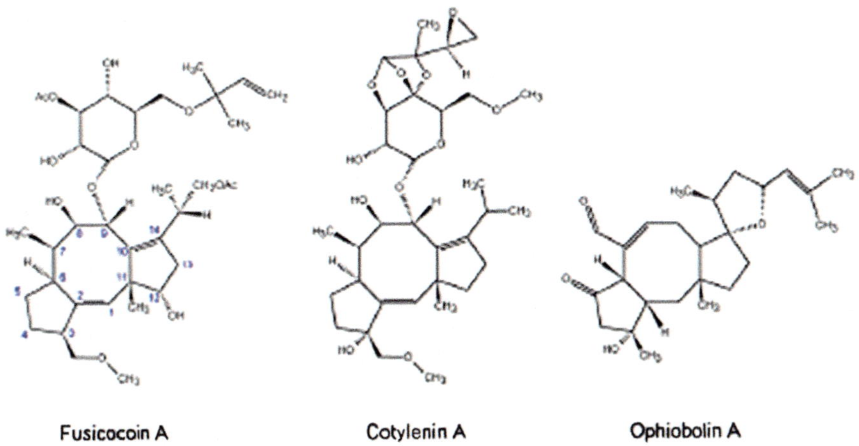

Fig. 2 Chemical structure of Fusicoccin-A, Cotylenin-A and Ophiobolin-A [44]

fusicoccanes are Fusicoccin-A (FC), Cotylenin-A (Cot-A), and Ophiobolin-A (OPH-A) (Fig. 2).

The fungus infects peach and almond trees, in which the secreted FC stabilizes the complex formed between the C-terminal tip of the plasma membrane H+-ATPase and a 14-3-3 protein. This stabilization irreversibly activates the H + -ATPase, that drives the opening of stomata resulting in uncontrolled water loss and wilting of the tree. Although the only known FC target in plants is the H+-ATPase, at the cell and organ level FC affects a range of different processes, like opening the stomata, breaking seed dormancy, inducing an apoptotic like cell death in sycamore cells, and reducing hydrogen peroxide and nitric oxide production in guard cells [45].

Many of these processes can be (in)directly explained by the activation of the H +-ATPase while others suggest the presence of another FC target. FC is not only active on all higher plants, but on other organisms as well.

It randomizes, for instance, the left-right asymmetry early in the Xenopus laevis embryonic development [46] and induces apoptosis in cancer cells derived from different origins, for which the efficacy of FC can be enhanced by combining it with the cytokine interferon (IFNα).

Cotylenin-A was first described by Sassa (REF) who discovered it in a fungal culture filtrate of Cladosporium sp and described its growth enhancing effect on plant cotyledons. The structure of Cot-A is rather similar to that of FC, and comparisons made on their effects in plants show that both molecules have similar physiological effects. In some cases FC is more potent (e.g. induction of seed germination and wilting of tomato cuttings), while Cot-A is more effective in others (e.g. stomatal opening and cotyledon elongation). Besides this, Cot-A has been shown to induce differentiation in human myeloid leukemia cells and inhibits the growth of various tumor types both in vitro and in vivo, when combined with IFNα or rapamycin.

Ophiobolins (OPHs) are produced by plant pathogenic fungi belonging to the Bipolaris species. Currently 23 analogs have been identified, to which various biological activities are ascribed that are all damaging plants either directly or indirectly. Besides being harmful for plants, OPHs also possess anti-microbial activities as well as toxicity towards animals and even cancer cells. The working mechanism of most OPHs is not completely clear and probably differs per analog. OPH-A, however, has been shown to mainly function through the specific inhibition of calmodulin, an important signalling molecule.

Anticancer effects have been described for several fusicoccanes, including FC, Cot-A and OPH-A. All show a clear growth reduction at relatively low concentrations (IC50-value mostly in the μM range, depending on the cell line). Although the tumor selectivity differs per molecule, with some being relatively cancer selective while others are indiscriminate and kill healthy cells as well as cancer cells, fusicoccanes form an interesting group of potentially new anticancer compounds. The working mechanism of most fusicoccanes in inhibiting the cancer cell growth is not clear yet. OPH-A most likely acts through the selective inhibition of calmodulin by binding covalently to calmodulin with its C7-aldehyde group [47]. FC and Cot-A both lack this aldehyde group, suggesting that these compounds function through a different cellular target. When combined with IFN, FC and Cot-A both induce apoptosis by the induction of the TRAIL pathway, however, this is most likely the final effect of the combined treatment and the primary target of both compounds is therefore still unknown.

2 Experimental Methods

2.1 *Preparation of Activated Carbons Based on Vegetable Raw Materials*

In this work, a series of experiments on the physical and chemical activation patterns of raw materials was performed. Physical activation of raw materials samples was performed in a rotating stainless steel reactor of 0.5 dm^3 at 19 rpm. Carbon dioxide gas, was chosen as an activating agent which is supplied from the bubbler into the reaction zone at a rate of 50 cm^3/min. Carbonization was carried out in a horizontal pyrolysis installation with adjustable electric heating (Fig. 8) in the temperature range from 400 to 900 °C, duration 1 h after carbonization yield at a predetermined temperature. For the experiment, crushed walnut shells, highlighting the sieving of the product of crushing the working fraction with a diameter of 2–4 mm [48].

2.2 Chromatographic Separation of Fusicoccin

For the experiment, 1 kg of seeds of spring wheat cultivar "Steklovidnaya-24" was taken. The seeds which were soaked for one day in a sterile cool tap water supplemented with 1 mM 6-BAP (6-benzyl aminopurine). 6-BAP was added as an inducer of the fusicoccin compounds. Then germinated seeds were homogenized in 3 l of 70% ethanol on the blade type homogenizer MPW-302 (Poland). The homogenate was centrifuged for 10 min at 10,000 × g. The resultant extract was purified by column nanostructured carbon sorbent.

In the presented work was used a column manufactured as follows (Fig. 3).

For a glass tube a rubber tube of the desired size was selected that it enters the tube only half or less. Then, with the narrow end of the plug blade we chose gentle concave surface, deepening the center plug, and then polishing the surface with sandpaper and the exact center of the plug was inserted into the plastic nozzle or syringe with a diameter of 1.5–2.5 mm. The nozzle must be inserted so that one end accounted for exactly the bottom surface of the funnel, and the other free exit from the wider end of the plug. If possible, it is desirable to have a plastic plug screw thread, which can be tightly screwed to various connecting plastic tubing or plug. After making the bottom plug with a hose, the narrow end of its snug nylon fabric (preferably from the children's bow) was inserted into the cork with a cloth in the lower end of the column. After the plug firmly, but not much comes to a column sticking out of the column pieces of nylon fabric carefully cut with a razor cap more tightly planted on the column and the column was checked for leaks. It was very

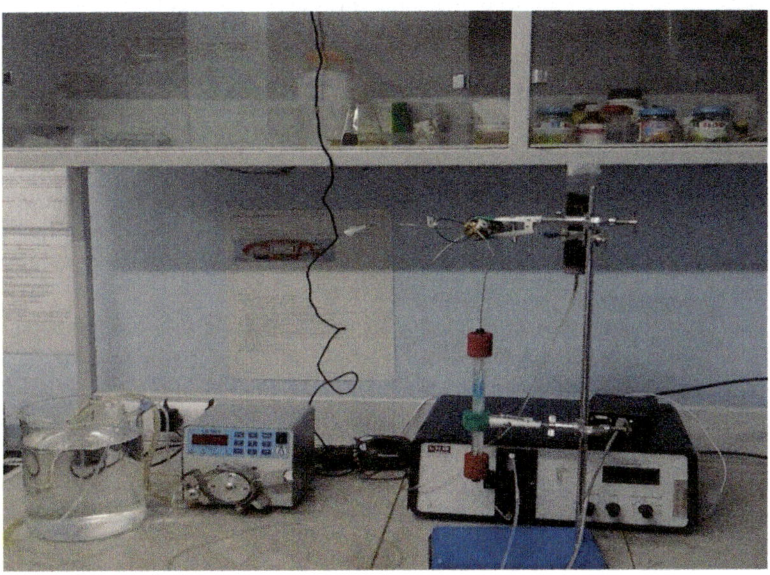

Fig. 3 The UV detecter connected with column chromatography

convenient to use the plug with the union in the upper part of the column. In this case, the cut surface of the funnel is not necessary, and the nozzle or needle is located on both sides of the plug. The top tube provides complete sealing of the column and continuity in the elution solution from a supply vessel with a minimum hydrostatic pressure [49].

2.3 Study of Biological Activity

2.3.1 Cytotoxic Activity

Cytotoxicity was assessed with the help of the survival test of brine shrimp larvae of *Artemia salina* (Leach). Experiments were carried out in 2-day-old larvae in cultivation conditions in vitro. The larvae were grown up by dipping of brine shrimp eggs *A. salina* (Leach) into the artificial sea water and by the 48 h incubation at a temperature of 37 °C. A weighed portion of the test sample was dissolved in 2 ml of methanol, and then it was taken from that solution by 500 μl (3 parallels), 50 μl (3 parallels), 5 μl (3 parallels). After evaporation of methanol 5 ml of artificial seawater were added to each vial. Thus, if the initial sample weight was 2 mg, the final sample concentrations were 100 μg/ml, 10 μg/ml and 1 μg/ml, respectively of each concentration in 3 replications. 10 2-day-old brine shrimp larvae *A. salina* were put into each vial containing the sample using the Pasteur pipette. Thereafter, all the vials were left at room temperature exposed to light for 24 h. After 24 h survived and dead larvae were counted. Then, using the obtained data of the upper and lower toxic limits a half-toxic sample dose was calculated.

2.3.2 Antimicrobial Activity

The study of antimicrobial activity of the samples mentioned above was performed pertaining to gram-positive bacteria strains *Staphylococcus aureus*, *Bacillus subtilis*, gram-negative strains *Escherishia coli* and *Candida albicans* yeast fungus by the method of diffusion into the agar wells. Comparator agents were gentamicin for bacteria and nystatin for *C. albicans* yeast fungus.

Cultures were grown in a fluid medium with pH 7.3 ± 0.2 at a temperature of 30–35 °C during 18–20 h. The cultures were diluted 1:1000 in a sterile 0.9% sodium chloride isotonic solution; then they were poured into cups with 1 ml of solution with relevant elective nutrient media clarify for the studied test-strains and inoculated according to the method of "continuous lawn". After drying on the surface of the agar wells of 6.0 mm size were formed, which were filled with the solution of the test sample, gentamicin and nystatin. Ethanol in equivoluminar quantities was used as control. Thus, the sample was tested in quantity of 1 μg and the comparator agent in quantity of 1 mg. The inoculations were incubated at 37 ° C, the accounting of growing colonies was performed after 24 h.

The antimicrobial activity of the samples was estimated by the diameter of test-strains growth inhibition zones (mm). The diameter of inhibition zones lower than 10 mm and the continuous increase in the Petri dish were estimated as the absence of the antibacterial activity, 10–15 mm—as a weak activity, 15–20 mm—as a moderate activity, over 20 mm—as an expressed one (REF). Each sample was tested in three parallel experiments. Statistical reporting was performed by methods of the parametric statistics with the calculation of arithmetic mean and standard error.

2.3.3 Analgesic Activity

The experimental objects were studied in the dose range from 25 to 100 mg/kg when administered intragastrically. The comparator agent diclofenac sodium was tested in the dose range from 25 to 100 mg/kg. The experimental objects and comparator agent were administered 30 min before administration of the 0.75% acetic acid solution.

2.3.4 Phagocytosis-Stimulating Activity

The comparator agent "Immunorm" (juice of Echinacea purpurea in alcohol, "Merkle", Germany) which was tested at a dilution of 0.9% sodium chloride solution in the final concentration of 8%.

Blood sampling was taken from a healthy donor in the fasting state from the ulnar vein into heparinized test tubes. As an object of the phagocytosis the daily culture *S. aureus* (strain 209) was used. At the microscopic examination (the magnification 15 × 90, oil immersion) the number of phagocytic cells (the phagocytic index, PI) out of 100 neutrophils (quantitative indicator) and the number of staphylococci, absorbed by one neutrophil—the phagocytic number (PN, qualitative indicator of the phagocytosis) were counted after 1 h of study, Table 5.

75 mg of the dry sample with account of its solubility were diluted in 0.2 ml of 96% alcohol until fully dissolved and made up to 2 ml of 0.9% sodium chloride solution. The substances were tested at a concentration of 1 mg/ml in three parallel experiments. The smears of the blood incubated with the dilution medium (96% alcohol and saline 1:9) were served as controls.

Statistical results reporting was performed by methods of the non-parametric statistics with the calculation of the arithmetic mean (M) and its standard error (m) [50].

3 Results

3.1 Computer Modeling of Activated Carbon for Fusicoccin

We have retrieved an X-ray structure of fusicoccin from the Brookhaven Protein Database (entry 3IQV) and optimised its geometry using molecular mechanics (MMFF) followed by semiempirical calculations (PM3). The size of the molecule (to its van der Waals limit) is about 1.9 × 1.5 × 1.1 nm. As it can rotate freely in solution it is unlikely to be bound within pores that are smaller than the compound's largest dimension. Overlaying the molecule on a model of a graphene sheet and removing carbon atoms until fusicoccin just fits leaves a pore just over 2.1 nm in diameter. To allow for any solvent and rapid molecular rotation, we would suggest a pore size of 2.5 nm as the minimum that could be expected to bind the molecule.

According to the results of computer modeling, cleaning composite components of fusicoccin using microporous carbon adsorbents not suitable as the size of the molecule of fusicoccin more than micropores.

And in the following figure (Fig. 4) the optimum pore size for purification of constituents of fusicoccin was determined by computer simulation.

As the results of computer simulation in this work study show that the most effective recognized mesoporous activated carbons.

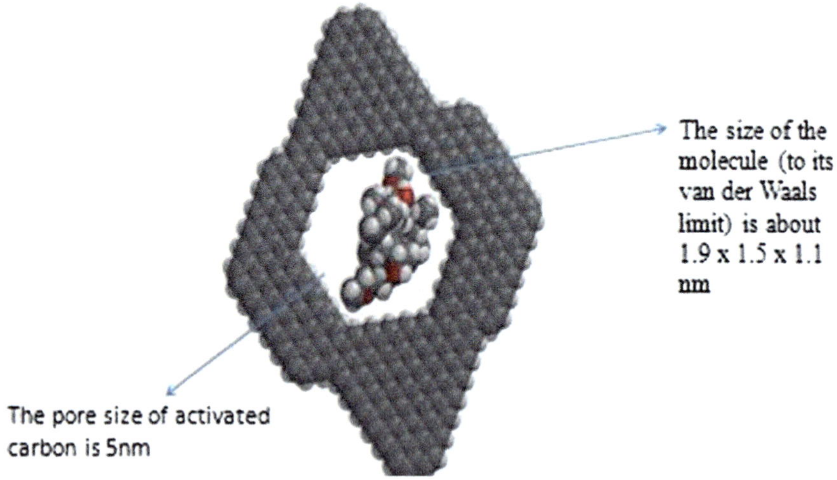

Fig. 4 The optimal pore size of carbon for column chromatography

3.2 *Extract of Phytohormone of Fusicoccin Containing Components*

Fusicoccin was obtained by a technique developed in the ICP, as part of the bouquet of organic compounds. In this connection, there arose the task of separation of biologically active substances. To solve this task, the technique has been used with liquid chromatography sorbent made of Greek walnut (GW). A distinguishing property of the selected sorbent is that it contains carbon and silica in its structure, this leading to the presence of both hydrophobic and hydrophilic properties. To control the chromatographic separation, a UV monitor type Uvicord S II manufactured by LKB (Sweden) was used.

To release columns from unadsorbed substances, the column was washed with 10% ethanol to completely remove them, and then bounded with sorbent phytohormone was eluted with 50% ethanol.

The spectroscopic study of purified GW was conducted on spectrophotometer Ultrospec +1100 pro company of Amersham Biosciences (UK) in the ultraviolet and visible regions of the spectrum.

Fot the test, GW samples obtained at temperature of 750 °C were chosen. To compare the specific characteristics, nanomaterial organic gel Octyl Sepharose CL-4B (Sweden) used in the world today was taken.

The liquid containing FC was passed through a separation column with the experimental material. The results of the chromatographic separation are shown in Fig. 5.

The first peak (Fig. 5) contained substances that do not bind with sorbent. For release the column from unadsorbed substances, the column was washed with 10% ethanol prior to complete removal, and then the phitohormone bound with the sorbent was eluted with 50% ethanol (2nd peak). Analysis of the figures of separation suggests that the synthesized sorbent has separation characteristics which are not inferior to the world analogues. However, the greatest advantage of this material is that it has a high resistance with respect to the microbiological media and the lack of parasitic sorption, unlike organic gel Octyl Sepharose CL-4B (b), having agarose in its structure [51].

3.3 *Mass Spectrometric Analysis. Fragmentation of Fusicoccin Using MS/MS*

Figures 6 and 7 show the spectra obtained when the polarity of the voltage applied in the electrospray (ESI) (source coupling the LC and MS) was positive and negative, respectively.

The full scan spectra in the preferential formation of a sodium adduct of fusicoccin [fusicoccin + Na]$^+$. Other species without such adduct were not detected. Multiple low intense peaks appeared, when infusing fusicoccin with ESI (−).

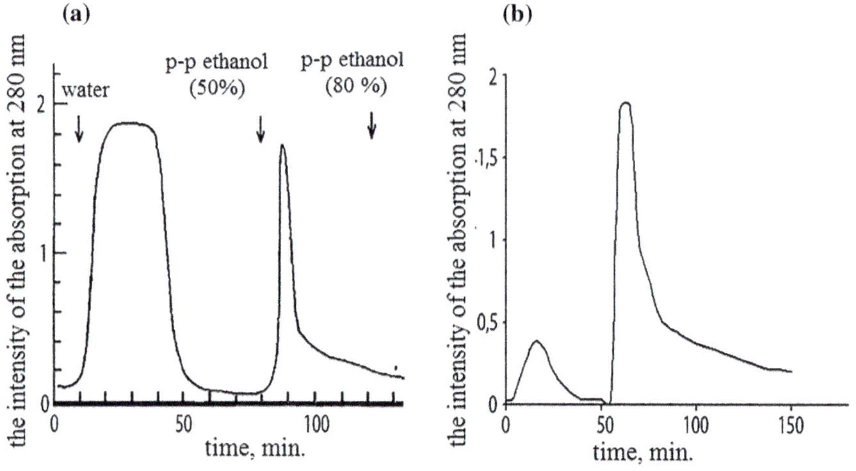

Fig. 5 Curves of chromatographic separation of fusicoccin on the column

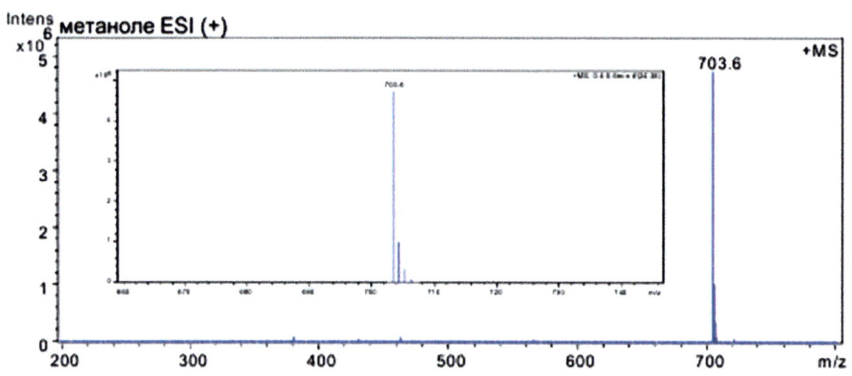

Fig. 6 Full scan spectra of fusicoccin in methanol ESI (+)

ESI (+) was chosen as the optimal conditions to monitor fusicoccin given that ESI (−) caused low intense peaks that we could not relate to fusicoccin.

The full scan spectra in F the preferential formation of a sodium adduct of fusicoccin [fusicoccin + Na]$^+$. Other species without such adduct were not detected. The most abundant product ions are 633 and 431. Fragmentation (MS/MS) of the ion m/z 703.3 with the amplitude applied in the trap shown in Fig. 8.

The optimal fragmentation amplitude (0.7 V) was chosen as the one causing the highest intensity of product ions keeping at least 5% of precursor ion (Fig. 9).

Fragmentation MS3 of m/z 703 [fusicoccin + Na]$^+$ and analysis of the resultant ions by product ion scan mode shown in Fig. 10.

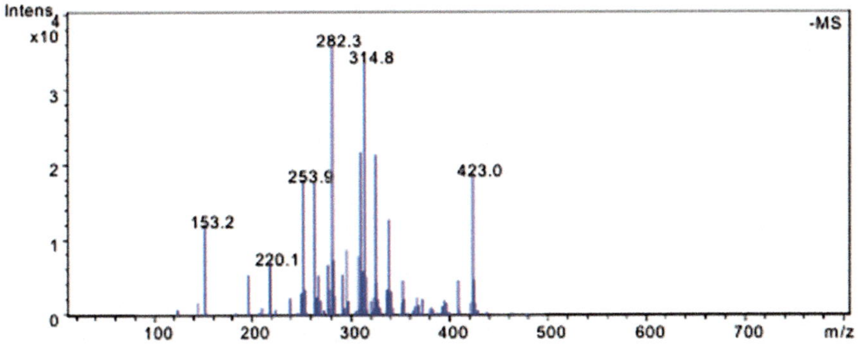

Fig. 7 Full scan spectra of fusicoccin in methanol ESI (−)

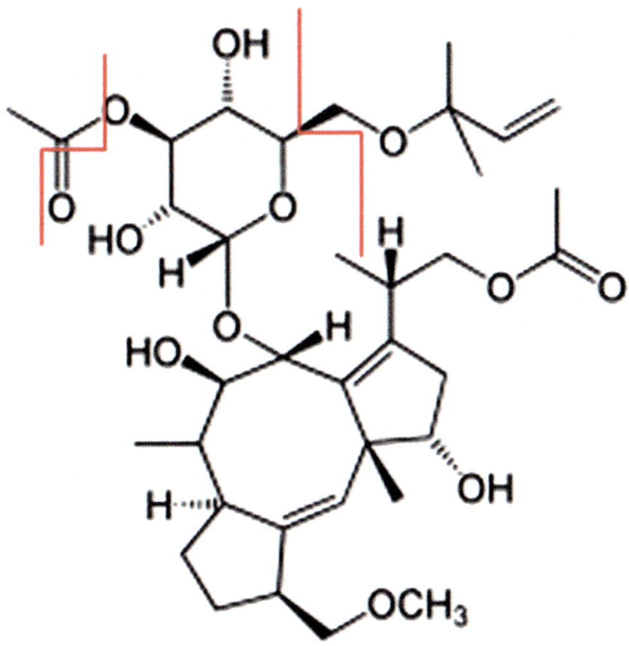

Fig. 8 Fragmentation (MS/MS) of the ion m/z 703.3 with the amplitude

Respectively, in Fig. 11 shown fragmentation MS^3 of m/z 703 $[fusicoccin + Na]^+$ (fragmentation route 2) and analysis of the resultant ions by product ion scan mode.

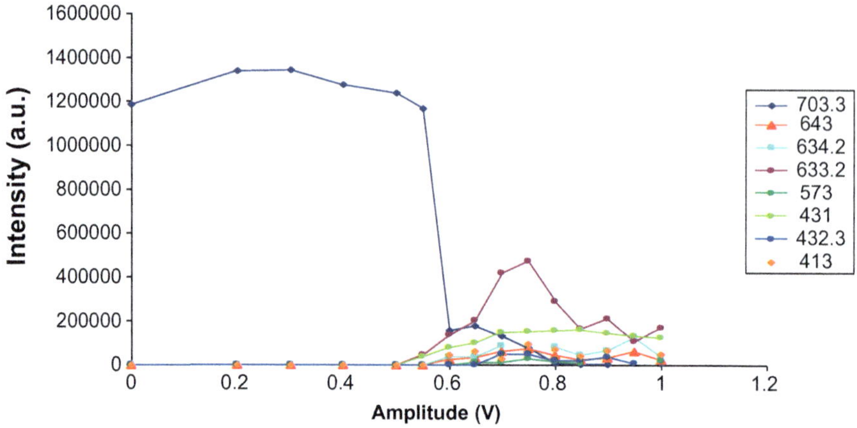

Fig. 9 Fragmentation (MS/MS) of the ion m/z 703.3 with the amplitude

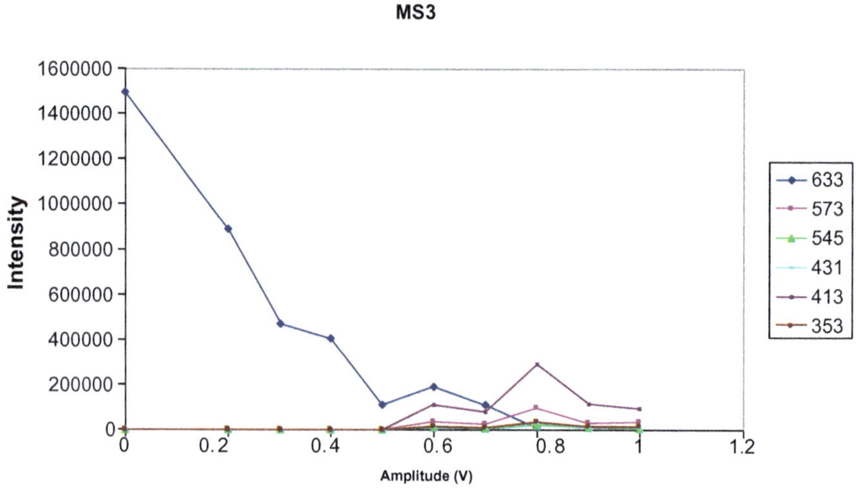

Fig. 10 Fragmentation MS3 of m/z 703 [fusicoccin + Na]$^+$ and analysis of the resultant ions by product ion scan mode

3.4 Analysis of Samples on Chromatogram

Ethanol extracts obtained from the elution of the retained fraction of wheat samples from a carbon column. In addition, there was a solid extract that was extracted at Brighton University by mixing it with 5 ml methanol and sonication (30 min), Fig. 12.

In Fig. 12 shown the full scan analisys from the solid extract sample.

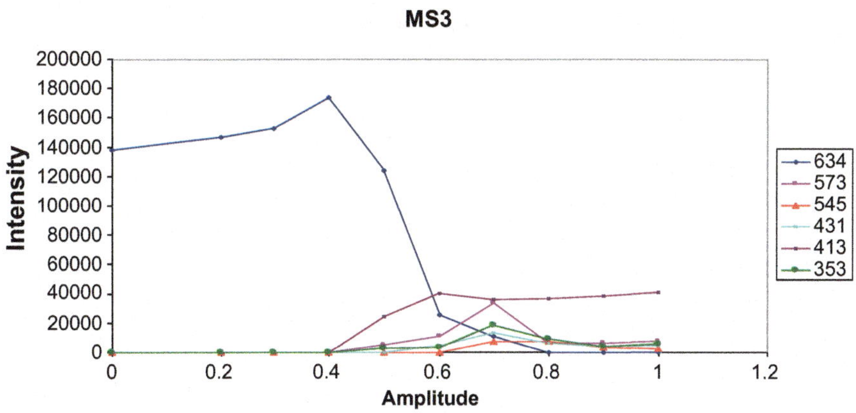

Fig. 11 Fragmentation MS3 of m/z 703 [fusicoccin + Na]$^+$ (fragmentation route 2) and analysis of the resultant ions by product ion scan mode

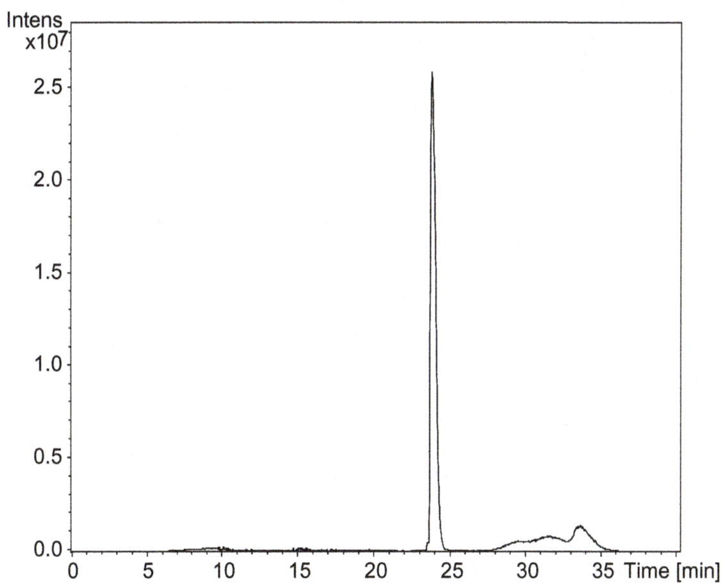

Fig. 12 4A spiked at concentration 0.5 ug/g

Two intense peaks can be detected, the first one with m/z 701 as most abundant ion (see spectra below), which also contains m/z 703 as isotopic ion, and in the second one, m/z 703 (rt 30–35 min) can also be detected (fusicoccin m/z).

3.5 *Determination of the Chemical Composition of the Samples by Liquid Chromatography*

Objects of the study: extract obtained by elution with 50% ethanol in the separation of 70% ethanol extract of milk-ripe wheat and extract obtained by elution with 10% ethanol in the separation of 70% ethanol extract of milk-ripe wheat, Figs. 13, 14 and 15.

The fraction extract obtained by elution with 10% ethanol in the separation of 70% ethanol extract of milk-ripe wheat shown in Fig. 14.

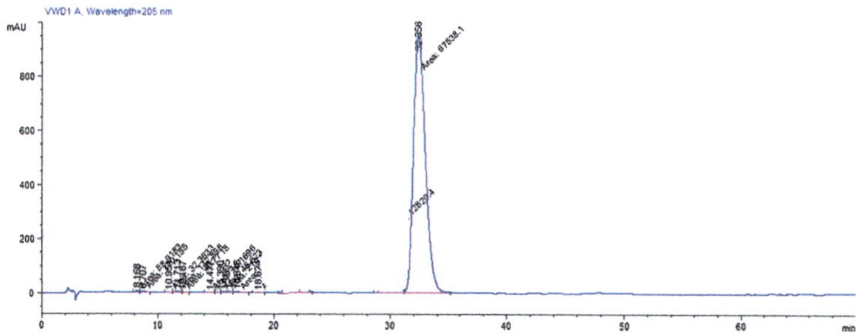

Fig. 13 Pure Fusicoccin's peak

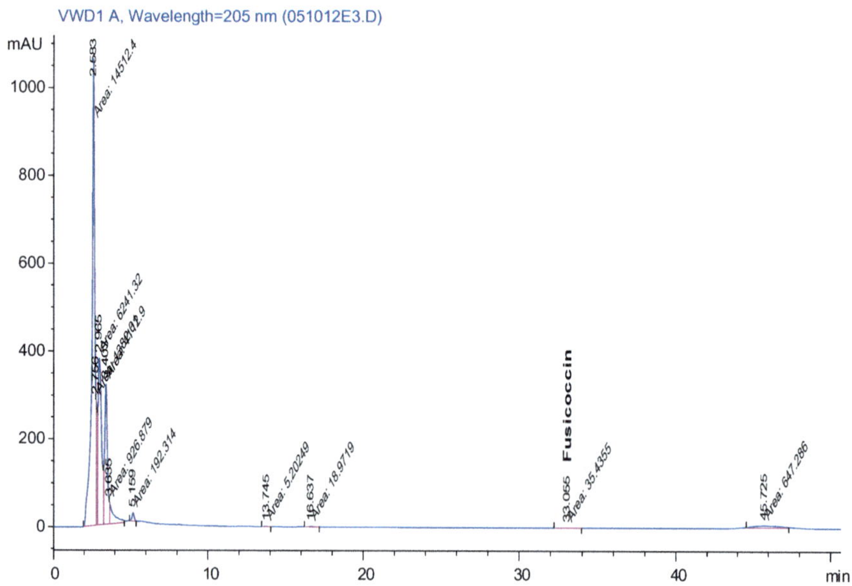

Fig. 14 The fraction extract obtained by elution with 10% ethanol in the separation of 70% ethanol extract of milk-ripe wheat

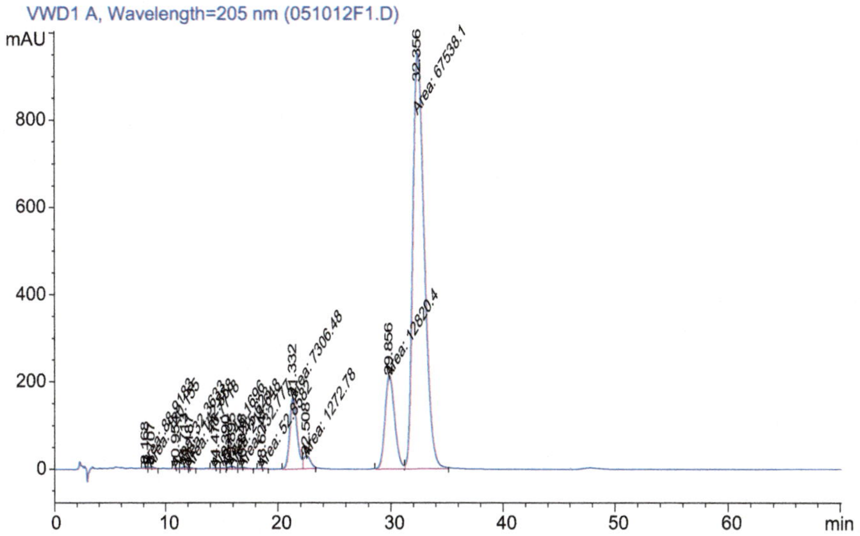

Fig. 15 The fraction obtained by elution with a 50% ethanol in the separation of 70% ethanol extract of milk-ripe wheat

Table 1 Result in fractions after preparative separation

Samples	The samples consists of fusicoccin or no?
The fraction obtained by elution with a 50% ethanol in the separation of 70% ethanol extract of milk-ripe wheat	Yes
The fraction extract obtained by elution with 10% ethanol in the separation of 70% ethanol extract of milk-ripe wheat	No

The fraction obtained by elution with a 50% ethanol in the separation of 70% ethanol extract of milk-ripe wheat shown in Fig. 15.

The results of the quantitative content of fusicoccin in fractions after preparative separation are shown in Table 1.

The fraction extract obtained by elution with 10% ethanol in the separation of 70% ethanol extract of milk-ripe wheat don't have fusicoccin.

3.6 The Examination of the Cytotoxic Activity of the Obtained Fractions

Objects of the study were 8 sample substances for the presence of cytotoxic activity pertaining to brine shrimp larvae Artemia salina (Leach) under cultivation conditions in vitro.

Samples designation (name):

1. A0—crude extract;
2. A1—fraction after the preparative separation;
3. A2—fraction after the preparative separation;
4. A3—fraction after the preparative separation;
5. A4—fraction after the preparative separation;
6. A5—fraction after the preparative separation;
7. A6—fraction after the preparative separation;
8. A7—fraction after the preparative separation.

Statistical results reporting was conducted using FNI computer program.
Findings of investigations.

The results of the testing of the cytotoxic activity of the samples A0, A1, A2, A3, A4, A5, A6, A7 pertaining to brine shrimp larvae *A. salina* (Leach) under cultivation conditions in vitro are listed in the Table 2.

As the table shows the crude extract A0 and the following fractions after the preparative separation (A1, A4, A5, A6, A7) exhibit the cytotoxic activity pertaining to brine shrimp larvae *A. salina* (Leach).

3.7 Antimicrobial Activity Tests

Materials and methods: Objects of the study were 8 samples of fusicoccin compounds for the presence of antimicrobial activity.

Samples designation (name):

Findings of investigations. The results of the samples' antimicrobial activity testing are listed in Table 3.

It was stated that all presented samples exhibit an antibacterial activity pertaining to the gram-positive microorganisms *S. aureus* and *B. subtilis*, and the gram-negative test-strain *E. coli*.

Table 2 The cytotoxic activity of the samples A0, A1, A2, A3, A4, A5, A6, A7 pertaining to brine shrimp larvae *Artemia salina* (Leach) under cultivation conditions in vitro

Substance name	Concentration (μg/ml)	Quantity of survived larvae			LD 50 (μg/ml)	95% confidence interval	Activity
		1st parallel	2nd parallel	3rd parallel			
1	2	3	4	5	6	7	8
A0 crude extract	1	10	10	10	140.39	60.04–932.24	has
	10	10	7	9			
	100	7	6	4			
A1 fraction after the preparative separation	1	10	10	10	39.22	20.63–96.74	has
	10	6	6	8			
	100	4	3	4			
A2 fraction after the preparative separation	1	10	10	10	–	–	does not have
	10	10	9	9			
	100	9	7	8			
A3 fraction after the preparative separation	1	10	10	10	–	–	does not have
	10	8	9	8			
	100	8	6	8			
A4 fraction after the preparative separation	1	9	10	9	32.62	15.92–91.15	has
	10	7	7	6			
	100	4	3	3			
	10	7	10	7			
	100	6	3	5			
A5 fraction after the preparative separation	1	10	10	10	95.55	40.17–559.30	has
	10	8	7	7			
	100	7	5	7			
A6 fraction after the preparative separation	1	10	10	10	135.57	56.08–984.11	has
	10	7	8	10			
	100	5	6	6			
A7 fraction after the preparative separation	1	10	10	10	76.11	37.34–263.85	has

Table 3 The antimicrobial activity of the samples

Substance name	*S. aureus* 505	*Bac. Subtilis*	*E. coli* M-17	*C. albicans*
(0) Crude extract	14.0 ± 0.3	16.0 ± 0.1	14.0 ± 0.3	13.0 ± 0.1
(1) Fraction after the preparative separation	16.0 ± 0.1	16.0 ± 0.2	15.3 ± 0.3	13.0 ± 0.2
(2) Fraction after the preparative separation	14.0 ± 0.3	15.0 ± 0.3	12.0 ± 0.2	15.0 ± 0.2
(3) Fraction after the preparative separation	16.0 ± 0.2	14.0 ± 0.1	14.0 ± 0.3	13.0 ± 0.3
(4) Fraction after the preparative separation	15.0 ± 0.2	16.0 ± 0.4	14.0 ± 0.2	14.0 ± 0.1
(5) Fraction after the preparative separation	14.0 ± 0.2	14.0 ± 0.2	15.0 ± 0.3	13.0 ± 0.3
(6) Fraction after the preparative separation	15.0 ± 0.2	16.0 ± 0.3	14.0 ± 0.1	12.0 ± 0.2
(7) Fraction after the preparative separation	14.0 ± 0.2	14.0 ± 0.4	15.0 ± 0.2	13.0 ± 0.3
Gentamicin	26 ± 1	24 ± 1	23 ± 2	–
Nystatin	–	–	–	22

Table 4 The analgesic activity of naturally occurring compounds and their derivatives

Substance name, dose					
		10 min	15 min	20 min	30 min
Fraction after the preparative separation (5)	Control	28 ± 1.4	42 ± 3.2	55 ± 3.5	71 ± 4.5
	Diclofenac	10 ± 2.4	17 ± 2.6	23 ± 2.8	34 ± 1.7
	25 mg/kg	18 ± 3.6	35 ± 5.5	47 ± 5.9	66 ± 6.5
	50 mg/kg	16 ± 2.4	28 ± 3.2	36 ± 4.5	44 ± 5.3
	100 mg/kg	17 ± 2.1	33 ± 4.5	43 ± 5.8	50 ± 4.9
Fraction after the preparative separation (6)	Control	32 ± 3.7	42 ± 4.3	51 ± 4.1	70 ± 3.3
	Diclofenac	16 ± 2	23 ± 1.6	30 ± 2.2	34 ± 1.8
	25 mg/kg	28 ± 5	41 ± 4.7	53 ± 4.8	66 ± 7
	50 mg/kg	18 ± 2.8	29 ± 4.3	37 ± 5	46 ± 6.6
	100 mg/kg	22 ± 4.2	33 ± 5.3	39 ± 6.3	50 ± 8.3
Fraction after the preparative separation (7)	Control	20 ± 1.5	35 ± 3.5	48 ± 3	69 ± 2,5
	Diclofenac	7 ± 1.3	12 ± 1.9	19 ± 1.5	25 ± 2.3
	25 mg/kg	21 ± 2	31 ± 2.1	38 ± 2.1	45 ± 1.5
	50 mg/kg	12 ± 0.6	26 ± 1.5	36 ± 1,6	45 ± 2.4
	100 mg/kg	11 ± 2.4	19 ± 2.3	25 ± 2.5	35 ± 4.2

3.8 The Study of the Analgesic Activity of Naturally Occurring Substances and Their Derivatives

The screening results of the analgesic activity of the crude extract and its fractions after preparative separation are listed in the Table 4. The analgesic effect of the submitted samples was determined by the ability to reduce the number of “writhes” (in %) calculated for 10, 15, 20 and 30 min and compared to the corresponding figures in animals in the control group, Table 4.

The study revealed that the fraction after preparative separation (7) has an expressed analgesic effect at a dose of 100 mg/kg, the remaining samples showed a weak analgesic activity.

3.9 Testing for the Phagocytosis-Stimulating Activity

The results of the screening study of the blood cells phagocytic activity are listed in the table (Table 5).

Based on data from the table the comparator agent “Immunorm” has an expressed phagocytosis-stimulating action pertaining to both quantitative and qualitative indicators of the blood neutrophils phagocytosis.

Table 5 The influence of the samples (0), (1), (2), (3), (4), (5), (6), (7) on the phagocytic activity of blood neutrophils, (M ± m)

Substance name	Neutrophils	
	PI (%)	PN, unit
Control with dilution medium	27.0 ± 1.5	4.1 ± 0.4
Immunorm	51.3 ± 4.8	8.2 ± 1.2
(0) Crude extract	24.1 ± 3.6	3.3 ± 0.23
(1) Fraction after the preparative separation	25.3 ± 1.3	3.1 ± 0.37
(2) Fraction after the preparative separation	29.7 ± 2.4	3.4 ± 0.2
(3) Fraction after the preparative separation	22.7 ± 1.8	4.1 ± 0.1
(4) Fraction after the preparative separation	17.1 ± 1.5	4.5 ± 1.3
(5) Fraction after the preparative separation	24.3 ± 5.5	3.9 ± 0.4
(6) Fraction after the preparative separation	23.1 ± 2.3	3.8 ± 1.1
(7) Fraction after the preparative separation	32.7 ± 2.9	3.4 ± 0.4

4 Conclusion

For the first time a method of determining the optimum pore size of sorbents for the selection of fusicoccin was developed using the program Spartan Graphical interface and found that as optimal sorbent with mesoporous structure for cleaning fusicoccin was used walnut shell. A sensitive and selective method is developed in samples of milky ripe wheat by analysis of fusicoccin by liquid chromatography and mass spectrometry. As a result of the investigation of the cytotoxic activity and biological activities of the extracts fusicoccin.

References

1. Dubinin, M. M. (1978). Adsorbents, their preparation, properties and applications. -M, 4–22.
2. Aygun, A., Yenisoy-Karakas, S., & Duman, I. (2003). Production of granular activated carbon from fruit stones and nutshells and evaluation of their physical, chemical and adsorption properties. *Microporous and Mesoporous Materials, 66,* 189–195.
3. Guo, Y., Yang, S., Yu, K., Zhao, J., Wang, Z., & Xu, H. (2002). Carbon materials for sorption. *Materials Chemistry and Physics, 74,* 320.
4. Daud, W. M., & Ali, W. S. (2004). Comparison on pore development of activated carbon produced from palm shell and coconut shell. *Bioresource Technology, 93,* 63–69.
5. Benaddi, H., Bandosz, T. J., Jagiello, J., Schwarz, J. A., Rouzaud, J. N., Legras, D., et al. (2000). Surface functionality and porosity of activated carbons obtained from chemical activation of wood. *Carbon, 38,* 669–674.
6. Asma, B. M. (2000). Apricot production. *Malatya Evin Ofset, Turkey*, 240 p.
7. Smisek, M., & Cerny, S. (1970). Active carbon manufacture, properties and aplications (370 p). New York: Elsevier Publishing Company.

8. Khezami, L., Chetouan, A., Taou, B., & Capar, R. (2005). Production and characterisation of activated carbon from wood components in powder: Cellulose, lignin, xylan. *Powder Technology, 157,* 48–56.
9. Hayashi, J., Kazehaya, A., Muroyama, K., & Watkinson, A. P. (2000). Preparation of activated carbons from lignin by chemical activation. *Carbon, 38,* 1873–1878.
10. Jandosov, J. M., Shabanova, T. A., Shamalov, M., Biysenbaev, M. A., & Mansurov, Z. A. (2010). Preparation of carbon materials with high specific surface area. *Combustion and Plasma Chemistry, 8*(3), 257–261.
11. Klijanienko, A., Grabowska, E. L., & Gryglewicz, G. Z. (2008). Development of mesoporosity during phosphoric acid activation of wood in steam atmosphere. *Bioresource Technology, 99,* 7208–7214.
12. Jibril, B., Houache, O., Maamari, R. A., & Rashidi, B. A. (2008). Effects of H_3PO_4 and KOH in carbonization of lignocellulosic material. *Journal of Analytical and Applied Pyrolysis, 83,* 151–158.
13. Azat, S. (2013). Synthesis of carbonized nano mesoporous sorbents based on vegetable raw materials. *Nanoscience and Nanoengineering International Journal, 1*(1), 41–44.
14. Azat, S., Pavlenko, V. V., Kerimkulova, A. R., & Mansurov, Z. A. (2012). Synthesis and structure determination of carbonized nano mesoporous materials based on vegetable raw materials. *Advanced Materials Research,* 535–537. Online available since Jun 14, 2012 at www.scientific.net.
15. Fierro, V., Fernandez, V. T., Montane, D., & Celzard, A. (2008). Adsorption of phenol onto activated carbons having different textural and surface properties. *Microporous and Mesoporous Materials, 111,* 276–284.
16. Cookson, J. T. (1980). Carbon adsorption handbook. In P. N. Cheremisinoff & F. Ellerbusch (Eds.) (pp. 241–279). Michigan: Ann Arbor Sci.
17. Keltsev, N. V. (1976). Fundamentals of adsorption technology. *Moscow Chemistry*, 512.
18. Greg, S., & Singh, K. (1970). Adsorption, surface area, porosity. M.: Mir, 259.
19. Mansurov, Z. A., Zhylybaeva, N. K., Ualieva, P. S., & Mansurova, R. M. (2002). Obtaining procedure and properties of the sorbents from plant raw material. *Chemistry for Sustainable Development, 3,* 321–328.
20. Mansurov, Z. A., Digel, I., Biisenbaev, M. A., Savitskaya, I., Kistaubaeva, A., Akimbekov, N., et al. (2012). *Composites and their applications*. INTECH 2012 Chap. 11, 271–295.
21. Mansurov, Z. A., Shabanova, T. A., & Mansurova, R. M (2004). The morphology of micro nano particles of carbonized plant materials. *Bulletin of KazNU. Series of Chemical, 2*(34), 129–135.
22. Basso, M. C., Cerrella, E. G., & Cukierman, A. L. (2002). Activated carbons from a rapidly renewable biosource for removal of cadmium(II) and nickel(II) ions from dilute aqueous solutions. *Industrial & Engineering Chemistry Research, 41,* 180–189.
23. Jia, Y. F., & Thomas, K. M. (2000). Adsorption of cadmium ions on oxygen surface sites in activated carbon. *Langmuir, 16,* 1114.
24. Azat, S., Kerimkulova, A. R., & Mansurov, Z. A. (2012). Synthesis and structure determination carbonated nanomaterials based on vegetable raw materials. In *VII International Symposium "Physics and Chemistry of Carbon Materials/Nanoengineering*", Almaty (pp. 124–126). On Sept 19–21.
25. Mansurov, Z. A. (2008). Synthesis of carbon nanomaterials and their applied aspects. *Herald A Series of Chemical, 2*(50), 16–31.
26. Azat, S. Mansurov, Z. A. (2011). Wastewater treatment using for carbonized nanosorbents. *Vestnik KazNU Chemical Series, 1*(61), 166–169.
27. Emuranov, M., Yu, S., Zhylybaeva, N. K., Biysenbaev, M. A., Shabanova, T. A., Ryabikin, Yu., et al. (2006). Multifunctional nanostructured carbonized sorbents. *Bulletin of National Academy of Sciences of Kazakhstan A Series of Chemical, 4,* 35–41.
28. Bevla, F. R., Rico, D. P., & Gomis, A. F. (1984). Activated carbon from almond shells. Chemical activation. Activating reagent selection and variables influence. *Industrial Engineering Chemistry Product Research and Development, 23,* 266–269.

29. Aygun, A., Yenisoy-Karakas, S., & Duman, I. (2003). Production of granular activated carbon from fruit stones and nutshells and evaluation of their physical, chemical and adsorption properties. *Microporous and Mesoporous Materials, 66,* 189–195.
30. Razvigorova, M., Budinova, T., Petrov, N., & Minkova, V. (1998). Purification of water by activated carbons from apricot stones, lignites and anthracite. *Water Research, 32,* 2135–2139.
31. Ballio, A., Chain, E. B., DeLeo, P., Erlanger, B. F., Mauri, M., & Tonolo, A. (1964). Fusicoccin: A new wilting toxin produced by Fusicoccum Amygdali Del. *Nature, 203*(4642), 297.
32. Babakov, A. V., Abramycheva, N. Y., Bilushi, S. T., & Shevchenko, V. P. (1990). Research fusicoccin interaction with the plasma membrane of higher plants. *Biology Membrane, 7.2. M,* 107–112.
33. Tajima, N., Nukina, M., Kato, N., & Sassa, T. (2004). Novel fusicoccins R and S, and the fusicoccin S aglycon (phomopsiol) from Phomopsis amygdali Niigata 2-A, and their seed germination-stimulating activity in the presence of abscisic acid. *Bioscience, Biotechnology, and Biochemistry, 68*(5), 1125–1130.
34. Aducci, P., Ballio, A., Fogliano, V., Fullone, M. R., Marra, M., & Proietti, N. (2003). Purification and photoaffinity labeling of fusicoccin receptors from maize. *Development, 130* (20), 4847–4858.
35. De Boer, A. H., Watson, B. A., & Cleland, R. E. (1989). Purification and identification of the fusicoccin binding protein from oat root plasma membrane. *Plant Physiology, 89*(1), 250–259.
36. Trofimova, M. S., Smolenskaya, I. N., Drabkin, A. V., Galkin, A. V., & Babakov, A. V. (1997). Does plasma membrane H+ATPase activation by fusicoccin involve protein kinase. *Physiologia Plantarum, 99*(2), 221–226.
37. Olivari, C., Meanti, C., De Michelis, M. I., & Rasi-Caldogno, F. (1998). Fusicoccin binding to its plasma membrane receptor and the activation of the plasma membrane H + -ATPase IV. Fusicoccin induces the association between the plasma membrane H + -ATPase and the fusicoccin receptor. *Plant Physiology, 116*(2), 529–537.
38. Oecking, C., Eckerskorn, C., & Weiler, E. W. (1994). The fusicoccin receptor of plants is a member of the 14-3-3 superfamily of eukaryotic regulatory proteins. *FEBS Letters, 352,* 163–166.
39. Olivari, C., Albumi, C., Pugliarello, M. C., & De Michelis, M. I. (2000). Phenylarsine oxide inhibits the fusicoccin-induced activation of plasma membrane H + -ATPase. *Plant Physiology, 122*(2), 463–470.
40. Baunsgaard, L., Fuglsang, A. T., Jahn, T., Korthout, H. A. A. J., De Boer, A. H., & Palmgren, M. G. (1998). The 14-3-3 proteins associate with the plant plasma membrane H + -ATPase to generate a fusicoccin binding complex and a fusicoccin responsive system. *The Plant Journal 13*(5), 661–671.
41. De Vries-van Leeuwen, I. J., Kortekaas-Thijssen, C., Nzigou Mandouckou, J. A., Kas, S., Evidente, A., & De Boer, A. H. (2010). Fusicoccin-A selectively induces apoptosis in tumor cells after interferon-a priming. *Cancer Letters, 293*(2), 198–206.
42. Sassa, T., Tajima, N., Sato, M., Takahashi, A., & Kato, N. (2002). Fusicoccins P and Q, and 3-epifusicoccins H and Q, new polar fusicoccins from isolate Niigata 2-A of a peach fusicoccum canker fungus. *Bioscience, Biotechnology, and Biochemistry, 66*(11), 2356–2361.
43. Come, C., Laine, A., Chanrion, M., Edgren, H., Mattila, E., Liu, X., et al. (2009). CIP2A is associated with human breast cancer aggressivity. *Clinical Cancer Research, 15*(16), 5092–5100.
44. Wurtele, M., Jelich-Ottmann, C., Wittinghofer, A., & Oecking, C. (2003). Structural view of a fungal toxin acting on a 14-3-3 regulatory complex. *EMBO Journal, Chemical structure of Fusicoccin-A, Cotylenin-A and Ophiobolin-A, 22*(5)(C), 987–994.

45. Korthout, H. A. A. J., & De Boer, A. H. A. (1994). fusicoccin binding protein belongs to the family of 14-3-3 brain protein homologs. *The Plant Cell, 6*(11), 1681–1692.
46. Sassa, T., Tajima, N., Sato, M., Takahashi, A., & Kato, N. (2002). Fusicoccins P and Q, and 3-epifusicoccins H and Q, new polar fusicoccins from isolate Niigata 2-A of a peach fusicoccum canker fungus. *Bioscience, Biotechnology, and Biochemistry, 66*(11), 2356–2361.
47. Bunney, T. D., De Boer, A. H., & Levin, M. (2003). Fusicoccin signaling reveals 14-3-3 protein function as a novel step in left-right patterning during amphibian embryogenesis. *Development, 130*(20), 4847–4858.
48. Pavlenko, V. V., Anurov, S. A., Mansurov, Z. A., Biysenbaev, M. A., Konkova, T. V., Azat, S., et al. (2014). Preparation of microporous activated carbons based on carbonized shell apricot. *Vestnik KazNU, Chemical Series, 3*(75), 103–113.
49. Mansurov, Z. A., Jandosov, J. M., Kerimkulova, A. R., Azat, S., Zhubanova, A. A., Digel, I. E., et al. (2013). Nanostructured carbon materials for biomedical Use. *Eurasian Chemico-Tecnological Journal, 15*(3), 209–217.
50. Mansurov, Z. A., Azat, S., Adekenova, A. S., Kerimkulova, A. R., Ivasenko, S. A., Shulgau, Z. T., et al. (2013). Extraction fusicoccin from wheat seeds using nanocarbon sorbents. *Advanced Materials Research, 647,* 67–70.
51. Azat, S., Adekenova, A. S., Ivasenko, S. A., Seydahmetova, R. B., Kerimkulova, A. R., & Mansurov, Z. A. (2012). Development of technology for drug fusicoccin on nanocarbon sorbents and the study of the biological activity. *Pharmaceutical Bulletin Scientific Journal, 2–3*(164), 57–60.

Author Biography

Professor Zulkhair Mansurov Ph.D. He is a General Director of the institute of combustion problems of the Ministry of Education and Science of the Republic of Kazakhstan. Zulkhair Mansurov successfully defended his Ph.D and Doctor of Sciences dissertations in1973 and 1990, correspondingly. In 1981 Zulkhair Mansurov got 10-months scientific probation at the University College of London, Great Britain. His scientific work includes investigations of the kinetics and mechanisms of hydrocarbon combustion, soot formation, nanotechnology and petro-chemistry. He is editor-in-chief "Combustion and Plasmochemistry Journal" and of the "Eurasian Chemico-Technological Journal" (English), indexed at scopus.

Part III
Physics

The Fourth Phase of Water: Implications for Energy, Life, and Health

Gerald H. Pollack

1 Introduction

Can drinking water supply energy? Is water merely another kind of food? How does water relate to health?

Answering these and related questions requires an understanding of the nature of water. Many presume that water must be completely understood, given its simplicity and pervasiveness through nature, but in fact precious little has been known about how water molecules organize themselves and interact with one another—until recently.

Students learn that water has three phases: solid, liquid and vapor. But there is something more: in our laboratory at the University of Washington we have uncovered a *fourth* phase. This phase occurs next to water loving (hydrophilic) surfaces. It is surprisingly extensive, projecting out from surfaces by up to millions of molecular layers. And it exists almost everywhere throughout nature, including your body. In fact, it fills your cells.

This newly identified phase of water has been described in a recent (2013) book: *The Fourth Phase of Water: Beyond Solid, Liquid and Vapor* [5] www.ebner-andsons.com. The book documents the basic finding and presents many applications beyond the ones mentioned above. It also deals with water's well-recognized anomalies, turning those anomalies into easily explained features.

The existence of a fourth phase may seem unexpected. However, it should not be entirely so: a century ago, the physical chemist Sir William Hardy argued for the existence of a fourth phase; and many authors over the years have found evidence for some kind of "ordered" or "structured" phase of water. Fresh experimental evidence not only confirms the existence of such an ordered, liquid-crystalline

G. H. Pollack (✉)
University of Washington, Seattle, WA 98195, USA
e-mail: ghp@u.washington.edu
URL: http://faculty.washington.edu/ghp/

G. M. Artmann et al. (eds.), *Biological, Physical and Technical Basics of Cell Engineering*, https://doi.org/10.1007/978-981-10-7904-7_13

phase, but also details its properties. It is more viscous, dense, and alkaline than H_2O and has relatively more oxygen since its formula is H_3O_2. As a result, it has a negative charge, and like a battery, can hold energy as well as deliver that energy when needed. These properties explain everyday observations and answer questions ranging from why gelatin desserts hold their water, to why teapots whistle.

The presence of the fourth phase carries many implications. Here, I will outline some basic features of this phase, and then deal with several of those implications including energetic aspects. We obtain energy from food; however we can also get energy from water. I will touch briefly on atmospheric science implications because everyone has interest in the weather, and then focus on some biological and health applications.

2 Does Water Transduce Energy?

The energy for building water structure comes from the sun. Radiant energy converts ordinary bulk water into ordered water, building this ordered zone. We found that all wavelengths ranging from ultraviolet, through visible, to infrared can build this ordered water. Near-infrared energy is the most capable. Water absorbs infrared energy freely from the environment; it uses that energy to convert bulk water into liquid crystalline water (fourth phase water)—which we also call "exclusion zone" or "EZ" water because it profoundly excludes solutes. Hence, buildup of EZ water occurs naturally and spontaneously from environmental energy. Additional energy input creates additional EZ buildup.

Of particular significance is the fourth phase's charge: commonly negative (Fig. 1). Absorbed radiant energy splits water molecules; the negative moiety constitutes the building block of the EZ, while the positive moiety binds with water

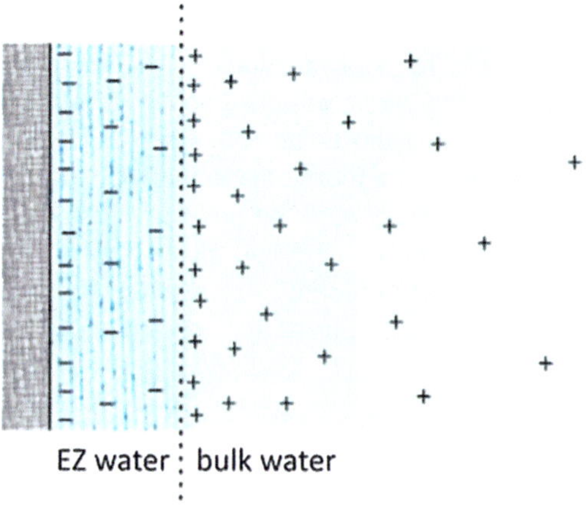

Fig. 1 Diagrammatic representation of EZ water, negatively charged, and the positively charged bulk water beyond. Hydrophilic surface at left

molecules to form free hydronium ions, which spread diffusely throughout the water. Adding additional light (radiant energy) stimulates more charge separation.

This process resembles the first step of photosynthesis. In that step, energy from the sun splits water molecules. Hydrophilic chromophores catalyze that splitting. The process considered here is similar but more generic: any hydrophilic surface may catalyze the splitting of water. Some surfaces work more effectively than others.

The separated charges resemble a battery. That battery can deliver energy in a manner similar to the way the separated charges in plants deliver energy. Plants, of course, comprise mostly water, and it is therefore no surprise that similar energy conversion takes place in water itself.

The stored electrical energy in water can drive various kinds of work, including flow. An example is the axial flow through tubes. We found that immersing tubes made of hydrophilic materials into water produces flow through those tubes, similar to blood flow through blood vessels (Fig. 2). The driving energy comes from the radiant energy absorbed and stored in the water. Nothing more. Flow may persist undiminished for many hours, even days. Additional incident light brings faster flow. This is not a perpetual motion machine: incident radiant energy drives the flow—in much the same way that it drives vascular flow in plants and powers water from the roots to nourish trees taller than the length of a football field.

3 Applications in Natural Science

The water-based energy conversion framework is rich with implication for many systems involving water. These systems may range from biology and chemistry all the way to atmospheric science and engineering. The fourth phase appears nearly everywhere: all that's needed is water, radiant energy, and a hydrophilic surface. The latter can be as large as a slab of polymer and as small as a dissolved molecule. The liquid crystalline phase inevitably builds—and its presence plays some integral role in the system's behavior.

Let me provide a few representative examples.

One example is…you. By volume, two thirds of your cells' content is water. However the water molecule is so small that making up that two-thirds volume requires 99% of all your molecules. Modern cell biology considers that 99% molecular fraction as mere background carriers of the "important" molecules of life

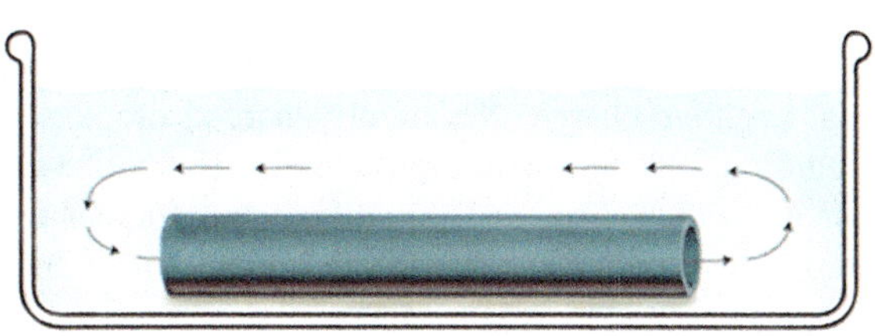

Fig. 2 Practically incessant flow occurs through hydrophilic tubes immersed in water

such as proteins and nucleic acids. Conventional wisdom asserts that 99% of your molecules don't do very much.

However, EZ water envelops every macromolecule in the cell. Those macromolecules are so tightly packed that the EZ water largely fills your cells. In other words most of your cell water is EZ water. This water plays a central role in everything the cell does—as elaborated in my earlier book, *Cells, Gels and the Engines of Life* [4] www.ebnerandsons.com.

What's new is the role of radiant energy: incident radiant energy powers many of those cellular functions. An example is the blood flowing through your capillaries. That blood eventually encounters high resistance: capillaries are often narrower than the red blood cells that must pass through them; in order to make their way through, those red cells need to contort. Resistance is high. You'd anticipate the need for lots of driving pressure; yet, the pressure gradient across the capillary bed is negligible. The paradox resolves if radiant energy helps propel flow through capillaries in the same way that it propels flow through hydrophilic tubes. Radiant energy may constitute an unsuspected source of vascular drive, supplementing cardiac pressure.

Why you feel good after a sauna now seems understandable. If radiant energy drives capillary flow and ample capillary flow is important for optimal functioning, then sitting in the sauna will inevitably be a feel-good experience. The infrared energy associated with heat should help drive that flow. The same if you walk out into sunlight: we presume that the feel-good experience derives purely from the psychological realm; but the evidence above implies that sunlight may build your body's EZs. Fully built EZs around each protein seem necessary for optimal cellular functioning.

A second example of the EZ's central role is weather. Common understanding of weather derives from two principal variables: temperature and pressure. Those two variables are said to explain virtually everything we experience in terms of weather. However, the atmosphere also contains water—micrometer-scale droplets commonly known as aerosol droplets or aerosol particles. Those droplets make up atmospheric humidity. When the atmosphere is humid, the many droplets scatter considerable light, conferring haze; you can't see clearly through that haze. When the atmosphere contains only few droplets, you may see clearly, over long distances.

The Fourth Phase book presents evidence for the structure of those droplets. It shows that EZ water envelops each droplet, while hydronium ions occupy the droplets' interior. Repelling one another, those hydronium ions create pressure, which pushes against the robust shell of EZ water. That pressure explains why droplets tend toward roundness.

How do those aerosol droplets condense to form clouds? The droplets' EZ shells bear negative charge. Negatively charged droplets should repel one another, precluding any condensation into clouds. Those like-charged aerosol droplets should remain widely dispersed throughout the atmosphere. However, droplets *do* condense into clouds, and the question is how that can happen.

Fig. 3 Like-charged entities attract because of an intermediate of opposite charge

The reason they condense is because of the unlike charges that lie in between the droplets. Richard Feynman, the legendary Nobel Prize physicist of the late 20th century understood the principle, opining that: "like-likes-like because of an intermediate of unlikes." The like-charged droplets "like" one another, so they come together; the unlike charges lying in between those droplets constitute the attractors (Fig. 3).

The like-likes-like principle has been widely appreciated, but also widely ignored: after all, how could like charges conceivably *attract*? A reason why this powerfully simple concept has been ignored is that the source of the unlike charges has been difficult to identify. We now know that the unlike charges can come from the splitting of water—the negative components building EZ shells, while the corresponding positive components provide the unlike attractors. With enough of those attractors, the negatively charged aerosol droplets may condense into clouds.

These two phenomena, radiant energy-induced biological function and like-likes-like cloud formation, provide examples of how water's energy can account for phenomena not otherwise explained. The fourth phase is the key building block that allows for construction of an edifice of understanding.

4 Practical Applications

Beyond scientific, the discovery of the fourth phase has practical applications. They include flow production (already mentioned), electrical energy harvesting, and even filtration. I briefly mention the latter two applications.

Filtration occurs naturally because the liquid crystalline phase massively excludes solutes and particles in much the same way as does ice. Accordingly, fourth phase water is essentially solute free. Collecting it provides solute-free and bacteria-free water. A patented working prototype has confirmed this expectation. Purification by this method requires no physical filter: the fourth phase itself does the separation, and the energy comes from the sun.

Energy harvesting seems straightforward: light drives the separation of charge, and those separated charges constitute a battery. Harvesting electrical energy should be realizable with proper electrodes. This (patented) technology development has the potential to replace standard photovoltaic systems with simpler ones based on water. More detail on these practical applications can be found in the Pollack laboratory homepage: http://faculty.washington.edu/ghp/.

5 Applications in Medical Science

Practical applications also exist within our bodies, and I present two of them: why your joints don't squeak: and why dislocated or sprained joints will swell *within seconds*.

Joints are sites at which bones press upon one another (Fig. 4). The bones may also rotate, as during deep-knee bends and push-ups. You'd think that rotation under pressure might elicit squeaky frictional resistance, but joint friction remains remarkably modest. Why so?

The ends of bones are lined with cartilage. Those cartilaginous materials do the actual pressing. Hence, the issue of joint friction reduces to the issue of the cartilaginous surfaces and the synovial fluid lying in between. How does this system behave under pressure?

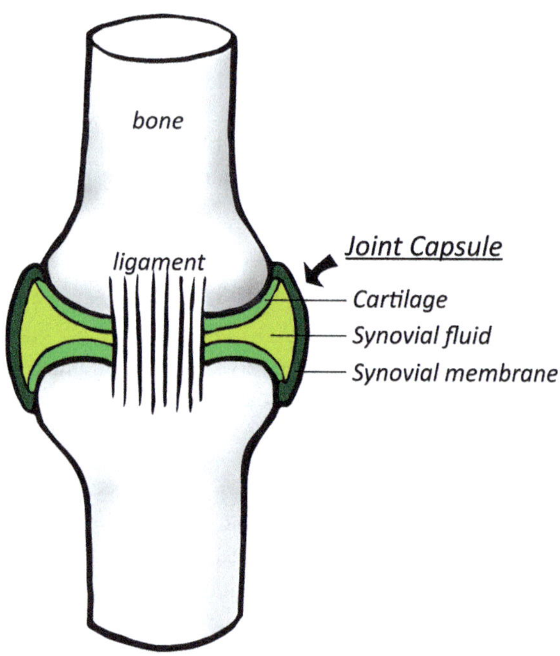

Fig. 4 Enveloping the joint, the capsule ensures that the fluid's hydronium ions don't disperse. The concentrated hydronium ions repel, keeping surfaces apart and assuring low friction

Cartilage is made of classic gel materials: highly charged polymers and water; therefore, cartilage is a gel. Gel surfaces bear EZs, so cartilage surfaces should likewise bear EZs. The splitting of water associated with EZ buildup creates many hydronium ions in the synovial fluid in between. Additional hydronium ions come from the molecules within that fluid, creating their own EZs and protons. Thus, many hydronium ions will lie in the area where two cartilaginous surfaces lie across from one another. The repulsive force coming from those hydronium ions should keep the cartilage surfaces apart—some investigators maintain that the cartilage surfaces never touch, even under heavy loads. That separation means that any rough spots, or asperities, will never come into contact with one another as the respective surfaces shear past one another; and that in turn means low friction.

For such a mechanism to actually work, some kind of built-in restraint should be present to keep the repelling hydronium ions in place. Otherwise, those hydronium ions may be forced out of the local region, thereby compromising lubrication. Nature provides that safety net: a structure known as the joint capsule envelops the joint. By constraining the dispersal of hydronium ions, that encapsulation ensures low friction. That's why your joints don't ordinarily squeak.

Regarding swelling, the second issue under consideration here, osmosis evidently plays a role. Since the cell is packed with negatively charged proteins, the cytoplasm should generate an osmotic draw similar to the osmotic draw generated by diapers or gels. Physiologists know that it does.

A peculiar feature of cells, however, is their relatively modest water content. Compared to 20:1 or higher for many common gels, the cell's water-to-solids ratio is only about 2:1. The many negatively charged macromolecules of the cell should generate a strong osmotic draw; yet the water content in the cell remains surprisingly low. That limited water content may come as a consequence of the macromolecular network's stiffness: cellular networks typically comprise tubular or multi-stranded biopolymers tightly cross-linked to one another. The resultant stiffness prevents the network from expanding to its full osmotic potential.

If those cross-links were to disrupt, however, then the full power of osmotic draw would take effect; the tissue could then build many EZ layers and therefore hydrate massively, bringing huge expansion (Fig. 5). That's what happens when body tissues are injured, especially with dislocations. The injury disrupts fibrous macromolecules and cross-links, eliminating the restraining forces that keep osmosis at bay; EZ buildup can then proceed virtually unimpeded.

The reason why swelling can be so impressive is that the cross-link disruption occurs progressively. Breaking one cross-link results in higher stress on neighboring cross-link; so disruption progresses in a zipper-like fashion. When that happens, the osmotic rush of water into the tissue can continue practically without restraint, resulting in the enormous immediate swelling that is often seen. The tissue will return to normal only when cross-links repair and the matrix returns to its normally restraining configuration.

Fig. 5 Example of post-injury swelling

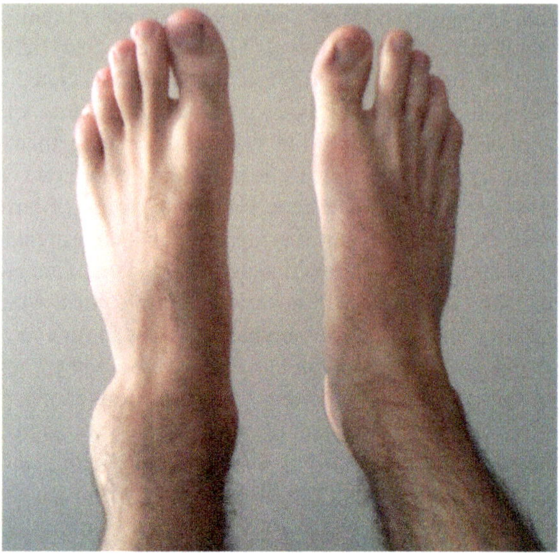

6 Water and Healing

During childhood illness, grandmothers and doctors will often advise: "drink more water." In his now-classical book, sub-titled *Your Body's Many Cries for Water: You Are Not Sick, You Are Thirsty,* [1] the Iranian physician Fereydoon Batmanghelidj confirms the wisdom of this quaint advice. The author documents years of clinical practice showing reversal of diverse pathologies simply by drinking more water. Hydration is critical.

Batmanghelidj's experience meshes with evidence of healing from special waters such as those from the Ganges and Lourdes. Those waters most often come from deep underground springs or from glacial melt. Spring waters experience pressure from above; pressure converts liquid water into EZ water because of EZ water's higher density. EZ water differs from bulk water in that it absorbs light in the UV region of 270 nm on spectrometry. The more light in this sector that is absorbed, the higher the concentration of EZ water in the sample. Specimens from the above and certain spring waters show a peak in this 270-nanometer zone, suggesting that relatively high EZ concentrations could contribute to their therapeutic benefits.

The same for mountain water: it too should have high EZ content. Our studies have shown that ice formation requires an EZ intermediate; i.e., bulk water does not convert directly to ice; it converts to EZ, which then converts to ice. Similarly for melting: melting ice forms EZ, which subsequently converts to bulk water. Fresh ice melt contains abundant EZ water.

For spring water and fresh ice melt, then, the high EZ content may explain the recognized health benefits. EZ water should rehydrate tissues better than ordinary water because of its higher dipole moment. To appreciate this argument, picture a bean with positive charge localized at one end, negative at the other. The positive end of that dipole orients toward the negatively charged cell, which then strongly draws in that dipole. The larger the dipole, the stronger will be the draw. Since EZs contain masses of separated charges, or large dipoles, EZ water should hydrate cells better than ordinary water. That's why EZ water may particularly promote good health.

7 Negative Charge and Anti-oxidants

Humans are considered neutral, but I suggest that we bear net negative charge.

Physical chemists reasonably presume that all systems tend toward neutrality because positive charge attracts negative charge. The human body being one of those "systems," we assume that the body must be neutral.

Not all systems are neutral, however. The earth bears net negative charge, while the atmosphere bears net positive charge. Water itself can bear charge: Anyone watching MIT professor Walter Lewin's stunning demonstration of the Kelvin water dropper http://www.youtube.com/watch?v=oY1eyLEo8_A&feature=related, where separated bodies of water eventually discharge onto one another, will immediately see that bodies of water *can* bear net charge. If any doubt remains, then the experience of getting an electric shock from touching certain kinds of drinking water (which my colleagues and I have personally experienced) should eliminate that doubt.

Charges can remain separated if input energy keeps them separated—something like recharging your cell phone battery and creating separated negative and positive terminals. Since we constantly absorb external energy from the environment, the theoretical possibility exists that we may bear net charge.

Consider the arithmetic. Cells make up some 60% of your body's mass, and they are negatively charged. Extracellular tissues such as collagen and elastin are next in line, and those proteins bear negative charge and adsorb negatively charged EZ water. Only some of the smaller compartments are positively charged with protons (low pH), and they commonly *expel*: urine, gastrointestinal system; sweat, and expired air (containing hydrated CO_2 or carbonic acid). They help *rid* the body of positive charge.

So, the arithmetic shows not only that our body bears net negative charge, but also that the body makes every effort to maintain that negativity by ridding itself of protons. It is as though maintaining negativity is a "goal" of life. Plants do it easily: they connect directly to the negatively charged earth; animals need to struggle a bit more to maintain their body's charge, in exchange for greater mobility.

How does our body's negative charge relate to the benefits of anti-oxidants?

Answering this question returns us to basic chemistry. Recall that "reduction" is the *gain* of electrons, while "oxidation" means electron *loss*. Oxidation strips molecules of their negative charge, working against the body's attempt to maintain high negativity. To guard against that loss we employ *anti*-oxidants. Anti-oxidants may keep us healthy simply by maintaining proper negativity.

8 Information in Water?

The question of the "memory of water" first became widely known from the studies of the late Jacques Benveniste, once a world-class French immunologist. In mid-career, Benveniste inadvertently turned from orthodoxy to controversy. He had been studying basophils, a type of white blood cell that secretes histamine when exposed to a particular antibody. Someone in his lab found an odd result: even when that antibody suspension had been diluted so extensively that not even a single antibody molecule could theoretically remain, exposure of that extremely diluted suspension produced the same response as did the original. Pouring what amounted to pure water elicited the same response as pouring the antibody.

Benveniste made the tragic mistake of labeling the phenomenon "water memory." The essentially pure water appeared to have retained "memory" of the molecules with which it had previously had contact; otherwise how could the water have elicited so specific a response? However, water molecules are known to jitter randomly many times each nanosecond; how possibly could actively dancing molecules retain information? Clearly, they cannot.

Before I recount the Galileo-like saga that befell Benveniste, I should mention that his results have been confirmed in multiple laboratories [2]. Furthermore, a possible physical-chemical basis for understanding how water could hold information is now evident: in may lie in water's liquid-crystalline fourth phase. The idea of water memory is no longer a scientific joke but, among some groups, a phenomenon ripe for exploration and one that is now being actively pursued.

In 1989, however, water memory was heresy. Benveniste's attempts to publish his lab's findings in the respected journal, *Nature*, were thwarted multiple times by the editor, Sir John Maddox. Finally, under pressure to publish a collective submission by several groups reporting the same result, Maddox relented. He'd publish the submission under one condition: he'd send a committee of peers to look over the shoulders of those French scientists to see what they were really about. The committee of peers would then report their findings to the readers of *Nature*. Seeing vindication on the horizon, Benveniste accepted Maddox's offer.

Several weeks later the committee arrived in Paris. Not exactly a committee of "peers," the visitors consisted of a threesome: Maddox himself, a journalist with limited experience in biology; Walter Stewart, a professional fraud sleuth from the National Institutes of Health; and "The Amazing Randi," a world-class magician. Randi's fame came partly from his own genius at magic and partly from his ability to uncover the basis of other magicians' tricks.

The committee's makeup sent signals of Maddox's intent. Since "water memory" seemed nigh unto impossible, clearly the French were engaging in some kind of duplicity, and what better committee than the one assembled could uncover the nature of their trick.

Although the "trick" was never specifically identified, the committee managed to find what it was looking for and the subsequent report to *Nature*'s readers dubbed water memory a "delusion." The impact was practically instantaneous. Water memory became a scientific joke, and Benveniste became the community's laughing stock—having trouble remembering? Why not drink some of Benveniste's memory water?

Benveniste never recovered. While pressing on with experiments showing that the stored information could be transmitted even over the Internet, he found himself unable to secure funds to support his work. His laboratory soon collapsed. He became demoralized, and finally succumbed following a routine surgical procedure.

Nevertheless, Benveniste opened the field of information storage in water, including the therapeutic use thereof, as championed by Nobelist Montagnier [3]. A possible key to understanding may lie in the fourth phase water, which is liquid crystalline. Crystals are stable. Molecules don't bounce around in the same way as ordinary liquid molecules. Indeed the ordered array of oxygen molecules resembles the ordered array of molecules in digital memories. In the latter case, molecules may take on either of two states, denoting zero or one. The oxygen molecules of water can do even better—theoretically assuming any of the five possible oxidation states, from negative two all the way to positive two. Hence the capacity of water memory mechanism may be superior to that of the standard digital memory.

It remains to be seen whether the water memory phenomenon will assume a dominant role in medical therapy and diagnosis, or will be rejected by a community that has already decided that the phenomenon cannot be true and is therefore unworthy of exploration.

9 The Future

Water's centrality for health is nothing new, but it has been progressively forgotten. With the various sciences laying emphasis molecular, atomic, and even sub-atomic approaches, we have lost sight of what happens when the pieces come together to form the larger entity. The whole may indeed exceed the sum of its parts. 99% of those parts are water molecules. To think that 99% of our molecules merely bathe the "more important" molecules of life ignores centuries of evidence to the contrary. Water plays a central role in all features of life.

Until recently, the understanding of water's properties has been constrained by the common misconception that water has three phases. We now know it has four. Taking into account this fourth phase allows many of water's "anomalies" to vanish: those anomalies turn into predictable features. Water becomes more understandable, and so do entities made largely of water, such as oceans, clouds, and, indeed, human beings.

References

1. Batmanghelidj, F. (1997). Your body's many cries for water: You are not sick, you are thirsty. Don't treat thirst with medications. Falls Church: Global Health Solutions.
2. Belon, P., Cumps, J., Ennis, M., Mannaioni, P. F., Sainte-Laudy, J., Roberfroid, M., et al. (1999). Inhibition of human basophil degranulation by successive histamine dilutions: Results of a European multi-centre trial. *Inflammation Research, 48*(Suppl 1), S17–8. https://doi.org/10.1007/s000110050376.
3. http://www.youtube.com/watch?v=XVBEwn6iWOo.
4. http://www.youtube.com/watch?v=JnGCMQ8TJ_g&list=PLwOAYhBuU3Ufr53AnJv9RLousgVN4G_uX&index=8.
5. http://youtu.be/i-T7tCMUDXU.
6. Montagnier, L., Aissa, J., Montagnier, J.-L., & Lavallee, C. (2009). Electromagnetic signals are produced by aqueous nanostructures derived from bacterial DNA sequences. *Interdisciplinary Sciences, 1*(2), 81–90. https://doi.org/10.1007/s12539-009-0036-7.
7. Pollack, G. (2001). *Cells, gels and the engines of life: A new unifying approach to cell function*. Ebner and Sons: Seattle.
8. Pollack, G. (2013). *The fourth phase of water: Beyond solid, liquid, and vapor*. Ebner and Sons: Seattle.

Author Biography

Professor Gerald H. Pollack (USA), Ph.D. He received his Ph.D. in biomedical engineering from the University of Pennsylvania in 1968. He then joined the University of Washington faculty and is now professor of Bioengineering. He is also Founding Editor-in-Chief of the journal, WATER, convener of the Annual Conference on the Physics, Chemistry and Biology of Water, and Executive Director of the Institute for Venture Science. His interests have ranged broadly, from biological motion and cell biology to the interaction of biological surfaces with aqueous solutions. His 1990 book, Muscles and Molecules: Uncovering the Principles of Biological Motion, won an "Excellence Award" from the Society for Technical Communication. His 2001 book, Cells, Gels and the Engines of Life, and his newest book, The Fourth Phase of Water: Beyond Solid, Liquid, and Vapor won that Society's "Distinguished Award," their highest distinction. The latter book went on to receive the World Summit Excellence Award. Pollack received an honorary doctorate in 2002 from Ural State University in Ekaterinburg, Russia, and was more recently named an Honorary Professor of the Russian Academy of Sciences, and foreign member and Academician of the Srpska Academy. He received the Biomedical Engineering Society's Distinguished Lecturer Award in 2002. In 2008, his colleagues chose him as the recipient of his university's highest annual distinction: the UW Faculty Lecturer Award. Pollack is a Founding Fellow of the American Institute of Medical and Biological Engineering and a Fellow of both the American Heart Association and the Biomedical Engineering Society. He received an NIH Director's Transformative R01 Award. He was the 2012 recipient of the Prigogine Medal for thermodynamics of dissipative systems, and in 2014 he received the Scientific Excellence Award from the World Academy of Neural Therapy, as well as the Dinsdale Prize from the Society for Scientific Exploration. He has presented two TEDx talks on water. In 2015, he won the Brandlaureate Award, previously bestowed on notables such as Nelson Mandela, Hillary Clinton and Steve Jobs. In 2016 he was awarded the 1st Emoto Peace Prize. And, he appears briefly in the 2016 Travis Rice sports-action film, The Fourth Phase, named after his recent book.

Emergence of Organisms from Ordered Mesoscopic States of Water (Liquids)—Physical Instead of Chemical Origin of Life

Igor Jerman

1 Introduction

The development of science is not a linear way of steady progression; it is more akin to a living organism and knows not only ebb and flow, but also critical bifurcation points. These are points of major decisions to be taken by the majority of scientists of a particular discipline. Once such decision is taken, the following way of scientific progress is separated from a part of previous achievements, explanations, models and findings as they are no more in harmony with the newly chosen paradigm. The science of biology is no exception to the rule. It passed through a very important decisive point during the period of Lamarck (the second half of the 18th century), where the majority of biologists—the moment coincided with the beginning of the biology as a separated natural science—decided that organisms follow different principles from the inanimate nature. Another important decisive point moved biology again away from its autonomous realm into the area of molecular reductionism, where life with all its richness of levels and ramifications is seen, studied and modeled as a complex chemical process individualized in organisms that are subject to natural selection; the latter being seen as a major evolutionary "force". Here, physics and the theory of dynamic systems have only a limited access, while there is no room for genuine biological laws. Since we still lack even a good and generally accepted definition of life, not to mention the lack of a deep and comprehensive theory of the living process, the last decision within biological community bereaved biology of very fruitful ideas, conceptions, views and considerations that were kept alive only by a small community of biological thinkers (e.g. the ones of biological fields—see more thorough studies of this topics by Bischof [1], Tzambazakis [2]). Hopefully, there are good signs that in a not very distant future, a new turning point in biological thinking will take place. Towards

I. Jerman (✉)
Institute Bion, Ljubljana, Slovenia
e-mail: igor.jerman@bion.si

G. M. Artmann et al. (eds.), *Biological, Physical and Technical Basics of Cell Engineering*, https://doi.org/10.1007/978-981-10-7904-7_14

this event I devote this chapter concerning the origin of life, since the nature of life and the origin of it are only two facets of the same enigma. Namely, the effort to truly grasp the possibility of the origin of life inevitably leads us towards a deeper understanding of life and its laws, since the capability to identify the point of life's emergence is equal to understanding the life's very essence.

The mystery of the origin of life is tightly connected with another mystery that is becoming more and more topical with an increasing number of uncovered exo-planets that resemble the Earth: is there life anywhere else in cosmos? Our main premise will be that if life is understood and treated only as a molecular phenomenon, as the prevalent contemporary view holds, the possibilities of the life's origin are very minuscule indeed. However, if life is more then just chemistry and if it has strong, physically based, genuine aspects of orderliness, then its origin could have been much more probable then estimated by the majority of contemporary scientists and life should be widespread in cosmos, though it may chemically differ from the Earth bound (DNA-RNA-proteins) life.

2 Theories of the Origin of Life: Some Basic Problems and Their Possible Solution

The science of the origin of life also knew its crucial turning point that has been slowly preparing around 200 years, from Italian researchers Francesco Reddi (1626–1697) and Lazaro Spallanzani (1729–1799) until the great French scientist Louis Pasteur (1822–1895) with his famous experiment that yielded the proverb "Omne vivum ex vivo" (all living from the living). Until the experiment at least the unicellular organisms were seen as a creation of spontaneous generation from the inanimate matter. The perception and understanding was that matter had spontaneous organizational powers to generate at least the simplest organisms out of its inanimate chaos. The Pasterur's experiment shattered this notion, opened an abyss of unanswered questions, since it was also (and still is) logical that (a) there was time in the past of the Earth when there was no life at all and that (b) just under certain conditions once in the history of the Earth life emerged. A logical conclusion from the Pasteur's experiment and these two established facts was that contemporary matter on the Earth

(a) has no power to spontaneously generate life and that
(b) once it should have had it.

Therefore, in principle the Pasteur's experiment did not take away "creative" powers from the inanimate matter, but only limited them to some unknown conditions in the past. All scientific history of the origin of life enigma since "Omne vivum ex vivo"[1] concerned the identification of those special conditions. Mostly

[1] As we could see this proverb is problematic as it presupposes that life has been present since the origin of the universe.

they were sought for in some special complex chemistry, possible soon after the cooling of the Earth, but most certainly not possible now. The basic problem of these origin of life approaches stems from the fact that contemporary life, even in its simplest forms, is highly complex and ordered. In terms of informational theory, it has a relatively high informational content that should be stably reproduced from generation to generation. Therefore, we have two problems: chemical and informational. The first one demands identifying and reproducing a stable path from simple molecules (like ammonium ion, formaldehyde, hydrogen etc.) to macromolecules that would resemble the ones preset in contemporary organisms. This demand is spiced up with the condition that many primordial chemicals for the living chemistry should be chiral (optically active)—and we know that in nature spontaneously generated chemicals are ever represented by both enantiomers in equal proportion (i.e. optically inactive). Through time various partial scenarios for the synthesis of this or that essential chemical (like amino acids, organic bases etc.) were proposed. Some of them were also successfully empirically proved [3, 4]. However, there still exists a huge gap between these partial approaches and the orchestrated complex biochemistry of even the most primitive living forms. It may be said that it is difficult if not impossible to envision (and later to reproduce) a stable path that leads from the afore mentioned simple chemicals to the first living being. From the standpoint of the theory of dynamical systems it would mean it is most unlikely to find a stable (i.e. at least under certain conditions probable) trajectory from the chemistry of the ancient pre-life Earth to the evidently stable attractor of our first ancestor.

The second aforementioned problem concerns the problem of stable complex information that is capable of continuous reproduction. As it was smartly formulated in a relatively recent consideration by Walker and Davies [5], the starting point for the emergence of life should mean an establishment of causal powers from the side of organized information over molecules. Here information in living systems is not seen as coded in a straight line (as in DNA), but in a diffuse, non-localized, way. In other words, the causal powers of life are viewed as trans-material, whereby no supernatural factor is introduced, only dynamically organized, diffuse and active, operational information. In present day life this dispersed, organized and operational (active) information only partially resides in the DNA, it is also widespread in the whole living state of the organism—something perceived also by the structuralist line of organicism [6]. From experiments it is well known that *Acetabularia mediteranea* can reproduce its cap even without the nucleus, therefore with no DNA [7]. Consequently, and in line with Walker and Davies, we may safely abandon to understand organisms only as trivial duplicators with errors (i.e. like copying machines), which is the mainstream view of contemporary established biology. On the contrary, they should be perceived as non-trivial duplicators, systems which organized information is capable of multiplying the organization of energy and matter within the system in a complex and widely dispersed way and simultaneously actively process information and express a high degree of flexibility (variability).

According to this advanced view of the living process that is sufficiently abstract to enable capturing also some strange forms of life in larger cosmos, life can be defined as *dynamic and highly organized information, expressing itself through material systems and capable of further potentially limitless enrichment of its content (complexity) and its organization.* The origin of life then means only the emergence of such information together with its upholding material system. With the origin of life information as an active principle of nature gains the ability of causation from top to bottom (from the whole to its parts, from the cell—or the organism—to its molecules). It is behind the wholeness of organisms. This notion of information, even if sharing some basic features with the Shannon's information theory [8], at the same time far exceeds it. If the Shannon's theory concentrates on quantitative aspect of information, the biological concept should cover also its content (semantic aspect) together with its function and operational power. In this view, information as an active principle behind the living process arises as a powerful natural principle that works within matter and at the same time steering, directing its processes. According to this outlook, life has its special principle that surpasses ordinary (inanimate) physical and chemical world, but is on the other hand still a **fully natural principle**, like matter, time, space etc. This principle resides in inanimate matter as a more or less remote potential—and emerges under suitable conditions.

The theory of organized information sounds attractive, but has also its prosaic difficulties. In his inventive theory of the origin of life, Manfred Eigen tackled the problem of the emergence of the organized information at the molecular level that could lead to the origin of the simplest organisms. He analyzed the behavior of a special class of self-reproducing chemical entities called *autocatalytic cycles* and proved that simple autocatalytic cycles that would emerge on the primordial Earth as something similar to contemporary t-RNAs (i.e. self reproducing relatively short RNAs) would have a very limited evolution, in other words they would end in a shortest possible self-reproducing molecule [9]. Therefore, in simple self reproducing molecules we could achieve organized information, however on a pre-life level, incapable to engender a true living process with in principle limitless evolutionary perspectives. To cope with this short dead end of molecular evolution based on autocatalytic cycles Eigen invented entities called hypercycles [9]. The latter are interesting entities, but at the same time not very probable.[2]

Long after the triumphant Miller's experiment (dated on 1952), where he exposed simple molecules to electrical discharge in a chemically closed system and got many amino acids (though optically inactive) the solutions of the prebiotic chemistry leading to the origin of life is still far from convincing. It is similar with organizational (organized information) aspect, even if we jump from Eigen to Kauffman and his autocatalytic (still purely chemical) nets. He made sophisticated computer simulations of the possibility of spontaneous organization of information

[2]Eigen proposed the possibility of a direct polypeptide coding from a strand similar to contemporary t-RNAs without any genetic coding system.

and came further form Eigen, yet he still had to confess that the nets are only a necessary but not the sufficient condition for the origin of life [10]. This all speaks in favor of the idea that the exclusively chemical outlook on life may not be a fruitful attitude; it may be just a cul-de-sac that will never explain life itself nor its origin. Therefore, to regain a real chance to solve the mystery of the origin of life, we should make a bold step into supramolecular realms and try to identify a vital factor that

(c) would have a high probability to play an important role in biological processes, even if not yet acknowledged by the majority of biological science circles,
(d) could maintain a high level of orderliness even in highly diversified chemical systems, therefore systems with no "memory" molecule and with no genetic code,
(e) would be able to increase its informational content (complexity) through time (evolution),
(f) would be capable of specific interactions with a wide range of molecules, with a capability to direct their transformations similar to catalysts,
(g) would be at least physically feasible, if not probable within a supposed conditions of pre-biotic Earth.

Since this factor is not chemical (even if it should fully interact with molecules—as stated in the fourth line above), from its very beginning life should have had at least two levels: chemical and some other (let's say physical) with organized information working at both levels simultaneously and also connecting them.

3 Views of Life and Water That Promises Fresh Possibilities to Disentangle the Mystery of the Origin of Life

3.1 (Bio)Physical Basis of Life

Since the reductionist biology has begun to establish itself (this process started during the second half of the 19th century), the minority of biologists opposed to the more and more "molecularized" outlook on life. They sought a unifying principle that could explain the immense orderliness and complexity found even in the simplest organisms. They were accused of vitalism from the established majority and at least with the start of the DNA era in biology certain lines of vitalistic thought were completely banned from biology. However, the organistic, supramolecular outlook on life that tried to identify, research and understand the living process also in the light of a special field, never stopped; during the 20th century it simply moved from a more or less mystical perception to a more scientific one. The physical nature of the field was unclear, but at least through countless cases of observational and experimental embryogenesis its fingerprint became evident. Many such processes found an apt explanation through the terminology and mathematics of fields whose nature holds some similarity to physical fields [6, 11]. Mostly it was named *morphogenetic* field. The biggest difference to the known physical fields was that the morphogenetic field (sometimes named also biofield [12, 13]) functioned in a very dynamic way. It should have a very complex chreod (i.e. a dynamic attractor, stable path) from the zygote to the fully grown individual, pierced by many points of symmetry breaking.

If the first biological field approach stems mostly from embryologists the second one stems from a purely physical reasoning. In the sixties of the past century, a British biophysicist Herbert Fröhlich found that due to a high electric membrane field the living state at the level of cells rests on rhythmic and coherent[3] oscillations of polar (macro)molecules (called also polarons, see [14]) and that these oscillations can bring much order even into cellular biochemistry [15, 16]. Fröhlich's ideas were later elaborated by an Italian group of quantum physicists (Del Giudice, Vitiello, Preparata) on the level of quantum field theory. The theory that could be called the theory of quantum bio-electrodynamics [17–19] argues that life at the cellular level (biochemistry, molecular movements) is highly organized through coherent (laser like) electromagnetic (EM) field that can behave as a particle of a very small mass and would have a rather peculiar form, namely, the one of very thin filaments. Endogenous EM field therefore builds a fine filamentous network within cells, which is highly stable and should persist even some time after destruction of the underlying order of molecular dipoles. The essential thing here is that the filamentous EM field in organisms is not an epiphenomenon of chemical processes, an unimportant collateral entity, but an indispensable one that ensures order and coordination of chemical and (micro)mechanical processes within cells. This field is supposed to represent a basis for the synthesis and disintegration of one of the most important cellular organelles—microtubules and other microfilaments that build the cellular skeleton and locomotory organelles. If the frequency of an EM filament resonantly matches a neighboring molecule, the latter is attracted to its outer surface and is oriented at the same time. In the filamentous field, chemical interactions are therefore ordered and interconnected through the resonant induction. The field is also important from the thermodynamic standpoint: the output energy of a chemical reaction is not dispersed since it continues travelling as a polarization wave, the so called soliton. In harmony with Fröhlich, Del Giudice's group also postulates that the coherent field may leak only from some defects occurring on internal cell's surfaces and that this emission should be coherent at all frequencies.

The biological electrodynamic system is therefore abstract as to the specific molecular composition of the cell. It should only fulfill certain dynamic systemic conditions for attaining the coherent regime. If we look at the contemporary life, many articles devoted to endogenous cellular coherent regime argue that there is a strong connection between coherent oscillations of the endogenous EM field and the orchestration of chemical reactions. This would account for the fact that within

[3]Coherent means that all oscillations are in the same phase like with lasers.

cells a huge number of chemical reactions occur with well defined time sequences without mistakes and at rates much higher than in vitro. The high rate and the absence of mistakes imply that the underlying chemical dynamics should not be based so much on random molecular collisions (Brownian motion) but more on the mutual long range attraction of the molecules in a selective way that can be attributed to the coherent cellular regime, see also [20]. Further considerations on this line are given at length by Meijer and Geesink [14]. There are many empirical confirmations of such electrodynamic coherent endogenous cellular field and it was also directly measured [21, 22].

The third supramolecular approach to the orderliness of the living process does not originate from the research or considerations of organisms but from some quantum field theoretical and also empirical researches of water. This is especially important since water is regarded as a cradle of life. Regarding ordinary liquid water there are two basic scientific outlooks:

(a) "chaotic": conventional, strongly established, that sees the liquid water in a state of an entirely chaotic movement resulting in Brownian motion of colloidal particles;
(b) "ordered": unconventional, though also based on science and ramified in various theoretical streams. They see water partially in a state of chaotic motion and partially as a long-range ordered state (depending on temperature and some other conditions, like the concentration of polar molecules or ions).

It is the second one that interests us here. A long-range order water should stem from phase transitions of water from the chaotic vapor-like phase to an ordered oscillatory phase where all molecules fluctuate in unison, laser like, which is called also as *coherence*. They oscillate in tune with a self-trapped EM field within minuscule regions called *coherent domains* (CDs), spanning around 100 nm, see [17, 19, 20, 23]. This basic water order was extensively researched by Preparata and Del Giudice. Coherent regime stabilizes CDs and makes them resistant to thermal noise, i.e. to chaos that still characterizes bulk water encompassing CDs. The order of the latter is *long range long term* order. *Long range* means that not only neighboring but also distant molecules within a CD are moving or oscillating in harmony with one another. *Long term* means that the coherent vibration or better said, the vibrational pattern of molecules within a CD has a very long persistence.

As it is well known, water in organisms and cells is not pure. It is imbued with many ions, organic molecules, colloidal particles and what is also very important, it mainly borders to countless hydrophilic surfaces. And the water bordering to hydrophilic surfaces was found even more highly organized than ordinary (pure) water with its CDs.[4] This is evident especially from an extensive work by Pollack and coworkers [24, 25]. This water has many signs of long range long term order that could have static as well as dynamic characteristics. Since it tends to exclude

[4]In pure water at room temperature around 40% of water molecules should reside in CDs and 60% in the bulk water.

solutes it was called "exclusion zone water" or simply EZ water and may span up-to 0.5 mm away from the surface. As already said, in organisms and various organic vesicles the relative amount of the interfacial water is very high.

Therefore, we have identified three biological, physical or physicochemical phenomena that speak in favor of a supramolecular entity that together with appropriate chemistry makes life possible. They could be further reduces to two entities: an emergent[5] EM field (morphogenetic or Fröhlich's field of coherent EM oscillations) and partially long range long term ordered water (coherent domains) surrounded by non-coherent (chaotic) bulk water. We may assume they are both real and have an essential role in living processes. In the following we shall discuss their essential role in the origin of life process.

4 Field and Ordered Water Approach to the Origin of Life

If chemical experiments into the origin of life proved to be only partially successful, actually a disappointment after Miller's triumphant experiments, the situation was much more successful with various cell-like structures that resembled simple organisms. Even from basic text books of biology it is known that Alexander Oparin succeeded in creating so called coacervates, quasi cellular structures, actually aggregates of colloidal droplets within water [26]. They resemble living unicellular organisms with no sharply defined form, they have a boundary (a quasi membrane), they can change shape like amoebas, can produce and release vacuoles, can merge, divide etc. Sidney Fox went even further, his so called proteinoid microspheres [27], synthesized from amino acids, showed even more striking features, something similar to the resting membrane potential and action potentials, i.e. a sort of excitability [28]. The basic question here is whether these dynamic structures only superficially resemble organisms or do they tell us something important regarding the origin of life. The answer from the molecular reductionist standpoint is clear: various quasi cellular structures, even if having electrical activity, reproduction etc. are only superficial "counterfeits" of organisms, a short evolutionary route of colloidal systems with dead end regarding the possibility of the origin of life, in this respect similar to the already mentioned Eigen's auto-catalytic cycles.

But from the deeper view of life that I disclosed above, the story may be quite different. Assuming countless researches and theoretical considerations concerning the most probable scenario for the origin of life on the young Earth, first living organisms should emerge in water systems that were

[5]This means the field is physical, but in a complex biological system it assumes new (i.e. emergent) characteristics. It is akin to emergent properties of superfluid He in comparison to the normal gas. The wave equation if the former cannot be derived from equations of the latter.

(a) rich in various hydrophilic surfaces,
(b) rich in minerals and organic molecules, many of them polar,
(c) abundant in free energy through appropriate molecules, e.g. polyphosphates.

If we take into account a relatively strong electric field surrounding hydrophilic surfaces [29], a lot of free energy (strong UV radiation, heavy storms, frequent volcano eruptions) and at least moderate concentration of longer polar molecules than according to the aforementioned research we may be pretty sure that such primordial water systems resulted in coacervate like forms, i.e. colloidal quasi cellular structures that have in principle large hydrophilic surfaces and thereby a high level of orderliness. From their very beginning they should have

(a) organized water on at least two or even more already postulated levels (basic and extended—higher level—coherent domains of various kinds) including exclusion zones and
(b) since all conditions (see [15]) are fulfilled for the emergence of coherent oscillations of polar molecules, we may expect the emergence of a Fröhlich-like regime even in these supposedly[6] non-living systems.

In other words, both regimes that are recognized as underlying contemporary organisms at the supramolecular level are very feasible even in the most probable pre-biotic water systems. But were they capable to connect with the molecular substance in a feed-back manner and were such systems capable to harbor dispersed active organized information?

If we try to answer the first question we may say in addition to the already disclosed considerations of quantum electrodynamics that biological macromolecules flowing in the bulk water between CDs are subject to influences from the coherent EM (so called evanescent) fields on the borders of CDs as they protrude from the latter. According to the general theorem of electrodynamics, molecules able to oscillate at the same frequency to the one of the CD field are strongly attracted and therefore able to react chemically [30, 31]. Hence, in an extended coherent region, i.e. the region that comprises many CDs and the space in between (see also next paragraph), the diffusive, Brownian motion of molecules is replaced by a selective dynamic regime, where molecules recognize and interact with one another via frequency matching. Here, biological dynamics appears as a mutual interplay between electromagnetism and biochemistry, where fields enable molecules to interact through resonance, and molecules are able to regulate the field frequency through their reaction energies.

According to the general scheme, water molecules in bulk are predicted to give rise to CDs having size of 0.1 μ [32]. In biological environments, however, these CDs are presumed to give rise to extended domains, i.e. domains of basic CDs together excited by the cellular metabolic energy flow, and may extend to the size

[6]As we have no generally accepted definition of life and we are tackling systems close to the emergence of life, it is not possible to be sure about the true place in the line living—non-living dimension.

of molecular complexes, or even to a whole cell [33]. Since the excitable spectrum of a CD is very rich [34], a variety of extended domains can emerge that may assume fractal (nested) architecture, as analyzed by Vitiello [35]. Extended domains entail two important consequences, namely a defined size of the coherent system, and the appearance of geometrical shapes [33]. In order to have a precise frequency matching, the relative positions of reacting molecules must assume a specific spatial configuration, corresponding to biological structures.[7] The concept of extended domains where the latter cover also the space between basic CDs connects the water orderliness with the already described Fröhlich's like regime of polar (molecular) oscillations. Therefore, from now on we will speak only of the *coherent regime* that would signify the concept of the extended coherent domain, implying also a possibility (if not likelihood) of Fröhlich's type of oscillations.

Taking all these together, there is plenty of supporting theoretical researches and considerations arguing that a coherent regime of water systems based on rich chemical grounds, especially if there is a constant flow of free energy, results in a mutual association between the coherent EM field and chemistry. The answer to our first question is therefore affirmative.

Now, what about the operational active information that should be the gist of life? The emergence of this sophisticated information would announce the origin of life, so let's repeat its basic characteristics: it should be stable, dispersed (non-localized), it should have control over chemical reactions and capable of increasing its own complexity. All what we know about coherent states of water and polar macromolecular systems and was spoken about at length above, speaks in favor of capability of such systems to give rise and to sustain it. Long term vibrational continuity enables its stability, long range phase correlations of oscillations enable its non-localized functioning. Through resonance and induction the vibrational states interact with countless molecules as well as with their ceaseless transformations and can direct the latter.

Therefore, we may safely assume that taking into account primordial conditions of the Earth (not going into much vaster possibilities of the origin of life elsewhere in cosmos, for instance on comets) systems with previously enumerated characteristics shouldn't be uncommon, difficult or improbable to find. But, would they be already alive? Where there is no true and generally accepted theory of life, the answer is difficult (of course it is easy and negative if we indentify life with DNA, proteins, metabolism etc., but then we loose generality and depth). According to the here presented definition of life (see the second chapter) we may claim that such organized systems are not yet alive if they do not have a clear possibility (actually a stable trajectory in a complex evolutionary phase space) to evolve into something more complex. With no such possibility, even if highly organized through the

[7]The recent experiments on photosynthetic systems provide an example where spatial arrangement of the component molecules within a coherent system must be carefully controlled in order to optimize the flow of energy [36].

coupling between coherent modes (field level) and chemistry (substance level), they would be similar to Eigen's autocatalytic cycles or Kauffman's autocatalytic nets.

5 From Abiotic Water Systems to Life

5.1 Expected Stages

Up to now we tried to understand life as a coupled pair of two mutually interacting and communicating levels: the field level and the one of molecular substance. The third element of life is non-material, namely emergent, operative and ordered information directing the living process and capable of further and potentially limitless evolution in the sense of increasing the system's complexity. Non living systems would then be the ones that would not fit into this scheme (the trio of molecular substance, coherent field and active information) and would have a very limited reach of a possible evolutionary "jump". The systems that would (a) include all three components and (b) would have only a limited reach of the next possible evolutionary jump could be treated as being somewhere between the living and the

Table 1 Three stages of evolution of matter-energy-information systems towards (and including) life

	A-biotic phase	Pre-biotic phase	Primordial life
Organization complexity	Simple coherent water domains	Quasi cellular structures (coacervates, microspheres, nanobacteria, …)	Evolving cellular structures
Evolution[a] = increasing complexity	Very limited reach, if any at all	Limited reach, but more or less frequently occurring	Potentially endless
The nature of molecular substance and processes	Water solutions of ions and polar molecules	Most probably complex and organized in autocatalytic cycles	Highly complex and organized in a huge autocatalytic network
The nature of field	Coherent domains of water systems are not in organized feed-back relations with molecular interactions of their own molecular level	Extended coherent domains are in partial feed-back relations with molecular interactions of their own molecular level	Extended coherent domains of cells are in full feed-back relations with molecular interactions of their own molecular level
Active highly complex information	Non existent	Gradually forming	Actual, full sway

[a]Biological evolution does not mean necessarily an increase in complexity, since many organisms evolved in a direction of a reduction of organs or tissues. But for the purpose of this article evolution is understood as an increasing complexity

non-living ones. Consequently, on the line between a chaotic non living systems and the assured highly organized life we may demarcate the following more or less three distinct stages (see Table 1):

With the help of this table we can assume that the origin of life took place between the second and the third stage, when the already complex systems with all three components and a limited evolutionary reach "mutated" into a new one which evolutionary capabilities became limitless, at least in principle, as it holds for contemporary life. We may hold the premise that "Natura non facit saltus" (the nature doesn't make jumps), which in our case would mean that the second stage was subject to a gradual evolution concerning less and less limited evolutionary reach (more and more remote mutational possibilities that are fitter from the predecessor systems).

5.2 Various Entities that Could Represent the Passage from Inanimate Systems to Living Beings

Now, let us move to real structures that could represent the first and the second stage of evolution of information-energy-matter systems towards life. The representatives of the first stage are not difficult to identify; these may be even pure water or various water solutions. With more complex polar molecules they are expected to contain more sophisticated coherent domains. Countless water systems (not necessarily water—many other polar solvents would behave in a quite similar way) bordering to a polar or ionized hydrophilic surface would also add up to this stage systems, maybe nearing the second one. Primordial Earth should possess many of them.

Are there any systems that would resemble the second stage? Yes; in the first place, these are the systems that were found and researched as possible candidates for the origin of life before these efforts were concentrated almost exclusively on the biochemical and molecular genetic level. We may speak of the already mentioned Oparin's coacervates and Fox's proteinoid microspheres. There are also a little more exotic entities, like Cairns Smith's clay systems, Reich's bions, Bahadur's jeevanus, Bechamp's microzymas, Naessen's somatids, nanobacteria, nanobes, biomorphs, nanovesicles etc.

Of all the above interesting creatures nanobacteria or more universally, nanoparticles[8] or nanovesicles (NVs in short), are the most examined. Nanobacteria are formations that are some 5–100 times smaller from ordinary bacteria ($\sim 1\ \mu$). They were first identified in Italian hot springs in 1993 [39] resembling also microstructures found in the Allan Hills meteorite remnants from Mars that were (in 1996) hypothesized to represent Martian primitive life forms [40]. They were also

[8]In [37] Kajander defines nanobacteria within the human body as calcifying nanoparticles. Martel et al. [38] extended this view since they treat nanobacteria as mineralo-organic nanoparticles.

found to accompany human life since they were discovered in all our bodily fluids and connected to many of our diseases, mostly to ones involving calcification (kidney stones, arteriosclerosis, tooth scale etc.).

While they were soon regarded as too small to be able to function as fully living beings by the majority of microbiologists, nanobacteria nevertheless demonstrated capability to contaminate cell cultures and even more, they proved to be resistant to ordinary sterilizing procedures like heating even over 90 °C and to many antibiotics [38, 39, 41]. They may form colonies larger than 1 mm and may exhibit budding and may propagate like microspheres and coacervates, even if at a much slower rate than Escherichia coli at optimal conditions [41]. Mostly, they appear as reproducing calcifying macromolecular vesicles [37] containing a sort of a wall and filamentous surface projections. By one of the most versed group of researchers (Martel, Young et al.) they are treated only as complex inorganic–organic systems [38, 39, 42].

The same group succeeded to make nanobacteria like creatures also artificially and on the basis of other cations than calcium, like Mn, Cu, Fe, Ni, Sr, Na etc. and some animal proteins, too [42]. They were called *bions*. Bions also form aggregates and colonies and are mostly (but not all of them) round shaped and of various sizes. Being produced absolutely artificially, they have no trace of genetic material and are classified as non-living, bio-mimetic complex chemical systems [42]. Biomimetic forms may be achieved also by solely inorganic chemicals, like silica, barium and calcium and are called biomorphs; they were aptly described by the group around Garcia-Ruiz [43].

Taken all these facts together, the behavior of these tiny bodies strongly resembles living organisms (their highly biomimetic nature, at least). Their complex multiplication indicates that **they do possess active organized information**. And since in general they do not possess the genetic apparatus distinctive for ordinary living beings and even viruses, their still very complex organization together with variable chemistry indicates a highly organized coherent electrodynamic field that we assumed as a basic condition for the origin of life. But at which point of the second stage they are? Are they only biomimetic or do they have a real potentials for the origin of life? If we return to nanobacteria there are authors claiming that they are definitely living creatures and criticize the previously mentioned estimation (see [37, 41]). A confirmation for their position they see in the nanobacteria' need for oxygen for their successful propagation. In their view nanobacteria are mostly infectious and harmful agents, usually where mineralization of tissues or teeth is at stake, but nevertheless they could be only opportunistic companions of our life, since they were found also in healthy humans. Moreover, they even see a possibility to heal some illnesses via appropriate chemicals together with nanobacteria [41, 44].

Therefore, the NVs are very interesting systems that, as already stated, roughly correspond to our second stage of the origin of life process. Various synthetic bions and biomorphs may truly represent only very complex forms of precipitation (crystallization) of various chemicals and could represent only an advanced first stage or a passage from the first to the second one. Then we may have various bions and perhaps also some nanobacteria that would represent the "middle" of the second stage. On the other hand, at least some nanobacteria—perhaps the ones found in bodily fluids—may represent systems that border the third stage or even trespass this boundary. This would mean a thrilling prospect that we do not need to go into a distant past, speculating a lot about the most probable path leading to the first living being and trying laboriously to reproduce the conditions that would have lead to it. No, we may be encircled by systems that are on various stages towards life at this very moment. To be sure, they are most probably quite different from the ones in the past or from the ones elsewhere in cosmos, yet they may be, according to the here presented definition still on the verge of life or even beyond it. Some of them may even be our companions within our own bodies or in the nature. If living, they could have a quite different biochemistry from our well known one, but they should have a capability for adaptation and in principle limitless evolutionary potential. These creatures may be called robust living beings with not so well defined biochemistry as the well known organisms. Therefore, if we take into account the here offered definition of life that is bound to specifically organized information expressing itself through a close and mutual interplay of the coherent field and countless molecular interactions, we may safely assume that our known and unambiguous (DNA-RNA-proteins) life (chemically highly specialized, S-life) is surrounded and even interspersed by a parallel life being much more robust (chemically robust, R-life). Its robustness is proved by its already described high resistance against many for ordinary organisms highly detrimental chemicals and physical conditions (extending even to a high gamma radiation, see [41]. This robustness indicates that the R-life is not organized on a detailed molecular informational matrix as is the case with the S-life, but is rather much more slack, only roughly defined in its chemistry. Nevertheless, its forms are still relatively highly organized owing to their specific physical characteristics discussed in the fourth chapter.

Even if we accept that at least some nanobacteria or bions are living beings, the question still remains, whether these forms may transit to the S-life. The answer to this question would demand an intensive further research. On the other hand, it is very probable that the S-life may not only pass to the R-life, but may actually *produce* some of its forms. Perhaps the R-Life can also be formed by an appropriate mixture and conditions involving various inorganic and organic substances as the research of bions by Reich showed [38, 45]; the name "bions" used in the works by Martel and Young was taken just from Reich [38]. Reich speaks about moving creatures, generated from purely sterilized conditions after autoclaving some living or formerly living material. They resembled paramecia or amoebas, which was confirmed also by a long and systematic research by DeMeo [46]. Here, one of very important things to be ascertained is whether the amoeboid movements—observed also with Oparins' coacervates—are the result of purely random Brownian motion of nanovesicles and their content or do they show at least some deflections from pure randomness. The latter would be a vey important evidence of truly organized systems that would merit the name *living*. The experiments of this kind are in preparation.

6 Conclusion

The story of the origin of life—a tremendously organized and complex process amid a chaotic world of matter and energy—is still a mystery. In this chapter I wanted to present a new, promising path towards its dismantling. It does not disregard conventional chemical and physical aspects of life. It only tries to connect them into a deeper whole, integrated via active organized information. As an essential factor of the living process, a coherent field together with partially dynamically ordered water were presented. This physical orderliness may have a pivotal role in maintaining a high organizational level of the systems approaching the first organism, since DNA or even RNA could be either too unstable, would lack chirality (optical activity, e.g. D ribose), would not achieve sufficient concentration, would intermingle with other molecules in a pre-living chaotic chemical system etc. As a result, the order of coherent modes of pre-living water vesicles would function as a dispersed systemic memory, would sustain organized information and would replace contemporary DNA or RNA. No genetic code would be needed—instead, the code of resonance matching between coherent field oscillations and molecular reactions would be in place. Only gradually and presumably long after the first organisms (at least considered so through the eyes of the here presented definition of life) appeared, the natural selection together with some laws of the living state dynamics would result in the transference of the dispersed memory contained in coherent oscillations into the focused molecular one (DNA, RNA). An interesting support for such transference may be found in a recent publication, stating that optical activity, so essential for contemporary life was found in ordered water systems [47].

Most probably, the origin of life was a long, gradual process, where various systems grew in their complexity, however, they did not have the open path (full potential) to limitless evolution. Many of such system are still here with us and are known as nanobacteria or more generally as nanovesicles. They may be quite natural or artificial, in any case they testimony that for nature it is not difficult to organize matter and energy into ordered dynamic structures so much resembling present day organisms that scientists still quarrel as to their living or non-living status. There is a real possibility that the origin of life was not an almost miraculous moment in some distant geological past—it can be continually happening all the time here with us, as a permanent passage of certain vesicular systems from a lower to a higher complexity. The message of this chapter is that the origin of life is not so unfavorable, so an improbable event as it is usually portrayed in the established biology. It may be quite widespread also on other planets and may there have a profoundly different (bio)chemistry from the Earth life. Any serious exobiology should seriously take this possibility into account.

Bibliography

1. Bischof, M. (1998). Holism and field theories in biology. *Biophotons* (pp. 375–394). Netherlands: Springer.
2. Tzambazakis, A. (2015). The evolution of the biological field concept. In D. Fels M. Cifra & F. Scholkmann (Eds.), *Fields of the cell* (pp. 1–28).
3. Groth, W. (1975). Photochemical formation of organic compounds from mixtures of simple gases simulating the primitive atmosphere of the earth. *BioSystems, 6,* 229–233.
4. Miller, S. L., & Orgel, L. E. (1974). *The origins of life on the Earth.* Englewood Cliffs: Prentice Hall Inc.
5. Walker, S. I., & Davies, P. C. W. (2013). The algorithmic origins of life. *Journal of the Royal Society Interface, 10,* 20120869. https://doi.org/10.1098/rsif.2012.0869.
6. Goodwin, B. C. (1985). Developing organisms as self-organising fields. In: *Mathematical essays on growth and the emergence of form* (pp. 185–200). Edmonton: The University of Alberta Press.
7. Goodwin, B. C., Skelton, J. L., & Kirk-Bell, S. M. (1983). Control of regeneration and morphogenesis by divalent cations in *Acetabularia mediterranea. Planta, 157*(1), 1–7.
8. Jaynes, E. T. (1957). Information theory and statistical mechanics. *Physical Review, 106*(4), 620.
9. Eigen, M., & Schuster, P. (1979). *The hypercycle.* Berlin: Springer.
10. Hordijk, W., Kauffman, S. A., & Steel, M. (2011). Required levels of catalysis for emergence of autocatalytic sets in models of chemical reaction systems. *International Journal of Molecular Sciences, 12*(5), 3085–3101.
11. Goodwin, B. C. (1984). A relational or field theory of reproduction and its evolutionary implications. *Beyond NeoDarwinism* (pp. 219–241). London: Academic Press.
12. Jerman, I., Leskovar, R. T., & Krašovec, R. (2009). Evidence for biofield. In: *Philosophical insights about modern science* (Vol. 9, pp. 199–216). Hauppauge, NY: Nova Science Publishers.
13. Rubik, B. (2002). The biofield hypothesis: Its biophysical basis and role in medicine. *The Journal of Alternative & Complementary Medicine, 8*(6), 703–717.
14. Meijer, D. K. F., & Geesink, H. J. H. (2016). Phonon guided biology. *NeuroQuantology, 14* (4), 718–755. https://doi.org/10.14704/nq.2016.14.4.985.
15. Fröhlich, H. (1975). The extraordinary dielectric properties of biological materials and the action of enzymes. *Proceedings of the National Academy of Sciences, 72,* 4211–4215.
16. Fröhlich, H. (1988). Theoretical physics and biology. In H. Frohlich (Ed.), *Biological coherence and response to external stimuli* (pp. 1–24). Berlin: Springer.
17. Jerman, I. (2016). The origin of life from quantum vacuum, water and polar molecules. *American Journal of Modern Physics. Special Issue: Academic Research for Multidisciplinary, 5*(4), 34–43.
18. Fröhlich, H. (1978). Coherent electric vibrations in biological systems and the cancer problem. *Microwave Theory and Techniques, IEEE Transactions, 26,* 613–618.
19. Del Giudice, E., Preparata, G., & Vitiello, G. (1988). Water as a free electric dipole laser. *Physical Review Letters, 61,* 1085–1088.
20. Del Giudice, E., De Ninno, A., Fleischmann, M., Mengoli, G., Milani, M., Talpo, G., et al. (2005). Coherent quantum electrodynamics in living matter. *Electromagnetic Biology & Medicine, 24,* 199–210.
21. Jelínek, F., Cifra, M., Pokorný, J., Vanis, J., Simsa, J., Hasek, J., et al. (2009). Measurement of electrical oscillations and mechanical vibrations of yeast cells membrane around 1 kHz. *Electromagnetic Biology and Medicine, 28*(2), 223–232.
22. Pollock, J. K., & Pohl, D. G. (1988). Emission of radiation of active cells. *Biological coherence and response to external stimuli* (pp. 139–147). Berlin: Springer.
23. Bono, I. Del, Giudice, E., Gamberale, L., & Henry, M. (2012). Emergence of the coherent structure of liquid water. *Water, 4,* 510–532.

24. De Ninno, A. (2017). Dynamics of formation of the exclusion zone near hydrophilic surfaces. *Chemical Physics Letters, 667,* 322–326.
25. Pollack, G. H. (2013). *The fourth phase of water.* Seattle, USA: Ebner and Sons Publishers.
26. Oparin, A. I., & Gladilin, K. L. (1980). Evolution of self-assembly of probionts. *BioSystems, 12,* 133–145.
27. Brooke, S., & Fox, S. W. (1981). Compartmentalisation in proteinoid microspheres. *BioSystems, 9,* 1–22.
28. Ishima, Y., Przybylski, A. T., & Fox, S. W. (1981). Electrical membrane phenomena in spherules from proteinoids and lecithin. *Biosystems, 13,* 243–251.
29. Ovchinnikova, K., & Pollack, G. H. (2009). Can water store charge? *Langmuir: The ACS Journal of Surfaces and Colloids, 25*(1), 542.
30. Del Giudice, E., et al. (2005). Coherent quantum electrodynamics in living matter. *Electromagnetic Biology and Medicine, 24,* 199–210.
31. Del Giudice, E., Spinetti, P. R., & Tedeschi, A. (2010). Water dynamics at the root of metamorphosis in living organisms. *Water, 2,* 566–586.
32. Arani, R., Bono, I., Giudice, E. D., & Preparata, G. (1995). QED coherence and the thermodynamics of water. *International Journal of Modern Physics B, 9,* 1813–1842.
33. Del Giudice, E., & Tedeschi, A. (2009). Water and autocatalysis in living matter. *Electromagnetic Biology and Medicine, 28,* 46–52.
34. Del Giudice, E., & Preparata, G. (1998). Electrodynamical like-charge attractions in metastable colloidal crystallites. *Modern Physics Letters B, 12,* 881–886.
35. Vitiello, G. (2009). Coherent states, fractals and brain waves. *New Mathematics and Natural Computation (NMNC), 5,* 245–264.
36. Scholes, G. D. (2010). Quantum-coherent electronic energy transfer: Did nature think of it first? *The Journal of Physical Chemistry Letters, 1,* 2–8.
37. Kajander, E. (2006). Nanobacteria—propagating calcifying nanoparticles. *Letters in Applied Microbiology, 42*(6), 549–552.
38. Martel, J., Peng, H. H., Young, D., Wu, C. Y., & Young, J. D. (2014). Of nanobacteria, nanoparticles, biofilms and their role in health and disease: Facts, fancy and future. *Nanomedicine, 9*(4), 483–499.
39. Young, J. D., & Martel, J. (2010). The rise and fall of nanobacteria. *Scientific American, 302* (1), 52–59.
40. Martel, J., Young, D., Peng, H.-H., Wu, C.-Y., & Young, J. D. (2012). Biomimetic properties of minerals and the search for life in the Martian meteorite ALH84001. *Annual Review of Earth and Planetary Science, 40,* 167–193.
41. Yaghobee, S., Bayani, M., Samiei, N., & Jahedmanesh, N. (2015). What are the nanobacteria? *Biotechnology & Biotechnological Equipment, 29*(5), 826–833.
42. Wu, C.-Y., Young, L., Young, D., Martel, J., & Young, J. D. (2013). Bions: A family of biomimetic mineralo-organic complexes derived from biological fluids. *PloS One, 8*(9), e75501.
43. García-Ruiz, J. M., Melero-García, E., & Hyde, S. T. (2009). Morphogenesis of self-assembled nanocrystalline materials of barium carbonate and silica. *Science, 323,* 362.
44. Lin, Y., Zheng, R., He, H., Du, H., & Lin, Y. (2009). Application of biomimetic mineralization: A prophylactic therapy for cracked teeth? *Medical Hypotheses, 73*(4), 493–494.
45. Reich, W. (1938). The bion experiments on the origin of life. *Amazon,* ISBN-13: 978-0374514464.
46. DeMeo, J. http://www.orgonelab.org/DeMeoBionsColor.pdf.
47. Elia, V., Yinnon, T. A., Oliva, R., Napoli, E., Germano, R., Bobba, F., et al. (2017). Chiral micron-sized H_2O aggregates in water: Circular dichroism of supramolecular H_2O architectures created by perturbing pure water. *Water, 8,* 1–29.

Author Biography

Professor Igor Jerman, Ph.D. He finished his studies in biology in 1980. In 1984 he defended his doctorate on biological holism. Afterwards, he undertook postdoctoral studies on biological structuralism and bioelectromagnetics with Prof. Goodwin at the Open University, Milton Keynes, UK. In 1987 he became an assistant professor of theoretical biology at the Biotechnical Faculty in Ljubljana, Slovenia, becoming a full professor in 1999. He is also one of the founders of the BION Institute where he acts as its professional and general director. His research encompasses theory of life and its origin, ordered states of water and measurements of ultra-weak emissions. He has a large scientific publication list both nationally and internationally.

Magnetic Resonance in Studying Cells, Biotechnology Dispersions, Fibers and Collagen Based Tissues for Biomedical Engineering

Victor V. Rodin

1 Introduction

Molecular, cellular and tissue engineering establishes the ways to improve the health of many people by restoring and maintaining cell and tissue functions [7, 13, 15, 23] as well as generating alternative tissues for reconstructive surgery. It is important to understand a mechanism of cellular interactions, a role of surface contacts and predict how cell growth, and how cell/tissue differentiation is affected by the environment [32, 34]. Cellular surroundings, both chemical and physical, can have strong effects on behavior, growth, differentiation and storage of cells [7, 32, 34, 37, 39, 54, 69]. The interdisciplinary approach of cell and tissue engineering provides a real basis for the investigation and fabrication of new biomedical devices with a broader perspective on quantifying efficacy and establishing clinical applications [7, 13, 23, 34, 37, 69]. Studying physical factors or creating special environmental conditions for the cells and tissues became very important.

Many physical methods have been applied and developed to restore, maintain, and to supply the cell and tissue functions. Nuclear magnetic resonance (NMR) methods are also applied in this area [12, 15, 17, 20, 21, 26, 36, 38, 39, 42, 53, 54]. It is important to understand the NMR achievements, apply those for biomedical engineering and estimate development further to unveil the details of recovering cell and tissue functions [21, 38, 39, 42, 54, 69]. Nowadays the NMR methods study the processes inside living cells [12, 17, 26]. The techniques clarify the intracellular protein-protein interactions responsible for most biological functions [7, 26, 39, 54]. NMR now can analyze living prokaryotic cells in details [12, 17, 26, 54].

V. V. Rodin (✉)
Institute of Organic Chemistry, Johannes Kepler University Linz,
Altenbergerstraße 69, 4040 Linz, Austria
e-mail: victor.rodin@jku.at

G. M. Artmann et al. (eds.), *Biological, Physical and Technical Basics of Cell Engineering*, https://doi.org/10.1007/978-981-10-7904-7_15

This chapter discusses mostly those NMR methods, which give the information about translational dynamics of molecules and macromolecules in complex systems including natural porous biomaterials. NMR relaxation, pulse field gradient (PFG) NMR, NMR imaging and multiple-quantum-filtered (MQF) NMR techniques are studying normal and disease cells and tissues, regulation of cell behaviour and fermentation processes resulting in special cell/tissue properties based on dynamics of molecules and diffusion characteristics [10, 16, 31, 38, 39, 41, 42, 51, 57]. Giving non-invasive characterization of engineered connective tissues, these tools consider a possibility to tabulate differences in the MR properties of tissue-engineered constructions [16, 21, 38, 57].

PFG NMR, NMR relaxation and MQF NMR methods are potential techniques in a study of molecular dynamics and diffusion properties in the anisotropic systems [31, 42]. An anisotropic molecular tumbling does not average all dipolar interactions to zero. The residual proton-proton dipolar interactions (RDIs) between macromolecular protons and water protons produce second-rank tensor that is responsible for the observed double-quantum-filtered (DQF) spectrum [16, 42, 51, 57]. There is no DQF signal in isotropic systems. The DQF spectra of connective tissues may represent a sum of spectra arising from a number of different non-interacting sites. If to consider local residual interaction at each site and orientation of local symmetry axis relative to the external field, this gives effective characterization of the total water motion in each site. DQF technique collects the information about local and macroscopic order in the systems with effective anisotropic motion of molecules [10, 31]. Proton and sodium MQF spectroscopy and proton spin-lattice (T_1) and spin-spin (T_2) relaxation times were used to study the chondrogenesis and osteogenesis of collagen contained tissues [21]. The apparent diffusion coefficient (D_{app}) was also applied in studying tissue engineering materials [20, 21, 38]. PFG experiment can be carried out at different diffusion times or applying the gradients in two orthogonal directions to natural-oriented tissues. This can clarify anisotropy and restriction diffusion and estimate the restricted distance and permeability coefficient [42, 49]. D_{app} values and NMR relaxation parameters (T_1, T_2) correlate well with the growth of engineered tissues and help to characterize the scaffolds [21]. MQF spectroscopy shows that the tissue-engineered cartilage has no order or preference in collagen orientation [21].

The chapter considers also two-dimensional correlation NMR spectroscopy as the distributions of diffusion coefficients in two orthogonal directions in the anisotropic systems. These 2D NMR methods can reveal the correlation of the diffusion motion of molecules along either collinear or orthogonal directions of applied pulse gradients of magnetic field. The approach is promising in studying tissues with anisotropic structure revealing microscopic local anisotropy in the presence of global isotropy.

2 Theory and Methods. Main NMR Techniques and Experiments

In order to apply MR parameters for biomedical engineering properly, more MRS/MRI experiments with different combination of scaffolds and cells and understanding the details of measurements are needed [20, 21, 38]. The efforts to develop NMR tools to the tissue characterization are depending on how cell and tissue engineers are involved in MR experiments, as well as how deep is their understanding the basics of MR methods. As discussed in [21] and functional MRI/MRS for regenerative medicine meetings, the collaboration between MR community and regenerative medicine community is important for successful transfer of biomedical engineering achievements to practice. Therefore additional explanation of MR parameters and factors affected those should be presented in more details.

This section introduces NMR methods and explains basic experimental approaches. Some models and experimental techniques are described to help in the realisation and interpretation of NMR relaxation and NMR diffusion experiments in one- and two-dimensions. A technique of double-quantum-filtered (DQF) NMR is presented for the study of molecular anisotropic motion.

The magnetic properties of atomic nuclei, basic principles of NMR phenomenon and NMR applications in various fields are described in publications [1, 4–6, 11, 19, 40, 42]. Electrons, protons and neutrons in atoms can be imagined as spinning on their axes. In some atoms for example, ^{12}C these spins are paired against each other and the nucleus of the atom has no overall spin. However, in other atoms, such as ^{1}H, ^{13}C, and ^{31}P, the nucleus does possess an overall spin. The nuclei with spin experience NMR phenomenon, which occurs when the nuclei with nonzero spin are placed in a static magnetic field and a second oscillating magnetic field is orthogonally applied [1, 4, 11, 19].

According to quantum mechanics, this nuclear spin is characterised by a nuclear spin quantum number, I [1, 5, 6, 40]. A nucleus of spin I will have $2I + 1$ possible orientations in magnetic field B_0. For a spin-half nucleus ($I = 1/2$) an interaction with a magnetic field results in two energy levels [1, 5, 19, 40]. When external magnetic field $B_0 = 0$ these orientations are of equal energy. In the case of nonzero B_0, the energy levels split. Each level is given by a magnetic quantum number, m, which is restricted to the values from $-I$ to I in integer steps. Thus, for a spin-half nucleus, there are only two values of magnetic quantum number m, $+1/2$ and $-1/2$ [11, 19, 40].

The energy state with $m = +1/2$ is denoted as α and is characterized with the lowest energy. This state is often described as 'spin up' notation. The state with $m = -1/2$ is denoted β, and it is described as 'spin down' notation. The m-values ($+1/2$, $-1/2$) express the parallel and antiparallel orientations of the nuclear spin with respect to the static magnetic field. The details of energy levels for two or more spins in molecule can be found elsewhere [4, 11, 19]. The initial populations of the energy levels of the nucleus in a magnetic field are determined by thermodynamics according to the *Boltzmann* distribution. The lower energy level will contain slightly more nuclei than the higher level when nuclei are in the state of equilibrium. With action of an electromagnetic radiation, it is possible to excite the nuclei from the low level into the higher level. The difference in energy between the energy levels $\Delta E = E_{\beta} - E_{\alpha}$ determines the frequency of radiation, which is needed for this excitation. When positive charged nucleus is spinning, this generates a small magnetic field. The nucleus therefore possesses a magnetic moment μ. This magnetic moment is proportional to its spin I, Planck's constant h, and the constant γ, which is called the *gyromagnetic ratio* [1, 19]. γ for nucleus is a ratio of magnetic dipole moment to its angular momentum. γ is a fundamental nuclear constant which has a different value for every nucleus [4, 40]. The energy E_m of each level m is proportional to the strength of the magnetic field at the nucleus B_0, magnetic quantum number m and γh. The transition energy ΔE will be also proportional to B_0. When the static magnetic field B_0 is increased, the transition energy increases too. It is possible to imagine a nucleus ($I = 1/2$) in a magnetic field. This nucleus is in a base state. It is at the lower energy level and its magnetic moment does not oppose the applied field. The nucleus is spinning on its axis. In the presence of a magnetic field, this axis of rotation will precess around the magnetic field. The frequency of precession is called the *Larmor* frequency. The *Larmor* frequency is identical to the transition frequency. The angle of precession, φ, (the angle between the direction of the applied field and the axis of nuclear rotation) will change when energy is absorbed by the nucleus. The absorption of energy leads to a higher energy state. Due to the difference in population of levels in the state of equilibrium, there is a macroscopic magnetization M. It is oriented along the direction of the static magnetic field B_0 (Z-axis).

NMR-studies can be done in time-domain (for example, studying translational dynamics) and in frequency-domain using one-dimensional and two-dimensional NMR techniques (for example, studying structure of proteins) [5, 19, 40]. Fourier transformation is used to do transfer between time-domain and frequency-domain. The Fourier transform (FT) is a mathematical technique for converting data of time-domain to data in frequency-domain (giving a spectrum), and an inverse Fourier transform (IFT) converts data from the frequency-domain to the time-domain.

The alternative magnetic field $B_1(t) = B_{1m}.\cos(\omega t)$ is applied perpendicularly to constant magnetic field B_0. Behaviour of the spin system in crossing magnetic fields is considered on the base of movement of magnetization vector M. Basic equations for the vector of the bulk magnetization M in outer magnetic field are the differential *Bloch* equations with the spin-lattice and spin-spin relaxation terms [5, 11, 19, 40]. These equations are much easier in a rotating coordinate system $X'Y'Z'$, which rotates with frequency ω around B_0 (Z, Z'-axes) in the direction of nuclear precession [11, 40]. In rotation frame, vector M rotates around magnetic field B_1 (X'-axis) with angle frequency $\omega = \gamma B_1$. In the absence of magnetic field B_1, magnetization M is along the direction of outer magnetic field B_0. The angle θ for rotation of magnetization M during precession time t_p is presented as $\theta = \gamma B_1 t_p$ [11]. During time t_p, if vector M turns by $\theta = 90°$, then this rotation is called as a 90°- or

$\pi/2$-pulse (duration time t_p is called the pulse length). When $\theta = 180°$ then this rotation is called as a π-pulse.

An application of $\pi/2$-pulse to the magnetization M (using magnetic field B_1) results in vector M being along the Y'-axis and the intensity of measurable signal (along Y'-axis) has maximal value. During the time (because of relaxation processes) a projection of magnetization vector M to the Y'-axis will decrease. This detected signal is called as a *Free induction decay* (FID).

The signal intensity depends on the population difference between the two energy levels considered. The system is irradiated with a frequency, whose energy is equal to the difference in energy between these two energy levels. Upward transitions absorb energy and downward transitions release energy. The probability for transitions in either direction is the same. The number of transitions in either direction is determined by multiplying the initial level population by a probability. The nuclei, which are in a lower energy state can absorb radiation. At absorption of radiation, a nucleus jumps to the higher energy state. When the energy level populations are the same, the number of transitions in either direction will also be the same. When there is no further absorption of radiation, the spin system will be saturated. However, a possibility of saturation means that the *relaxation processes* would occur to return nuclei to the lower energy state.

The sample in which the nuclei are held is called the *lattice*. All nuclei in the sample that are not observable are considered as *lattice*. Nuclei in the lattice are in vibrational and rotational motion, which creates a complex magnetic field. The magnetic field caused by motion of nuclei within the lattice is called the *lattice field*. The components of the lattice field, which are equal in frequency and phase to the *Larmor* frequency of the considered nuclei can interact with nuclei in the higher energy state, and cause them to lose energy and return back to the lower (equilibrium) state. The equilibrium is considered as a state of polarization with magnetization M_0 directed along the longitudinal magnetic field B_0. The restoration of the equilibrium is named longitudinal (spin-lattice) relaxation.

The relaxation time constant, T_1 describes the average lifetime of nuclei in the higher energy state. T_1 is typically in the range of 0.1–20 s for protons in non-viscous liquids and other dielectric materials at room temperatures. A larger T_1 indicates a slower or more inefficient spin relaxation [1, 4, 11]. The efficiency of spin-lattice relaxation depends on factors that influence molecular movement in the lattice, such as viscosity and temperature.

The longitudinal relaxation times are often measured using the inversion recovery (IR) pulse sequence ($180° - t - 90°$) at repetition time $\geq 5T_1$ [40, 42, 47, 48, 52]. The period of $5T_1$ is waited in order to ensure that the fully recovered signal was acquired. When t is varied then experimental dependence of magnetization $M_t = f\ (t)$ can be obtained and fitted by the law $M_t = \sum M_{0i} \cdot [1 - 2 \cdot \exp(-t/T_{1i})]$, where i is the number of relaxation component. An aim is to model experimental data as a sum of several relaxation components or distribution of NMR-relaxation times. To fit the data inverse Laplace transform (ILT) in one direction is applied [42, 53]. The kernel function in ILT is associated with M_t data from IR NMR experiments. Figure 1 shows T_1 distributions obtained by one-dimensional ILT on

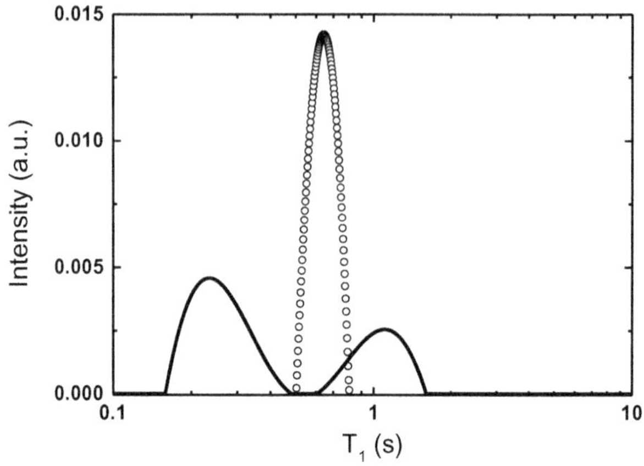

Fig. 1 T_1 distribution of ^1H in the suspension of *E. coli* with concentration of 0.07 g dry mass/g H_2O (open circles) and in the same sample with xenon at 1.2 MPa (solid line) [53]. T = 280 K, ν = 90 MHz. All intensities have been normalized per maximal signal

the data sets of IR NMR experiments for bacterial suspension of *E. coli* and *E. coli* under xenon conditions. In original suspension, an exchange between intra- and extracellular water is too fast on NMR scale and gives one T_1 component. The formation of xenon hydrates affected exchange dynamics of water between intra- and extracellular water compartments resulted in apparent two T_1 components/ peaks.

An interaction between neighbouring nuclei with identical frequencies but differing magnetic quantum states is described by spin-spin (transverse) relaxation. These nuclei can exchange quantum states. A nucleus in the lower energy level will be excited, while the excited nucleus relaxes to the lower energy state. The populations of the energy states do not change, but the average lifetime of a nucleus in the excited state will decrease. This can result in broadening resonance lines. More details are presented in publications [1, 4, 11, 19, 40]. Transverse relaxation is characterized by time constant T_2. Spin-spin relaxation is a process whereby nuclear spins come to thermal equilibrium between themselves. At room temperatures, the T_2 values for different materials are usually in the range from 10 μs to 10 s. It is always true for $T_2 \leq T_1$. Transverse relaxation is unlike the longitudinal relaxation. It is sensitive to the interaction, which results in dephasing nuclear spins [39, 42].

For measurement of T_2, it is possible to use *Han* pulse sequence ($90° - \tau - 180°$). This sequence creates a signal of spin echo at $t = 2\tau$ [11, 19, 40]. When interval 2τ is varied, an experimental dependence $M_{2\tau} = f(2\tau)$ can be obtained and modelled to calculate the T_2 [42]. The modification of *Han* method is *Carr-Purcell* (CP) pulse sequence ($90° - \tau_{cp} - 180° - 2\tau_{cp} - 180° - 2\tau_{cp} - 180° - \ldots$) which contains π-pulses train [11]. The *Carr-Purcell-Meiboom-Gill* (CPMG) sequence is the modification of CP-method: uses phase shift of π-pulses (by 90°) with respect to initial 90° pulse [40, 52]. CPMG decays are modelled by a sum of relaxation components [39, 43]. Alternatively, a distribution of T_2 components, which could reconstruct spin-echo decay from CPMG experiment, should be found. To extract the function $f(T_2)$, the inverse Laplace transform is used. The probability density $f(T_2)$ is calculated from the Eq. (1) for the spin-echo signal M_t presented according to Refs. [2, 6]:

$$M_t = \int_0^{+\infty} \mathrm{e}^{-t/T_2} f(T_2) dT_2 \quad (1)$$

Biological tissue represents a class of porous materials with complex internal structure. The physical structure is critical to many aspects of the physiological functions of tissues additionally to their biological and chemical properties [10, 20, 31, 41, 57]. For a material considered as porous media [29, 56], T_2 is found to be enhanced by surface relaxation. It has been used to measure pore sizes [56, 61]. Some examples on the geometrical properties of the porous bone parameter for osteoporosis and other bone metabolic diseases are presented in publications [20, 38]. Reference [38] considered the ratio T_2/T_1 as an indication of the water environment: the smaller ratio for osteogenic tissues is a sign of complex or more solid osteogenic constructs when compared to the chondrogenic constructs.

More examples are on collagen tissues [42], where T_1 relaxation was studied. The T_2 results provided additional information on relaxation mechanisms and the properties of bound water in biomaterials. They were analyzed in terms of water distribution models in natural biopolymers [3, 14, 16, 42, 52, 63]. Hydration water is essential for the function of the biomolecules, folding and secondary structures, dynamics and hydrophobic interactions [14, 34, 69]. The molecules in the hydration layers exhibit much slower dynamics [3, 52]. The ^{1}H T_2 relaxation time of water might be used to monitor the healing process of collagen structures of the tendon non-invasively [63].

Mostly biological systems exhibit heterogeneity [16, 20, 29, 42]. Therefore, one expects a distribution of relaxation times. An accurate characterization of the true relaxation time distributions in tissue are still open because independent assessment of this relaxation time distribution and tools for that are developing [14, 16].

Water in a biological tissue exists in a variety of different environments with an exchange occurring between some of them [3, 61, 63]. For clarification of the exchange details, it is possible to carry out the $T_2 - T_2$ correlation experiment resulting in 2D map of T_2 distribution [6, 42]. To produce 2D $T_2 - T_2$ map, two-dimensional Laplace inversion is used. Figure 2 presents the T_2-store-T_2 2D ^{1}H NMR spectrum of intra pore water within collagen tissue and discovers three environments as diagonal spots. Natural collagens and elastin make up much native extracellular matrix (ECM) of organisms. The ECM provides structure of the tissues. It also communicates with the cellular components and regulates daily cellular processes and wound healing. Collagens, elastin and silk fibroin are used in the creation of bioactive ECM analogues in tissue engineering [55, 68].

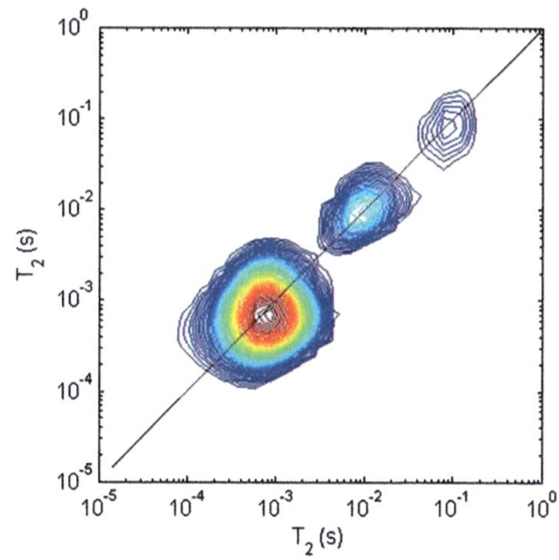

Fig. 2 Experimental T_2-store-T_2 2D correlation spectrum for collagen of the aponeurosis of 1.25 years old calf. The storage time is 20 ms. Frequency ν = 20 MHz. T = 298 K. HL = 0.51 g H_2O/g dry mass

Scaffolds manufactured from collagen materials promise better clinical functionality in tissue engineering application [68]. Understanding the conditions of an exchange between different environments in tissue engineered collagen is important.

As tissues grow, morphogenesis alters their physical and chemical properties. Changes in tissue hydration, macromolecular content, or water/lipid ratios are reflected in MR parameters. Measuring FIDs can result in estimation of hydration/dehydration of tissue-engineered materials and characteristics of scaffolds. Figure 3 shows FID for *B. mori* silk [42, 48] with fitting the data to the equation characterizing summary signal of silk and water protons. This gives an estimation of MR parameters for calculation of second moment M_2 and correlation times of molecular motion. Silk has remarkable strength and excellent biocompatibility. Silks have been used in biomaterials based on using the natural fibers or silk fibroin (SF) solutions [24, 25, 58, 60, 70]. Silk-based materials have been reported to

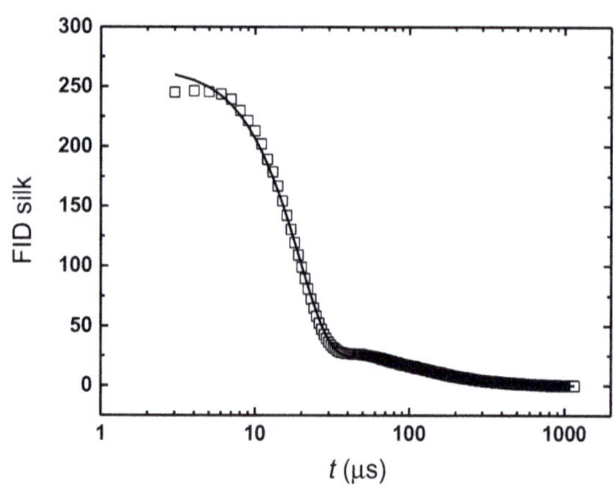

Fig. 3 ^{1}H FID in *B. mori* silk (open squares) at 0.07 g H_2O/g dry matter, T = 298 K, ν = 20 MHz. Solid line is the fit to an equation composed of an exponential function (water protons) and a Gaussian-sinc function (silk protons) [42, 52]. $S_{silk} = S_{s0} \times \exp(-a^2t^2/2) \times \mathrm{sinc}(bt)$, where S_{s0} is the intensity at $t = 0$ and a, b are the parameters used in the equation: $M_2 = a^2 + b^2/3$ [40, 43]

support attachement and differentiation of various cell types including mesenchymal stem cells [25, 55, 58, 60]. They have been also utilized for engineering of ligaments, bone, cartilage, tendon [24, 55, 58, 60, 70].

A diffusion coefficient is targeted MR parameter for correlation with the growth of engineering tissues [20, 21, 38]. PFG studies can be then realized in one- and two-dimensions. In one-dimension case, the three pulse stimulated-echo (STE) sequence [40, 42, 49] (Fig. 4, *top*) is used with storage times τ_2 and encoding times $\Delta = \tau_1 + \tau_2$ (where τ_1 is the gap between the 1st and 2nd and τ_2 is the interval between the 2nd and 3rd 90°-RF pulses). When the PFG technique is applied to heterogeneous systems, it is often necessary to take into account the existence of internal magnetic field gradients because of the heterogeneities in magnetic susceptibility of the sample [8, 42, 49]. The effects of internal magnetic field susceptibility gradients can be negated, for example, by the *Cotts* sequences [8], which introduce bipolar gradient pairs separated by additional 180°-pulses in place of each gradient pulse of the STE sequence. For example, the '13-intervals

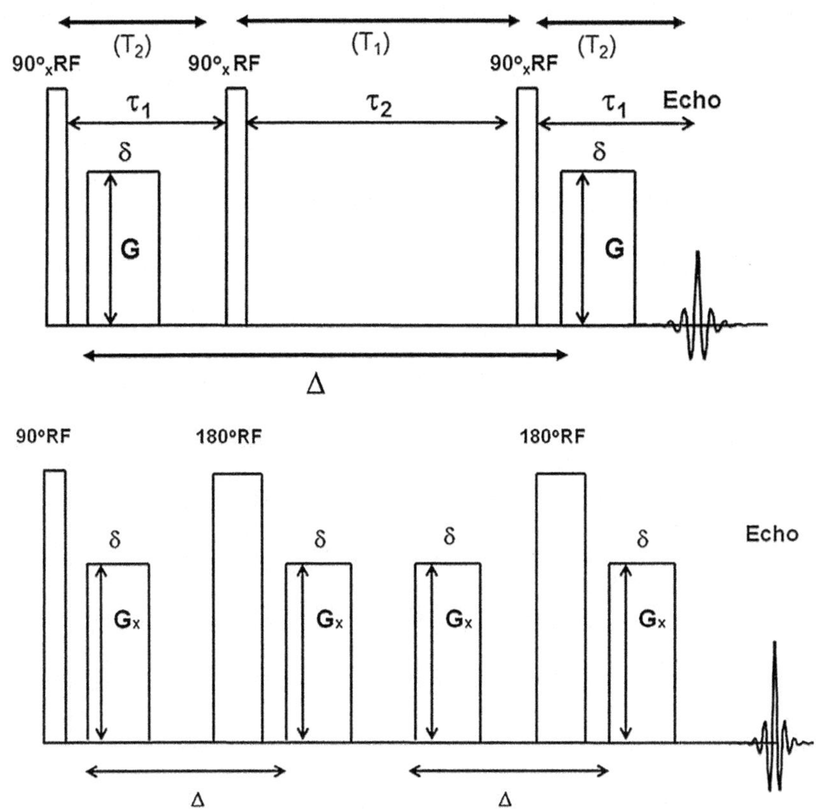

Fig. 4 One-dimensional PFG STE experiment *(top)*; 2D DDCOSY pulse sequence with two pairs of collinear gradient pulses G_x *(bottom)*

sequence' of *Cotts* [8] using a bipolar pair of 1 ms gradient pulses in place of each 2 ms pulse has been implemented in collagen studies [49] to compare the results of STE and *Cotts* pulse sequence on the same samples.

If the NMR echo intensity is recorded in presence of a pair of field gradient pulses of strength G, duration δ and separation Δ then it is smaller than the echo intensity in the absence of the gradients because of diffusive attenuation. According to the published works, the amplitude of echo depends on the gradient factor $b = q^2(\Delta - \delta/3)$ and the echo time $(\tau_2 + 2\tau_1)$ as: $I(\tau_1, \tau_2, b)/I(0) = \exp(-\tau_2/T_1) \times \exp(-2\tau_1/T_2) \times \exp(-bD)$ [5, 42, 43, 49]. For molecules diffusing in liquids the additional echo attenuation $I(G)/I^*(0)$ is given in Eq. (2), where $q = \gamma G\delta$ and γ is the (^{1}H) nuclear *gyromagnetic ratio.*

$$\frac{I(G)}{I*(0)} = \exp\left[-q^2\left(\Delta - \frac{\delta}{3}\right)D\right] \tag{2}$$

In the case of restricted diffusion in the system [8, 64, 65], the diffusion coefficient D in Eq. (2) should be replaced by an apparent diffusion coefficient D_{app}. It is sometimes considered that the total water signal in diffusion studies has a small portion of the water signal, which is contributed by the water molecules that are bound to macromolecules by strong hydrogen bonding. The diffusion of this bound water is so slow relative to the diffusion time Δ. Therefore, one considers that its signal is a constant in diffusion experiment [5, 49].

In STE diffusion experiment, longitudinal relaxation has a place during diffusion interval between the 2nd and 3rd RF pulses (Fig. 4, *top*). Because of T_1 could be affected by cross-relaxation, STE diffusion experiment (for example, in collagen tissue or *B. mori* silk fibers [42, 49]) can also experience this phenomenon. Cross-relaxation is driven by the difference between the deviations from equilibrium in both phases (biopolymer and water), and this phenomenon is considered on the base of two phase model [48]. R_{1w} and R_{1c} are the rates of longitudinal (spin-lattice) relaxation in each phase (water protons and collagen protons, respectively). k_w and k_c are the cross-relaxation rates, $p_w = k_c/(k_w + k_c)$, $p_c = k_w/(k_w + k_c)$ are the populations of water and collagen, respectively. $p_w + p_c = 1$. The rate of cross-relaxation will have q-dependence and will consequently depend on gradient G. The equations, which include exchange terms in *Bloch* equations for Z-magnetization of both phases, are presented anywhere [4, 5, 19, 40, 42]. Additional experiments (e.g., based on Goldmann-Shen pulse sequence) could be carried out to find cross-relaxation parameters [42, 48, 52]. When the parameters k_w, R_{1w}, k_c and R_{1c} are known, the diffusion coefficient can be determined by a nonlinear regression analysis fitting the echo-attenuation curve to derived expression. The diffusion coefficients measured in collagen tissues or silk based scaffolds should be calculated taking into account the cross-relaxation factor [42, 49].

Two-dimensional diffusion-diffusion correlation experiments (DDCOSY) [6, 42, 49] comprising of two spin echo pulse sequences with two pairs of PFG were carried out with encoding times of 6 ms and gradients in the x-directions (Fig. 4, *bottom*) or in both x- and y-directions. Echo decays are inverted using the inverse

Laplace transform (ILT) algorithm according to [66]. To monitor the changes in molecular diffusion of water in normal and regenerating tendon properly, as well as in the tissue engineered materials, where anisotropic diffusion could be a characteristics of the tissue growth or healing process, simulations of 2D DDCOSY experiments for isotropic and anisotropic environments should be performed. Figure 5 shows the results of these simulations for the experiments with the collinear pairs and with the orthogonal pairs of magnetic field gradient pulses [49].

The fibers of natural silk and collagen, tendons, cartilage and other tissue-engineered materials with anisotropic motion of bound water have been studied by the MQF NMR techniques [10, 16, 20, 21, 31, 38, 41, 42, 51, 57]. Filtering the NMR signal with the aid of these techniques enables the observation of the contributions to the FID that originate exclusively from anisotropic media and the ordered structures [10, 21, 41, 51, 57, 59]. DQF spectroscopy is a sensitive method for depicting biological tissues, in which the residual dipolar interaction may be of local or macroscopic nature. The DQ coherences are formed as a result of interaction with the collagen fibers [16, 42, 51, 59]. A dependence of the DQF signal on the creation time and the value of creation time for a maximal DQF signal may be considered as the characteristics of each tissue [10, 57]. Therefore, comparing DQF signals from collagen fibers in normal or regenerating tissues, and in scaffolds or tissue-engineered materials would result in understanding DQF findings for different tissues [10, 31, 41, 57].

DQF spectra of water molecules were detected in oriented collagen fibers from the aponeurosis of young and adult cows [42, 51] indicating the residual proton-proton dipolar interaction and anisotropy of water molecules. The dependence of the DQF signal on the creation time (τ_{creat}) of double quantum coherence

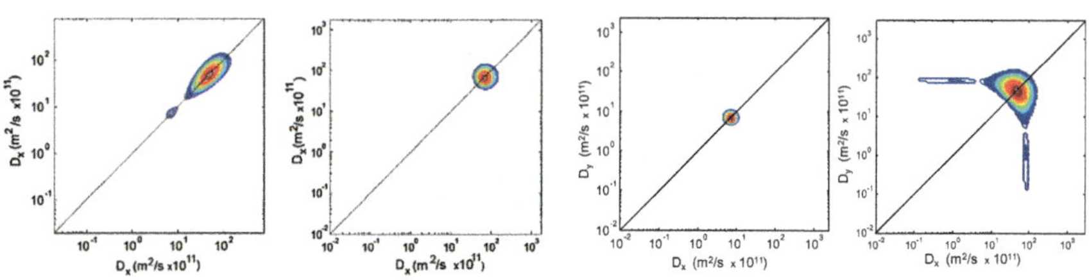

Fig. 5 Simulated 2D (D_1, D_2) spectra in modelled DDCOSY experiments with the collinear pairs (1st, 2nd *left*) and with the orthogonal pairs (1st, 2nd *right*) of the magnetic field gradient pulses: $D_2 = 5D_1 = 10^{-9}$ m^2/s (1st *left*), $D_2 = D_1 = 10^{-9}$ m^2/s (2nd *left*), $D_2 = D_1 = 10^{-10}$ m^2/s (2nd, *right*), $D_2 = 10D_1 = 10^{-9}$ m^2/s (1st, *right*). Diffusion anisotropy is characterised by two extended spots on the diagonal in model with collinear pairs (1st *left*) and by two off-diagonal "wings" and main spot on the diagonal in model with orthogonal pairs of gradient pulses (1st *right*). Isotropic diffusion is highlighted as one round spot on the diagonal (2nd *left*, 2nd *right*)

observed maximal signal at $(\tau_{creat})_{max}$ and gave a possibility to compare characteristics of anisotropic motion of water molecules in the collagen tissues of different types. The following pulse sequence: $90°_{\varphi} - (\tau_{creat}/2) - 180°_{\varphi 1} - (\tau_{creat}/2) - 90°_{\varphi} - t_1 - 90°_{\varphi 2} - t_2$ (acquisition) [31, 51, 59] has been used (Fig. 6).

As a result of the residual dipolar interaction, double rank spherical tensor operators ($T_{2,+1}$, $T_{2,-1}$) are created during the creation time τ_{creat} [31]. The 180°-pulse refocuses frequency shifts and magnetic field inhomogeneities. For description of the processes during this pulse sequence (Fig. 6) irreducible spherical tensor operator basis $T_{l,p}$ (l is a rank of the tensor; p is a coherence of the tensor with $-l \leq p \leq l$) is used [10, 51]. Residual dipolar interactions result in the formation of tensors of high rank ($T_{l,\pm 1}$, $l = 2, \ldots$) which can be detected by MQF NMR methods [10, 16, 41, 51, 57]. These tensors are presented as combination of known spin operators I_z and $I_{\pm}=I_x \pm iI_y$ [31]. For example, for two ½ spins, spherical tensor $T_{1,0}$ is presented as $I_z^a + I_z^b$, tensors $T_{1,\pm 1}$ as $-(\pm 2^{-1/2})\cdot(I_{\pm}^a+I_{\pm}^b)$, tensors $T_{2,\pm 1}$ as $-(\pm ½)\cdot(I_z^a I_{\pm}^b+I_{\pm}^a I_z^b)$, and tensors $T_{2,\pm 2}$ as $½\cdot(I_{\pm}^a I_{\pm}^b)$.

The second 90°-pulse transforms the second rank tensors into double quantum coherence tensors ($T_{2,+2}$, $T_{2,-2}$) which are selected using cycling of the phase, φ. The third 90° pulse transforms the coherence back to -1, so it may be detected during acquisition. The detected DQ signal of water results from anisotropic molecular motion. The DQF NMR signals were measured by varying the DQ evolution time, t_1 (from 10 μs to 5 ms). The intensity of DQF NMR signal

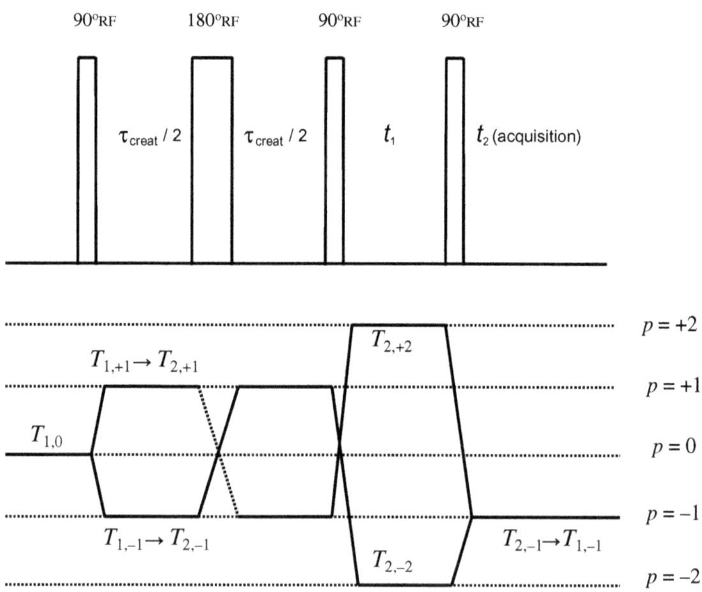

Fig. 6 DQF pulse sequence and the coherence-transfer pathway diagram. $T_{l,p}$ are spherical tensors of rank l and coherence p. Each pathway is chosen by the corresponding phase cycle [16, 31, 51, 57]

decreased with increase of evolution time (at fixed τ_{creat}) [42, 51]. The DQF evolution time t_1 was kept as short as possible to avoid lost of signal due to relaxation in experiments with change of creation time. Typical DQF experiment with pulse sequence of Fig. 6 had phase cycle according to published data [16, 51, 57]. The resulting DQF signal for detectable transverse magnetization is given by [27, 35]:

$$M(\tau_{creat}, t_1, t_2) = 0.15[\exp(-s_2\tau_{creat}) - \exp(-s_1\tau_{creat})]\exp(-s_{dq}t_1)[\exp(-s_2t_2) - \exp(-s_1t_1)] \tag{3}$$

In Eq. (3) s_{dq} is DQ transverse relaxation rate, and s_1, s_2 are two single-quantum transverse relaxation rates. At constant values of t_1 and t_2, the detectable transverse magnetization signal was dependent only on creation time and relaxation rate constants s_1 and s_2 [27, 57] according to Eq. (4):

$$M_{xy} = A[\exp(-s_2\tau_{creat}) - \exp(-s_1\tau_{creat})] \tag{4}$$

The time dependent free precession signal in DQF experiments is the difference between two exponentials rather than the weighted sum obtained by single-pulse acquisition. The corresponding absorption spectrum is the difference between two *Lorentzians* with different widths [31, 51, 52]. The relative line widths of the broad and narrow components depend on the ratio of the two relaxation rates. For example, the broad component may not be detected if the contribution of narrow resonance dominates ($s_1 >> s_2$) [27]. Results from experiments at series of τ_{creat} values therefore enable calculation of the relaxation rate constants s_1 and s_2 [27, 35] using nonlinear least-squares fit of the DQF signal dependence on the creation time. The τ_{creat} values, which result in maximal intensity of DQF сигналов in the dependences of intensity $I_{DQF} = f(\tau_{creat})$, can be then calculated as $\tau_{max} = \ln(s_1/s_2)/(s_1 - s_2)$ [42, 51].

3 Magnetic Resonance in Studying Cells and Nanoparticles in Biotechnology

When cells recover their function after damage/disease, many processes have a place. NMR methods are used to study the processes inside living cells [17, 26, 54]. In-cell NMR spectroscopy is a multidimensional NMR study of macromolecules, in particular, nucleic acids and proteins in living cells. The aim is to analyze their structures, functions, and cell membrane properties [12, 39, 54]. In order to selectively derive signals of macromolecules of interest from a large number of other signals of cell the techniques of isotope-aided solution NMR spectroscopy are used.

The most intensively investigated prokaryotic model organism is the *Escherichia coli* (*E. coli*), K-12 strain [62]. It is considered as very important representative of cellular suspensions in biotechnology. *E. coli* is a common gram-negative rod-shaped bacterium, found in normal human (and other animals) bacterial flora. It is about 2.0 μm in length and its diameter is 0.25–1.0 μm [62]. These microorganisms are facultative anaerobic bacteria, which can be easily grown in a laboratory.

For the in-cell approach, bacterial suspensions of *Escherichia coli* have been used as "host cells" so far. Proteins to be analyzed are over-expressed in cells grown in isotope-enriched media, i.e., the proteins are semi-selectively labelled with NMR active nuclei such as ^{15}N and ^{13}C [26]. The whole cells are then NMR studied to detect only the signals of labeled molecules.

In a number of cases, the T_1 and T_2 relaxation times might inform about cell state and recovery of cell functions quite effectively. These are also informative about restoration or damage of bacterial membranes and their permeability as indicator to the factors of environmental changes [39, 53]. The relaxation rates of the water molecules in cell suspensions are governed by several important factors, in particular, the strength of local magnetic interactions between water nuclei, the molecular motion and proton exchange rates. Nuclear magnetic inter- and intramolecular dipole-dipole coupling describes the interactions between water nuclei [5, 6, 19, 40]. The magnetic interactions are partially averaged within the hydration layer [67] by the processes depending on the interactions of water with macromolecules/cell surface. These are proton transfer, dynamic orientation and diffusion of water molecules through regions of different orientations [42, 53]. The NMR relaxation times are sensitive to molecular motions in the range of 10^{-8} to 10^{-12} s [43, 47, 48, 52, 53, 67]. In liquid solutions, water molecules tumble at a correlation time of about 10^{-12} s [39]. This motion is considerably slowed down when water molecules interact with biological macromolecules/cell surface. For example, the correlation time is of the order of 10^{-12} s for the water molecules non-involved in the association through hydrogen bonds with surface of cells. The rotational motion of water molecules associated via hydrogen bonds with polar groups of the macromolecules is reduced so that their correlation time is even sometimes of the order of 10^{-6} s [42]. Under the conditions of rapid exchange between the hydration and bulk water in suspensions and macromolecular solutions, single relaxation rate ($1/T_1$) is observed [53]. Although two types of compartments are obviously presented in bacterial suspensions of *E. coli*, single exponential relaxation of protons in suspensions of *E. coli* (Fig. 1) confirms fast exchange between intracellular and intercellular water on the NMR time scale.

The permeability of the cell membrane to water is quite high (exchange time of the order of 10 ms [53]). Water proton relaxation times in *E. coli* bacteria showed one exponential behavior, i.e., two sites (two compartments: intracellular and extracellular spaces) are in a fast exchange limit. Thus, a researcher needs to reduce one of the relaxation times to less than exchange time in order to remove this condition and to generate two exponential behavior [39, 53]. In the presence of paramagnetic ions of manganese in the extracellular water shortening the relaxation times outside the cells, it was possible to record two separate NMR signals in the population of cells with non-damaged cellular membranes [50]. Another approach to affect the conditions of fast exchange limit in *E. coli* suspensions is associated with use of inert gas xenon in bacterial suspension (Fig. 1). This produces clathrate hydrates and changes the cell permeability resulting in registration of two different T_1 components (Fig. 1).

For the purposes of cell engineering, some works focus on advanced magnetic resonance spectroscopy and MRI methods to use those in metabolic studies [13, 15, 32, 69]. Metabolism is central to neuroimaging [15]. It can reveal pathways by which glial and neuronal cells use nutrients for their growth and function. MR spectroscopy on nuclei of ^{17}O and ^{31}P provide rates of oxygen use and ATP synthesis inside mitochondria whereas ^{19}F MRS performs measurement of cytosolic glucose metabolism. MRS on nuclei of ^{13}C is the method used to measure glutamatergic neurotransmission and cell-specific energetics for signaling and non-signaling purposes [15]. NMR methods are used for imaging brain metabolism. They demonstrate translational potential to understand brain disorders better and guide diagnosis and treatment.

Many biotechnological products are based on the correct realization of grown biomass. Lignocellulosic biomass is applied as a renewable resource for the production of fuels, chemicals, and materials [12]. In order to realize the potential of the biomass fully, an improved understanding of the chemical and molecular structures within biomass must be developed. NMR can carry out structural characterization of lignocellulosic biomass without causing structural modification [12]. This type of analysis results in enhancing plant-bioengineering actions.

Nanotechnology is rapidly growing area with broad classes of nanoparticles applied for different targets and destinations. Nanoparticles can be constructed of different sizes and shapes [9, 28, 46]. Some of their basic properties are currently used in biotechnology, others are still developed for future applications. Different materials can be presented in nanoparticles resulting in various chemical and surface properties [18, 30, 44, 45].

There are various classes of nanoparticles applied for drug delivery, gene transfection, semiconductor nanocrystals and other destinations in biotechnology. For example, some classes, which have been studied by NMR methods quite effectively [9, 28, 30, 44–46], are liquid crystals, liposomes, superparamagnetic nanoparticles and nanoshells. Currently they have lead to the development of many new areas of research, in particular, new knowledge about nanoparticle interactions within cells is now rapidly growing and giving new progresses in biotechnology for new nanoparticles and discovery of new applications [18, 28, 30].

Nuclear magnetic resonance gives many techniques to investigate the properties of liquid crystalline materials, in particular, molecular ordering, phase structure, and dynamics in complex anisotropic phases [9, 28]. Pharmaceuticals based on liquid crystals are composed of organic liquid crystal materials that mimic proteins and lipids, i.e., naturally-occurring biomolecules. They are considered quite safety for drug delivery and can target specific areas in the body. These are especially suitable in the cases when tissues are inflamed, or even when tumours are present.

The NMR data obtained for liquid crystals lead to the discoveries of interesting new soft materials made to address the fundamental physical and chemical phenomena. This is also developing the potential applications in sensor and display devices [9].

^{1}H and ^{31}P NMR spectroscopy was applied in studying liposomes [30, 44–46, 67], which are used effectively in the cosmetic and pharmaceutical industries. Liposomes are lipid-based liquid crystals [28]. Liposomes are the engineered nanoparticles used for drug delivery. NMR can provide the details of interaction of protectors with liposomes [44] as well as an evidence for involvement of surface group of bilayer, phosphoryl groups and metal cations into interaction of liposomes with polyA:polyU duplex [45, 46]. The lipid bilayer of hydrated biological membranes is characterized by a liquid-crystalline state, which is highly dynamic one. If cooling or dehydration is applied to liposomes, the membranes undergo a cooperative transition to a more-ordered, gel phase [28]. This characteristic phase transition is very important in studies of freezing-tolerant organisms. The techniques of NMR allow for the detection and characterization of the phase transitions over a wide temperature range. The study [28] reported how the freezing point depression affects the lipid phase transition, causing reduction in the lipids' melting temperature. NMR unveils this phenomenon studying membrane-associated water phases as well as cryoprotectants for membrane [44].

Superparamagnetic molecules are those that are attracted to a magnetic field. They do not retain residual magnetic properties after removal of the field. Nanoparticles of iron oxide with diameters in the 5–100 nm range have been used for selective magnetic bioseparations. Superparamagnetic iron oxide nanoparticles (SPION) are applied for drug delivery and gene transfection [33]. SPIONs are used as contrast agents for MRI of inflammatory processes. For the in vivo MRI detection of macrophage infiltration contrast agents are applied on the base of SPIONs [33].

Polymers are often used for covering nanoparticles in order to change surface properties [18]. The study [18] showed a novel approach to synthesize thiolated sub-100 nm organosilica nanoparticles from 3-mercaptopropyltrimethoxysilane (MPTS). The formation of MPTS nanoparticles proceeds through the condensation of methoxysilane groups and simultaneous disulfide bridging caused by partial oxidation of thiol groups. Excellent colloidal stability in dilute aqueous dispersions has been showed for these nanoparticles. However, at higher concentrations, these MPTS nanoparticles could further self-assembly into chains and necklaces [18]. They exhibited very good ability to adhere to ocular mucosal surfaces, i.e., could be applied in drug delivery. The thiolated nanoparticles can be easily modified through PEGylation resulting in a loss of their mucoadhesive characteristics [33]. PEGylated nanoparticles have attracted the attention of pharmaceutical researchers due to the numerous advantages of these systems for drug delivery. Possible location of exchangeable (and other) protons is inside the particles. Figure 7 shows the PFG STE NMR data for PEGylated nanoparticles in D_2O-H_2O at different diffusion times (big delta = 18–800 ms).

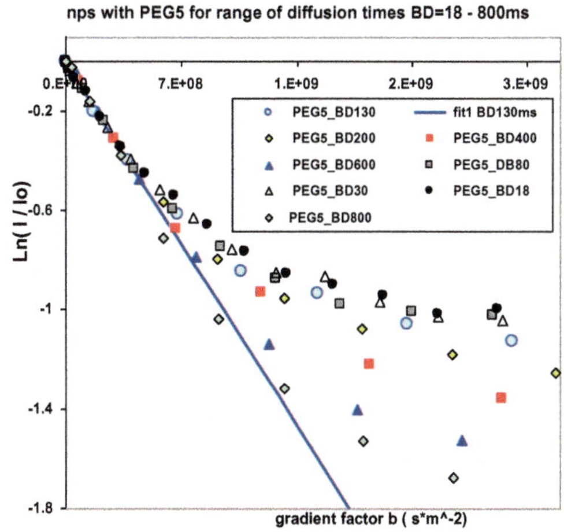

Fig. 7 Relative echo intensities (normalized to maximal signal) vs gradient factor $b = (\gamma G\delta)^2(\Delta - \delta/3)$ at varying diffusion times (18–800 ms) in dispersions of PEGylated nanoparticles (PEG 5000). The slope of echo curve at small b-values did not depend on diffusion time giving apparent diffusion coefficient of solvent (HDO): $D_{HDO} = 1.8 \times 10^{-9}$ m²/s. The rest (tail) signals (at large b-values) depended on diffusion time and characterised the protons of nanoparticles. T = 295 K, ν = 400 MHz

4 Magnetic Resonance in Studying Fibers and Collagen Based Tissues

The changes in water-biomolecules interactions in tissue-engineered materials are handled by the degree of cross-links in collagen nano-scaffolds. NMR details hydration properties of collagens with different cross-links, and water interactions in scaffolds. A description of water diffusion in collagen fibers with different cross-links should lead to understanding of internal interactions and knowledge of the structure of porous media. The apparent translation diffusion coefficients at two directions of applied gradient (along the direction of static magnetic field B_0 and perpendicular to B_0) have been studied in oriented collagen fibers (along the static magnetic field B_0) [42, 49]. The diffusion time dependence of D_{app} showed for both directions of applied gradient a restriction for motion of water molecules. A model of equally spaced plane parallel permeable barriers for NMR diffusion studies have been used to estimate the restricted distance and permeability coefficient [47, 49, 64, 65]. Diffusion anisotropy of water protons (D_{app}) was observed in two collagen tissues with different level of cross-links. One example of the analysis of NMR diffusion data is presented in Fig. 8 for collagen tissue of 8 years old steer (C8y, high level of cross-links).

The restricted distance and permeability coefficient could be monitored as important MR parameters during tissue growth and in collagen based scaffolds. In the studies [20, 21, 38], D_{app} has been measured, but there was no any information

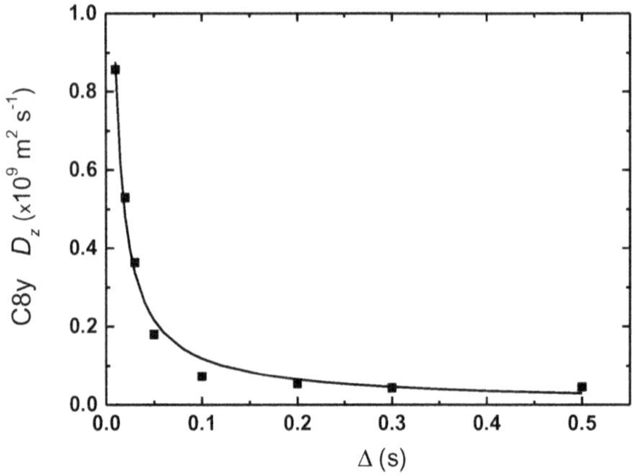

Fig. 8 Dependence of water diffusion coefficient D_{app} on the diffusion time for C8y fibers oriented along static magnetic field B_0 (*Z*-direction). Gradient was applied along *Z*-direction. D_{app} values included the cross-relaxation factor [42, 48, 49]. Solid line is the fit according to the model of equally spaced plane parallel permeable barriers developed by Tanner [64, 65] to estimate the size of restricted distance. $a = 9.89$ μm, and permeability coefficient $p = 0.00034$ cm s^{-1}. T = 298 K, ν = 400 MHz, HL = 0.6 g H_2O per g dry mass

about the details of the diffusion in tissue: cross-relaxation, restriction, anisotropy or permeability of porous media. These additional characteristics could clarify a lot of valuable information about tissue-engineered materials. Ref. [49] presented the results of 2D ILT applied to the data of 2D DDCOSY NMR experiment on the hydrated collagen fibers C8y. The data showed two peaks on the diagonal of 2D diffusion map, i.e., it was similar to the case of simulation studies at $D_2 \gg D_1$ (Fig. 5). These results (two separated peaks on the diagonal of the square) reflect diffusion anisotropy with different D_1 and D_2 diffusion coefficients. An agreement between the theory and the experiment was good testifying that 2D DDCOSY experiments on tissue-engineered materials can produce additional valuable information about diffusion anisotropy of water in collagen scaffolds. These approaches can potentially be used to diagnose and monitor treatment of collagen based tissues at diseases that involve macromolecular reorganization and associated changes in cross-relaxation and diffusion anisotropy. The use of 2D Laplace inversion algorithm in fast and efficient producing *D*-*D* correlation spectra enables to apply two-dimensional diffusion correlation spectroscopy for the study of molecular dynamics at collinear/orthogonal pairs of pulse gradients.

DQF NMR has been applied to study collagen tissues with different quantity of covalent intermolecular cross-links, which increase during the life of animal (15-months old calf and 8 years old steer) [51]. It was observed that intensities of DQF signals in fibers oriented along outer static magnetic field exceed significantly the DQF signals in nonoriented fibers. The registered ^{1}H DQF NMR signals were responsible for RDI between protons of water molecules interacting with oriented collagen fibers [10, 31, 41].

The DQF signals in oriented collagen fibers reached their maximum intensities at creation times of $(\tau_{creat})_{max}$ = 0.165 and 0.18 ms for C15m and C8y, respectively [42]. For randomly distributed collagen fibers, $(\tau_{creat})_{max}$ = 0.100 ms (C8y) and $(\tau_{creat})_{max}$ = 0.101 ms (C15m) [49]. According to the known results [10, 31, 41, 57], ^{1}H DQF NMR signals as a function of τ_{creat} are responsible for important NMR parameters such as the RDI (related to the degree of order in the sample) and the proton exchange rate. The velocity of the exchange diminishes with decrease of temperature [31]. The differences in the τ_{creat} dependence of the DQF signals could be observed for C8y and C15m collagen at decrease of temperature from 298 to 233 K [16, 51] (C15m data are in Fig. 9). The increase of DQF signal for these collagens with decreasing temperature to 233 K was in line with other DQF data obtained at changing temperature and depending on proton exchange rate [31, 42]. Fitting the shape line (at 233 K) as a difference of two *Lorentzians* also showed that the line width increased at decreasing temperature to 233 K and became ~7290 Hz (negative peak) and ~1730 Hz (positive peak) for C8y and ~3923 Hz (negative peak) and ~1214 Hz (positive peak) for C15m, respectively.

According to the DQF studies in tendon [16, 31], the line shape and the amplitude of the ^{1}H DQF spectra are affected by the ratio of the residual ^{1}H–^{1}H dipolar interactions and the proton exchange rate, that is, the intensity of DQF signal declines when the proton exchange rate increases. References [10, 31, 57] showed that for tendon, the RDI is temperature independent. The proton exchange rate changed by more than a factor of 3 at changing the temperature

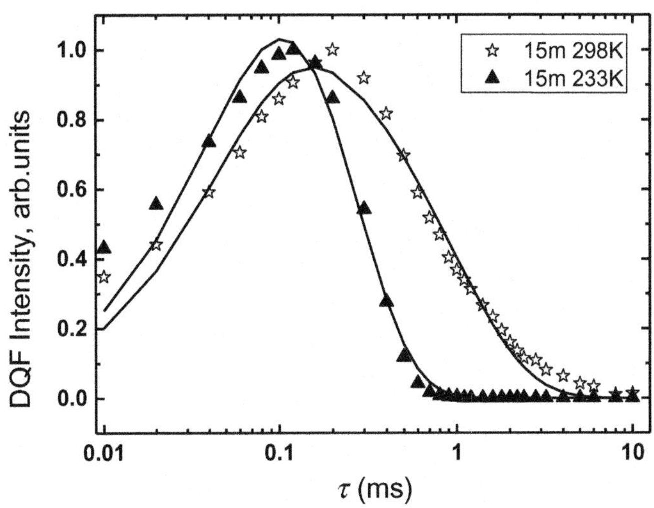

Fig. 9 ^{1}H DQF NMR signal intensity (normalized per maximum) versus the creation time for C15m collagen fibers oriented along the magnetic field B_0 at T = 298 K (*stars*) and T = 233 K (▲). ν = 400 MHz, HL = 0.6 g H_2O per g dry mass. Solid lines are fits to the Eq. (4) [27, 35]. Relaxation constants: s_1 = 19.92 ms^{-1}, s_2 = 1.08 ms^{-1} at T = 298 K (*stars*) and s_1 = 14.17 ms^{-1}, s_2 = 6.18 ms^{-1} at T = 233 K (▲) [42]

by 40° (from 313 to 274.5 K). In the DQF studies of C8y and C15m fibers [42, 51], the changes in temperature are larger (by 65°). Therefore, it is obvious that the decrease of the proton exchange rate was main impact in the increase of the DQF signal intensity at low temperature. Other additional reasons and the factors affecting the creation of DQ coherence could be clarified from the DQF relaxation experiments [characteristics of DQF signal with varying evolution time t_1 (Fig. 6)] [16, 31, 42] and from comparison of DQF results with the T_2 characterisation taken from normal CPMG measurements [4, 39, 40, 63]. Understanding relaxation processes with contributions from all/main sources as well as correct proton density measurements lead to proper analysis of MR parameters in tissue-engineered materials [10, 42, 59].

Reference [59] describes the different contributions to T_2 relaxation in the articulate cartilage emphasizing that ignoring the two exponentials can produce serious errors in the proton density. This study showed also how to slow T_2 relaxation, to minimize its angular dependence, and how to apply ^{2}H DQF MRI technique correctly to measure the density and orientation of collagen fibers in articular cartilage. The attachment of the cartilage to the bone has a stabilizing effect on the collagen matrix [59]. The collagen fibers in cartilage have different thickness. The largest one is near the bone. It decreases continuously towards the surface [59]. The bone supports the collagen fiber matrix. There is a calcified zone near the bone. Osteoarthritis is a disease, which involves the erosion of an articular cartilage [59]. ^{1}H and ^{2}H DQF NMR studies could provide essential progress in understanding the physicochemical origin of changes in cartilage properties and disruption of its major constituents: collagen and proteoglycans. This could improve tissue engineering process. ^{1}H DQF NMR was used for four weeks in bovine chondrocytes seeded in alginate beads and chondrocytes pellets [20]. The data showed that the average residual proton dipolar coupling (called sometimes as splitting [10, 31, 41, 57]) is lower in chondrocyte pellets ($\sim$20 Hz) and chondrocytes seeded in alginate beads ($\sim$70 Hz) as compared to the native cartilage ($\sim$few kHz). This dipolar coupling indicates the partial alignment of molecules leading to incomplete averaging. Thus, lower dipolar coupling in chondrocyte pellets indicates that the collagen molecules in engineered tissues does not have a preferred direction. They mostly exhibit a random isotropic spatial distribution [20].

In addition to DQF NMR/MRI and NMR relaxation methods (on nuclei ^{1}H and ^{2}H) the applications of sodium relaxation and sodium triple-quantum filter spectroscopy (TQF) for the study of tissue growth dynamics in three different cartilage tissue engineering constructs have been investigated [22]. This study showed that the motion of sodium ions and their relaxation rates were dependent on the amount of macromolecules, high cell density and the anisotropy of the cartilage tissue engineering constructs. It was found that average quadrupolar couplings were lower in the engineered tissue as compared to the native tissues. This is due to the lack of order in collagen accumulated in the engineered tissue [22]. Thus, the use of sodium triple-quantum filter spectroscopy provides a reliable tool to investigate anisotropy and growth dynamics of cartilage tissue engineered constructs.

5 Concluding Remarks

The chapter considered NMR&MRI methods as the sensitive characterization tools, which can provide different parameters related to changes in the cell/tissue microstructure. The MR techniques are available for different needs, different cells and tissues. These techniques are differed of many other methods that require sectioning, slicing and decalcification of the tissue. The NMR methods have the advantage that the physicochemical properties of the cells and tissues can be studied non-invasively on intact samples giving a characterization at conditions of growth and restoring/storage. The specialists in cell and tissue engineering need to pair up with NMR specialists to take and realise all advantages, which MR techniques can offer. Hopefully, the material of the chapter can establish a bridge between researchers specialised in some particular MR techniques and cell and tissue engineers. These collaborations can result in developing new tools, which can be extremely useful for quantification of extracellular matrix components and for the details of molecular dynamics. Those would be an added advantage for the biomedical engineering purposes.

References

1. Abragam, A. (1961). *The principles of nuclear magnetism* (p. 599). Oxford: Clarendon Press.
2. Berman, P., Levi, O., Parmet, Y., Saunders, M., & Wiesman, Z. (2013). Laplace inversion of low-resolution NMR relaxometry data using sparse representation methods. *Concepts in Magnetic Resonance, Part A, 42A*(3), 72–88.
3. Blinc, R., Rutar, V., & Zupancic, I. (1995). Proton NMR relaxation of adsorbed water in gelatine and collagen. *Applied Magnetic Resonance, 9*(2), 193–216.
4. Blümich, B. (2005). *Essential NMR (for scientists and engineers)*. Springer.
5. Callaghan, P. T. (2001). *Principles of nuclear magnetic resonance microscopy* (p. 492). Oxford: Clarendon Press.
6. Callaghan, P. T. (2011). *Translational dynamics and magnetic resonance* (p. 547). Oxford: Oxford University Press.
7. Castellanos, I., Balteanu, B., Singh, T., & Zderic, V. (2016). Therapeutic modulation of calcium dynamics using ultrasound and other energy-based techniques. *IEEE Reviews in Biomedical Engineering*. https://doi.org/10.1109/RBME.2016.2555760.
8. Cotts, R. M., Hoch, M., Sun, T., & Markert, J. T. (1989). Pulsed field gradient stimulated echo method for improved NMR diffusion measurements in heterogeneous system. *Journal of Magnetic Resonance, 83,* 252–266.
9. Dong, R. Y. (Ed.). (2010). *Nuclear magnetic resonance spectroscopy of liquid crystals*. World Scientific Publishing Co. Pte. Ltd, p. 464. ISBN: 978-981-4273-66-4.
10. Eliav, U., Komlosh, M., Basser, P. J., & Navon, G. (2012). Characterization and mapping of dipolar interactions within macromolecules in tissues using a combination of DQF, MT and UTE MRI. *NMR in Biomedicine, 25*(10), 1152–1159.
11. Farrar, T. C., & Becker E. D. (1971). *Pulse and fourier transform NMR: Introduction to theory and methods* (p. 118). Academic Press.

12. Foston, M., Samuel, R., He, J., & Ragauskas, A. J. (2016). A review of whole cell wall NMR by the direct-dissolution of biomass. *Green Chemistry, 18*(3), 608–621.
13. Gupta, V., Khan, Y., Berkland, C. J., Laurencin, C. T., & Detamore, M. (2017). Microsphere-based scaffolds in regenerative engineering. *Annual Review of Biomedical Engineering, 19*. https://doi.org/10.1146/annurev-bioeng-071516-044712.
14. Henkelman, R. M., Stanisz, G. J., Kim, J. K., & Bronskill, M. J. (1994). Anisotropy of NMR properties of tissues. *Magnetic Resonance in Medicine, 32*(5), 592–601.
15. Hyder, F., Rothman, D. L. (2017). Advances in imaging brain metabolism. *Annual Review of Biomedical Engineering, 19*. https://doi.org/10.1146/annurev-bioeng-071516-044450.
16. Ikoma, K., Takamiya, H., Kusaka, Y., & Yoshiteru Seo, Y. (2001). ^{1}H double-quantum filtered MR imaging of joints tissues: Bound water specific imaging of tendons, ligaments and cartilage. *Magnetic Resonance Imaging, 19,* 1287–1296.
17. Inomata, K., Ohno, A., Tochio, H., Isogai, S., Tenno, T., Nakase, I., et al. (2009). High-resolution multi-dimensional NMR spectroscopy of proteins in human cells. *Nature, 458,* 106–109. https://doi.org/10.1038/nature07839.
18. Irmukhametova, G. S., Mun, G. A., & Khutoryanskiy, V. V. (2011). Thiolated mucoadhesive and PEGylated nonmucoadhesive organosilica nanoparticles from 3-mercaptopropyltrimethoxysilane. *Langmuir, 27,* 9551–9556.
19. Keeler, J. (2010). *Understanding NMR spectroscopy* (2nd ed., p. 511). Wiley.
20. Kotecha, M., Klatt, D., & Magin, R. L. (2013). Monitoring cartilage tissue engineering using magnetic resonance spectroscopy, imaging, and elastography. *Tissue Engineering Part B, 19* (6), 470–484.
21. Kotecha, M., Yin, Z., Magin, R. L. (2013). Monitoring tissue engineering and regeneration by magnetic resonance imaging and spectroscopy. *Journal of Tissue Science and Engineering S11*. http://dx.doi.org/10.4172/2157-7552.S11-007.
22. Kotecha, M., Ravindran, S., Schmid, T. M., Vaidyanathan, A., George, A., Magin, R. L., et al. (2013). The application of sodium triple-quantum coherence NMR spectroscopy for the study of growth dynamics in cartilage tissue engineering. *NMR in Biomedicine, 26*(6), 709–717. https://doi.org/10.1002/nbm.2916.
23. Lakin, D. A., Snyder, B. D., & Grinstaff, M. W. (2017). Assessing cartilage biomechanical properties: Techniques for evaluating the functional performance of cartilage in health and disease. *Annual Review of Biomedical Engineering, 19*.
24. Liu, H., Ge, Z., Wang, Y., Toh, S. L., Sutthikhum, V., & Goh, J. C. H. (2007). Modification of sericin-free silk fibers for ligament tissue engineering applications. *Journal of Biomedical Materials Research Part B, 82B,* 129–138.
25. Liu, H., Fan, H., Wang, Y., Toh, S. L., & Goh, J. C. H. (2008). The interaction between a combined knitted silk scaffold and microporous silk sponge with human mesenchymal stem cells for ligament tissue engineering. *Biomaterials, 29,* 662–674.
26. Luchinat, E., Banci, L. (2017). In-cell NMR: A topical review. *IUCrJ, 4*, Part 2, 108–118.
27. Lyon, R. C., Pekar, J., Moonen, C. T. W., & McLaughlin, A. C. (1991). Double-quantum surface-coil NMR studies of sodium and potassium in the rat brain. *Magnetic Resonance in Medicine, 18,* 80–92.
28. Mandal, A., & van der Wel, P. C. (2016). MAS ^{1}H NMR probes freezing point depression of water and liquid-gel phase transitions in liposomes. *Biophysical Journal, 111*(9), 1965–1973. https://doi.org/10.1016/j.bpj.2016.09.027.
29. Menon, R. S., Rusinko, M. S., & Allen, P. S. (1991). Multiexponential proton relaxation in model cellular systems. *Magnetic Resonance in Medicine, 20,* 196–213.
30. Moniz, T., de Castro, B., Rangel, M., & Ivanova, G. (2016). NMR study of the interaction of fluorescent 3-hydroxy-4-pyridinone chelators with DMPC liposomes. *Physical Chemistry Chemical Physics: PCCP, 18*(6), 5027–5033. https://doi.org/10.1039/C5CP05273D.

31. Navon, G., Shinar, H., & Eliav, U. (2001). Multiquantum filters and order in tissues. *NMR in Biomedicine, 14*(2), 112–132.
32. Neumann, K.-H., Kumar, A., & Imani, J. (2009). *Plant cell and tissue culture—A tool in biotechnology, principles and practice*. Springer: Berlin, Heidelberg. Chapter 12: Cell division, cell growth, cell differentiation, p. 235.
33. Neuwelt, A., Sidhu, N., Hu, C. A., Mlady, G., Eberhardt, S. C., Sillerud, L. O., et al. (2015). Iron-based superparamagnetic nanoparticle contrast agents for MRI of infection and inflammation. *AJR. American Journal of Roentgenology, 204*(3), W302–W313. https://doi.org/10.2214/AJR.14.12733.
34. Nooeaid, P., Salih, V., Beier, J. P., & Boccaccini, A. R. (2012). Osteochondral tissue engineering: Scaffolds, stem cells and applications. *Journal of Cellular and Molecular Medicine, 16*(10), 2247–2270.
35. Payne, G. S., & Styles, P. (1991). Multiple-quantum-filtered ^{23}Na NMR spectroscopy in model systems. *Journal of Magnetic Resonance, 95,* 253–256.
36. Pellecchia, M., Bertini, I., Cowburn, D., Dalvit, D., Giralt, E., Jahnke, W., et al. (2008). Perspectives on NMR in drug discovery: A technique comes of age. *Nature Reviews Drug Discovery, 7,* 738–745. https://doi.org/10.1038/nrd2606.
37. Ponomarev, A., Makeev, O., Gurevich, L., & Rodin, V. V. (2016). Different xenon conditions for preservation of viable tissue in order to use in regenerative medicine. *European Cells and Materials, 31*(Suppl. 1), 382. ISSN 1473–2262.
38. Pothirajan, P., Ravindran, S., George, A., Magin, R. L., & Kotecha, M. (2014). Magnetic resonance spectroscopy and imaging can differentiate between engineered bone and engineered cartilage. In *IEEE 36th Annual International IEEE EMBS Conference*, (pp. 3929–3932). https://doi.org/10.1109/embc.2014.6944483.
39. Rodin, V. V. (1997). *Characterization of the biotechnological dispersion systems using magnetic resonance methods*. Moscow State University, M., PhD Thesis, Doctor of Chem. Sciences: Macromolecular and Colloid Chemistry, 278 p.
40. Rodin, V. V. (2004). *Magnetic resonance methods* (p. 95). Moscow: Press MIPhT. ISBN: 5-7417-0228-7.
41. Rodin, V. V. (2014). NMR approaches in studying dynamic molecules in porous materials. *Journal of Material Science Engineering, 3*(3), 245.
42. Rodin, V. V. (2017). Methods of magnetic resonance in studying natural biomaterials. In Wang Z. (Eds.), *Encyclopedia of physical organic chemistry* (pp. 2861–2908, 1st ed., Vol. 4, part 4, chapter 53), Wiley. ISBN 978-1-118-47045-9. http://eu.wiley.com/WileyCDA/WileyTitle/productCd-1118470451.html.
43. Rodin, V. V, & Cosgrove, T. (2016). Nuclear magnetic resonance study of water-polymer interactions and self-diffusion of water in polymer films. *OALJ: Chemical Materials Science, 3*(10), 1–17. https://doi.org/10.4236/oalib.1103018.
44. Rodin, V. V., & Izmailova, V. N. (1994). Influence of disaccharides on the stability of PhL-vesicles to low temperature action by NMR-data. *Colloid Journal, 56*(6), 833–839.
45. Rodin, V. V., & Izmailova, V. N. (1995). Lipid-nucleic interactions in dispersions of PhL-vesicles by NMR-data. *Colloid Journal, 57*(2), 231–239.
46. Rodin, V. V., & Izmailova, V. N. (1998). Study of biocolloids by radiospectroscopy methods (interactions of phospholipid vesicles with polynucleotides). *ACH-Models in Chemistry, 134* (6), 833–844.
47. Rodin, V. V., & Knight, D. P. (2003). Self-diffusion of water into silk fibers from magnetic field pulse gradient data. *Biophysics, 48,* 429–435.
48. Rodin, V. V., & Knight, D. P. (2003). Natural materials by NMR data: Cross-relaxation in macromolecules of natural silk. *Materials Science, 10,* 16–21.
49. Rodin, V. V., & Nikerov, V. A. (2014). NMR relaxation and PFG NMR studies of water dynamics in oriented collagen fibres with different degree of cross-linking. *Current Tissue Engineering, 3*(1), 47–61. https://doi.org/10.2174/2211542003666140626211652.

50. Rodin, V. V., Isangalin, F Sh, Volkov, V., Volkov, V., & Ya, V. (1986). Investigation of formation of the xenon clathrates in *E. coli* suspensions using the NMR method of paramagnetic doping. *Biophysics, 31*(2), 274–277.
51. Rodin, V. V., Foucat, L., & Renou, J. P. (2004). The dipolar interactions and dynamics of water molecules in collagen fibres from two connective tissues of different ages by ^{1}H double-quantum-filtered NMR spectroscopy. *Materials Science, 5,* 2–10.
52. Rodin, V. V., Reznichenko, G. M., & Vasina, E. L. (2004). Properties of natural polymer fibers with low water content. *Materials Science, 1,* 34–42.
53. Rodin, V. V, Ponomarev, A., Gerasimov, M., & Gurevich, L. (2017). Xenon-water interaction in bacterial suspensions as studied by NMR. *International Journal of Biochemistry and Biophysics, 5*(1), 26–36. http://www.hrpub.org/download/20170330/IJBB4-10890467.pdf.
54. Selenko, P., & Wagner, G. (2006). NMR mapping of protein interactions in living cells. *Nature Methods, 3,* 80–81. https://doi.org/10.1038/nmeth0206-80.
55. Sell, S. A., Wolfe, P. S., Garg, K., McCool, J. M., Rodriguez, I. A., Gary, L., et al. (2010). The use of natural polymers in tissue engineering: A focus on electrospun extracellular matrix analogues. *Polymers, 2,* 522–553. https://doi.org/10.3390/polym2040522.
56. Senturia, S. D., & Robinson, J. D. (1970). Nuclear spin lattice relaxation of liquid confined in porous solids. *SPE J., 10,* 237–244.
57. Seo, Y., Ikoma, K., Takamiya, H., Kusaka, Y., Tsoref, L., Eliav, U., et al. (1999). ^{1}H double-quantum-filtered MR Imaging as a new tool for assessment of healing of the ruptured Achilles tendon. *Magnetic Resonance in Medicine, 42,* 884–889.
58. Seo, Y.-K., Choi, G.-M., Kwon, S.-Y., Lee, H.-S., Park, Y.-S., Song, K.-Y., et al. (2007). The biocompatibility of silk scaffold for tissue engineered ligaments. *Key Engineering Materials, 342,* 73–76.
59. Shinar, H., & Navon, G. (2006). Multinuclear NMR and microscopic MRI studies of the articular cartilage nanostructure. *NMR in Biomedicine, 19*(7), 877–893.
60. Soffer, L., Wang, X., Zhang, X., Kluge, J., Dorfmann, L., Kaplan, D. L., et al. (2008). Silk-based electrospun tubular scaffolds for tissue-engineered vascular grafts. *Journal of Biomaterials Science, Polymer Edition, 19,* 653–664.
61. Song, Y.-Q. (2012). Focus on the physics of magnetic resonance on porous media. *New Journal of Physics, 14,* 055017 (9 p).
62. Taj, M. K., Samreen, Z., Ling, J. X., Taj, I., Hassan, T. M., & Yunlin, W. (2014). *Escherichia coli* as a model organism. *International Journal of Engineering Research Science & Technology, 3*(2). ISSN 2319-5991.
63. Takamiya, H., Kusaka, Y., Seo, Y., Noguchi, M., & Ikoma, K. (2000). Characteristics of proton NMR T_2 relaxation of water in the normal and regenerating tendon. *The Japanese Journal of Physiology, 50,* 569–576.
64. Tanner, J. E. (1978). Transient diffusion in system partitioned by permeable barriers. Application to NMR measurements with a pulsed field gradient. *The Journal of Chemical Physics, 69*(4), 1748–1754. https://doi.org/10.1063/1.436751.
65. Tanner, J. E., & Stejskal, E. (1968). Restricted self-diffusion of protons in colloidal systems by the pulsed-gradient, spin-echo method. *The Journal of Chemical Physics, 49,* 1768–1777.
66. Venkataramanan, L., Song, Yi-Q, & Hurlimann, M. D. (2002). T_1-T_2 correlation spectra obtained using a fast two-dimensional Laplace inversion. *Journal of Magnetic Resonance, 154,* 261–268.
67. Wider, G. (1998). Technical aspects of NMR spectroscopy with biological macromolecules and studies of hydration in solution. *Progress in Nuclear Magnetic Resonance Spectroscopy, 32*(3), 193–275.
68. Zeugolis, D., Khew, S. T., Elijah, S. Y., Yew, E. S. Y., & Ekaputra, A. K. (2008). Electro-spinning of pure collagen nano-fibres—Just an expensive way to make gelatin? *Biomaterials, 29,* 2293–2305.

69. Zhang, A., Sun, H., Xu, H., Qiu, S., & Wang, X. (2013). Cell metabolomics. *OMICS: A Journal of Integrative Biology, 17*(10), 495–501.
70. Zhou, J., Cao, C., & Ma, X. (2009). A novel three-dimensional tubular scaffold prepared from silk fibroin by electrospinning. *International Journal of Biological Macromolecules, 45,* 504–510.

Author Biography

Victor V. Rodin (Austria), Ph.D., Institute of Organic Chemistry, Johannes Kepler University Linz. His research interests are in NMR methods, including new methodologies and applications to complex mixtures and materials. He graduated from Moscow Institute of Physics and Technology. He received Ph.D. in Biophysics (Research Institute of Biophysics) and DSc (Ph.D.) in Macromolecular and Colloid Chemistry (MSU). He has worked in excellent NMR centers: USA (University of California), France (INRA), UK (University of East Anglia, University of Bristol, University of Surrey), Austria (Johannes Kepler University of Linz). His research experience focuses on the development of NMR methods and analysis to study biomaterials, drugs, biotechnological systems, xenon, synthetic polymers, natural biopolymers, and other materials. Based on obtained results he delivered lectures to students on physical methods, chemistry, biophysics, mathematics. He wrote 2 books (2004, 2018) and 1 chapter for Encyclopedia of Physical Organic Chemistry (2017) on NMR methods.

Novel MRI Contrast from Magnetotactic Bacteria to Evaluate In Vivo Stem Cell Engraftment

Ji-Hye Jung, Yuko Tada and Phillip C. Yang

1 The Limited Stem Cell Engraftment Reduces Therapeutic Efficacy

The tremendous potential of human pluripotent stem cells (hPSCs) for regenerative medicine today is a topic of intense investigation to scientists worldwide. Since the first report of human embryonic stem cells (hESCs) in 1998, steady progress has been made as the cornerstone of stem cell research [1]. In 1998, Thomson and colleagues succeeded in isolating the hESCs from the inner cell mass of blastocysts. The pluripotency of the hESCs are considered to be a promising source to achieve any cell replacement therapy. They exhibit unlimited self-renewal and are enable to differentiate into every cell type in the adult body. The hESCs offer the possibility of cell therapy for many incurable and degenerative diseases. They hold great promise for tissue engineering and drug discovery applications. However, a major limitation of hESC research is the constant dispute regarding immune rejection, tumorigenesis and ethical concerns. However, in 2007, Shinya Yamanaka of Japan reported a landmark discovery, the human induced pluripotent stem cells (iPSCs), which has broadened the horizons of regenerative medicine. iPSCs are generated when the Yamanaka Factors (sex determining region Y-box 2 (Sox2), Krüppel-Like Factor 4 (Klf4), V-Myc Avian Myelocytomatosis Viral Oncogene Homolog (cMyc) and POU Class 5 Homeobox 1 (Oct3/4)) are transfected into adult somatic cells.

J.-H. Jung · Y. Tada · P. C. Yang (✉)
Stanford Cardiovascular Institute, Stanford, USA
e-mail: phillip@stanford.edu

Y. Tada
e-mail: ytada@stanford.edu

J.-H. Jung · Y. Tada · P. C. Yang
Division of Cardiovascular Medicine, Center for Clinical Science Research (CCSR); 3115C, Stanford University School of Medicine, 269 Campus Drive, Stanford, CA 94305, USA

G. M. Artmann et al. (eds.), *Biological, Physical and Technical Basics of Cell Engineering*, https://doi.org/10.1007/978-981-10-7904-7_16

The cells are reprogrammed to embryonic stage with pluripotent properties [2]. The discovery of the iPSCs led to revolutionary changes in stem cell research. This breakthrough not only addressed the ethical limitations of hESCs but also advanced the development of personalized medicine. Astounding progress has been made, leading to a variety of pharmacogenomics approaches. Disease-specific iPSCs have also offered great promise with profound insights into the pathophysiology of previously incurable diseases [3–6].

Although the challenges of the immune rejection and ethical issues of the hESCs were obviated by the advent of somatic cell reprogramming, significant challenges remain in clinical translation. One of the major problems is to establish safe and effective methods to differentiate the hPSCs into a pure population of specific lineage in vitro. The proper manipulation of hPSCs is not completely understood even though several essential differentiation factors have been identified. Another major hurdle is the limited cell engraftment in vivo following delivery. Although researchers today are able to generate a specific cell lineage, it is difficult to reach therapeutic benefit when the injected cells do not engraft into the recipient's organ of interest [7].

Stem cell therapy should prioritize patient safety and tolerance. Stem cells are associated with number of risks and require successful integration of the transplanted cells in the desired microenvironment of the target tissue. In this chapter, we will focus on the challenges of stem cell engraftment in the heart and discuss a novel in vivo MRI contrast agent based on magnetotactic bacteria, which enables high live cell specificity.

2 Technological Development Is Required to Assess Stem Cell Engraftment

Myocardial infarction (MI) usually results in irreversible myocardial cell loss and heart failure due to the disruption of blood supply. Despite recent advances in the standard of care of MI, injured myocardium and scar tissue still cannot be restored [8]. It has been shown that stem cell therapy in the acute phase of MI attenuates cardiomyocyte apoptosis and local inflammatory response while promoting local neoangiogenesis and myocardial perfusion [9, 10]. In the late phase of MI, cell therapy may replace the dead myocardium with viable cardiomyocytes, smooth muscle cells, and endothelial cells to reduce scar formation [11]. These discoveries have been followed by clinical trials with transplantation of adult somatic stem cells in patients with acute MI [12].

In one of the first reported pre-clinical studies of human pluripotent stem cell-derived cardiomyocytes (CMs), Kehat and colleagues demonstrated that the hESC-derived CMs (hCMs) transplanted into pigs have shown their potential to function as biological pacemakers in electrophysiologically silenced or atrioventricular (AV) blocked hearts [13]. One of the technical challenges in the implementation of hCMs was the purity and yield of the differentiated hCMs. Efficient and reproducible methods of mice and hESC differentiation have been introduced by manipulating the cardiac-specific signaling pathways in embryonic development [14]. Various differentiation methods including, genetic modifications, cytokines, and small molecules have been conducted to obtain a homogeneous and functional hCMs. Recently, hCMs, successfully generated on a large scale, demonstrated reliable engraftment and restoration of damaged heart tissue in a primate MI model [15]. These results seemed promising; however, clinical translation of is still limited by tumorigenesis, immune rejection, genetic instability, ventricular arrhythmia, and ethical considerations.

Similar to the hESCs, iPSCs are pluripotent and generate all three germ layers. The cardiogenic potential of the iPSC population has been studied in both mice [16] and human derivatives [17]. The iPSCs have been differentiated into CMs (iCMs) and other cardiovascular cells such as smooth muscle, fibroblast, and endothelial cells.

Since the advent of stem cell transplantation, intense efforts have been made to track the transplanted stem cells. However, the inability to localize and assess the viability of the cells delivered into the heart is recognized as one of the major limitations for clinical translation of cell therapy [8, 18]. Such information is important to evaluate the engraftment of iCMs and their therapeutic efficacy in vivo. These challenges necessitate novel imaging technologies to conduct reliable translational investigation of any hPSC derivatives. An ideal non-invasive platform will enable high sensitivity, optimal spatial and temporal resolution, and exquisite tissue contrast. Longitudinal visualization of live cell specificity to confirm the engraftment of the delivered cells within the host myocardium is necessary.

3 Imaging Technology Monitors In Vivo Stem Cell Engraftment

3.1 Imaging Modalities Used for In Vivo Stem Cell Tracking

At present, no available imaging technology satisfy the ideal properties of in vivo stem cell tracking [19]. Multiple techniques are used currently for in vivo monitoring of labeled cells, including magnetic resonance imaging (MRI), single photon emission computed tomography (SPECT), positron emission tomography (PET), and optical imaging such as fluorescence imaging (FLI) and bioluminescence imaging (BLI). Each modality has individual advantages and drawbacks related to spatial resolution, sensitivity, clinical availability, safety, and labeling method. Cell labeling methods are largely divided into direct and indirect labeling. In direct labeling, tracers are bound to the cell surface or transported inside the cells by diffusion, endocytosis, phagocytosis, or active transport. While direct labeling needs no gene editing, one of the disadvantages is that the produced signals can be diluted by cell division or proliferation. Another critical problem is that the signal may persist regardless of the cell viability because of the uptake of dead cells by phagocytizing cells [20]. In contrast, in indirect labeling, reporter genes such as firefly luciferase (Fluc) and/or herpes simplex virus thymidine kinase (HSV-tk)) are transfected into the target cells to express specific proteins that react with the administered substrates, leading to the emission of signals [21]. Thus, the detected signal strength correlates with cell viability [22]. However, genetic modifications are not desirable for clinical translation because of the problems including inflammatory toxicity and carcinogenesis of the virus vectors.

In BLI, bioluminescent gene, Fluc, are inserted into the genome of the target cell. Expressed enzyme catalyzes the conversion from administered luciferin into oxiluciferin, releasing photon energy. In FLI, fluorescence signals from the fluorescent protein such as the enhanced green fluorescent protein (eGFP) are detected. Although their sensitivity is excellent and long-term cell monitoring is possible, the application of BLI and FLI is limited to the small animals because of the limited signal penetration. PET/SPECT generally utilize direct radioscintigraphic cell labeling of high-energy gamma-emitting radiotracers. Radionucleotides such as 111In-oxine, which passively diffuse into the cytosol, are used for SPECT cell tracking. PET achieves better contrast and spatial resolution than SPECT. The advantage of PET/SPECT is the exquisite signal sensitivity and transmission in addition to the ease of clinical translation. Although tracers with different half-life and decay profiles such as FDG, Cu, Zr, and Mn are used, the short half-life and the radioactive tracers makes PET unsuitable for long-term cellular tracking [23]. On the other hand, MRI has an exquisite spatial resolution compared to PET/SPECT or BLI/FLI; however, the sensitivity of MRI is relatively low. Superparamagnetic iron oxide particles (SPIONs), improves the capability of MRI to detect stem cells in the myocardium longitudinally and offers the only suitable agent for clinical translation of stem cell tracking [24, 25]. Higher sensitivity makes SPIONs feasible for molecular MRI [26] while also combining with precise evaluation of cardiac function and myocardial tissue characterization [27]. This capability enables precise cell localization and assessment of their specific regional effects in the different areas of myocardial injury to correlate cell engraftment with therapeutic efficacy [28, 29]. Stem cell labeling with magnetic particles can also be combined with magnetically-targeted stem cell delivery, which utilize external magnet devices manipulate cells to guide cells to the target lesion sites [30, 31].

3.2 *Superparamagnetic Nanoparticle Labeled Cells Enable In Vivo Monitoring with MRI*

The SPIONs are the magnetic nanoparticles most commonly used to label cells for MRI tracking. These nanoparticles have been widely used to detect inflammation since they are taken up by phagocytizing cells after their administration into the body [32] (Fig. 1). Supraparamagnetic property of SPIONs accelerates transverse relaxation of magnetized protons by producing local field inhomogeneity, called T2* relaxation effect. SPION containing tissues are shown as signal defect on T2*-weighted gradient echo sequence consisting of long TR/TE and low flip angle [33]. In general, exact cell quantification using MRI is difficult [20]. However, accumulation of the SPION-labeled cells in the tissue can be quantified by acquiring several gradient echo sequences at different TEs and calculating T2* decrease from the T2* decay curve [34].

The structure of SPIONs consists of the functional core, coating, and surface properties, affecting the efficiency of cellular uptake, distribution, metabolism and potential toxicity. The functional core with superparamagnetic property is a single-domain iron oxide molecule (<10 nm) consisting generally of Fe_3O_4 (magnetite), gamma-Fe_2O_3 (maghematite), or alpha-Fe_2O_3 (hematite) [35]. SPIONs are coated with a biocompatible polymer such as dextran and carboxydetran polymer to prevent aggregations, structural changes, and degradation when exposed to the biological system [36, 37]. SPIONs usually label cells in a direct manner. Clathrin mediated endocytosis serves as the main mechanism for the most cell types including stem cells without phygocytizing capacity, which is facilitated by cationic compounds such as poly-L-Lysine and protamine [38–40]. Magnetoeloctroporation and magnetosonoporation are alternatively used to allow rapid incorporation into freshly isolated cells [41, 42]. In certain culture conditions, mesenchymal stem cells (MSCs) are known to regain its innate phagocytic capacity and take up SPIONs without facilitators [43, 44]. Antibody mediated targeting such as anti-CD34 antibody coated magnetic particles for hematopoietic stem cells could be another promising method to label target stem cells specifically in vivo [45, 46].

SPIONs are classified into standard SPION (SSPIO, 50–150 nm), ultra-small SPION (USPIO, 10–50 nm), and very-small SPION (VSPIO, <10 nm) based on

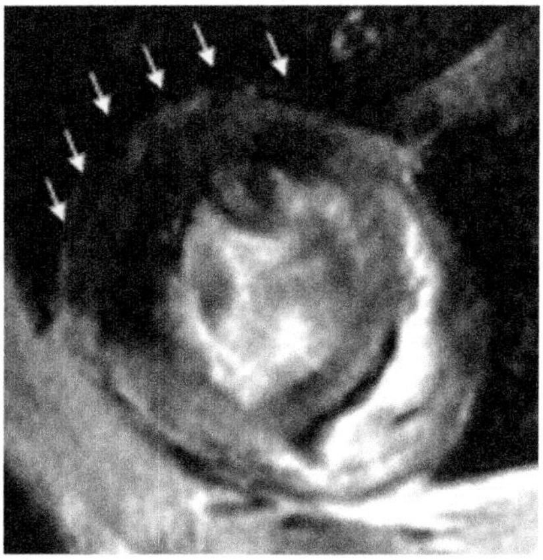

Fig. 1 Myocarditis detected by ferumoxytol. The short axis cardiac MRI was taken from the rat induced with autoimmune myocarditis. Myocardial inflammation where ferumoxytol accumulates is clearly detected as signal void on GRE image (arrows)

Table 1 Examples of clinical SPION

Generic name	Brand name	Classification	Coating	Diameter (nm)	Status
Ferumoxide	Feridex/ Endorem	SSPIO	Dextran	120–180	Discontinued
Ferumoxtran-10	Combidex/ Sinerem	USPIO	Dextran	15–30	Discontinued
Ferucarbotran	Resovist/ Cliavist	SSPIO	Carbodextran	60	Discontinued
Ferucarbotran	Supravist	USPIO	Carbodextran	21	
Feruglose	Clariscan	USPIO	Pegylated starch	20	Discontinued
Ferumoxytol	Faraheme	USPIO	Carboxymethyl dextran	30	FDA approved

their hydrodynamic diameters [47]. Optimal particle size of SPIONs for intravenous administration is 10–100 nm [48]. SPIONs are clinically used as contrast agents for evaluating blood volume fraction, perfusion, and cancer metastasis (Table 1). Unfortunately, most of them were discontinued due to the safety reason or infrequent use. Ferumoxytol is the only clinically available SPION approved by FDA as an iron replacement for the patients with renal anemia. In addition, various non-clinical SPIONs are available for experimental use. For instance, Molday (BioPAL, Inc) is an USPIO MRI contrast agent with a magnetic core and hydrodynamic sizes of around 8 and 35 nm, respectively, which is designed for cell labeling and requires no transfection reagents.

The major limitation of SPIO-labeling is the maintenance of in vivo signal after cell death. There is no reliable MRI viability signal from the transplanted cells [49, 50]. They generate false positive signals even after transplanted cell death. Persistence of significant MRI signal derived from ferumoxide-containing macrophages is seen despite only few viable stem cells 3 weeks after transplantation, indicating that the MRI of ferumoxide-labeled stem cells does not demonstrate long-term stem cell engraftment signal in the heart [51]. Another limitation is the decrease in the SPIO signal as the transplanted stem cells divide, presenting difficulty in tracking the cells in longitudinal studies [52]. SPION has generally low toxicity to the cells [40]. Nevertheless, some in vitro experiments suggested the possibility that iron oxide or transfection facilitators could cause cellular stress (mitochondrial dysfunction, ROS generation), alterations in gene expression and cellular differentiation (promoting chondrogenic, osteogenic, and adipogenic differentiation of mesenchymal stem cells), decrease in cell proliferation, and promotion of pro-inflammatory environment [25].

4 Magnetotactic Bacteria Evaluate Live Cell Specificitiy of Engrafted Stem Cells

It is important to evaluate stem cell engraftment for clinical translation of stem cell therapy. The yield of the cell engraftment in the myocardium characterizes the most essential process of post-transplantation biology. With the potential emergence of cell therapy for heart failure, accurate in vivo evaluation of stem cell engraftment and the resultant effects on the regional myocardial viability and function may represent a critical measure of therapeutic efficacy [53, 54]. Thus, live cell specific technology is necessary to advance regenerative medicine.

4.1 Development of Magnetotactic Bacteria

Magnetotactic bacteria (MTB) are a polyphyletic group of gram-negative prokaryotes. Members of this group are characterized by their ability to align themselves and exist along a magnetic field. They owe this ability to the magnetosome, a specific intracellular structure that contains crystals of magnetic metals (usually magnetite or greigite) bounded by a membrane [55].

The magnetosome, the most important and defining intracellular component of the MTB, is a membrane-bound inorganic magnetic iron crystal. Since its discovery, researchers have recognized it as a true prokaryotic organelle because of its distinct complexity that rivals that of eukaryotic counterpart. Despite the universal presence of the magnetosomes within the MTB, there are inter-species difference in their size, composition and morphology. Morphologically, a few different forms of magnetosomes have been observed across species by electron microscopy (Fig. 2). In most strains of MTB, the magnetosomes are fixed in several chains that are in turn bound to the cell membrane in order to limit magnetostatic energy. The most commonly used minerals in the magnetosomes are, as mentioned previously, magnetite or greigite. However, many other iron sulfides have also been found within sulfide producing MTBs, such as cubic FeS and tetragonal FeS [57]. The researchers recognized that the magnetosome formation has few different processes. The most recent proposed model lists three discrete steps. The first step consists of the creation of magnetosome membrane, which invaginates out of the cytoplasmic membrane (primed by GTPase), as well as the lining up of these vesicles along cytoskeletal filaments in order to form a chain structure. Second, iron is transported into the cell. This transport is strictly regulated by an oxidation-reduction system because of the potentially dangerous effects of having too much intracellular iron. In the third and final step, magnetosome proteins regulate morphology and trigger magnetite crystal nucleation (Fig. 3) [58].

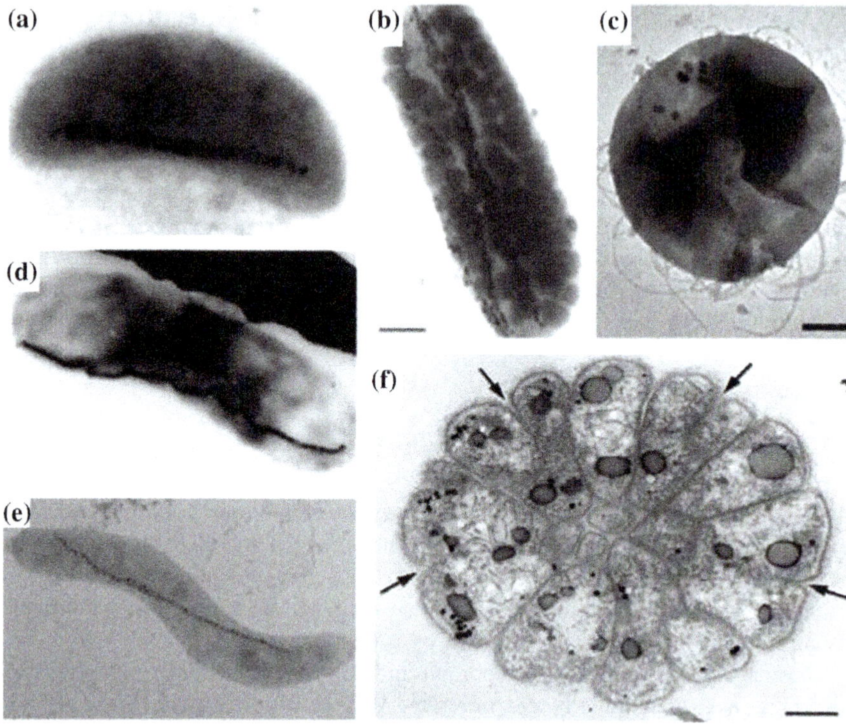

Fig. 2 Various morphology of MTB. **a** vibrios; **b** and **d** rods (Bar = 1.0 μm) and **c** coccoid (Bar = 200 nm); **e** spirilla; and **f** multicellular organism (Bar = 1.0 μm). Reprinted with permission from Yan et al. [56]

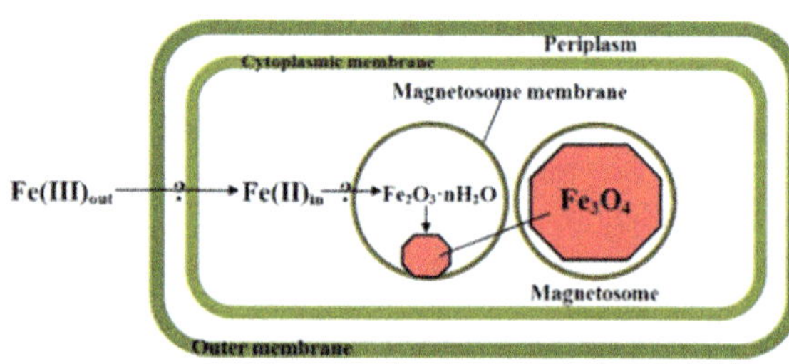

Fig. 3 A simplified proposed model of magnetite biomineralization. Reprinted with permission from Yan et al. [56]

4.2 *Development of Novel MRI Agent for In Vivo Tracking*

Since the discovery of the magnetatotic bacteria, the biomedical application such as the evaluation of cell engraftment, employing this technology has been attempted. However, the major hurdles to the widespread application of magnetosomes involve the fastidiousness of MTB growth, which makes them difficult to culture on a large scale and requires full investigation of the genetic and environmental control over magnetosome synthesis. These limitations are constantly being addressed by numerous scientists [59]. To address these issues, the novel magneto-endosymbionts (MEs) contrast agent was developed by biotech company (Magnelle® reagents Bell Biosystems, Inc., CA), which are derived from the magnetotactic bacteria Magnetospirillum magneticum strain AMB-1 (AMB-1) (Fig. 4a). AMB-1 coordinates over 100 genes to synthesize membrane bound magnetic nanoparticles in magnetosomes, which are highly effective MRI contrast agents [61]. Importantly, MEs are removed by macrophages and cleared from the tissue, eliminating any confounding MRI contrast as evidenced by the absence of non-specific dephasing signal. Hence, the MEs are maintained only in the viable cells to provide accurate longitudinal engraftment signal of the transplanted stem cells. The transfection of MEs into any recipient cells have no effects on the cell viability and cytotoxicity in vitro. Propidium iodide (PI) is a membrane

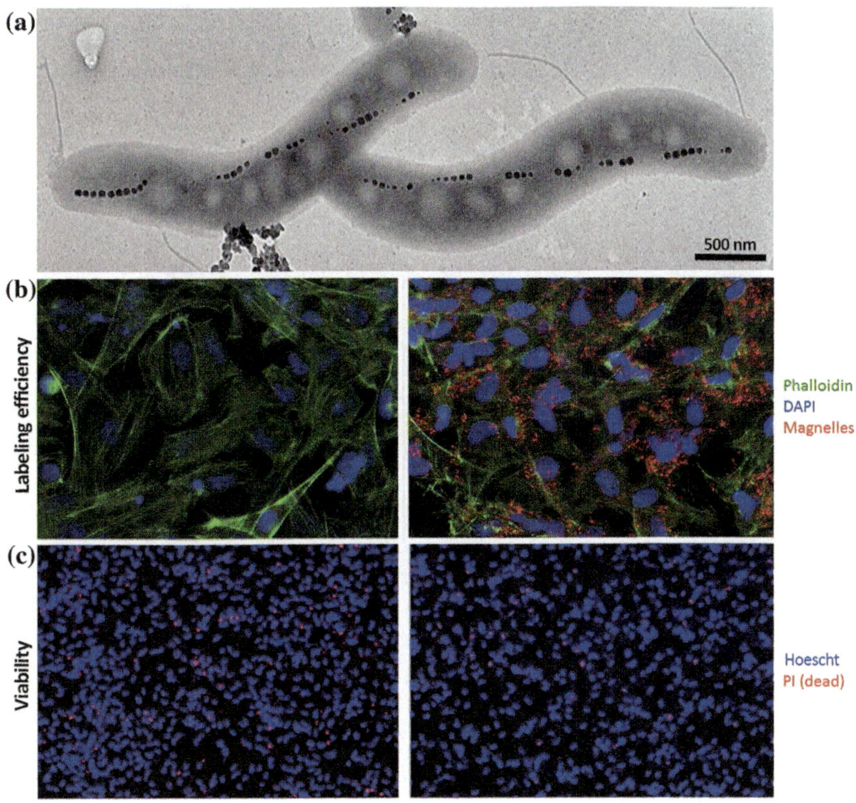

Fig. 4 **a** TEM image showing the magnetosome structure within MEs; **b** fluorescent images of the unlabeled (left panel) and ME labeled (right panel) iCMs stained with ME anti-body in red, phalloidin in green and DAPI in blue, showing internalization of MEs in the iCMs; and **c** viability assessment of unlabeled (left panel) and ME labeled (right panel) iCMs using propidium iodide (PI) assay, dead cells are shown with a red signal while blue Hoescht staining corresponds to the total cell population Reprinted with permission from Mahmoudi et al. [60]

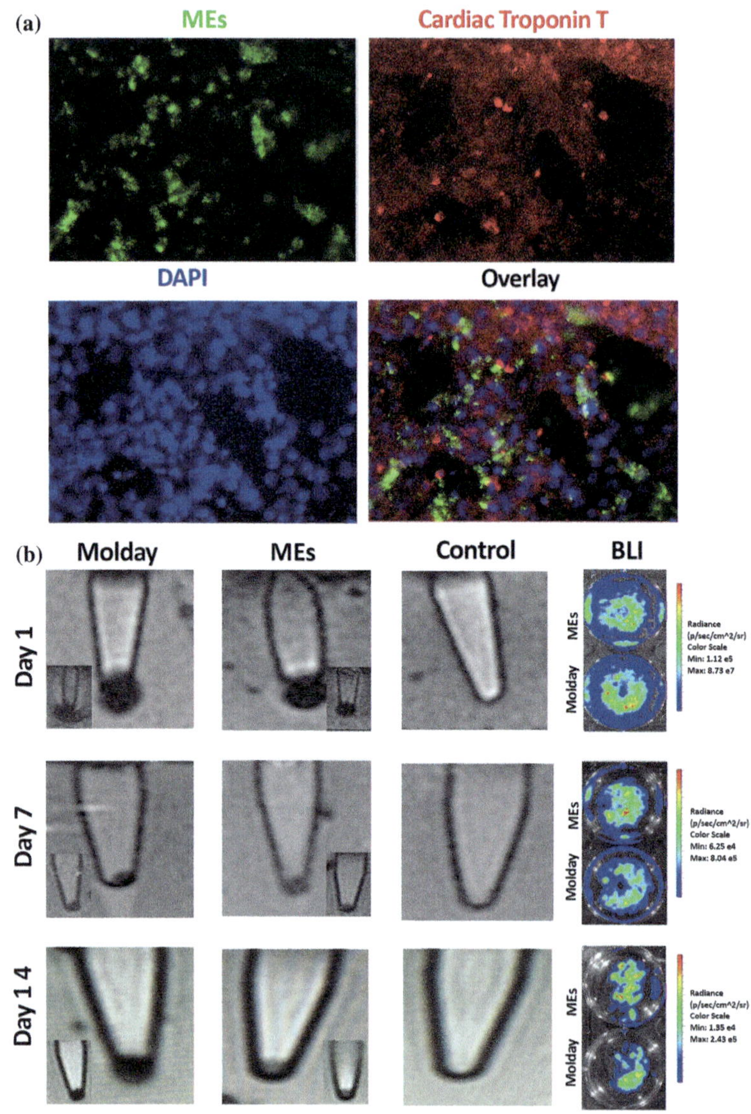

Fig. 5 **a** Fluorescent images of the ME-labeled iCMs used for the evaluation of potential ME effects on cardiac marker. MEs are stained in green and cardiac troponin T in red. Nuclei are counterstained with DAPI in blue and **b** MRI images (T2*) of the bottom of Eppendorf tubes of Molday-labeled iCMs, ME-labeled iCMs, and PBS only control sample at different time points (days 1, 7, and 14 after labeling); the bottom panel shows the replicate MRI images for Molday- and ME-labeled cells. Right panel shows the corresponding BLI signal, indicating viability of the Molday- and ME-labeled iCMs at days 1, 7, and 14 post labeling (BLI experiments were performed in triplicate). Reprinted with permission from Mahmoudi et al. [60]

impermeable dye and does not enter viable cells with intact membranes. As demonstrated by the PI staining (Fig. 4c), there was no significant difference between control unlabeled cells, which had a viability of 88% ± 2, and ME-labeled cells which had 93% ±3 cell viability ($p > 0.05$). No significant decrease in cell viability was seen after labeling with MEs (Fig. 4b, c). Moreover, we have demonstrated that the ME-labeled iCMs maintain their unique character such as contractility and strong expression of the sarcomeric protein (i.e., cardiac troponin T). This experimental demonstration indicates that the MEs do not affect the cardiac cell properties after magnetic labeling (Fig. 5a). The signal generated from the in vitro and in vivo bioluminescence imaging (BLI) and MRI studies showed corresponding evidence of robust contrast to track the iCMs based on our in vitro MRI signal on days 1, 7, and 14 when compared to the Molday ION labeled iCMs. The phantom experiment showed that the strong T2* signal of the labeled cells (with both Molday IONs and MEs) on days 7 and 14 after labeling are seen by the dark dephasing signal. The viability of the cells at different time points was monitored by BLI (Fig. 5b). After the injection of ME-labeled iCMs into a live mouse heart, both BLI and MRI signal were detected seven days after cell injection. Mice #2 and #3 show BLI signal clearance after a week while mouse #1 retains BLI signal through 2 wks. As expected, the Molday-labeled cells indicated no BLI signal by day 14; however, the cells continued to generate dephasing MRI signal consistent with the persistence of the nanoparticle signal in the mouse heart. In contrast to the Molday-labeled iCMs, the ME-labeled cells demonstrated significant correlation between BLI and MRI signals in vivo (Fig. 6). From these experiments, the mice exhibited strong MRI signal in both ME- and Molday-labeled iCMs. This finding, indicates the major limitation of synthetic iron oxide nanoparticles for in vivo monitoring of live cells in the heart as shown previously while the ME-labeled iCMs demonstrate live cell specificity [49, 50].

5 Future Perspectives

Stem cell technology represents an attractive approach to realize regenerative medicine. iPSC technology represents a landmark discovery to deliver this promise. Adult stem cells, such as MSCs, may also represent an alternative cell source to treat degenerative diseases as there are no ethical concerns or tumorigenicity potential. The development of novel stem cell therapy will continue to evolve. Various stem cell types from specific patients with unique disease process, novel gene-editing/modification, or stem cell secretomes such as the exosomes may replace damaged tissue and prevent organ failure. Thus, reliable and precise in vivo stem cell tracking represents one of the key essential technology in advancing the field of stem cell therapy. However, there is no reliable in vivo imaging method to confirm the therapeutic contribution of the transplanted stem cells. In order to obtain high resolution and reliable signal of the transplanted cells, sensitive detection of the cell engraftment signals is necessary.

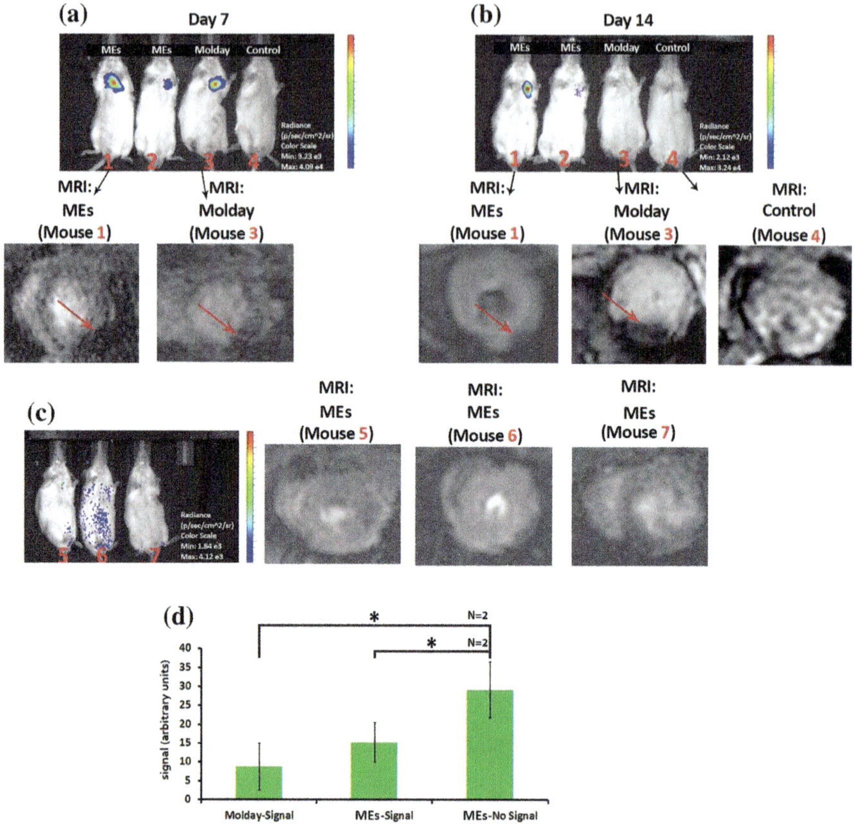

Fig. 6 Representative BLI images of the mice injected with ME- and Molday-labeled reporter gene-transduced iCMs or PBS control after **a** 7 days; **b** 14 days following cell injections (top) and their corresponding in vivo MRI images of the murine hearts (below). Red arrows show the signal from the injected labeled cells; **c** representative in vivo BLI images of the selected mice with dead cardiomyocytes at day 14 of ME labeled cells and the corresponding in vivo MRI images; **d** signal intensity (SI) for Molday and ME (dephasing signal) and ME (no dephasing signal). A significant absence of signal of ME-labeled dead cells compared to other samples with positive signal ($p < 0.05$). Reprinted with permission from Mahmoudi et al. [60]

High spatial and temporal resolution imaging of the organ function and morphology is imperative for in vivo cell tracking. MRI represents a promising imaging modality as it combines the chemical sensitivity of nuclear magnetic resonance with exquisite tissue contrast and high spatial and temporal resolution. These specifications provide optimal technical features to assess organ physiology. However, MRI lacks the sensitivity to detect the transplanted cells. The high sensitivity provided by the magnetic nanoparticles for cell tracking applications addresses this issue; however, in vivo monitoring of the cell viability has not been feasible due to the confounding signal of the nanoparticles in the tissue or the macrophages after the death of transplanted stem cells. The novel ME approach reviewed here leverages the rapid response of the immune system, which may detect and eliminate the biological component of the ME derived from MTB. Existence of biological impurities (particularly proteins, nucleic acids, and polysaccharides) of the MEs exposed by dead stem cells allow rapid clearance by the immune system.

Recent advances in regenerative medicine demonstrate the vast potential of cell therapies for a wide range of therapeutic needs in neurodegenerative disease, cardiomyopathy, oncology, and immunology. However, a robust method to track these cells after delivery had been a limiting factor for rapid clinical translation. Our group demonstrated the first use of MEs to track the iCMs in vivo by MRI. The study showed the use of ME as a novel contrast agent for MRI-based cell tracking, which allowed longitudinal visualization of cell survival in a murine model. The use of the ME enabled live cell sensitivity (LCS), which may have significant advantage over commercially available IONs, such as the Molday IONs. This novel technology will allow researchers to effectively track live cells in vivo to monitor their engraftment and biological distribution. Several issues, including the sensitivity of MRI and dosing of ME, should be addressed to optimize the reliability and precision in cell tracking applications for future cell-based investigation. The highly novel MTB technology represents one of the viable stem cell tracking method to advance regenerative medicine.

References

1. Thomson, J. A., et al. (1998). Embryonic stem cell lines derived from human blastocysts. *Science, 282,* 1145–1147.
2. Takahashi, K., et al. (2007). Induction of pluripotent stem cells from adult human fibroblasts by defined factors. *Cell, 131,* 861–872.
3. Tanaka, A., Yuasa, S., Node, K., & Fukuda, K. (2015). Cardiovascular disease modeling using patient-specific induced pluripotent stem cells. *International Journal of Molecular Sciences, 16,* 18894–18922.
4. Maehr, R., et al. (2009). Generation of pluripotent stem cells from patients with type 1 diabetes. *Proceedings of the National Academy of Science U. S. A., 106,* 15768–15773.
5. Richard, J.-P., & Maragakis, N. J. (2015). Induced pluripotent stem cells from ALS patients for disease modeling. *Brain Research, 1607,* 15–25.
6. Payne, N. L., et al. (2015). Application of human induced pluripotent stem cells for modeling and treating neurodegenerative diseases. *New Biotechnology, 32,* 212–228.
7. Chidgey, A. P., Layton, D., Trounson, A., & Boyd, R. L. (2008). Tolerance strategies for stem-cell-based therapies. *Nature, 453,* 330–337.
8. Jung, J.-H., Fu, X., & Yang, P. C. (2017). Exosomes generated from iPSC-derivatives: New direction for stem cell therapy in human heart diseases. *Circulation Research, 120,* 407–417.
9. Hynes, B., et al. (2013). Potent endothelial progenitor cell-conditioned media-related anti-apoptotic, cardiotrophic, and pro-angiogenic effects post-myocardial infarction are mediated by insulin-like growth factor-1. *European Heart Journal, 34,* 782–789.
10. Valina, C., et al. (2007). Intracoronary administration of autologous adipose tissue-derived stem cells improves left ventricular function, perfusion, and remodelling after acute myocardial infarction. *European Heart Journal, 28,* 2667–2677.

11. Rota, M., et al. (2008). Local activation or implantation of cardiac progenitor cells rescues scarred infarcted myocardium improving cardiac function. *Circulation Research, 103,* 107–116.
12. Hare, J. M., et al. (2009). A randomized, double-blind, placebo-controlled, dose-escalation study of intravenous adult human mesenchymal stem cells (prochymal) after acute myocardial infarction. *Journal of the American College of Cardiology, 54,* 2277–2286.
13. Kehat, I., et al. (2004). Electromechanical integration of cardiomyocytes derived from human embryonic stem cells. *Nature Biotechnology, 22,* 1282–1289.
14. Sumi, T., Tsuneyoshi, N., Nakatsuji, N., & Suemori, H. (2008). Defining early lineage specification of human embryonic stem cells by the orchestrated balance of canonical Wnt/β-catenin, activin/nodal and BMP signaling. *Development, 135,* 2969–2979.
15. Chong, J. J. H., et al. (2014). Human embryonic-stem-cell-derived cardiomyocytes regenerate non-human primate hearts. *Nature, 510,* 273–277.
16. Mauritz, C., et al. (2008). Generation of functional murine cardiac myocytes from induced pluripotent stem cells. *Circulation, 118,* 507–517.
17. Zhang, J., et al. (2009). Functional cardiomyocytes derived from human induced pluripotent stem cells. *Circulation Research, 104,* e30–e41.
18. Passier, R., van Laake, L. W., & Mummery, C. L. (2008). Stem-cell-based therapy and lessons from the heart. *Nature, 453,* 322–329.
19. Frangioni, J. V., & Hajjar, R. J. (2004). In vivo tracking of stem cells for clinical trials in cardiovascular disease. *Circulation, 110,* 3378–3383.
20. Li, Z., et al. (2008). Comparison of reporter gene and iron particle labeling for tracking fate of human embryonic stem cells and differentiated endothelial cells in living subjects. *Stem Cells (Dayton, Ohio), 26,* 864–873.
21. Parashurama, N., et al. (2016). Multimodality molecular imaging of cardiac cell transplantation: Part II. In vivo imaging of bone marrow stromal cells in swine with PET/CT and MR imaging. *Radiology, 280,* 826–836.
22. Parashurama, N., et al. (2016). Multimodality molecular imaging of cardiac cell transplantation: Part I. Reporter gene design, characterization, and optical in vivo imaging of bone marrow stromal cells after myocardial infarction. *Radiology, 280,* 815–825.
23. von der Haar, K., Lavrentieva, A., Stahl, F., Scheper, T., & Blume, C. (2015). Lost signature: Progress and failures in in vivo tracking of implanted stem cells. *Applied Microbiology and Biotechnology, 99,* 9907–9922.
24. Lewin, M., et al. (2000). Tat peptide-derivatized magnetic nanoparticles allow in vivo tracking and recovery of progenitor cells. *Nature Biotechnology, 18,* 410–414.
25. Jasmin, et al. (2011). Optimized labeling of bone marrow mesenchymal cells with superparamagnetic iron oxide nanoparticles and in vivo visualization by magnetic resonance imaging. *Journal of Nanobiotechnology, 9*(4).
26. Cunningham, C. H., et al. (2005). Positive contrast magnetic resonance imaging of cells labeled with magnetic nanoparticles. *Magnetic Resonance in Medicine, 53,* 999–1005.
27. Arai, T., et al. (2006). Dual in vivo magnetic resonance evaluation of magnetically labeled mouse embryonic stem cells and cardiac function at 1.5 t. *Magnetic Resonance in Medicine, 55,* 203–209.
28. Dash, R., et al. (2011). Dual manganese-enhanced and delayed gadolinium-enhanced MRI detects myocardial border zone injury in a pig ischemia-reperfusion model. *Circulation: Cardiovascular Imaging, 4,* 574–582.
29. Hung, T.-C., et al. (2008). Multimodality evaluation of the viability of stem cells delivered into different zones of myocardial infarction. *Circulation: Cardiovascular Imaging, 1,* 6–13.
30. Nishida, K., et al. (2006). Magnetic targeting of bone marrow stromal cells into spinal cord: through cerebrospinal fluid. *NeuroReport, 17,* 1269–1272.
31. Vandergriff, A. C., et al. (2014). Magnetic targeting of cardiosphere-derived stem cells with ferumoxytol nanoparticles for treating rats with myocardial infarction. *Biomaterials, 35,* 8528–8539.

32. Uchida, M., et al. (2008). A human ferritin iron oxide nano-composite magnetic resonance contrast agent. *Magnetic Resonance in Medicine, 60,* 1073–1081.
33. Nitz, W. R., & Reimer, P. (1999). Contrast mechanisms in MR imaging. *European Radiology, 9,* 1032–1046.
34. Bos, C., et al. (2004). In vivo MR imaging of intravascularly injected magnetically labeled mesenchymal stem cells in rat kidney and liver. *Radiology, 233,* 781–789.
35. Santoyo Salazar, J., et al. (2011). Magnetic iron oxide nanoparticles in 10–40 nm range: Composition in terms of magnetite/maghemite ratio and effect on the magnetic properties. *Chemistry of Materials,* 23, 1379–1386.
36. Moraes, L., et al. (2012). Neuroprotective effects and magnetic resonance imaging of mesenchymal stem cells labeled with SPION in a rat model of Huntington's disease. *Stem Cell Research, 9,* 143–155.
37. Bull, E., et al. (2014). Stem cell tracking using iron oxide nanoparticles. *International Journal of Nanomedicine, 9,* 1641–1653.
38. Hillaireau, H., & Couvreur, P. (2009). Nanocarriers' entry into the cell: Relevance to drug delivery. *Cellular and Molecular Life Sciences CMLS, 66,* 2873–2896.
39. Cores, J., Caranasos, T. G., & Cheng, K. (2015). Magnetically targeted stem cell delivery for regenerative medicine. *Journal of Functional Biomaterials, 6,* 526–546.
40. Suzuki, Y., et al. (2007). In vitro comparison of the biological effects of three transfection methods for magnetically labeling mouse embryonic stem cells with ferumoxides. *Magnetic Resonance in Medicine, 57,* 1173–1179.
41. Qiu, B., et al. (2010). Magnetosonoporation: instant magnetic labeling of stem cells. *Magnetic Resonance in Medicine, 63,* 1437–1441.
42. Walczak, P., Kedziorek, D. A., Gilad, A. A., Lin, S., & Bulte, J. W. M. (2005). Instant MR labeling of stem cells using magnetoelectroporation. *Magnetic Resonance in Medicine, 54,* 769–774.
43. Khurana, A., et al. (2013). Iron administration before stem cell harvest enables MR imaging tracking after transplantation. *Radiology, 269,* 186–197.
44. Liu, L., et al. (2016). A new method for preparing mesenchymal stem cells and labeling with ferumoxytol for cell tracking by MRI. *Scientific Reports, 6,* 26271.
45. Chen, J., et al. (2013). Guidance of stem cells to a target destination in vivo by magnetic nanoparticles in a magnetic field. *ACS Applied Materials & Interfaces, 5,* 5976–5985.
46. Chung, J., et al. (2011). In vivo molecular MRI of cell survival and teratoma formation following embryonic stem cell transplantation into the injured murine myocardium. *Magnetic Resonance in Medicine, 66,* 1374–1381.
47. Singh, N., Jenkins, G. J. S., Asadi, R., & Doak, S. H. (2010). Potential toxicity of superparamagnetic iron oxide nanoparticles (SPION). *Nano Reviews, 1.*
48. Elias, A., & Tsourkas, A. (2009). Imaging circulating cells and lymphoid tissues with iron oxide nanoparticles. *Hematology American Society Hematology Education Program* 720–726. https://doi.org/10.1182/asheducation-2009.1.720.
49. Chen, I. Y., et al. (2009). Comparison of optical bioluminescence reporter gene and superparamagnetic iron oxide MR contrast agent as cell markers for noninvasive imaging of cardiac cell transplantation. *Molecular Imaging and Biology (MIB) Official Publication of the Academy of Molecular Imaging, 11,* 178–187.
50. Suzuki, Y., et al. (2008). In vivo serial evaluation of superparamagnetic iron-oxide labeled stem cells by off-resonance positive contrast. *Magnetic Resonance in Medicine, 60,* 1269–1275.
51. Terrovitis, J., et al. (2008). Magnetic resonance imaging overestimates ferumoxide-labeled stem cell survival after transplantation in the heart. *Circulation, 117,* 1555–1562.
52. Kim, J. A., Åberg, C., Salvati, A., & Dawson, K. A. (2011). Role of cell cycle on the cellular uptake and dilution of nanoparticles in a cell population. *Nature Nanotechnology, 7,* 62–68.
53. Hendry, S. L., et al. (2008). Multimodal evaluation of in vivo magnetic resonance imaging of myocardial restoration by mouse embryonic stem cells. *The Journal of Thoracic and Cardiovascular Surgery,* 136, 1028–1037.e1.

54. Kim, P. J., et al. (2015). Direct evaluation of myocardial viability and stem cell engraftment demonstrates salvage of the injured myocardium. *Circulation Research, 116,* e40–e50.
55. Blakemore, R. (1975). Magnetotactic bacteria. *Science, 190,* 377–379.
56. Yan, L., et al. (2012). Magnetotactic bacteria, magnetosomes and their application. *Microbiological Research, 167,* 507–519.
57. Schüler, D., & Frankel, R. B. (1999). Bacterial magnetosomes: Microbiology, biomineralization and biotechnological applications. *Applied Microbiology and Biotechnology, 52,* 464–473.
58. Arakaki, A., Nakazawa, H., Nemoto, M., Mori, T., & Matsunaga, T. (2008). Formation of magnetite by bacteria and its application. *Journal of the Royal Society, Interface, 5,* 977–999.
59. Araujo, A. C. V., Abreu, F., Silva, K. T., Bazylinski, D. A., & Lins, U. (2015). Magnetotactic bacteria as potential sources of bioproducts. *Marine Drugs, 13,* 389–430.
60. Mahmoudi, M., et al. (2016). Novel MRI contrast agent from magnetotactic bacteria enables in vivo tracking of iPSC-derived cardiomyocytes. *Scientific Reports, 6.*
61. Komeili, A. (2012). Molecular mechanisms of compartmentalization and biomineralization in magnetotactic bacteria. *FEMS Microbiology Reviews, 36,* 232–255.

Author Biography

Ji-Hye Jung (USA) Mrs. Jung is a Postdoctoral Fellow in the Yang Lab at the Stanford University School of Medicine. She completed her Ph.D. at the Korea University Graduate School of Medicine in Seoul, Korea, where she studied the basic biological mechanisms of human embryonic stem cells (hESCs), induced pluripotent stem cells (iPSCs) and adult stem cells. Currently, at Stanford University, she is investigating cardiac tissue repair mechanisms after ischemic injury by identifying novel therapeutic biologics using iPSC technology. Her research is published in numerous SCI journals.

Conformational Motions of Disordered Proteins

Andreas M. Stadler

Molecular dynamics in proteins animate and play a vital role for biologically relevant processes of these biomacromolecules. Quasielastic incoherent neutron scattering (QENS) is a well-suited experimental method to study protein dynamics from the picosecond to several nanoseconds and in the Ångström length-scale. In QENS experiments of protein solutions hydrogens act as reporters for the motions of methyl groups or amino acids to which they are bound. Neutron Spin-Echo spectroscopy (NSE) on the other hand offers the highest energy resolution in the field of neutron spectroscopy. It enables the study of slow collective motions in proteins up to several hundred nanoseconds and in the nanometre length-scale. In the following chapter I will present recent experimental studies that demonstrate the relevance of molecular dynamics for protein folding and for conformational transitions of intrinsically disordered proteins (IDPs). During the folding collapse the protein chain is exploring the accessible conformational space via molecular motions. A large flexibility of partially folded and unfolded proteins, therefore, is mandatory for rapid protein folding. IDPs on the other hand are a special case as they are largely unstructured under physiological conditions in their native states. A large flexibility of IDPs is a characteristic property of the proteins as it allows, for example, the interaction with various binding partners or the rapid response to different conditions.

A. M. Stadler (✉)
Jülich Centre for Neutron Science, JCNS and Institute for Complex Systems ICS, Forschungszentrum Jülich GmbH, 52425 Jülich, Germany
e-mail: a.stadler@fz-juelich.de

G. M. Artmann et al. (eds.), *Biological, Physical and Technical Basics of Cell Engineering*, https://doi.org/10.1007/978-981-10-7904-7_17

1 Flexibility as a Requirement for the Biological Function of Proteins

At physiological temperature and in the solution state, in which they are biologically active and fulfil their respective functions, biological macromolecules display a dynamic variety of motions. Motions occur on various length-scales, from a fraction of a nanometre to several nanometres, and on a wide range of time-scales, from fast motions on the femtosecond to very slow folding processes on the time scale of a second. Proteins are long polypeptide chains that fold into specific three-dimensional structures, which are essential for biological function. Interactions with the aqueous environment are highly important for correct protein folding. Folding energies are relatively weak, comparable with thermal energy at room temperature, making protein structures very sensitive to temperature and solvent composition. There are two necessary conditions that a folded protein structure must fulfil: its time-average must be stable under particular physiological conditions and it must display the motions around this time-average that allow its activity. The solvent plays an essential role in protein folding, stabilization and biological activity. Scattering methods are an excellent tool to study the structure of proteins in solution [10, 13, 20], the properties of crowded protein solutions [27] or whole cellular objects [16]. Neutron spectroscopy is particularly useful to investigate the dynamic nature of proteins in solution [24, 25, 28]. In the following chapter I will describe some illustrative examples of experimental studies using neutron spectroscopy that have been performed recently.

1.1 *Protein Folding and the Case Study of Apomyoglobin*

Protein folding occurs from the disordered and unfolded state along the folding funnel towards the fully folded conformation that is biologically active and functional. A schematic illustration is shown in Fig. 1. For some proteins local minima exist on the free energy surface that can trap the protein transiently in partially folded intermediate states. After a certain time spent in the local energy minimum the protein escapes and finally reaches the global free energy minimum, where it folds into its native conformation. The partially folded intermediates are also termed molten globules (MGs) as the protein chain is in a compact conformation, but its three-dimensional structure is still not fully formed and remains locally unstructured and flexible. Throughout the folding transition the unfolded and partially folded protein chain is sampling the accessible conformational space in order to reach the correctly and fully folded conformation at the end of the folding funnel. This illustrates directly the high relevance of molecular dynamics for protein folding [24–26].

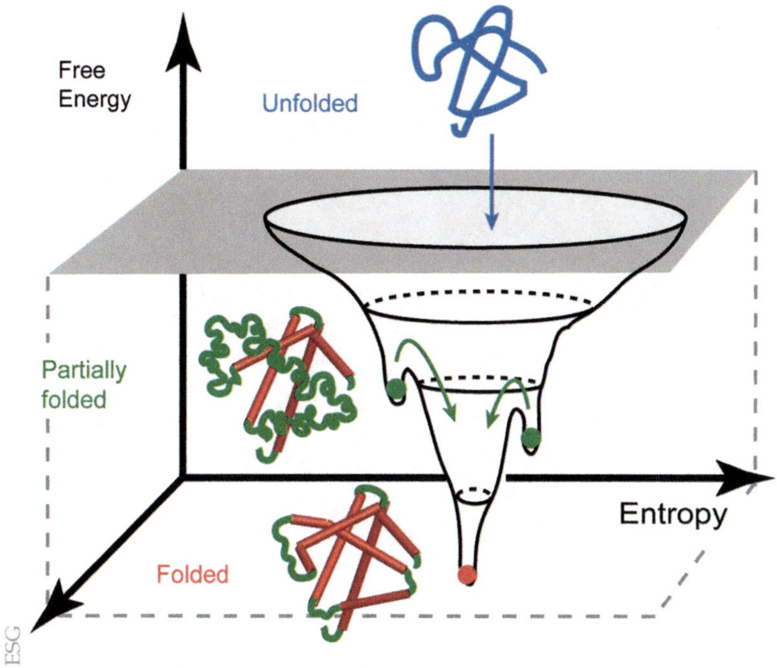

Fig. 1 Schematic illustration of the myoglobin folding transition. The fully unfolded protein chain rapidly folds via several kinetic folding intermediates towards the fully folded native structure in the global minima of the folding funnel. The red cylinders mark folded helices of the folded and of the partially folded conformation. Loops and disordered elements are shown in green of the native state and of the intermediate

In thermodynamic terms it is the difference of the free energy

$$\Delta G = G_{unfolded} - G_{folded} = \Delta H - T \cdot \Delta S \quad (1)$$

between the unfolded and the folded state that is driving the folding transition. The difference in ΔG between the unfolded and folded state consists of the enthalpic contribution ΔH and the entropic contribution ΔS. Major contributions to the enthalpic term are hydrophobic, Van-der-Waals interactions and hydrogen bonding, whereas the entropic term is related to changes in conformational entropy between the unfolded protein and the folded complex. It is important to note here that both terms ΔH and ΔS contain the contribution of the protein and of the hydration shell.

Myoglobin (Mb) serves as a model system to study basic physical properties of protein folding [7] and protein dynamics [12]. The fully folded protein consists of eight α-helices that are interconnected by loops. The protein acts as a scaffold for a haeme-group that can bind one O_2 molecule reversible, see Fig. 2 for the crystal structure of Mb. Mb with bound haeme-group is termed holo-Mb, while the

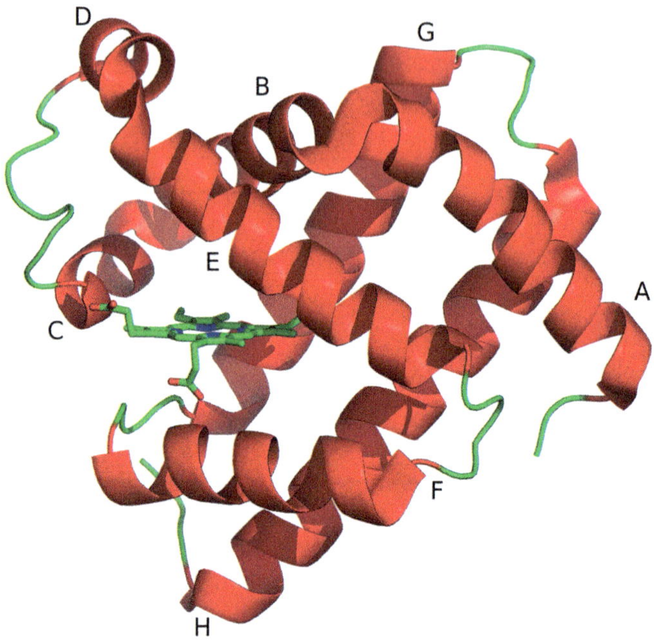

Fig. 2 Folded structure of myoglobin as determined by X-ray crystallography (pdb code: 2V1 K). The folded protein is formed of eight α-helices (drawn in red) that are connected by turns (shown in green). The protein carries a haeme-group (drawn with lines) that can bind one oxygen molecule. The long G and H helices form the hydrophobic core to which the smaller helices A and B are bound in the partially folded intermediates

haeme-free protein is called apo-Mb. The ancient Greek words 'holo' and 'apo' mean 'complete' and 'without', which refers to the presence or absence of the haeme-group.

The overall structure of native apo-Mb is generally well defined, but the haeme-binding pocket formed is unstructured and flexible [9]. During the folding process the long helices G and H, see Fig. 2, are formed first via a hydrophobic collapse [7]. In one molten globule state (MG I1) that occurs during the kinetic folding process the A helix is bound to the G and H helices [19]. In a second molten globule state (MG I2) that occurs after the MG1 state during the folding process the helices A, G, H are further stabilized and the helix B is additionally bound [19]. Therefore, the secondary structure content of the MG I2 state is higher than in the MG I1 state. Folding times of the apo-Mb molten globules are in the order of some micro- to milliseconds, while the full assembly of the native protein takes place within a few seconds [7]. Neutron spectroscopy experiments are quite time consuming and typically one has to measure the protein sample for a few hours up to several days, depending strongly on the spectrometer being used. Therefore, it is technically not feasible to investigate the dynamics of the transient folding intermediates during the kinetic folding process by neutron spectroscopy in real-time.

The apo-Mb system, however, has the tremendous advantage that the partially folded molten globules MG1 and MG2 can be stabilized under equilibrium conditions by choosing specific solvent conditions [25]. Furthermore, a third partially folded molten globule can be stabilized that has less secondary structure content than the MG I1 and MG I2 states [25]. These properties enable neutron spectroscopy experiments of the various folded conformations of apo-Mb and make it an ideal model system for the investigation of the role of molecular dynamics for protein folding.

1.2 The Intrinsically Disordered Myelin Basic Protein

For many years since the first protein structures were determined, biological activities of proteins were associated with the folding of the polypeptide chain into a well-defined three-dimensional structure. However, this structure/function paradigm was challenged with the discovery that many proteins are fully disordered or contain disordered regions under physiological conditions [6]. As disorder is a characteristic property of these proteins, they have been termed intrinsically disordered proteins (IDPs) or also intrinsically unfolded proteins. These proteins are, however, not fully unfolded, but contain a large amount of disordered regions. The biological role of IDPs is believed to be highly conformational adaptive, which would be important for association with multiple binding partners, or to respond rapidly to different environmental conditions [8, 30]. IDPs which are involved in molecular recognition processes can even fold into different structures when bound to their different physiological target proteins, reviewed for example in [33]. Crystallographic structures of IDPs do not exist yet, due to the existence of a large number of conformational states. However, the structure of IDPs at low resolution in solution can be well investigated by small-angle X-ray and neutron scattering (SAXS/SANS) [5], while neutron spin-echo spectroscopy (NSE) has proven to be the method of choice to study polymer dynamics [23] and protein motions [4, 28] up to several hundred ns.

Myelin basic protein (MBP) is a major component of the myelin sheath in the central nervous system. MBP is not an integral membrane protein, but due to its positive charge interacts with the lipid head groups and stabilizes the membranes [17]. From a medical aspect MBP has significant importance as there are many neurological disorders, e.g. multiple sclerosis, which are related to MBP misfolding and malfunction [17]. In aqueous solution and in the lipid-free form MBP is primarily unstructured and is classified as intrinsically disordered [17], as judged from the net charge versus hydrophobicity scheme by Uversky et al. [31, 32]. From a biophysical point of view MBP is an interesting system to study the basic physical properties of disordered or partially unfolded proteins in solution.

2 Protein Dynamics Investigated by Neutron Spectroscopy

Energy-resolved neutron scattering is amongst the few methodologies that have provided experimental data in the area of biological molecular dynamics. In quasielastic incoherent neutron scattering (QENS) experiments it is the incoherent scattering of H atoms that is mainly observed and analysed. In the time-scales examined, H atom motions reflect those of the chemical groups to which they are bound. They are, therefore, indicators of internal and global dynamics of proteins in solution [14, 15, 21, 29] on the picosecond to nanosecond time-scale.

2.1 Picosecond Dynamics of Unfolded, Partially Folded and Native Proteins

In a first set of experiments fast dynamics of apo- and holo-Mb on the picosecond time-scale have been studied by QENS [25]. The experiment has been performed on the neutron time-of-flight spectrometer IN6 located at the Institut Laue-Langevin. The instrument was used with an energy resolution of 100 μeV allowing probing fast protein motions on the picosecond time-scale. Sample temperature was 16 °C throughout the experiment. Different folded, partially folded and fully unfolded conformations of apo- and holo-Mb have been investigated. Choosing specific solvent conditions allowed the modification of the secondary structure content of the protein. Folding intermediates can be stabilized by the addition of salt or pD 4, while the acid denatured state at pD 2 was used as a reference for the unfolded state. QENS experiments are performed in heavy water buffer as it allows to suppress the contribution of the solvent to measured signal. Table 1 summarizes the solvent conditions, the general state of the protein and the secondary structure content of the protein.

The recorded QENS spectra contain the relevant information on internal protein dynamics. For a full description of the theory and application of QENS in biology see e.g. [11]. Mean square displacements (MSDs) $\langle x^2 \rangle$ that describe the flexibility of the protein have been calculated from the recorded QENS spectra. It was evident

Table 1 Solvent conditions, unfolded (U), molten globule (MG), folded (F) states of the protein and secondary structure content [25]

Protein	State	α-helical content (%)
apo-Mb at pD 2	U	6
apo-Mb at pD 2, 100 mM NaCl	MG	22
apo-Mb at pD 4	MG I1	43
apo-Mb pD 2, 20 mM NaTCA	MG I2	49
apo-Mb at pD 6	F	54
holo-Mb pD 6	F	66

that the partially folded MGs have a higher flexibility than the folded states, while the denatured and fully unfolded chain has largest flexibility. The difference in conformational entropy in proteins can be determined from the MSDs as suggested by [22] according to

$$\Delta S_{conf} = 3R\ln\left(\sqrt{\frac{\langle x_u^2\rangle}{\langle x_f^2\rangle}}\right) \quad (2)$$

where $\langle x_u^2\rangle$ and $\langle x_f^2\rangle$ are the MSDs of the unfolded state and the folded/partially folded conformation at the same temperature, and $R = 8.3144$ J/K/mol is the gas constant. In our work the acid denatured state was considered as a reference for the unfolded state. The determined values of ΔS_{conf} at 16 °C are shown in Fig. 3 as a function of the measured α-helical content of the proteins.

We find a linear increase of ΔS_{conf} with the helical structure content. The thermodynamic entropy difference $\Delta S = \Delta S_{\mathrm{conf}} + \Delta S_{\mathrm{hydr}}$ at 16 °C calculated from known thermodynamic parameters, see references in [25], is also shown in Fig. 3. The thermodynamic entropy difference consists of the conformational entropy difference of the protein ΔS_{conf} and of the entropy difference of the hydration water ΔS_{hydr}. It is evident that the thermodynamic entropy difference of the MGs is only slightly larger than zero, while the folded native apo- and holo-proteins have negative values of ΔS. The physical reason for the difference between ΔS and ΔS_{conf} is the contribution of the hydration water to the thermodynamic measurements. The entropy content of the hydration water can be determined from the measured data by taking the difference $\Delta S_{\mathrm{hydr}} = \Delta S - \Delta S_{\mathrm{conf}}$, although there is a contribution of conformational motions on slower time scales, which cannot be detected by the neutron spectrometer used. Our results demonstrate that water molecules around the MGs are only slightly more dynamically disordered than in the unfolded state, whereas in the folded conformations the entropy content of the hydration water is significantly larger than in the MGs and in the unfolded state. Buried hydrophobic residues in the core of the protein are responsible for that effect, which become progressively more solvent exposed in the partially folded conformations than in the native states.

2.2 *High-Resolution Neutron Spectroscopy to Study Global Diffusion and Slower Internal Dynamics*

Based on the success of the previous study we furthermore investigated dynamics of the differently folded conformations of apo- and holo-Mb on the slower picosecond to nanosecond time-scale using high-resolution neutron time-of-flight and neutron backscattering spectroscopy [24]—Published by the PCCP Owner

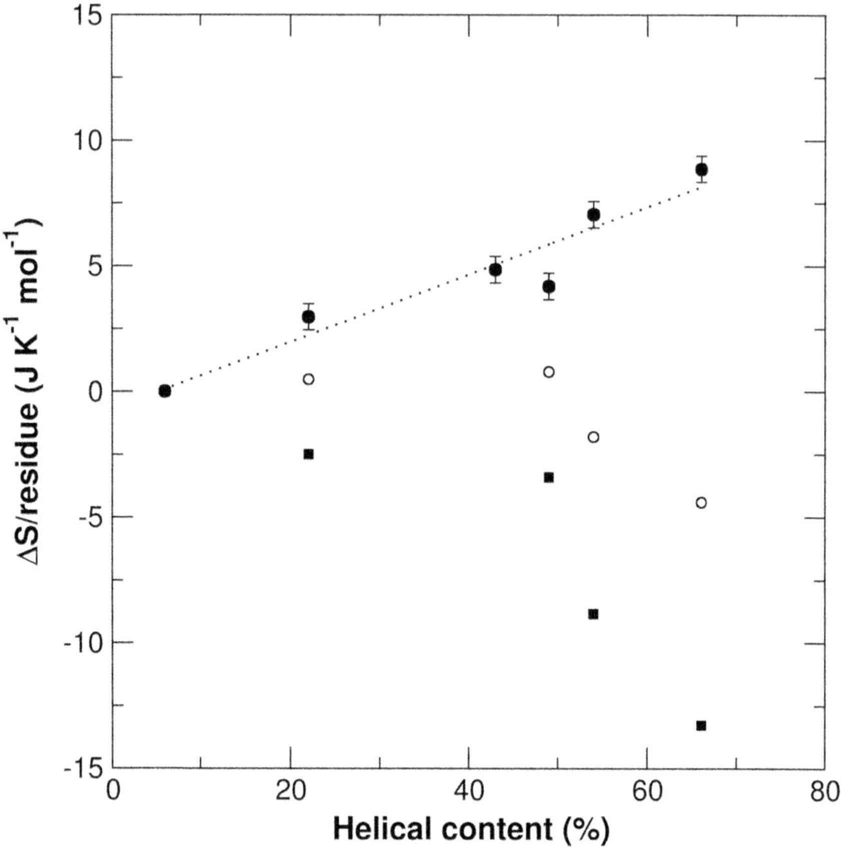

Fig. 3 Entropy difference per residue of the different partially and fully folded conformations of apo-Mb and holo-Mb. The difference of conformational entropy ΔS_{conf} measured by neutron scattering (filled circles), the thermodynamic entropy difference $\Delta S = \Delta S_{\mathrm{conf}} + \Delta S_{\mathrm{hydr}}$ calculated with the Gibbs-Helmholtz equation with the known thermodynamic parameters (empty circles), and the difference $\Delta S_{\mathrm{hydr}} = \Delta S - \Delta S_{\mathrm{conf}}$ (filled squares) due to the entropy content of the hydration water and slow moving protein dynamics [25]

Societies. Experiments were performed on the neutron time-of-flight spectrometer IN5 at the Institut Laue-Langevin and on the neutron backscattering spectrometers IRIS at the ISIS spallation source and on IN16 at the ILL. The energy resolutions of IN5 and IRIS were 12 and 18 μeV allowing to probe slower motions on the picosecond time-scale, whereas the high 0.9 μeV energy resolution of IN16 accessed slower dynamics up to the nanosecond time-scale. A compilation of the different samples their solvent composition and the protein concentration is given in Table 2.

The measured QENS spectra allowed the separation of internal protein dynamics and global protein diffusion. The QENS spectra could be described well by a sum of

Table 2 α-helical content measured by circular dichroism, unfolded (U), folded (F), molten globule (MG) states, solvent composition and protein concentration of the investigated protein solutions [24]

α-helical content (%)	State	Solvent composition	c (IN16) (mg/ml)	c (IRIS) (mg/ml)	c (IN5) (mg/ml)
6	U	apo-Mb pD 2	62	65	41
22	MG	apo-Mb pD 2, 0.1 M NaCl	54	84	–
43	MG	apo-Mb pD 4	54	–	32
54	F	apo-Mb pD 6	50	–	48
66	F	holo-Mb pD 6	56	56	

two Lorentzians. The narrow Lorentzian accounts for global protein diffusion, whereas the second broader Lorentzian describes the combination of global protein diffusion and internal protein motions. Exemplary QENS spectra of the unfolded apo-Mb at pD 2 (6% α-helical content) measured using IRIS and IN5 are shown in Fig. 4.

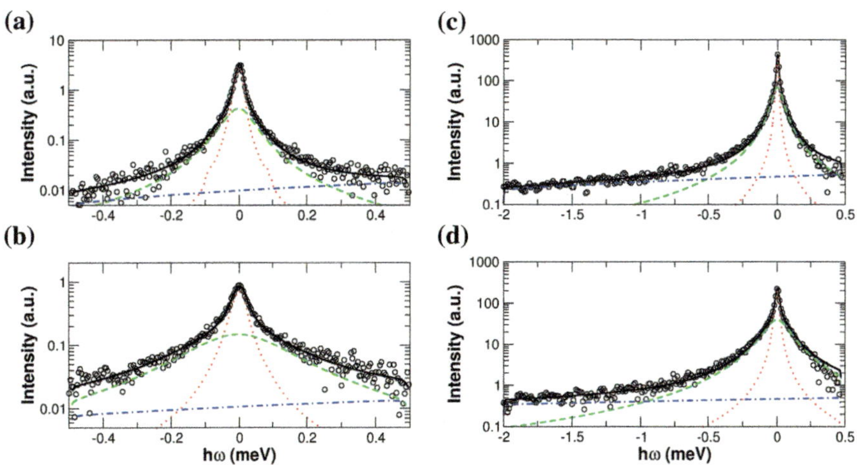

Fig. 4 QENS spectra of unfolded apo-Mb at pD 2 (6% α-helical content) measured on IRIS **a** at the scattering vectors $q = 0.50\ \text{Å}^{-1}$ and **b** $q = 1.44\ \text{Å}^{-1}$, and on IN5 **c** at $q = 0.54\ \text{Å}^{-1}$ and **d** $q = 1.10\ \text{Å}^{-1}$. The sample temperatures are 303 and 295 K for the shown IRIS and IN5 data, respectively. Empty symbols are the experimental data. For clarity only every forth data point is shown for IRIS spectra and every fifth point for data measured on IN5. The solid black line represents the total fit to the QENS spectra, the red dotted line is the narrow Lorentzian accounting for global protein diffusion, the green dashed line is the broad Lorentzian caused by the combination of global diffusion and internal protein dynamics, and the blue dashed-dotted line accounts for the linear background. All curves are convoluted with the instrumental resolution functions [24]

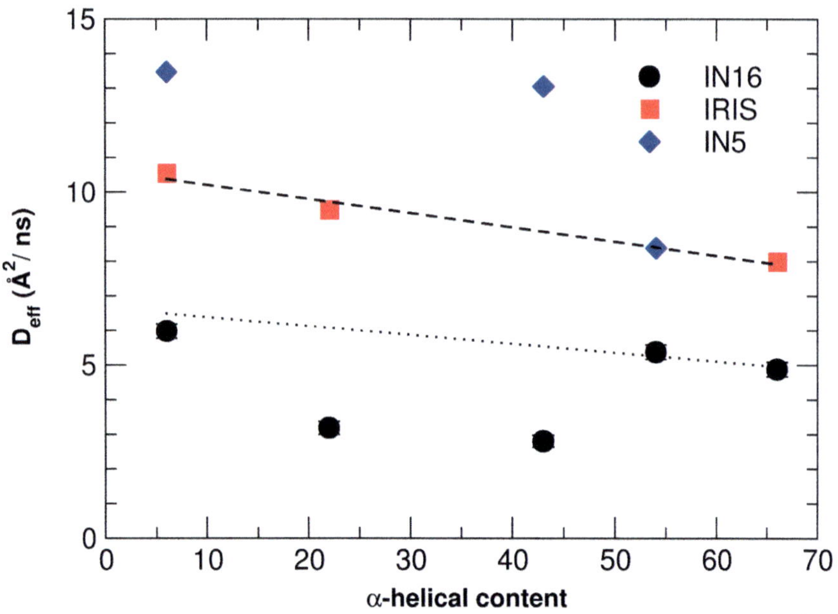

Fig. 5 Effective diffusion coefficients as a function of the degree of a-helical content at a temperature of 289.2 K. The dashed line is a linear fit to the IRIS and IN5 data excluding the more diluted solutions measured on IN5. The dotted line shows the scaled linear fit (dashed line) such that the IN16 value of holo-Mb at 66% a-helical content is matched. The scaled linear fit agrees with the IN16 results of the acid denatured and native apo-Mb states, but reveals deviations of the MGs due to gel formation [24]

Effective diffusion coefficients of global protein diffusion were extracted from the q^2-dependence of the measured line-widths of the narrow Lorentzian. The determined effective diffusion coefficients of global protein diffusion at 289.2 K are shown in Fig. 5.

Due to the lower protein concentrations of acid denatured apo-Mb at pD 2 (c = 3%) and of the MG state at pD 4 (c = 4%) measured by IN5 the determined diffusion coefficients are larger than the values of the other samples investigated by IRIS and IN5, and remain at a similar level irrespective of the degree of α-helical content. In the long-time limit of global protein diffusion observed by dynamic light scattering (DLS) unfolded and MG states of apo-Mb are characterized by slower protein diffusion due to the larger structural expansion as compared to the native folded states [25]. On the other hand, the short-time limit of global protein diffusion probed by IRIS and IN5 follows an inverse behaviour for proteins at higher concentration (c $\geq$ 5%): a reduced diffusivity with increasing degree of α-helical content. A physical explanation for that behaviour would be that repulsive interactions between charged proteins at pD 2 and 4 at high concentration increase protein diffusion in the short-time limit. Indeed, charged colloidal spheres diffuse slightly faster than non-charged spheres at the volume fractions explored here [2]. At lower protein concentration (c = 3–4%) charge interactions seem to play a minor role.

Concerning the D_{eff} measured on IN16 we obtain reduced absolute values as compared to the IRIS and IN5 results. Global protein diffusion of acid denatured apo-Mb is slightly faster than the folded conformations, which is in agreement with the observations described above. We find a D_{eff} value of holo-Mb measured on IRIS being a factor of 1.64 larger than the corresponding IN16 result. After rescaling the D_{eff} determined with the low-resolution spectrometers the values of acid denatured and folded apo-Mb coincide with the IN16 results (see dotted line in Fig. 5). The explanation here is that the slow internal process in the protein interferes with the narrow Lorentzian in the QENS spectra of the lower-resolution instruments, which cannot discriminate both processes accurately due to the lower energy resolution. As a result larger effective diffusion coefficients are obtained for IRIS and IN5.

A further remarkable observation is that the diffusivity of the MG states measured on the high-resolution spectrometer is strongly reduced as compared to the other protein solutions. The used MG states at high concentration had gel-like macroscopic viscosity. Previously, it was observed by high-resolution neutron spectroscopy of concentrated protein solutions around the denaturing temperature that protein denaturation results in the formation of an entangled gel-like network with significantly reduced global protein diffusion as compared to the native folded state [18]. Concerning the MG states the reduced D_{eff} values also seem to be the signature for the onset of protein entanglement of the partially folded proteins on the nanosecond time scale. Although the acid denatured state is even fully unfolded the higher charge of the protein increases protein repulsion, which appears to prevent protein entanglement and the reduction of global protein diffusion. Concerning the MG state with 22% α-helical content we find a significant discrepancy between the scaled IRIS and IN16 data. That observation indicates that on the faster picosecond time scale probed by IRIS the initial step of global protein diffusion is not sensitive to protein entanglement, while on the slow nanosecond time scale as seen by IN16 the individual proteins sense the presence of the neighbouring macromolecules, which reduces global protein diffusion.

MSDs of internal motions were calculated from the QENS spectra. The difference in conformational entropy between the acid denatured state and the partially or folded conformations was calculated according to Eq. 2. The experimentally determined $\Delta S_{conf}(T)$ profiles of the partially and fully folded conformations of apo- and holo-Mb are plotted in Fig. 6a as a function of temperature. In general, larger ΔS_{conf} values are found for more folded conformations. The difference of conformational entropy is close to or below zero for the MG states, which is an obvious result as the RMSDs of the acid denatured state were found to be similar or larger than the corresponding values of the MGs. A remarkable result is the parabolic shape of the $\Delta S_{conf}(T)$ profiles most clearly visible for the native holo- and apo-Mb states at pD 6. The parabolic profile is reminiscent of the parabolic shape of the thermodynamic free energy difference $\Delta G(T) = \Delta H - T\Delta S$ between the unfolded and folded states of proteins, where ΔH is the enthalpic component and ΔS is the contribution due to entropy. The maxima of $\Delta S_{conf}(T)$ are shifted to lower temperatures from holo-Mb via apo-Mb at pD 6 to the MG states, which is also the case

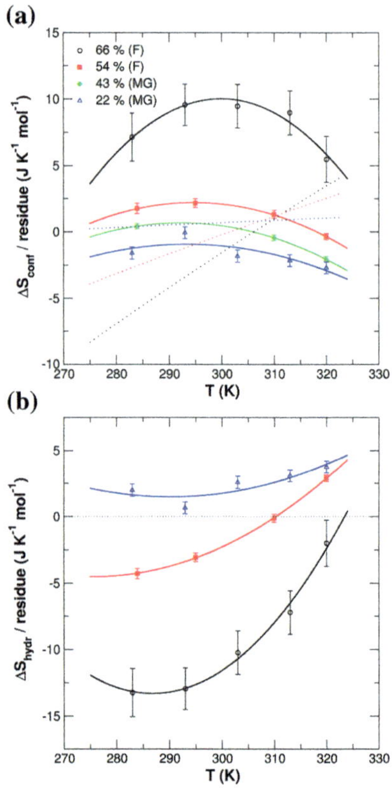

Fig. 6 **a** Difference in conformational entropy $\Delta S_{\rm conf}$ between the acid denatured state of apo-Mb and the protein in the investigated partially folded and folded states as obtained from Eq. 2. The α-helical content of the native folded (F), partially folded molten globule (MG), and unfolded (U) states is given in the legend (empty symbols IRIS, filled symbols IN5 data). The dotted lines are $\Delta S = \Delta S_{\rm conf} + \Delta S_{\rm hydr}$ calculated from thermodynamic parameters. **b** Difference in conformational entropy $\Delta S_{\rm hydr} = \Delta S - \Delta S_{\rm conf}$ related to water molecules of the hydration shell and protein motions that are out of the time window of the neutron spectrometers. The solid lines in **a** and **b** are polynomial fits of second order and serve as a guide to the eye [24]

for the thermodynamic $\Delta G(T)$ profiles [25]. Important to repeat here is that ΔG, ΔH and ΔS are thermodynamic quantities that contain the contributions of both the protein and the hydration layer, e.g. $\Delta S_{\rm conf}$ and $\Delta S_{\rm hydr}$ for the entropic contributions.

The $\Delta S(T)$ curves of folded holo- and apo-Mb (66 and 54% α-helical content) and of the MG state stabilised by 0.1 NaCl (22% α-helical content) are plotted in Fig. 6a in comparison to $\Delta S_{\rm conf}(T)$. The quantities ΔS and $\Delta S_{\rm conf}$ are different as $\Delta S_{\rm conf}$ is derived from the extent of protein fluctuations without the contribution of the hydration shell.

The entropy difference between the unfolded and the folded states that is related to the contribution of water molecules in the hydration layer of the protein can be determined by

$$\Delta S_{hydr}(T) = \Delta S(T) - \Delta S_{Conf}(T) \quad (3)$$

The ΔS_{hydr} values of the different structural conformations of apo- and holo-Mb are plotted in Fig. 6b. One should note that the experimentally determined ΔS_{hydr} values are not only related to hydration water, but also contain a contribution of slow protein motions that are out of the accessible time window of the neutron spectrometers.

Water molecules in the hydration layer of the folded proteins are found to be more disordered and mobile than in the unfolded state, which leads to negative ΔS_{hydr} values. See Fig. 7 for an illustration of disordered hydration water around fully folded native proteins, whereas hydration water around unfolded and partially disordered protein regions is less mobile and more ordered. The experimental results show that with decreasing degree of protein folding the hydration layer is getting less mobile and for the MG state is even more ordered than in the unfolded acid denatured apo-Mb, which is visible by the general increase of ΔS_{hydr}. The physical reason is that water molecules form ordered structures around solvent exposed hydrophobic amino-acids [1]. With decreasing α-helical content hydrophobic residues from the protein core are getting more solvent exposed and thus induce ordering of the hydration shell. Interestingly, we observe temperature dependent ΔS_{hydr} profiles. At the point $\Delta S_{\text{hydr}}(T_0) = 0$ the entropic contribution of hydration water of the folded conformation equals that of the hydration layer around the unfolded state and is even smaller at higher temperatures. The

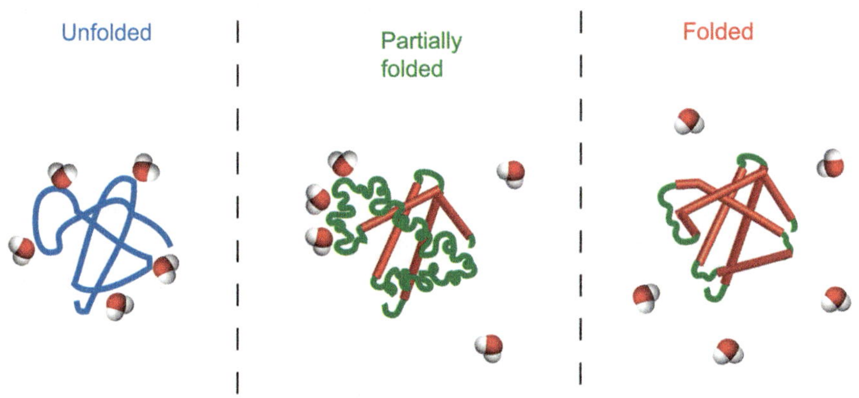

Fig. 7 Schematic illustration water in the close vicinity of the protein surface for unfolded, partially folded and folded native conformations. Water molecules close to hydrophobic regions of unfolded and partially folded regions that are solvent exposed are ordered and less mobile, while hydration water in close contact with hydrophilic regions of folded proteins is less ordered and more mobile

intersection temperature T_0 is shifted to lower temperatures for the less folded apo-Mb states. That behaviour is related to the interplay of two phenomena: (i) melting of the ordered water structures around the hydrophobic residues of the unfolded state at higher temperatures, and (ii) simultaneously an increase of ordered water molecules in the hydration shell of the folded states. An explanation for (ii) might be that water molecules can progressively interact with hydrophobic residues in the core of the folded structures at higher temperatures, which induces ordering of the hydration shell.

3 Dynamics of the Intrinsically Disordered Myelin Basic Protein

IDPs lack a well-defined folded structure and contain a high degree of structural freedom and conformational flexibility, which is expected to enhance binding to their physiological targets. In solution and in the lipid-free state MBP belongs to that class of proteins. In recent experiments we studied the protein as a biophysical model system for IDPs. Using SAXS and modelling by a coarse-grained structural ensemble gave indications for a compact core with flexible ends. NSE measurements revealed a large contribution of internal dynamics to the overall diffusion.

Small-angle scattering was measured to gain information about the structure of MBP as a prerequisite for the NSE experiments. Reverse Monte Carlo simulations were applied to generate a coarse-grained ensemble representing the structural characteristics of MBP. The EOM software program [3] was used to generate 10,000 random-like protein chains followed by the selection of a representative ensemble describing the measured SAXS data. The general features of the selected coarse-grained conformations indicate a central core region with flexible termini, while the overall shape appears to be slightly bent, see Fig. 8.

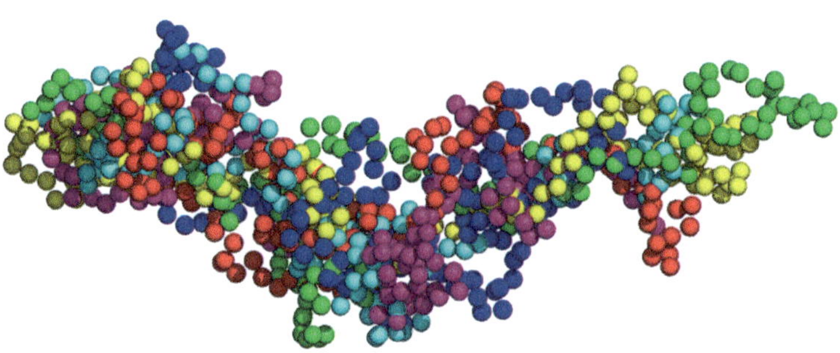

Fig. 8 Representative coarse-grained conformations of MBP as determined by using the EOM software [3]. The structures are rotated by 90° in the lower part of the figure. The blue structure is referred to as model A. Reprinted with permission from [28]. Copyright (2014) American Chemical Society

NSE was used to investigate slow collective dynamics of MBP up to 140 ns in the small-angle range, where internal dynamics and global protein diffusion are visible. A sample with a high protein concentration of 54 mg/ml was used that allowed to measure NSE spectra up to 140 ns at the compromise of the presence of hydrodynamic interactions and structure factor effects due to the high protein concentration.

We used the structural ensemble to interpret the full NSE spectra. The full NSE spectra of the 54 mg/ml solution were described an analytical model accounting for rigid body diffusion and internal protein dynamics [28]. Internal protein dynamics was described by an exponential decay with a q-independent rate Γ, and a q-dependent contribution of internal dynamics to the NSE spectra.

The resulting fit with a representative structural model (coarse-grained structure called model A in Fig. 8) to the NSE spectra is shown in Fig. 9. At shorter times the contribution of internal protein dynamics to the spectra becomes directly visible as compared to the long time limit. In general, the structure based model gives a good fit to the NSE data.

As we determined several possible structural models the NSE spectra were fitted with each model. Average values have been determined that describe the general dynamic features of the different possible protein conformations. On average a single relaxation time $1/\Gamma = 8.4 \pm 2.0$ ns of the whole structural ensemble could be determined from the NSE measurement. To interpret the observed internal protein dynamics in more detail the lowest lying collective excitations of the structural ensemble were calculated using normal mode (NM) analysis. We remind here that NMs 1–6 are the so-called trivial modes that account for translational and rotational motion of the particle. Information about internal dynamics is contained in the higher order modes. The calculated displacement patterns of the first and

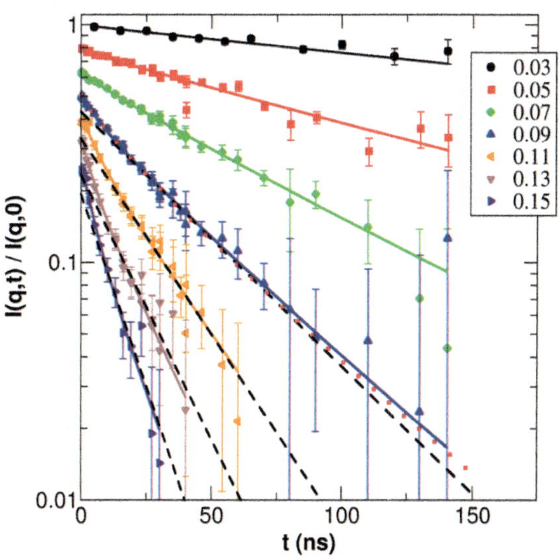

Fig. 9 Measured NSE spectra of MBP. All spectra start at unity but are shifted consecutively by a factor of 0.8 for clarity. The q-values in the legend are given in $Å^{-1}$. The solid lines are fits to the NSE data using structural model A. The dashes black lines indicate the long-time diffusion limit. The dotted red line is a stretched exponential fit. Reprinted with permission from [28]. Copyright (2014) American Chemical Society

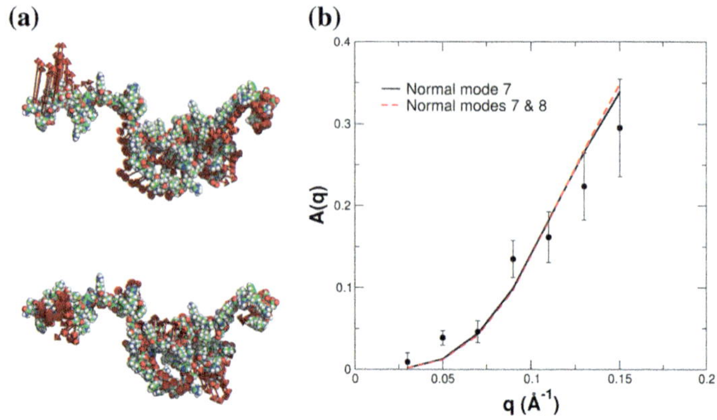

Fig. 10 **a** Calculated NMs of the representative MBP conformation model A. NM 7 in the upper part, NM 8 in the lower part of the figure. The lengths of the vectors are increased for better visibility. **b** Contribution of internal protein dynamics to the NSE spectra $A(q)$. The solid and dashed lines are fits to the data with NM 7 and the combination of NMs 7 and 8. Adapted with permission from [28]. Copyright (2014) American Chemical Society

second non-trivial NMs 7 and 8 describing internal collective dynamics of a representative conformation (model A) are shown in Fig. 10.

In general, the NMs 7 and 8 correspond to bending and stretching motions of all structural models. The NMs 7 and 8 were used to describe the measured contribution of internal dynamics $A(q)$ to the NSE spectra, see Fig. 10b. NM 7 is the dominating collective excitation and fully reproduces the measured $A(q)$. The RMSDs were calculated as average values of the whole structural ensemble and the errors as standard deviations. The RMSD of NM 7 was found to be 2.1 ± 0.3 Å, while the weighted combination of NMs 7 and 8 yielded a RMSD of 2.2 ± 0.3 Å. Including higher NMs does not modify the RMSD. The displacement vectors from NM analysis indicate that the centre part of MBP remains rather rigid, while the termini are more flexible. In general, the amplitudes of motion correspond to distances between the end termini of the coarse-grained structures. An intriguing idea would be that the observed dynamics represents movements with an average relaxation time of 8.4 ns between the different structures of the selected configurational ensemble.

4 Conclusions

Protein dynamics is of great importance for protein folding, as the polypeptide chain needs to explore the accessible conformational space during the folding transition. The most relevant thermodynamic quantity concerning protein folding that can be determined from the neutron scattering experiments is the difference in conformational entropy ΔS_{conf} between two specific structural states. Classic calorimetric experiments allow the determination of ΔS, which contains both the contribution of the protein (ΔS_{conf}) and that of the hydration shell (ΔS_{hydr}). Using thermodynamic parameters from the literature and the experimentally determined ΔS_{conf} values as input we could determine ΔS_{hyd} of the different folded states of the protein [24, 25]. Whereas the hydration layer of the partially folded state is found to be slightly more ordered than that of the acid denatured unfolded state, mobile and disordered water molecules in the hydration layer of the folded states make a significant contribution to the entropic stabilization of the folded structures. At higher temperatures the hydration layer around the folded states is getting less mobile and more ordered than in the unfolded state, which provides a source for the entropic stabilization of the unfolded state and facilitates protein unfolding.

In a second research project we investigated the structural and dynamic characteristics of an intrinsically disordered protein [28]. As a model system we chose the intrinsically disordered myelin basic protein. Structural aspects were investigated by SAXS and SANS, whereas NSE experiments have been performed to characterize the dynamic features of the protein. Computer based reconstructions based on the measured SAXS data provided us with a coarse-grained conformational ensemble. The different structures have been used as test-structures for the interpretations of the NSE results. The NSE experiments revealed a large contribution of internal dynamics to the overall diffusion. An analytical model based on global protein diffusion and internal protein dynamics was used successfully for the interpretation of the full NSE spectra. Displacement patterns along the first two normal modes demonstrated that collective stretching and bending motions of the flexible ends dominate the internal modes. The observed dynamics were interpreted as nanosecond conformational fluctuations within the reconstructed coarse-grained structural ensemble. In that way the MBP protein is able to explore large configurational space as it is generally believed to be a characteristic properties of IDP. Our results are important for a biophysical understanding of the nature and extend of large-scale conformational motions in IDPs.

Acknowledgements I would like to thank the Heinz Maier-Leibnitz Zentrum, the Institut Laue-Langevin, the ISIS Pulsed Neutron and Muon Source and the European Synchrotron Radiation Facility for provision of neutron and synchrotron radiation beam time. I also would like to thank my collaborators who appear as co-authors of the presented scientific articles.

References

1. Ball, P. (2008). Water as an active constituent in cell biology. *Chemical Reviews, 108,* 74–108. https://doi.org/10.1021/cr068037a.
2. Banchio, A. J., & Nägele, G. (2008). Short-time transport properties in dense suspensions: From neutral to charge-stabilized colloidal spheres. *The Journal of Chemical Physics, 128,* 104903. https://doi.org/10.1063/1.2868773.

3. Bernadó, P., Mylonas, E., Petoukhov, M. V., et al. (2007). Structural characterization of flexible proteins using small-angle X-ray scattering. *Journal of the American Chemical Society, 129,* 5656–5664.
4. Biehl, R., & Richter, D. (2014). Slow internal protein dynamics in solution. *Journal of Physics: Condensed Matter, 26,* 503103. https://doi.org/10.1088/0953-8984/26/50/503103.
5. Cordeiro, T. N., Herranz-Trillo, F., Urbanek, A., et al. (2017). Small-angle scattering studies of intrinsically disordered proteins and their complexes. *Current Opinion in Structural Biology, 42,* 15–23.
6. Dunker, A. K., Oldfield, C. J., Meng, J., et al. (2008). The unfoldomics decade: An update on intrinsically disordered proteins. *BMC Genomics, 9,* S1. https://doi.org/10.1186/1471-2164-9-S2-S1.
7. Dyson, H. J., & Wright, P. E. (2017). How does your protein fold? Elucidating the apomyoglobin folding pathway. *Accounts of Chemical Research, 50,* 105–111. https://doi.org/10.1021/acs.accounts.6b00511.
8. Dyson, H. J., & Wright, P. E. (2005). Intrinsically unstructured proteins and their functions. *Nature Reviews Molecular Cell Biology, 6,* 197–208.
9. Eliezer, D., & Wright, P. E. (1996). Is apomyoglobin a molten globule? Structural characterization by NMR. *Journal of Molecular Biology, 263,* 531–538. https://doi.org/10.1006/jmbi.1996.0596.
10. Endres, S., Granzin, J., Circolone, F., et al. (2015). Structure and function of a short LOV protein from the marine phototrophic bacterium Dinoroseobacter shibae. *BMC Microbiology, 15,* 30. https://doi.org/10.1186/s12866-015-0365-0.
11. Fitter, J., Gutberlet, T., & Katsaras, J. (Eds.). (2006). *Neutron scattering in biology—Techniques and applications*. Berlin: Springer.
12. Frauenfelder, H., McMahon, B. H., & Fenimore, P. W. (2003). Myoglobin: The hydrogen atom of biology and a paradigm of complexity. *Proceedings of the National Academy of Sciences, 100,* 8615–8617. https://doi.org/10.1073/pnas.1633688100.
13. Granzin, J., Stadler, A., Cousin, A., et al. (2015). Structural evidence for the role of polar core residue Arg175 in arrestin activation. *Scientific Reports, 5,* 15808. https://doi.org/10.1038/srep15808.
14. Grimaldo, M., Roosen-Runge, F., Hennig, M., et al. (2015). Hierarchical molecular dynamics of bovine serum albumin in concentrated aqueous solution below and above thermal denaturation. *Physical Chemistry Chemical Physics, 17,* 4645–4655. https://doi.org/10.1039/c4cp04944f.
15. Grimaldo, M., Roosen-Runge, F., Zhang, F., et al. (2014). Diffusion and dynamics of γ-globulin in crowded aqueous solutions. *The Journal of Physical Chemistry B, 118,* 7203–7209. https://doi.org/10.1021/jp504135z.
16. Guehrs, E., Stadler, A. M., Flewett, S., et al. (2012). Soft X-ray tomoholography. *New Journal of Physics, 14,* 13022. https://doi.org/10.1088/1367-2630/14/1/013022.
17. Harauz, G., Ishiyama, N., Hill, C. M., et al. (2004). Myelin basic protein-diverse conformational states of an intrinsically unstructured protein and its roles in myelin assembly and multiple sclerosis. *Micron, 35,* 503–542.
18. Hennig, M., Roosen-Runge, F., Zhang, F., et al. (2012). Dynamics of highly concentrated protein solutions around the denaturing transition. *Soft Matter, 8,* 1628. https://doi.org/10.1039/c1sm06609a.
19. Jamin, M., & Baldwin, R. L. (1998). Two forms of the pH 4 folding intermediate of apomyoglobin. *Journal of Molecular Biology, 276,* 491–504. https://doi.org/10.1006/jmbi.1997.1543.
20. Kaschner, M., Schillinger, O., Fettweiss, T., et al. (2017). A combination of mutational and computational scanning guides the design of an artificial ligand-binding controlled lipase. *Scientific Reports, 7,* 42592. https://doi.org/10.1038/srep42592.
21. Monkenbusch, M., Stadler, A., Biehl, R., et al. (2015). Fast internal dynamics in alcohol dehydrogenase. *The Journal of Chemical Physics, 143,* 75101.

22. Receveur, V., Calmettes, P., Smith, J. C., et al. (1997). Picosecond dynamical changes on denaturation of yeast phosphoglycerate kinase revealed by quasielastic neutron scattering. *Proteins, 28,* 380–387.
23. Richter, D., Monkenbusch, M., Arbe, A., & Colmenero, J. (2005). *Neutron spin echo in polymer systems*. Berlin: Springer.
24. Stadler, A. M., Demmel, F., Ollivier, J., & Seydel, T. (2016). Picosecond to nanosecond dynamics provide a source of conformational entropy for protein folding. *Physical Chemistry Chemical Physics, 18,* 21527–21538. https://doi.org/10.1039/C6CP04146A.
25. Stadler, A. M., Koza, M. M., & Fitter, J. (2015). Determination of conformational entropy of fully and partially folded conformations of holo- and apomyoglobin. *The Journal of Physical Chemistry B, 119,* 72–82. https://doi.org/10.1021/jp509732q.
26. Stadler, A. M., Pellegrini, E., Johnson, M., et al. (2012). Dynamics-stability relationships in Apo- and Holomyoglobin: A combined neutron scattering and molecular dynamics simulations study. *Biophysical Journal, 102,* 351–359.
27. Stadler, A. M., Schweins, R., Zaccai, G., & Lindner, P. (2010). Observation of a large-scale superstructure in concentrated hemoglobin solutions by using small angle neutron scattering. *The Journal of Physical Chemistry Letters, 1,* 1805–1808. https://doi.org/10.1021/jz100576c.
28. Stadler, A. M., Stingaciu, L., Radulescu, A., et al. (2014). Internal nanosecond dynamics in the intrinsically disordered myelin basic protein. *Journal of the American Chemical Society, 136,* 6987–6994. https://doi.org/10.1021/ja502343b.
29. Stadler, A. M., van Eijck, L., Demmel, F., & Artmann, G. (2011). Macromolecular dynamics in red blood cells investigated using neutron spectroscopy. *Journal of the Royal Society, Interface, 8,* 590–600. https://doi.org/10.1098/rsif.2010.0306.
30. Tompa, P. (2012). Intrinsically disordered proteins: A 10-year recap. *Trends in Biochemical Sciences, 37,* 1–8.
31. Uversky, V. N. (2002). Natively unfolded proteins: A point where biology waits for physics. *Protein Science, 11,* 739–756.
32. Uversky, V. N., Gillespie, J. R., & Fink, A. L. (2000). Why are "natively unfolded" proteins unstructured under physiologic conditions? *Proteins, 41,* 415–427.
33. Wright, P. E., & Dyson, H. J. (2009). Linking folding and binding. *Current Opinion in Structural Biology, 19,* 31–38. https://doi.org/10.1016/j.sbi.2008.12.003.

Author Biography

Andreas Stadler (Germany, France), Ph.D. He studied physics and biophysics in Munich and Paris. He continued his doctoral studies in the field of molecular biophysics using neutron spectroscopy with Dr. Giuseppe Zaccai at the Institut Laue-Langevin in Grenoble. He worked as postdoc with Prof. Georg Büldt at Forschungszentrum Jülich using X-ray holography to visualize cellular objects. Since 2011 he is staff scientist at Jülich Centre for Neutron Science at Forschungszentrum Jülich. Main research topics include the dynamics of partially and intrinsically disordered proteins and the application of neutron and X-ray scattering methods to biophysical questions.

Probing the Structure and Dynamics of Cells, Cell Components and Endogenous Nanoparticles Under Extreme Conditions with Neutrons

Judith Peters, Ruth Prassl and Philippe Oger

1 Introduction

Living cells are highly complex systems being composed of membranes (including lipids and proteins), cytosol containing among others mitochondria, lysosomes, ribosomes, Golgi apparatus, microtubules and the nucleus which comprises the DNA. Despite the difficulties related to such complexity, more and more efforts are undertaken to study and understand that complexity by means of different techniques such as modelling and simulations [1, 2], scattering (X-rays [3, 4] or neutrons [5, 6]) or imaging techniques (electron microscopy [7], fluorescence microscopy [8], etc.). In the present chapter, we will concentrate on neutron scattering methods to investigate living cells and natural nanoparticles, especially under extreme external conditions.

Neutrons represent an ideal probe for biological systems, as the neutron energies furnished by research reactors or spallation sources are of about a few meV and therefore match very well the order of magnitude of thermal energies present in these systems and typical binding energies [9]. Moreover, the corresponding wavelengths, which can be calculated through $\lambda = h/\sqrt{2mE}$ where m is the neutron

J. Peters (✉)
LiPhy, Univ. Grenoble Alpes, 38044 Grenoble, France
e-mail: peters@ill.fr

J. Peters
Institut Laue Langevin, 71 avenue des Martyrs, 38000 Grenoble, France

R. Prassl
Institute of Biophysics, Medical University of Graz, 8010 Graz, Austria
e-mail: ruth.prassl@medunigraz.at

P. Oger
INSA Lyon, CNRS, UMR 5240, Univ Lyon, 69621 Villeurbanne Cedex, France
e-mail: philippe.oger@insa-lyon.fr

G. M. Artmann et al. (eds.), *Biological, Physical and Technical Basics of Cell Engineering*, https://doi.org/10.1007/978-981-10-7904-7_18

mass and E its energy, h is Planck's constant and equal to 6.626×10^{-34} Js, are in the range of 0–20 Å, in good agreement with typical binding lengths in macromolecules. Neutrons have no electrical charge and penetrate easily into matter without damaging it. Consequently, structural and dynamical information can be obtained by various neutron scattering experiments even on live cells. This is of paramount importance as both structure and dynamics are connected and depend additionally on external conditions as the hydration state, temperature, salinity, pH value and pressure.

Cells are known to be able to adapt to various extreme conditions (see Fig. 1), although the mechanisms behind are not yet completely understood in all cases. For instance, some cells can bear temperatures below the freezing point while others require temperatures above the boiling point of water. Let us cite as examples the psychrophile *Aquaspirillum arcticum* living at 4 °C, the mesophiles *Escherichia coli* (*E. coli*) and *Proteus mirabilis*, used to 37 °C, the human body temperature, the thermophile *Thermus thermophilus*, adapted to 75 °C [5], and the hyperthermophile *Methanopyrus kandleri* living at up to 122 °C [10, 11]. The terms psychrophile, mesophile, thermo- or hyperthermophile give a classification of growth temperatures of cells from the lowest to the highest. The question how bacteria can adapt to such different temperature conditions was investigated in the past. For instance, concerning dynamics it was suggested that all of them had a similar flexibility at their physiological temperature, but the resilience increased with temperature, pointing to a higher stability of the thermo/hyperthermophile compared to the mesophile and to the psychrophile, while the flexibility is maintained in the limits required for the functioning of the bacteria [5].

Other species, such as the Archaeon *Halobacterium salinarum*, have evolved in an environment of high salinity, thus they are called halophiles. Most halophiles accumulate salt in their cytosol to counterbalance the osmotic pressure due to the high salt concentration outside their membrane. In this way, they are well adapted to salt and in contrast perceive low salt conditions as a stress. We found recently that such cells have the highest flexibility and lowest stability at 0 M salt and inversely a low flexibility and high stability in their native salinity at 4.2 M [12]. However, the cells present a pronounced response to salt concentrations only slightly lower than the native one. Otherwise, as dilution of the environment, for instance by rain, can change the salt concentration temporarily, it seems that such cells have to induce survival mechanisms quite early when the environmental salt concentration decreases.

Environments with extreme pH conditions are also encountered, for instance in the digestive tract. Therefore, for example *E. coli* has to tolerate them even if it is not an acidophile. Similar to salt, the external pH value can influence the intracellular pH and consequently the activity of many enzymes and the structure and flexibility of membranes [13]. A bacterium whose natural habitat is such an extremely acid environment as pH 2 is *Acidithiobacillus ferrooxidans*. Preliminary neutron results on membrane fragments originating from different bacteria seem to indicate a higher stability induced by a low pH value of samples not adapted to high salinity, although resilience is almost unchanged for the acidophile at pH 7 or pH 4.

Fig. 1 Examples of extremophiles: acidic waters (pH 2.2) of Rio Tinto harbouring bacteria that consume oxygen, salty stone at Dead Sea in Israel, *Escherichia coli*, silex spring in Yellowstone National Park, black smoker at 2400 m depth (courtesy of Wikimedia Commons)

These examples show that no general conclusions can be forwarded for the adaptation to different extreme conditions, although their study is of major importance for a better understanding of the diverse functioning of cells and possible applications towards cell engineering. In the following, we will show examples of microbes originating from the deep sea and living close to hot vents, therefore under combined conditions of high pressure and high temperature. Once we have introduced why pressure is even the better suited thermodynamic variable to probe molecular dynamics than temperature, we will present our conclusions about adaptation mechanisms to the environment close to hydrothermal vents. In a second step, we will introduce low density lipoproteins (LDL), which are endogenous nanoparticles transporting cholesterol in the blood, and outline how results from neutron experiments under high hydrostatic pressure can help to shed light on several properties of native lipoproteins and of a modified form of LDL mimicking a pathological state.

2 Neutron Scattering and Instruments

The quantity which is experimentally accessible by neutron scattering is the so called dynamic scattering function $S(Q, \omega)$, which is a function of both the momentum transfer Q and the energy transfer ω, in units of $\hbar$, between the neutron and the atomic nuclei of the sample. Neutron scattering comprises a coherent and an incoherent part, corresponding to a superposition of neutrons diffused by different atomic nuclei, thus to a collective effect, and to a superposition of neutrons diffused by the same atomic nucleus, but at different times. The corresponding scattering functions can be written as [14]:

$$S_{coh}(Q,\omega) = \frac{1}{N}\frac{1}{2\pi\hbar}\sum_{i,j}\int_{-\infty}^{\infty}\left\langle e^{-i\vec{Q}\,\vec{r}_i(0)}e^{i\vec{Q}\,\vec{r}_j(t)}\right\rangle e^{-i\omega t}dt \quad , \quad (1)$$
$$S_{inc}(Q,\omega) = \frac{1}{N}\frac{1}{2\pi\hbar}\sum_{i}\int_{-\infty}^{\infty}\left\langle e^{-i\vec{Q}\,\vec{r}_i(0)}e^{i\vec{Q}\,\vec{r}_i(t)}\right\rangle e^{-i\omega t}dt$$

where $\vec{r}_i$ are the positions of the nuclei and N their total number. The coherent scattering function gives access to structural information whereas the incoherent one permits to follow averaged individual motions of atomic nuclei within the sample, as the neutrons cannot distinguish between the particles. Although both parts are always present in neutron scattering experiments, the proportion of each is given by the corresponding cross section. The incoherent neutron scattering cross section of hydrogen atoms is about 40 times higher than that of any other atom present in biological samples, whereas it is small for the hydrogen's isotope deuterium [15]. Such fact permits an isotope labelling, e.g. an H/D exchange allowing to highlight certain parts of the sample, or the use of D_2O instead of H_2O as hydration agent to make the sample better visible. On the contrary, the coherent cross section of deuterium exceeds that of hydrogen, so that per-deuterated parts are better visible in the structure.

The scattering process might be elastic, when no energy is exchanged between the neutron and the sample, but the incident neutron changes its direction, or inelastic including the exchange of a certain amount of energy. If the energy transfer is small, it results in a broadening of the elastic peak called quasi-elastic scattering. The inelastic peaks are clearly separated from the elastic one and correspond to collective excitations as, for instance, phonons. The several types of scattering give access to different information about the sample.

Elastic incoherent neutron scattering (EINS) informs about small vibrational motions around the positions of equilibrium of the atomic nuclei, mainly of the H atoms, or of the molecular subgroups to which they are bound. However, it does not permit to distinguish the kinds of motions. Quasi-elastic incoherent neutron scattering (QENS) requests a more complex data analysis, but permits to distinguish between diffusional, rotational and/or confined motions and to quantify them by the self-diffusion coefficient D and the residence time τ, which is characteristic for jump-diffusion, where the particles perform a translational diffusive motion for a short time and then vibrate at their position for the time τ, after which they perform another jump. Inelastic neutron scattering will not be considered here.

The experiments were performed at the various spectrometers IN6 [16], IN5 [17] and IN13 [18] (see two examples in Fig. 2) of the Institut Laue Langevin (ILL) in Grenoble/France, giving access to different energy resolutions and to several Q-domains. The studies on lipoproteins were completed by measurements on the small angle neutron scattering instrument SANS-II [19] at the Paul-Scherrer-Institute (PSI) in Villigen, Switzerland, which gives access to low resolution structure information as the shape and the dimensions of the particle.

The spectrometer's time window can be calculated from the energy resolution using the Heisenberg uncertainty principle. It was about 20 ps for IN6 and about 100 ps for IN5 and IN13, on which we measured QENS and EINS. The different contributions to atomic motions (lattice and internal molecular vibrations, reorientations, translations) are assumed to be decoupled, because they occur on significantly different time scales, what was shown by molecular dynamics simulations [20] and quasi-elastic neutron scattering [21]. In this sense the instruments can be seen as a motion filter to focus on certain dynamics occurring within the window of a given spectrometer. All three spectrometers are indeed sensitive to internal motions, which correspond to mean square displacements of the order of 1 $Å^2$, but instead IN6 probes fast motions, typically of water molecules or other small molecular subgroups. The accessible Q-domains depend also significantly on the instrumental setup, ranging up to 1 $Å^{-1}$ on IN5, up to 2 $Å^{-1}$ on IN6 and up to 4.9 $Å^{-1}$ on IN13. The corresponding length scales can also be evaluated through Heisenberg's uncertainty principle. Table 1 summarises the instrumental characteristics.

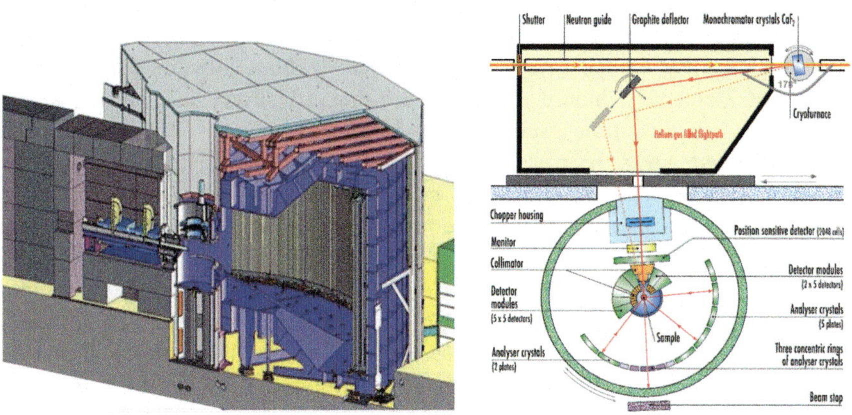

Fig. 2 The disk chopper time-of-flight spectrometer IN5 (left) and the thermal backscattering spectrometer IN13 (right) of the ILL (courtesy of ILL)

Table 1 Instrumental characteristics of the spectrometers IN5, IN6 and IN13 at the ILL

Spectrometer	IN5	IN6	IN13
Wavelength (Å)	10	5.1	2.23
Accessible Q-range ($Å^{-1}$)	0.3–1.0	0.4–2.0	0.2–4.9
Length scale (Å)	6–21	3–16	1–30
Energy resolution (μeV)	10	75	10
Time scale (ps)	100	15	100

2.1 High Hydrostatic Pressure

The typical energies always present in macromolecules are thermal energies corresponding to a few meV at ambient temperature. Temperature variation influences the thermal energy as well as the volume of a system

$$\Delta V = \Delta V^0 + \Delta\alpha(T - T^0) - \Delta\beta(p - p^0) \tag{2}$$

through the thermal expansion factor ($\Delta\alpha = \partial\Delta V/\partial T$) which makes it difficult to separate the two effects. Pressure is another thermodynamic variable which only affects the system volume via the compressibility factor ($\Delta\beta = \partial\Delta V/\partial p$). Thus, the energy variation $\Delta G = \Delta H - T\,\Delta S = \Delta U - p\,\Delta V - T\,\Delta S$ is better defined from a thermodynamic point of view if only pressure is changed (G: Gibbs free energy, H: enthalpy, U: internal energy, S: entropy). As we aim at testing dynamics which has its origin in the application of forces to test the flexibility of a system, it seems clear that pressure application, which is directly related to force application, is better adapted than temperature.

The existing lack of research in this domain is due to technical challenges of applying high hydrostatic pressure (HHP), especially in combination with neutron scattering, leading to the rather poor knowledge about pressure effects in the domain of biophysics. Indeed, high pressure requests strong and thick materials, whereas the use of neutrons excludes many materials which absorb too much or are easily activated. Over the past years, we developed a new HHP equipment for biological samples in solution in collaboration with the sample environment team of the ILL [22]. It consists of a pressure controller, which communicates with the instrument control software NOMAD, the high pressure stick [23] and the cell. Pressure is transmitted hydrostatically from the controller to the sample through a capillary using the Fluorinert™ liquid [24] that has a pour point of 178 K. As the stick is inserted into a cryostat to control temperature, one has to avoid that the liquid freezes at the cold point of the cryostat, therefore it must be heated by a wire and isolated thermally from its environment by a secondary vacuum.

The HHP cell is cylindrical, built of the high-tensile aluminium alloy 7075-T6, with an external diameter of 15 mm and an internal diameter of 6 mm. It withstands pressure up to 500 MPa (1 MPa = 10 bar). The sample solution is separated from Fluorinert by a separator on the top of the cell. As a sample thickness of 6 mm would lead to multiple scattering and thus to the loss of information about the

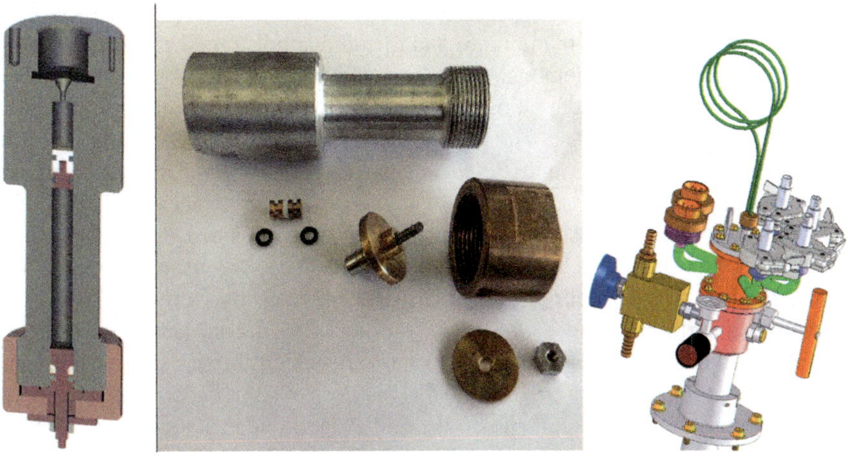

Fig. 3 500 MPa pressure cell providing Ø6 mm sample volume with in situ separator (left and middle) and top of the HHP stick allowing the heating of the pressure transmitting liquid (right) developed at ILL (see https://www.ill.eu/sane)

scattering angle, an Al cylinder of 4 mm diameter is inserted in the middle of the cell. The sample volume is about 1 ml. A drawing of the cell and stick is shown in Fig. 3.

The HHP equipment can be moved and installed on all ILL spectrometers as well as on the diffractometer D16. The experiments described below were performed thanks to the reliable functioning of it.

2.2 *Neutrons as Efficient Probes to Study Live Cells*

Cellular life is divided into three major branches in the "tree of life". One distinguishes one group of complex cells, the Eukaryotes, which have a nucleus containing the genome DNA and which often form multicellular organisms, and two groups of cells lacking a nucleus, the Prokaryotes including Bacteria and Archaea [25]. Although similar in shape and size, they present significant differences, in particular concerning the enzymes involved in transcription and translation or in terms of membrane lipids. The two groups of Prokaryotes also differ in terms of ability to colonize environments under extreme pH, temperature, salinity or oligotrophic conditions. Indeed, the most extremophilic Prokaryotes all belong to the Archaea [26]. This particular ability comes in part from the lipids composing their membranes. Archaeal membrane lipids are formed by ether bonds instead of ester bounds, linking the polar headgroups to isoprenoid chains instead of fatty acids [27]. Ether bonds have a higher chemical resistance than ester bonds, while isoprenoid chains have a tighter packing than aliphatic chains, ensuring the stability and function of the membrane under extreme conditions. Structural adaptation in Archaea is not restricted to membranes and membrane lipids. In the next paragraphs we will present studies demonstrating how neutron measurements under HHP allowed to decipher the adaptation strategies to high hydrostatic pressure in the order of *Thermococcales*. *Thermococcales* are divided in three genera: *Thermococcus*, *Pyrococcus* and *Paleococcus* exhibiting optimal growth temperatures ranging from 70 to 105 °C. They have been mostly isolated from hydrothermal vent systems ranging from the surface to the deepest part of the oceans (4500 m below sea level) [28]. In our study we used *Thermococcus kodakarensis*, a piezosensitive (an organism sensitive to pressure) Archaeon isolated from a surface solfatara of Kodakara island [29], Japan, and *Thermococcus barophilus,* the first true piezophile (an organism which thrives at high pressure) hyperthermophile described in 1999 [30]. This deep sea Archaeon was isolated in the hydrothermal vent Snakepit site on the mid-Atlantic ridge at a depth of 3550 m. Its optimal growth conditions are 40 MPa, 85 °C and 3% salinity, while those of *T. kodakarensis* are similar except for pressure (0.1 MPa). The two Archaea are genetically closely related, making them the ideal microbial system to monitor pressure adaptation mechanisms. Genetic and genomic studies have been unable to identify the structural adaptation in these organisms, which is a contrasting situation in comparison to other major environmental constraints, such as temperature, for which the structural adaptation of proteins is clearly visible in the genomes of the cells.

To be able to explain the adaptation in piezophiles, we passed to other techniques. As dynamics seemed to have a crucial influence on the good functioning of certain extremophiles, we studied *T. barophilus* and *T. kodakarensis* by EINS and QENS at pressures of 0.1 and 40 MPa in heavy water to optimise the contrast [31, 32]. As the Al high pressure cell does not support temperatures above 50 °C, all measurements were done at room temperature. The two organisms have very impermeable membranes at their optimal growth temperature of 85 °C. One of the advantages of our experimental setup was the low temperature of acquisition, which rendered the membranes even more compact and impermeable. Thus, we expected the H_2O/D_2O exchanges to be very limited between the extra- and intracellular spaces.

We found that bacterial cell pellets have around 70–80% of total water content and otherwise contain biomolecules (see Fig. 4) [33]. However, proteins and membranes have a hydration layer that differs significantly from the bulk water phase, the former type being slowed down with respect to the latter one [34–36]. These differences are visible in the QENS and EINS data provided proper modeling.

Living cells are functional systems per se, but since they are extremely complex, simple phase transitions that can be seen using single proteins, cannot be observed. The dynamics of the system varies nonetheless under different external conditions such as temperature or pressure, which allows to obtain the dynamics averaged over the whole cells. Obtaining averaged dynamics' values is an advantage when studying whole cells, which composition is not known. To validate the use of

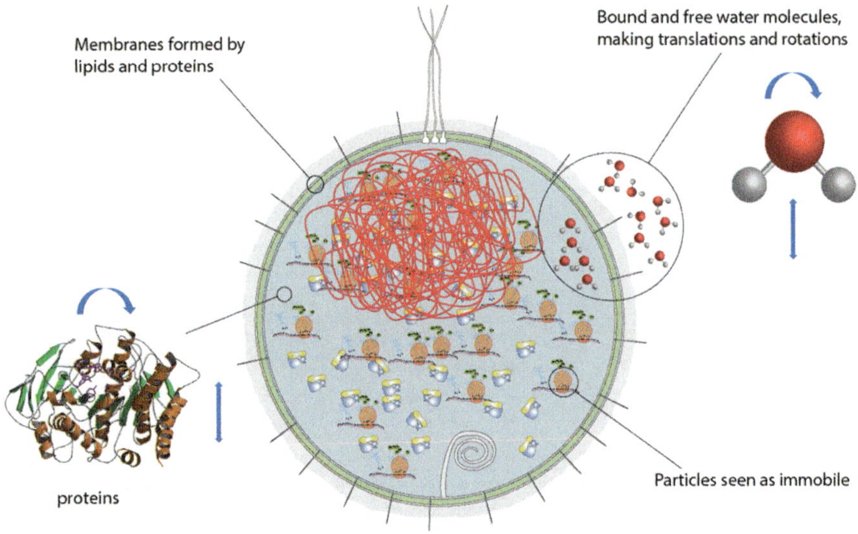

Fig. 4 Sketch of an archaeon with its components and their possible motions

incoherent neutron scattering to differentiate cell dynamics, we first measured the two *Thermococcales* and *E. coli* as mesophilic reference system in three consecutive years as function of HHP on the backscattering spectrometer IN13 at the ILL to check the reproducibility of the measures [31]. Figure 5 shows the neutron intensities summed over all scattering angles in arbitrary units and clearly demonstrates that the data points were agreeing quite well. The summed intensities being inversely proportional to the amplitudes of motion [37], one should notice that HHP decreases the amplitudes of motion, as predicted by Le Châtelier's principle: high pressure favors a decrease in volume and thus a damping of atomic motions. Whereas the summed intensity of *E. coli* seems to increase almost linearly with pressure, *T. kodakarensis* and *T. barophilus* exhibit a steep rise in intensity up to pressures of less than 10 and about 30 MPa, respectively, followed by a plateau. Interestingly, these values are close to the optimal growth pressure values of the respective species. Above the optimum pressure, the systems are therefore rather insensitive to pressure and very stable.

In a second step we performed QENS measurements of the two *Thermococcales* on the time-of-flight spectrometer IN5 at ILL with the same energy resolution of 10 μeV as on IN13 at 0.1 and 40 MPa, corresponding to the two molecular dynamics' regimes found with EINS [32]. If the samples were simple fluids, the incoherent scattering function would have the form of a Lorentzian curve for each Q value as function of the energy transfer ω, with its half-width-half-maximum (HWHM) proportional to $D\,Q^2$, where D is the translational diffusion coefficient [14]. If the sample is more complex and not a fluid, the van Hove theory predicts for $S(Q, \omega)$ a sum of Lorentzian functions, each of them representing a part of the

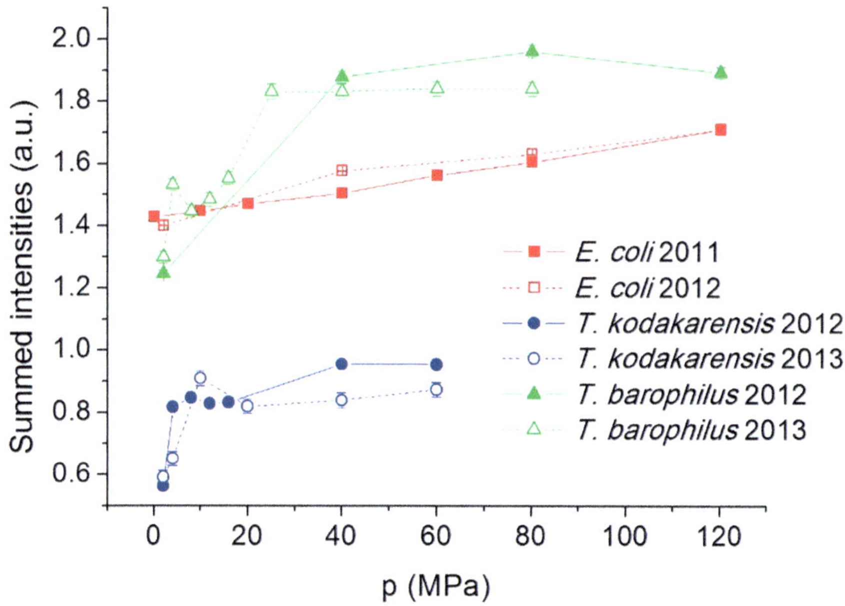

Fig. 5 Summed elastic incoherent neutron scattering intensities of *T. kodakarensis*, *T. barophilus* and *E. coli* measured in 2011, 2012 and 2013 on IN13

sample with its characteristic motions [38]. If nothing is known about the sample, such an analysis can be puzzling, as we will show in the last section. However, as the proportions of water and biomolecules in bacterial cell pellets, with around 70–80% of total water content, are well documented [33], we chose to apply a model that was previously used to analyse QENS data on neurological tissue which has a similar molecular composition as our cells [39]. In this context, four different components and a background $B(Q)$ are assumed and convolved with the instrumental resolution function $R(Q, \omega)$:

$$S(Q,\omega) \approx \begin{bmatrix} p_{elastic} \times \delta(\omega) + p_{bulk} \times S_1(Q,\omega,D_{T_{bulk}},\tau_{bulk},D_{R_{bulk}}) \\ + p_{hyd} \times S_2\left(Q,\omega,D_{T_{hyd}},\tau_{hyd},D_{R_{hyd}}\right) \\ + p_{bio} \times S_3\left(Q,\omega,\Gamma_{proteome}\right) + B(Q) \end{bmatrix} \otimes R(Q,\omega) \tag{3}$$

where $p_{elastic}$ is the elastic fraction describing protons that perform only atomic vibrations. $S_1(Q, \omega)$ and $S_2(Q, \omega)$ are the scattering intensities attributed to free water, which has properties very similar to bulk water, and of water molecules bound to the inner or outer surface of the cell, called hydration water. These two components include translational and rotational motions, described mathematically by a convolution of two Lorentzians, and they are characterized by a diffusion constant D_T, a rotational diffusion constant D_R and a residence time τ [40].

To obtain an optimal fit of the experimental data, a third broad Lorentzian function was necessary with the constant HWHM $\Gamma_{proteome}$ associated to the motions within the cell itself. As proteins represent indeed the major constituents in microorganisms [41], their dynamics dominates the component associated with biomolecules and we call it "proteome". p_{bulk}, p_{hyd}, $p_{proteome}$ (with $p_{elastic} + p_{bulk} + p_{hyd} + p_{proteome} = 1$) are, respectively, the fractions of atoms corresponding to the various components. The fits were performed by using the program mQfit [42] and the result of the best fit is given in Table 2.

The sum of the fractions of the two water populations is about 76% for the two samples, but the bulk water fraction is higher for *T. barophilus* and lower for *T. kodakarensis*. In total it agrees well with the values reported in the literature for the water content in cells [43]. On the contrary, the fractions of the elastic part and the one corresponding to the proteome are almost the same for the two samples and the two pressure values, as the number of scatterers is not influenced by pressure application. The bulk water diffusional coefficient D_{Tbulk} is constant for all four configurations. Water is indeed almost incompressible, what explains its insensitivity to pressure, and only slightly perturbed by the presence of the biomolecule, because 1–2 hydration water layers separate the cell surface and the bulk water [36]. The residence time τ of bulk water molecules, which is related to interactions between molecules [44], increases in *T. barophilus* under pressure application giving a hint for a stronger confinement effect in this state. The diffusion coefficient of hydration water molecules is not much affected for *T. kodakarensis*, but significantly reduced for *T. barophilus*, which seems to interact more strongly with its environment than the piezosensitive cell. The residence time could not be determined with sufficient accuracy for hydration water, because the corresponding HWHM curves did not reach a plateau value within the Q-range accessible by the instrument. The most surprising result was found for the proteome itself: whereas

Table 2 Best fit parameter values for the two samples at 0.1 MPa and at 40 MPa (reproduced from [32]

	T. barophilus		*T. kodakarensis*	
	0.1 MPa	40 MPa	0.1 MPa	40 MPa
$p_{elastic}$	0.01 ± 0.01	0.01 ± 0.01	0.03 ± 0.01	0.03 ± 0.01
p_{hyd}	0.10 ± 0.01	0.10 ± 0.01	0.17 ± 0.01	0.18 ± 0.02
p_{bulk}	0.67 ± 0.04	0.66 ± 0.03	0.60 ± 0.03	0.60 ± 0.03
$p_{proteome}$	0.23 ± 0.03	0.23 ± 0.03	0.23 ± 0.03	0.22 ± 0.03
D_{Tbulk} (10^{-5} cm^2 s^{-1})	1.98 ± 0.02	1.98 ± 0.02	1.98 ± 0.02	1.98 ± 0.02
τ_{bulk} (ps)	1.05 ± 0.11	1.55 ± 0.16	1.28 ± 0.13	1.15 ± 0.12
D_{Rbulk} (ps^{-1})	0.11 ± 0.01	0.09 ± 0.01	0.11 ± 0.01	0.11 ± 0.01
D_{Thyd} (10^{-7} cm^2 s^{-1})	5.17 ± 0.12	3.34 ± 0.07	4.86 ± 0.10	4.41 ± 0.09
D_{Rhyd} (ps^{-1})	0.06 ± 0.02	0.09 ± 0.03	0.05 ± 0.01	0.05 ± 0.01
$\Gamma_{proteome}$ (meV)	0.431 ± 0.004	0.451 ± 0.003	0.363 ± 0.005	0.347 ± 0.005

its dynamics was slightly decreased by pressure for *T. kodakarensis*, it increased for *T. barophilus* contrary to most other organisms' cells. We concluded that pressure adaptation consists at least partly in a mechanism which permits the biomolecules to remain flexible or even increase the flexibility in the physiological condition to assure the good functioning of the cell. Moreover, the interaction with water is also slightly modified leading to a reduced hydration shell which is slowed down in its motions. Pressure adaptation seems therefore to be operated partly through molecular dynamics including complex and sometimes counterbalancing effects.

The results can be summarized as presented in Fig. 6.

2.3 Neutron Scattering as Approach to Study Endogenous Nanoparticles

Another investigation with neutrons concerned low density lipoprotein (LDL) particles, which are natural quasi-spherical nanoparticles with a typical size of about 20 nm [45]. So far it was not possible to solve their structure at atomic level, which is mostly due to their large size and heterogeneous composition. However, neutron scattering techniques permitted to correlate their low-resolution structure obtained through small angle neutron scattering (SANS) with their averaged molecular dynamics, which would not be achievable by other means. Cholesterol, triglycerides and other lipids are transported in the blood by lipoproteins. Lipoproteins are associated with cardiovascular diseases such as

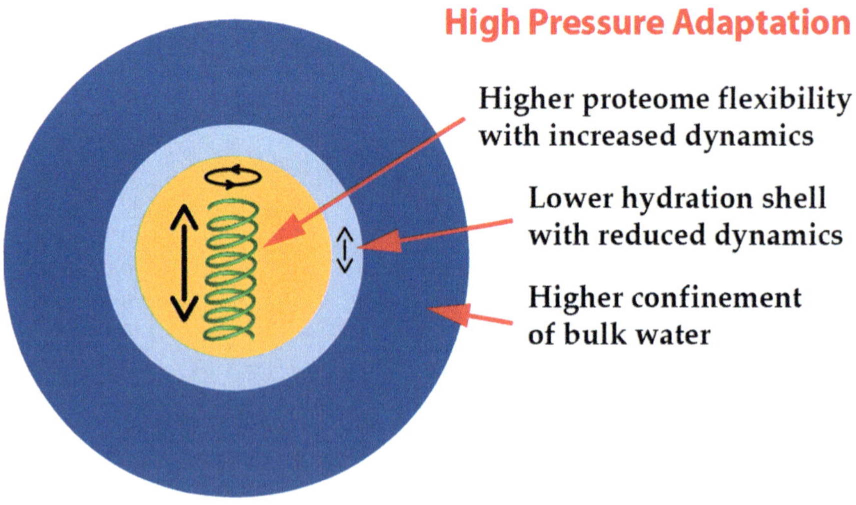

Fig. 6 Sketch summarizing the effects of HHP on *T. barophilus*. The dark blue surface represents bulk water, the light blue surface hydration water and the yellow kernel the cell. The green spring characterizes the proteome and its contributions (translation and rotation) to dynamics

atherosclerosis or stroke, which are among the most prevalent causes of death in developed countries [54]. The interaction of surface-located apolipoproteins with specific cellular receptors indeed determines whether triglycerides and cholesterol will be added or removed from the lipoprotein transport particles [46]. According to these functionalities, it is a common term to distinguish between "bad" and "good" cholesterol when designating LDL cholesterol and high density lipoproteins (HDL) cholesterol, respectively. However, certain blood levels of HDL and LDL are essential to maintain health. In our study, we concentrated on LDL, which is formed by a phospholipid monolayer on the surface, the huge apolipoprotein B100 (ApoB100) wound around the surface of the particle and an apolar oily core filled with cholesteryl esters and triglycerides (see Fig. 7). LDL's structure was investigated exhaustively by small angle scattering techniques and cryo-electron microscopy [47–52], giving rise to a rough representation of the nano-particle.

We were particularly interested in the native LDL (N-LDL) and a modified form of LDL, the triglyceride enriched LDL called TG-LDL hereafter, since the latter one mimics the pathological state of LDL found in a hyperlipidemic patient. Both particles were studied in D_2O solution under HHP up to 300 MPa by QENS on IN5 [17] and IN6 [16] at ILL and by SANS on SANS-II [19] at the PSI. We used such high pressures to make sure that effects could be distinguished. The samples were at 310 K, above the phase transition temperature of LDL [55]. As these particles are highly hydrophobic, we assumed that no water would be inside the core in contrast to living cells and we also did not expect a measurable amount of hydration water on the LDL's surface. Moreover, we subtracted the heavy water, which was measured in the same HHP cell, from the data to obtain the signal from the nanoparticles, only. No model was known so far to analyse the QENS data of such a complex system. Therefore, we combined QENS data taken on the same samples from two spectrometers and did a global fit of all data together to limit the number

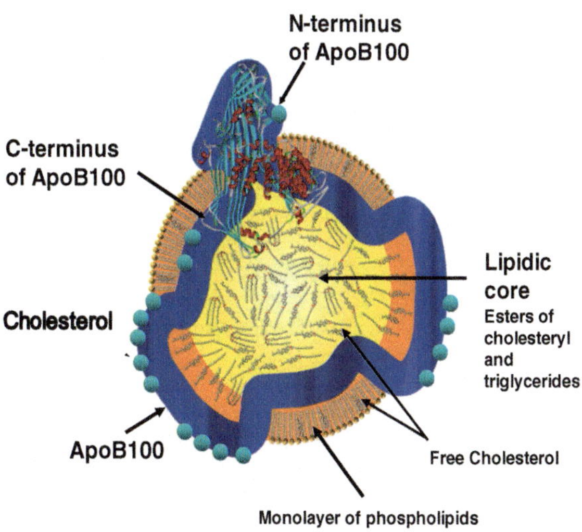

Fig. 7 Representation of LDL and its components (reprinted from [53] with permission)

of free fit parameters. We tried a so-called model-free analysis, where a sum of an elastic peak and Lorentzian functions was used to model the experimental points [38]. We found that three Lorentzians were sufficient to obtain a good fit (see Fig. 8a). The next step consisted in identifying the motions and the parts of the sample to which they could be associated. For that we plotted the full width half maximum (FWHM) of the Lorentzians against Q^2 to get additional information about the type of motion. If the FWHM increases linearly with Q^2 passing through the origin, one can assume free Brownian diffusion [14]. If the FWHM is independent on Q^2, it is the signature of rotational diffusion and if it increases first linearly tending then to a limiting value at high Q, it can be associated to jump-diffusion [40]. In the present case, we found all three kinds of motions (see Fig. 8b). The narrowest Lorentzian corresponded to slow translational diffusion, the medium contribution to jump-diffusion and the broadest curve to fast rotational diffusion. Surprisingly, the obtained fit parameters were identical for N-LDL and TG-LDL samples at low (2 MPa) and high hydrostatic pressure (300 MPa). For the translational diffusion we got a coefficient $D_T = (0.11 \pm 0.02) \times 10^{-5}$ cm^2/s, for the jump diffusion $D_{JD} = (2.1 \pm 0.1) \times 10^{-5}$ cm^2/s, for the residence time of the jump diffusion $\tau = (1.7 \pm 0.1)$ ps and for the FWHM of the rotational diffusion $\Gamma_{rotation} = (0.72 \pm 0.05)$ meV. Comparison with recent QENS studies on 1, 2-dimyristoyl-*sn*-glycero-3-phosphocholine (DMPC) lipids by Wanderlingh et al. [56] permitted to suggest that the translational diffusion could be ascribed to a centre-of-mass motion of the lipids, the jump-diffusion to confined motions within the lipids and the rotational diffusion to fast and small local motions, for instance to the rotation of CH_2 molecules.

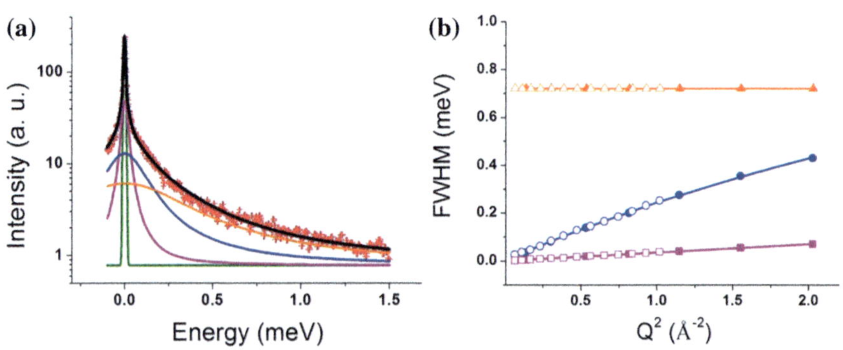

Fig. 8 QENS data and fits. Logarithmic plot of a typical QENS spectrum (IN5) of N-LDL at 310 K and 2 MPa measured at $Q = 1.00$ Å^{-1} (red points) and quasi-elastic contributions determined by the fit: the green line corresponds to the elastic contribution, the magenta, blue and orange lines to Lorentzian functions. The black line shows the final fit (**a**); FWHM values of the three Lorentzian distributions as function of Q^2 obtained from IN5 (smaller Q-range, open symbols) and IN6 (full symbols) and their fits according to a translational diffusive model (magenta points), a jump-diffusion model (blue points) and a rotational diffusive model (orange points). The estimated error is ±2 on the last given digit (**b**) (reprinted from [57])

On the contrary to the QENS contributions, we found an effect on the elastic peak of TG-LDL, which was increased at 300 MPa. The elastic peak is representative for motions which are not resolved within the time window of an instrument. The raise of the peak indicates therefore the slowing down of some molecular groups in TG-LDL under HHP. It means that the same kind of motions were present at ambient and high pressure, but not the same number of molecular groups participated to them. Similar results were also found by EINS on IN13: the amplitudes of motion were identical within error bars for N-LDL at 0.1 and 300 MPa, but they were clearly decreased at 300 MPa for TG-LDL with respect to 0.1 MPa [57]. The conclusion was that N-LDL coped surprisingly well with very HHP with almost no changes in the dynamical behaviour, whereas the modified particle TG-LDL had its number of molecular groups participating to dynamics reduced under HHP and was thus less able to withstand external forces.

We wanted to correlate these findings with structural information and performed small angle neutron scattering at the same pressures as used for the dynamics' studies on the instrument SANS-II [19] at the PSI. With this technique, the neutron intensity is recorded in forward direction (at small angles), so that structural details with dimensions $d \gg \lambda$ become resolvable. The experimental intensity curve is then compared through a fit to a theoretical curve supposing a certain geometry of the sample. The theoretical model can be more and more refined until convergence of the fit. As the momenta transferred by the neutrons were in the range 0.1–2.0 nm^{-1}, corresponding to dimensions of the order 3–60 nm, we had access to the overall shape of the LDL particles. We used an ellipsoidal model, including three different radii R_1, R_2 and R_3, to fit the data and got good fit results. In fact, we found variations for both samples under HHP but again much less pronounced for N-LDL than for TG-LDL, which seemed to be less stable under pressure. It appeared that the native particle has a rather elongated shape, irrespectively of pressure, whereas the modified form TG-LDL, is more spherical at ambient pressure and resembles the native LDL at high pressure.

We calculated that in both samples the particle volume decreased and the surface-to-volume ratio increased with pressure, but the changes were less pronounced for TG-LDL, which on the other hand became deformed. The surface-to-volume ratio change in LDL could be due to differences in the compressibility between shell and core. Other reasons could be a pressure dependent conformational change of the apolipoprotein, which could become more stretched or flattened by pressure to cover a larger surface area. Another effect could be a different equilibrium of unesterified cholesterol between the core and the phospholipid membrane, which could also lead to a change in surface-to-volume ratio. In conclusion, our results indicated that LDL particles withstand HHP quite well, however, a TG-rich core lipid composition seemed to have an impact on the molecular dynamics of LDL particles under pressure. Such a combined dynamical and structural investigation became possible due to the potential of neutron scattering techniques to give access to multivariate data.

3 Conclusions and Outlook

Incoherent neutron scattering techniques are a unique tool to get insights to molecular dynamics and structure, as neutrons can penetrate deeply into matter without damaging it. In contrast to X-rays, neutrons have a high sensitivity to H atomic nuclei, which represent about half of the atoms in biological systems. As we know today that dynamics are strongly correlated with structure and functionality of biomolecules, such measurements complete our picture about biological mechanisms. In the present study, we reported recent work on life cells, cell components and endogenous nanoparticles, in particular under high hydrostatic pressure conditions, to better understand their functions and interplays, what could even open perspectives for cell engineering. Pressure is indeed a thermodynamic variable much less used than temperature, mainly due to technological difficulties related to it, but which gives a more direct access to information about volume changes, compressibility, flexibility and stability of the samples.

In a first example, we presented studies on two hyperthermophile Archaea, one living at the surface of Earth and the other at 3550 m depth in the Atlantic Ocean, to better understand the mechanisms which lead to an adaptation to such harsh pressure conditions. We found indeed significant differences between the dynamics of both, but also the water in- and outside the cells seemed to be influenced by these organisms and contributed to their survival. However, we investigated the samples so far only at ambient temperature conditions although they use to live at 85 °C. Recently, we developed a new HHP cell in Titanium-Zirconium, capable of withstanding simultaneously high pressure and temperatures up to 100 °C [58]. Therefore, the next step will consist in studying the same organisms under physiological temperature conditions. The drawback consists in the fact, that the scattered neutron intensity decreases rapidly when temperature rises, in addition to the considerable absorption by the HHP cell. Therefore, long counting times are requested to guarantee a sufficient signal-to noise ratio for data analysis.

The second example concerned natural nanoparticles, LDLs, which transport cholesterol in the blood. It is a matter of fact, that LDL is related to diseases such as atherosclerosis or stroke. Therefore, we compared native LDL to a triglyceride enriched form, which is mimicking a state of hyperlipidemia and exposed both to HHP to test their stability and flexibility. As LDL is composed of more than 80% of lipids, we expected LDL to have a similar conformational landscape as lipid membranes, which are known to be among the most sensitive biosystems with respect to pressure. However, N-LDL proved to be very resistant against pressure application, concerning both dynamics and structure, while TG-LDL is perturbed by it. Our neutron scattering experiments do certainly not allow to conclude about the medical aspects of these findings, but it would be tempting to elaborate such connection.

Acknowledgements This work has been supported by the Austrian Science Fund (FWF Project No. I 1109-N28 to R. P.) and by two projects financed by the Agence Nationale de la Recherche (ANR; project number ANR 2010 BLAN 1725 01 Living deep and project number

ANR-12-ISV5-0002-01 LDLPRESS to J.P.). We thank the ILL for allocation of beamtime and the SANE group of ILL for their support to develop the HHP equipment and C. Payre and J. Maurice for their help to perform the experiments. The work is partially based on experiments performed at the Swiss spallation neutron source SINQ, Paul Scherrer Institute, Villigen, Switzerland. The work benefitted from SasView software, originally developed by the DANSE project under NSF award DMR-0520547. We are gratefully acknowledging the help of the local contacts on the various instruments, in particular B. Frick, M. M. Koza, J. Ollivier, J. Kohlbrecher and G. Nagy. We wish to thank our many co-workers for their help and fruitful discussions, in particular N. Martinez, G. Michoud, A. Cario, B. Franzetti, M. Jebbar, B. Lehofer and M. Golub.

References

1. Karr, J. R., Sanghvi, J. C., Macklin, D. N., Gutschow, M. V., Jacobs, J. M., Bolival, B., Jr., et al. (2012). A whole-cell computational model predicts phenotype from genotype. *Cell, 150,* 389–401. https://doi.org/10.1016/j.cell.2012.05.044.
2. Schulz, R., Lindner, B., Petridis, L., & Smith, J. C. (2009). Scaling of multimillion-atom biological molecular dynamics simulation on a petascale supercomputer. *Journal of Chemical Theory and Computation, 5,* 2798–2808. https://doi.org/10.1021/ct900292r.
3. Nam, D., Park, J., Gallagher-Jones, M., Kim, S., Kim, S., Kohmura, Y., et al. (2013). Imaging fully hydrated whole cells by coherent x-ray diffraction microscopy. *Physical Review Letters, 110*(9), 098103. https://doi.org/10.1103/PhysRevLett.110.098103.
4. Jiang, H., Song, C., Chen, C. C., Xu, R., Raines, K. S., Fahimian, B. P., et al. (2010). Quantitative 3D imaging of whole, unstained cells by using X-ray diffraction microscopy. *Proc Natl Acad Sci U S A, 107*(25), 11234–11239. https://doi.org/10.1073/pnas.1000156107.
5. Tehei, M., Franzetti, B., Madern, D., Ginzburg, M., Ginzburg, B. Z., Giudici-Orticoni, M. T., et al. (2004). Adaptation to extreme environments: macromolecular dynamics in bacteria compared in vivo by neutron scattering. *EMBO Reports, 5*(1), 66–70. https://doi.org/10.1038/sj.embor.7400049.
6. Jasnin, M. (2009). Atomic-scale dynamics inside living cells explored by neutron scattering. *Journal of the Royal Society, Interface, 6*(Suppl 5), S611–S617. https://doi.org/10.1098/rsif.2009.0144.focus.
7. Luef, B., Fakra, S. C., Csencsits, R., Wrighton, K. C., Williams, K. H., Wilkins, M. J., et al. (2013). Iron-reducing bacteria accumulate ferric oxyhydroxide nanoparticle aggregates that may support planktonic growth. *ISME Journal, 7*(2), 338–350. https://doi.org/10.1038/ismej.2012.103.
8. Hajj, B., Wisniewski, J., El Beheiry, M., Chen, J., Revyakin, A., Wu, C., et al. (2014). Whole-cell, multicolor superresolution imaging using volumetric multifocus microscopy. *Proc Natl Acad Sci U S A, 111*(49), 17480–17485. https://doi.org/10.1073/pnas.1412396111.
9. Gabel, F., Bicout, B. J., Lehnert, U., Tehei, M., Weik, M., & Zaccai, G. (2002). Proteins dynamics studied by neutron scattering. *Quaterly reviews of biophysics, 35*(4), 327–367.
10. Kurr, M., Huber, R., Konig, H., Jannasch, H. W., Fricke, H., Trincone, A., et al. (1991). *Methanopyrus kandleri,* gen. and sp. nov. represents a novel group of hyperthermophilic methanogens, growing at 110 °C. *Archives of Microbiology, 156,* 239–247.
11. Takai, K., Nakamura, K., Toki, T., Tsunogai, U., Miyazaki, M., Miyazaki, J., et al. (2008). Cell proliferation at 122 °C and isotopically heavy CH4 production by a hyperthermophilic methanogen under high-pressure cultivation. *Proceedings of the National Academy of Sciences of the United States of America, 105,* 10949–10954.
12. Vauclare, P., Marty, V., Fabiani, E., Martinez, N., Jasnin, M., Gabel, F., et al. (2015). Molecular adaptation and salt stress response of *Halobacterium salinarum* cells revealed by neutron spectroscopy. *Extremophiles, 19*(6), 1099–1107. https://doi.org/10.1007/s00792-015-0782-x.

13. Lund, P., Tramonti, A., & De Biase, D. (2014). Coping with low pH: Molecular strategies in neutralophilic bacteria. *FEMS Microbiology Reviews, 38*(6), 1091–1125. https://doi.org/10.1111/1574-6976.12076.
14. Bée, M. (1988). *Quasielastic neutron scattering: Principles and applications in solid state chemistry*. Adam Hilger, Philadelphia: Biology and Materials Science.
15. Sears, V. F. (1992). Neutron scattering lengths and cross sections. *Neutron News, 3,* 26–37.
16. http://www.ill.eu/instruments-support/instruments-groups/instruments/in6/description/instrument-layout/.
17. http://www.ill.eu/instruments-support/instruments-groups/instruments/in5/description/instrument-layout/.
18. Natali, F., Peters, J., Russo, D., Barbieri, S., Chiapponi, C., Cupane, A., et al. (2008). IN13 backscattering spectrometer at ILL: Looking for motions in biological macromolecules and organisms. *Neutron News, 19*(4), 14–18. https://doi.org/10.1080/10448630802474083.
19. Strunz, P., Mortensen, K., & Janssen, S. (2004). SANS-II at SINQ: Installation of the former Riso-SANS facility. *Phys B - Cond Matt, 350,* E783–E786.
20. Bée, M. (2003). Localized and long-range diffusion in condensed matter: State of the art of QENS studies and future prospects. *Chemical Physics, 292*(2–3), 121–141.
21. Doster, W. (2006). Dynamical structural distributions in proteins. *Physica B, 385–386,* 831–834.
22. Peters, J., Trapp, M., Hughes, D., Rowe, S., Demé, B., Laborier, J.-L., et al. (2011). High hydrostatic pressure equipment for neutron scattering studies of samples in solutions. *High Pressure Research, 32*(1), 97–102.
23. Lelièvre-Berna, E., Demé, B., Gonthier, J., Gonzales, J. P., Maurice, J., Memphis, Y., et al. (2017). 700 MPa sample stick for studying liquid samples or solid-gas reactions down to 1.8 K and up to 550 K. *Journal of Neutron Research, 19,* 77–84.
24. Sidorov, V. A., & Sadykov, R. A. (2005). Hydrostatic limits of Fluorinert liquids used for neutron and transport studies at high pressure. *Journal of Physics: Condensed Matter, 17,* 3005–3008.
25. Woese, C. R., Kandler, O., & Wheelis, M. L. (1990). Towards a natural system of organisms: proposal for the domains Archaea, Bacteria, and Eucarya. *Proceedings of the National Academy of Sciences U S A, 87*(12), 4576–4579.
26. Rampelotto, P. H. (2013). *Extremophiles and extreme environments. Life (Basel), 3*(3), 482–485. https://doi.org/10.3390/life3030482.
27. Albers, S. V., & Meyer, B. H. (2011). The archaeal cell envelope. *Nature Reviews Microbiology, 9*(6), 414–426. https://doi.org/10.1038/nrmicro2576.
28. Jebbar, M., Franzetti, B., Girard, E., & Oger, P. (2015). Microbial diversity and adaptation to high hydrostatic pressure in deep-sea hydrothermal vents prokaryotes. *Extremophiles, 19*(4), 721–740. https://doi.org/10.1007/s00792-015-0760-3.
29. Atomi, H., Fukui, T., Kanai, T., Morikawa, M., & Imanaka, T. (2004). Description of *Thermococcus kodakaraensis* sp. nov., a well studied hyperthermophilic archaeon previously reported as *Pyrococcus* sp. KOD1. *Archaea, 1,* 263–267.
30. Marteinsson, V. T., Birrien, J. L., Reysenbach, A. L., Vernet, M., Marie, D., Gambacorta, A., et al. (1999). *Thermococcus barophilus* sp. nov., a new barophilic and hyperthermophilic archaeon isolated under high hydrostatic pressure from a deep-sea hydrothermal vent. *International Journal of Systematic and Evolutionary Microbiology, 49*(2), 351–359. https://doi.org/10.1099/00207713-49-2-351.
31. Peters, J., Martinez, N., Michoud, G., Cario, A., Franzetti, B., Oger, P., et al. (2014). Deep sea microbes probed by incoherent neutron scattering under high hydrostatic pressure. *Zeitschrift für Physikalische Chemie, 228,* 1121–1133.
32. Martinez, N., Michoud, G., Cario, A., Ollivier, J., Franzetti, B., Jebbar, M., et al. (2016). High protein flexibility and reduced hydration water dynamics are key pressure adaptive strategies in prokaryotes. *Scientific Reports, 6,* 32816. https://doi.org/10.1038/srep32816.
33. Bratbak, G. (1985). Bacterial biovolume and biomass estimations. *Applied and Environment Microbiology, 49,* 1488–1493.

34. Dellerue, S., & Bellissent-Funel, M. C. (2000). Relaxational dynamics of water molecules at protein surface. *Chemical Physics, 258,* 315–325.
35. Schiro, G., Fichou, Y., Gallat, F. X., Wood, K., Gabel, F., Moulin, M., et al. (2015). Translational diffusion of hydration water correlates with functional motions in folded and intrinsically disordered proteins. *Nature Communications, 6,* 6490. https://doi.org/10.1038/ncomms7490.
36. Perticaroli, S., Ehlers, G., Stanley, C. B., Mamontov, E., O'Neill, H., Zhang, Q., et al. (2017). Description of hydration water in protein (green fluorescent protein) solution. *Journal of the American Chemical Society, 139*(3), 1098–1105. https://doi.org/10.1021/jacs.6b08845.
37. Peters, J., Martinez, N., Trovaslet, M., Scannapieco, K., Koza, M. M., Masson, P., et al. (2016). Dynamics of human acetylcholinesterase bound to non-covalent and covalent inhibitors shedding light on changes to the water network structure. *Physical Chemistry Chemical Physics: PCCP, 18*(18), 12992–13001. https://doi.org/10.1039/c6cp00280c.
38. Van Hove, L. (1954). Correlations in space and time and born approximation scattering in systems of interacting particles. *Physical Review, 95*(1), 249–262.
39. Natali, F., Dolce, C., Peters, J., Gerelli, Y., Stelletta, C., & Leduc, G. (2013). Water dynamics in neural tissue. *Journal of the Physical Society of Japan, 82*(Suppl. A), SA017.
40. Volino, F., & Dianoux, A. J. (1980). Neutron incoherent-scattering law for diffusion in a potential of spherical-symmetry: General formalism and application to diffusion inside a sphere. *Molecular Physics, 41*(2), 271–279. https://doi.org/10.1080/00268978000102761.
41. Martinko, M. T. M. J. M., & Parker, J. (2006). *Brock's biology of microorganism 11th.* New Jersey: Prentice Hall.
42. Martinez, N., Natali, F., & Peters, J. (2015). mQfit, a new program for analyzing quasi-elastic neutron scattering data. *EPJ Web of Conferences, 83*(3010), 3011–3014.
43. Stadler, A. M., Embs, J. P., Digel, I., Artmann, G. M., Unruh, T., Buldt, G., et al. (2008). Cytoplasmic water and hydration layer dynamics in human red blood cells. *Journal of the American Chemical Society, 130*(50), 16852–16853. https://doi.org/10.1021/ja807691j.
44. Jasnin, M., Stadler, A. M., Tehei, M., & Zaccai, G. (2010). Specific cellular water dynamics observed in vivo by neutron scattering and NMR. *Physical Chemistry Chemical Physics: PCCP, 12,* 10154–10160.
45. Laggner, P., & Müller, K. (1978). The structure of serum lipoproteins as analysed by X-ray small-angle scattering. *Quarterly Reviews of Biophysics, 11,* 371–425.
46. Krieger, M. (1998). The "best" of cholesterols, the "worst" of cholesterols: a tale of two receptors. *Proceedings of the National Academy of Sciences of the United States of America, 95,* 4077–4080.
47. Muller, K., Laggner, P., Glatter, O., & Kostner, G. (1978). The structure of human-plasma low-density lipoprotein B. An X-ray small-angle scattering study. *European Journal of Biochemistry, 82,* 73–90.
48. Maric, S., Lind, T. K., Lyngso, J., Cardenas, M., & Pedersen, J. S. (2017). Modeling small-angle X-ray scattering data for low-density lipoproteins: Insights into the fatty core packing and phase transition. *ACS Nano, 11,* 1080–1090.
49. Orlova, E. V., Sherman, M. B., Chiu, W., Mowri, H., Smith, L. C., & Gotto, A. M. (1999). Three-dimensional structure of low density lipoproteins by electron cryomicroscopy. *Proceedings of the National Academy of Sciences, 96,* 8420–8425.
50. Liu, Y., & Atkinson, D. (2011). Enhancing the contrast of ApoB to locate the surface components in the 3D density map of human LDL. *Journal of Molecular Biology, 405,* 274–283.
51. Liu, Y. H., & Atkinson, D. (2011). Immuno-electron cryo-microscopy imaging reveals a looped topology of apoB at the surface of human LDL. *Journal of Lipid Research, 52,* 1111–1116.
52. Vea, Kumar. (2011). Three-dimensional cryoEM reconstruction of native LDL particles to 16 angstrom resolution at physiological body temperature. *PLoS ONE, 6,* e18841.

53. Prassl, R., & Laggner, P. (2009). Molecular structure of low density lipoprotein: Current status and future challenges. *European Biophysics Journal, 38*(2), 145–158. https://doi.org/10.1007/s00249-008-0368-y.
54. Benjamin, E. J., et al. (2017). Heart disease and stroke statistics—2017 update: A report from the American Heart Association. *Circulation, 135,* e146–e603.
55. Deckelbaum, R. J., Shipley, G. G., Small, D. M., Lees, R. S., & George, P. K. (1975). Thermal transitions in human plasma low density lipoproteins. *Science, 190,* 392–394.
56. Wanderlingh, U., D'Angelo, G., Branca, C., Nibali, V. C., Trimarchi, A., Rifici, S., et al. (2014). Multi-component modeling of quasielastic neutron scattering from phospholipid membranes. *J Chem Phys, 140*(17), 174901. https://doi.org/10.1063/1.4872167.
57. Golub, M., Lehofer, B., Martinez, N., Ollivier, J., Kohlbrecher, J., Prassl, R., et al. (2017). High hydrostatic pressure specifically affects molecular dynamics and shape of low-density lipoprotein particles. *Scientific Reports, 7,* 46034. https://doi.org/10.1038/srep46034.
58. Peters, J., Golub, M., Demé, B., Gonthier, J., Payre, C., Maurice, J., Sadykov, R., Lelièvre-Berna, E. (2018) A new high pressure cell going up to 100 °C. Rev Sc Instr., To be submitted.

Author Biography

Professor Judith Peters (Germany, France), Ph.D. She graduated in Physics at the University Joseph Fourier Grenoble 1 and received her Ph.D. in Theoretical Physics in 1988. She then had a post-doctoral position at the University of Heidelberg working on molecular dynamics simulations. After one year as exchange scientist at the University of St. Petersburg, 7 years as assistant professor at the University of Applied Sciences in Berlin and 10 years as scientist at the Helmholtz Zentrum Berlin, she holds now a professorship of Physics at Université Grenoble Alpes. Her research interests comprise dynamical studies of biosystems by neutron scattering techniques, particularly under high pressure conditions.

Microbial Sampling from Dry Surfaces: Current Challenges and Solutions

Ilya Digel, Nuraly Sh. Akimbekov, Aida Kistaubayeva and Azhar A. Zhubanova

1 Introduction

> Surface sampling methods are not standardized. Different media are employed, and in the case of swabs, different results have been reported for wet and dry swab methods and contact plates. Replicate sample contact plates should be expected to give similar results under identical conditions, but rates of recovery have been reported to be both lower than expected and highly variable. In general, surface monitoring has been found to recover <50%, even when used with relatively high inoculum levels on standardized coupons. In actual production environments where organisms are stressed to varying degrees, recovery rates may be lower.
>
> United States Pharmacopeia (USP), Chapter 1116: "Microbiological Control and Monitoring of Aseptic Processing Environments", 2014.

Various types of dry (not submerged) surfaces such as plastics, stainless steel, glass, wood etc. used today in all branches of human activity, are all subject to contamination by microorganisms [1]. In general, surface colonization is beneficial for microorganisms in terms of growth and nutrient uptake so that most bacteria and fungi live attached to a solid surface—artificial, animal, mineral, or vegetable [2–4].

Many studies have emphasized the primary role of microbiological surface sampling in obtaining sufficient and representative information about the character and number of biological agents associated with materials. Microbiological public safety relies on sufficient recovery of significant microorganisms and following species/strain identification that together provide key evidences for decision-making.

I. Digel (✉)
Institute for Bioengineering, University of Applied Sciences Aachen, Jülich, Germany
e-mail: digel@fh-aachen.de

N. Sh. Akimbekov · A. Kistaubayeva · A. A. Zhubanova
Al-Farabi Kazakh National University, Almaty, Kazakhstan

G. M. Artmann et al. (eds.), *Biological, Physical and Technical Basics of Cell Engineering*, https://doi.org/10.1007/978-981-10-7904-7_19

Detection and evaluation of microbial colonization is required in many industrial facilities since the attached microflora greatly affects the rate of material corrosion. In medicine and tissue engineering, successful medical implants rely on the ability of the solid surface to avoid bacterial adhesion. Both in vivo and in vitro, adhered bacteria are main causes of infection for medical devices, including vascular grafts and catheters, cerebrospinal fluid shunts, artificial heart valves, orthopedic devices and cardiac pacemakers. Therefore, constant microbiological monitoring in hospitals is strictly required.

In microbial forensics, from a biological agent found at a crime scene, valuable inferences can be made. Collection, isolation and detection of surface associated microbe populations, including biological threat agents, is of vital importance in hygiene and sanitation issues, environmental health services, food safety, infection control, etc. The anthrax attack of 2001, when the Unites States Postal Service unknowingly distributed mail contaminated with anthrax spores, raised global awareness of the lack of adequate and standardized surface sampling techniques to discover and analyze bio-threats [5, 6]. Comprehensive sampling studies done in the past decade have revealed new possibilities for improving the detection and recovery of many biological threat agents, like *Bacillus anthracis* under different environmental conditions [7, 8].

Another significant motivation for in-depth investigation of surface-associated microorganisms and the search for appropriate sampling methods has arisen from space exploration. The spacecraft cleanliness is an issue of great concern for recent and future space missions, as stated by the United Nations Outer Space Treaty, Article IX: "Parties to the Treaty shall pursue studies of outer space, including the Moon and other celestial bodies, and conduct exploration of them so as to avoid their harmful contamination…" [9].

The constantly increasing number of studies reveals that the existing methods can recover frustratingly low percentages of surface-associated microorganisms. In fact, currently no available sampling method fully satisfies the practical demands. Although bioanalytical methods has made huge progress over the last twenty years, traditional CFU-assays are now greatly supported and even substituted by PCR, DNA-microarrays, ELISA and mass-spectroscopy, the sampling methods remain essentially incapable of providing adequate microbial recovery. The actual sampling outcome usually strongly depends on numerous factors, including characteristics of the surfaces (like porosity), amount, distribution and nature of the attached microorganisms, as well as environmental parameters (like temperature and humidity) and even on the individual sampling style of the executing person [10–12].

Some surfaces are especially challenging and troublesome for most of routinely applied sampling approaches. Even elaborated National Aeronautics and Space Administration standard swab protocols are admittedly ineffective on irregular or water absorbing surfaces [13]. The sanitarian and hygienic norms are therefore facing the problem of how to combine or modify the available methods in order to fulfill the requirements most competently.

The aim of this study was to overview theoretical and practical aspects of microbial attachment on dry solid surfaces and the problematics of development of a universal microbiological sampling method. In order to improve the recovery efficacy and sampling reliability, each existing sampling technique has to be carefully assessed in terms of suitable surfaces and optimal sampling parameters. Inherent advantages and drawbacks of sampling methods will be discussed in detail later in this chapter. This study was intended to help investigators and practitioners in gaining rational estimation of existing bio-sampling techniques. Thus, it would assist both in theoretical understanding and in practical implementation.

2 Physicochemical Aspects of Microbial Attachment and Detachment

The research of microbial adhesion is a large field covering different aspects of nature and human life, such as soil and plant ecology, biomedical and food industry, and most importantly, safety. Most kinds of adhesive bacteria and fungi have been shown to attach surfaces with non-covalent forces typically laying in the range of several piconewtons per contact site, that for the whole cell sum up into considerable adhesion strength [14]. Moreover, the working groups of Herman in 2014 and Dufrène in 2015 demonstrated that the specific force of even a single non-covalent bond can be remarkably strong, reaching several nanonewtons, which is equivalent to the covalent bond strength [15, 16]. Additionally, involvement of specialized adhesion molecules facilitates formation of covalent bonds between the contacting surfaces due to creating favorable local steric and chemical context. Expectedly, the sensitivity of any sampling method is significantly lower in recovering adsorbed cells from surfaces as compared to the free cells.

In 1986 Marshall and colleagues [17] proposed to describe surface colonization by microorganisms as consisting of two distinct stages, referred to as **reversible** and **irreversible** attachment, correspondingly. Although the boundary between these stages is quite vague, this view has been accepted by the majority of researchers [18–20].

The initial (reversible) attachment of microorganisms to material surfaces is the primary stage of adhesion, making the following molecular and cellular phases possible. The term “irreversible attachment” is used when microbes can no longer move perpendicularly away from the surface [21]. Among other terms used to describe microorganism-surface interactions, “adsorption” is sometimes defined as the initial, reversible stage of microbial attachment, in contrast to “adhesion”, which would imply additional irreversible and even specific bonding.

Despite the importance of cell-surface interactions and high scientific interest to this topic, the fundamental knowledge of the microbial adhesion mechanisms on dry surfaces contains many gaps. Likewise, the exact roles of physicochemical characteristics of the participating surfaces are still not fully understood. From the physicochemical point of view, a microbial cell can be simplistically modeled as a (typically negatively) charged particle of 0.5–10 μm size. Such an abstraction is seemingly expedient for modeling of microbial attachment in water solutions, allowing helpful predictions based on well-known physical and mathematical frameworks (Langmuir, DLVO) [17, 22, 23].

Thermodynamically, microbial attachment to dry surfaces results in overall decrease of Gibbs free energy, but in the case of electrostatic repulsion, a significant energy barrier has to be overcome. More broadly seen, initial microbial attachment in air can be described as a manifestation of interplay between various long-range (>100 nm) and short-range (2–50 nm) interactions, dominated by attractive Van der Waals forces, H-bonding and (often repulsive) electrostatic forces [24]. At shorter distances, water, which is always absorbed to the substratum (even on apparently dry surfaces), produces a significant potential barrier for closer attachment through solvation effects.

Most of the previously published reports were focused mainly on electrostatic effects in cell-surface association as the initial stage of adhesion but the applicability of the "electrostatic models" for dry adhesion has to be examined more thoroughly. The force on the particle as a function of the separation distance is proportional to the gradient of the potential energy. Indeed, a general theory predicts that interactions between oppositely charged groups, being relatively weak ("screened") in water due to its high dielectric constant, result in a strong, long-range attraction in air (Fig. 1a).

Although a favorable electrostatic situation greatly contributes to bacterial adhesion in air, other noncovalent intermolecular interactions are presumably involved as well. Hydrogen bonds (Fig. 1b), hydrophobic interactions (Fig. 1c), steric forces and various types of van der Waals bonding contribute relatively weakly as compared to electrostatic forces but, due to their great number, their combined effects can result in very strong bonding. In contrast to the water environment, where additional strong energetic effects may arise due to the entropy associated with solvation and orientation of water molecules surrounding the functional groups, in air the interaction between the polar hydrophilic groups leads to a relatively strong short-range molecular attraction. This strong attraction between dry surfaces indicates the importance of differential choice of a sampling technique depending on whether the surface is wet or dry.

Theoretical force-distance expressions can be used to interpret the data collected with the atomic force microscopy (AFM). Invention of AFM greatly facilitated investigations on the fundamental forces and bonds that control how a bacterium interacts with a solid surface but the small magnitudes of the individually involved forces makes the task of "force deconvolution" (distinguishing between the contributions of forces of different origin) extremely hard [16].

An additional difficulty in physicochemical modeling and experimental measurements of bacterial adhesion on air-exposed surfaces arises due to appearance of a "dried conditioning film". Various organic (lipids, proteins, polysaccharides) and inorganic (salt) molecules, once adsorbed from the medium at any occasional and even sporadic contact of the surface with water solutions, may significantly and uncontrollably modify physicochemical properties of the dried surface. Studies of microbial adhesion have been widely conducted for more than thirty years;

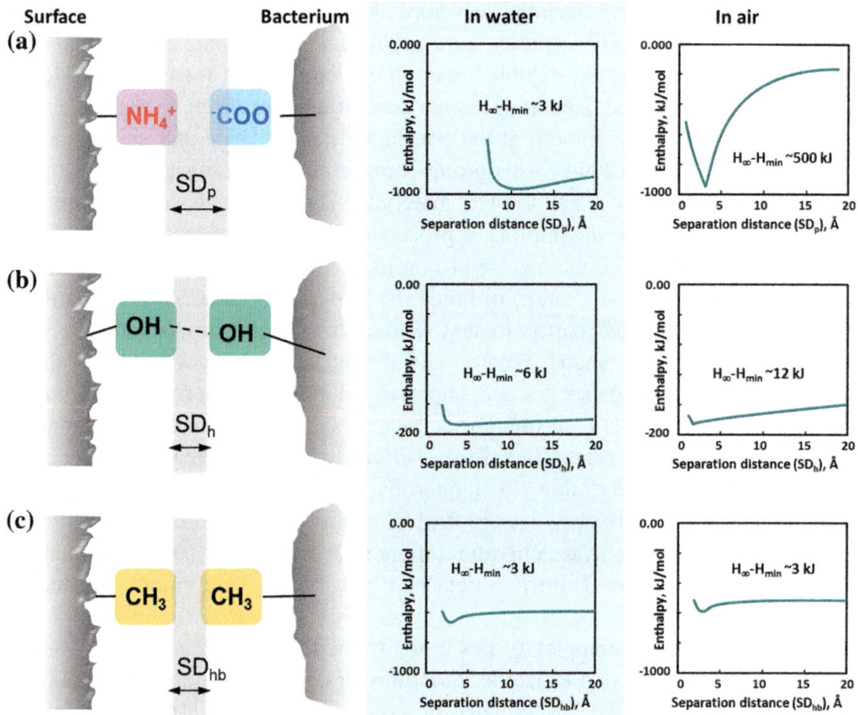

Fig. 1 Examples of three principal groups of intermolecular interactions and related enthalpic effects of molecular association in water and in air. **a** Electrostatic attraction; **b** hydrogen bonding; **c** hydrophobic interaction. Plots present simulated (using MatLab® software) entropy values as a function of separation distances (SD_p, SC_h and SD_{hb}, for electrostatic, hydrogen bond and hydrophobic interactions, respectively) in air and water. The approximate net adsorption enthalpy for each interaction type is shown as the difference between H_{min} and H_{∞} representing minimal enthalpy and the enthalpy at infinite separation, respectively

however, most of those works do not provide detailed molecular-scale information on organic and inorganic constituents involved in cell attachment. In particular, surface hydrophobicity and charge sometimes vary extraordinary, depending on numerous influencing factors, which can arise both from the environment and from the surface material involved.

These facts explain why the attempts to relate artificial surface physicochemical properties to bacterial adhesion undertaken by numerous research groups over the past decades have led to contradictory results. This can be illustrated by the ongoing debates concerning the impact of surface hydrophobicity on bacterial initial adhesion. On the one hand, hydrophobic interactions are believed to promote microbial adsorption [25–27], on the other hand, evidences have been collected indicating that bacteria do not have a strong preference for adhesion to hydrophobic or hydrophilic surfaces [28].

Another major concern related to sampling is the surface texture. In general, surfaces can be classified as porous, semi-porous and nonporous. Strictly speaking, virtually any surface would exhibit fractal-like irregularities; therefore, the definitions need to be adjusted depending on the context and according to the scale used. Nonporous surfaces are smooth at the considered length scale and usually do not appreciably absorb any fluids. This group is represented by polished metals, glass, and many painted and plastic-coated materials. Usually, such surfaces are well suitable for the traditional sampling approaches like swabbing.

According to our observations, semi-porous surfaces usually can slowly absorb moderate amounts (0.1–5.0 ml/g) of fluids. Examples are leather, rubber, hard and sealed soft wood, coated paper, treated fabric, etc. Porous surfaces can be characterized by pore size, shape, volume and depth; such surfaces typically absorb fluids readily. Rough paper, carpet, unprocessed wood, drapery, sponge, fabric, concrete, asphalt may serve as references. Since a sampling item dislodges attached microorganisms by creating a shear force, surface topography greatly influences the microorganism recovery due to numerous reasons. Firstly, microorganisms attaching in sheltered sites are protected from shear hydrodynamic forces. Secondly, due to local surface curvature, more attachment sites between contacting surfaces can be created. Thirdly, a detached cell may remain entrapped, floating inside a cavity or a pore.

A further source of complexity lies in the tremendous heterogeneity of bacteria at the single cell level due to lateral variations in composition and physical properties, like intrinsic shape asymmetry or polar pili location [16]. Bacterial cell surfaces are generally hydrophobic, but the cells can possess hydrophilic appendages [16, 29–32].

Summarizing, physicochemical aspects of initial microbial "in-air" adhesion together with many important clues in terms of both experimental observations and theoretical interpretations, still remains to be a source of confusion [33]. The main reason for this is that early bacterial adhesion is strongly multifactorial, depending on the nature and topology of the adhesive substrate, the conditioning films forming on the surface, the characteristics of the environment and other various properties of the cell surface. Moreover, all these properties vary with time, influencing each other by still indistinct interrelations.

3 Specialized Attachment Structures of Microorganisms

Microbes are not inert inanimate particles and can quickly respond to changing environmental conditions on both metabolic and structural levels [34]. Once in initial contact with a surface, microbes develop different types of attachment behaviors. Pure physical models of cellular adhesion do not take into account the complex self-regulating properties of living cells and the existence of specialized adhesion mechanisms. Understanding of the mechanisms determining biological response is a key element for maximally adequate description, reliable prediction and control of cellular adsorption.

As mentioned above, microbial adhesion to a material surface can be described as a two-phase process: (a) initial, instantaneous, and reversible physical phase (**phase one**) and (b) large-time dependent and irreversible molecular and cellular phase (**phase two**). The exact mechanism of a two-stage nature microbial attachment is still under discussion. Many authors believe that the second stage is undoubtedly mediated by new-synthesized adhesion molecules, creating strong chemical bounds between contacting surfaces. This point of view has found support in several studies that revealed cooperativeness in cellular attachment, as well as increased expression of specific gene groups prior to the second stage [35–38].

The phase two of microbial binding to biotic surfaces (wood, skin, paper) is mainly accomplished by biochemical interactions based on high specific affinity such as enzyme-substrate, lectin-carbohydrate, and antigen-antibody pair forming. The binding of cellulolytic *Clostridia* to cellulose-containing substrates induces formation of a "cellulosome" containing cellulose binding domains [39]. Adhesion to abiotic surfaces (plastics, glass, metals) also involves specialized biological structures, though presumably to lesser extent. Firmer adhesion of bacteria to an abiotic surface is accomplished by the bridging function of bacterial surface polymeric structures, which include capsules, fimbriae or pili, and slime.

In the case of highly specialized interactions, variations in physical and chemical parameters (temperature, local pH, etc.) play a relatively unimportant role in the adsorption process. The strength and specificity of these interactions are mainly affected by the macromolecular composition, charge and conformational state of corresponding binding sites [40]. Molecular, biological and biochemical models of cell adhesion focus predominantly on identification, isolation and structural analysis of phase-two-related attachment-responsible biological molecules and their genetic determinants. Below, some of the adhesion-related structures are briefly discussed.

3.1 Adhesins

All bacteria have a surface made of various biological polymers. **Adhesin** can be defined as a substance produced by microorganisms and responsible for specific adhesive activities. The molecules responsible for attachment are of different chemical nature and include surfactants, various carbohydrates, glycoproteins, lectins or lectin-like proteins, teichoic and lipoteichoic acids, etc. [39]. Adhesins can be identified as the functional part of cell walls, capsules, fimbriae or pili, and slime. Even previously "unsuspected" molecules, such as extracellular DNA or lipids, have been shown to play a role in the process of adhesion or biofilm maturation [41, 42].

Microbial adhesion-related machinery is often sophisticated, combining binding domains with catalytic units as well as hydrophobic domains, which act as linkers between the catalytic units and other domains, integrating the molecule into peptidoglycan-teichoic acid lattice. In general, adhesins display more complex domain features when adhesive bacteria do not express microscopically recognizable structures like fimbriae.

Microorganisms may possess multiple adhesins for different surfaces (different receptors) involved in close contacts between the bacterial cell and the substrate. While most adhesins bind to surfaces non-covalently, the streptococcal surface protein SfbI was reported to mediate covalent interaction with fibrinogen using unusual internal thioester bonds formed as a "chemical harpoon" between Cys and Gln residues [43]. Thioester-containing domains are surprisingly prevalent in pilins of clinically relevant gram-positive bacteria, indicating the importance of further study of intra-molecular isopeptide bonds and their role in adhesion.

Biologically defined expression level of adhesins can dramatically vary even inside of the same microbial population [44, 45]. This results in significant population heterogeneity in respect to cell surface hydrophobicity, charge, adhesiveness etc. Depending on their localization, microbial adhesins can be classified as **fimbrial** (located on fimbria or pili) and **non-fimbrial** (directly anchored to the outer cell membrane). Fimbrial adhesins will be briefly discussed below, together with fimbria. One of the most studied groups of non-fimbrial adhesins involved in cell adhesion to abiotic surfaces is a multi-domain *Bap*-family (biofilm-associated proteins). *Bap*-proteins are abundant in both gram-positive and gram-negative bacteria and share a specific N-terminal secretion signal for transport from the cytoplasm, glycine-rich C-terminal domain and a core domain composed of highly repeated motifs [46].

Filamentous hemagglutinins (non-fimbrial) expressed by both human and plant pathogens have been shown to attach cells to abiotic surfaces, thereby facilitating formation of air-exposed biofilms [47]. The members of the non-fimbrial adhesin protein family are called trimeric auto-transporters and mediate attachment to a vast number of both biotic and abiotic surfaces [48].

3.2 Fimbriae (Attachment Pili)

It has been recently argued by Renner et al. that the physical appendages of bacteria (flagella, fibrillae and pili) may help to overcome the repulsive intermolecular forces [42]. Bacterial fimbriae (also called pili) are a group of rigid, straight, filamentous appendages on a bacterial surface 4–7 nm in diameter. They arise from the outer surface of the outer membrane and may extend for considerable distances (up to 2 μm). A bacterial cell usually possesses several hundred to one thousand fimbriae [39]. Fimbriae are major adhesive structures on a bacterial surface found in many gram-negative and some gram-positive bacteria and are primarily composed of identical subunits of protein fimbrillin (pilin), which can act itself as fimbrial adhesin. For instance, *Pseudomonas aeruginosa* uses type IV pilin for binding to *stainless* steel, polystyrene, polyvinylchloride and other abiotic surfaces [49].

Pilus-deficient strains of *P. aeruginosa* do not bind to steel, whereas purified type IV pili bound with high affinity in a concentration-dependent, saturable manner.

Fimbriated bacteria are often more hydrophobic, which indicates the possible role of hydrophobicity in mediating surface adhesion. Finally, the appendages may contribute in covalent consolidation of the bacteria-surface bond by stimulating chemical reactions such as oxidation and hydration. Fimbriae may mediate bacterial adhesion also by adhesins FimH, F71, F8, SS142, K88 and F13 associated with fimbriae [46]. FimH is specific for mannose-containing receptors that can bind to abiotic surfaces in a mannose-dependent manner. Mannose was found to specifically inhibit biofilm formation in *E. coli* on a variety of abiotic surfaces, including various plastic polymers and glass [50].

3.3 Flagella and Fibrillae

Bacterial flagellum is typically a long helical 20 nm-thick hollow tube made up of flagellin. Flagella are anchored in the membranes by a basal body and a filamentous hook composed of several protein rings. Due to their flagella, microorganisms like uropathogenic *E. coli* can attach to surfaces irreversibly, while retaining active motility by mechanisms known as gliding, swarming, twitching, swimming, darting and sliding. Another important pathogen, *Vibrio cholerae* first utilize the flagella to spread across the surface, and then anchor onto the surface with pili and possibly outer membrane proteins [35, 51, 52].

Fibrillae are amorphous surface adhesive structures, which appear to lack the regular filamentous forms of fimbriae. Fibrillary structures have been observed at the surface of various streptococci and they are thought to contribute to the better adhesion onto hydroxyapatite substrata like dental materials [53].

3.4 Capsules and Slime

In addition to the different adhesins described above, many bacterial species produce extracellular polysaccharides that promote adhesion. These polysaccharides can be firmly associated with the cell surface, forming a **capsule**, or be loosely associated or even released from the cell surface, forming **slime**. Both capsules and slime are composed of (predominantly) polysaccharides and proteins that can act as adhesins. For example, a self-generated viscous slime largely composed of polysaccharide adhesins (PS/A) and cins (PIA) helps *S. epidermidis* to colonize various abiotic surfaces [54].

To summarize, numerous studies ascertained that microorganisms successfully adapt and adhere to virtually all-natural and manufactured materials. Virtually any biological molecule made by the bacterium can be used to support or modulate its adhesion to surfaces (Fig. 2). Recent discoveries in the regulation and the

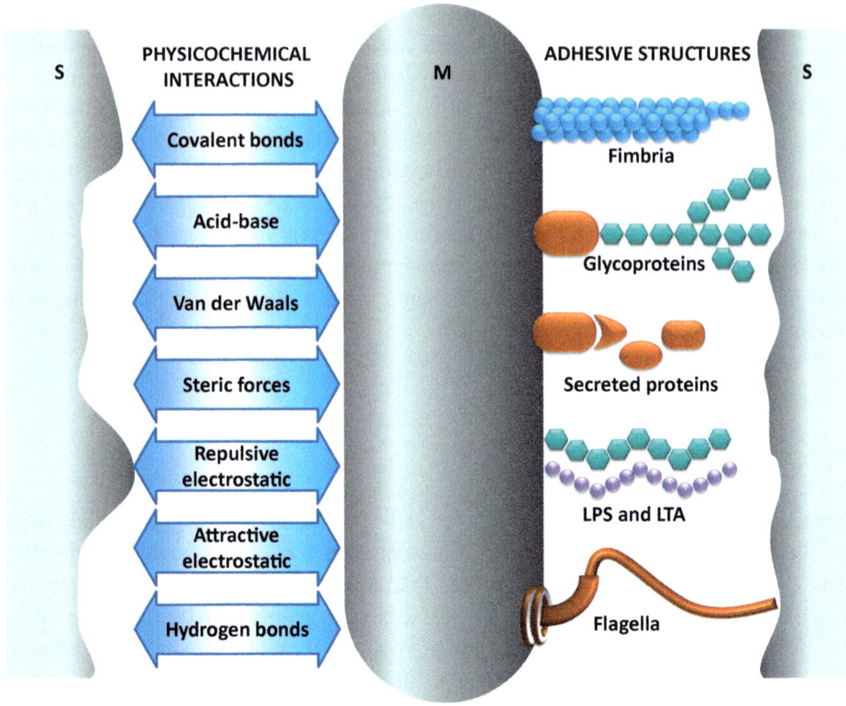

Fig. 2 Simplified overview of universal interactions (left) and specialized structures that mediate adhesion of microorganisms (M) on dry surfaces (S)

modulation of adhesion suggest that this process is much more complex and dynamic than originally anticipated, and especially for dry surfaces, has a huge influence on the microbiological sampling efficacy.

4 Methodological Aspects of Sampling

Extensive efforts were made to develop rapid, simple, accurate and reliable methods to collect and analyze surface-associated microorganisms. To date, numerous approaches, procedures and tools have been proposed together with conventional ones in an attempt to satisfy the demands of hygiene standards and conditions: swabbing, sponge-spatula, wiping, bulk, scraping, rinsing, vacuuming, tape-lifting, sonication and contact-plating just to mention some (Fig. 3). Some methods are better suited for inspecting flat surfaces, while others are supposed to be applied for irregular interfaces.

Virtually all presently available solid surface sampling methods have been recently reviewed by Rached Ismaïl et al. who classified them as destructive

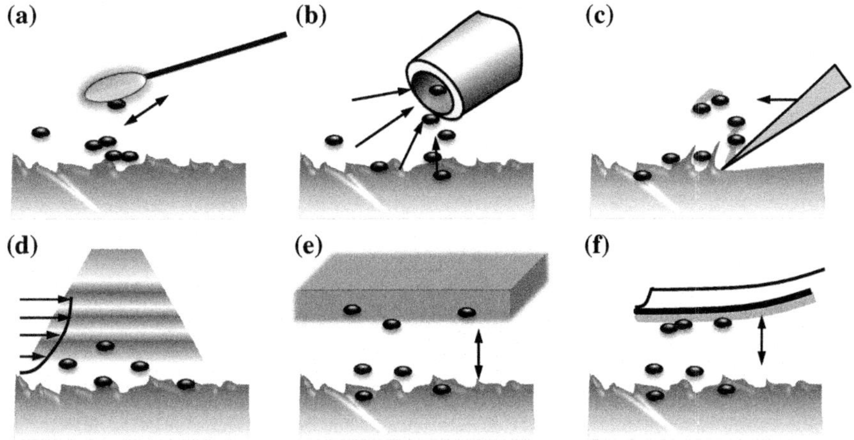

Fig. 3 Overview of dry surface sampling techniques. **a** Swabbing; **b** vacuuming and rinsing; **c** scraping and scratching, **d** sonication; **e** contact-plating; **f** tape-lifting

(rinsing, sonication, scraping) and non-destructive (swabbing, scrubbing and printing) [1].

In Table 1, most of existing microbial sampling methods are listed according to the tool/device used and surface types to be tested respectively. We believe that such a categorization could facilitate quick selection of an appropriate sampling type (Table 1).

Table 1 Currently available microbial sampling techniques for dry surfaces.

As it was pointed out by Erin E. Silvestri and coworkers, recovery rate of the microbial bioload from the sampled dry surface varies greatly, depending on numerous factors: the surface type to be tested, the size of the sampling surface area, the type of the sampling tool/device, etc. [55, 56]. Moreover, careful sampling organization as well as a comfortable working space also promotes safe and effective sampling. From the organizational point of view, a sampling process may be divided into several stages, including supervision/preparation, recovery, processing, analyzing and interpretation (Fig. 4).

Traditional microbiological methods require long-term routine CFU-counting procedures before receiving results. Together with the recovery of microbes, their rapid detection by using fluorescence and hybridization techniques becomes progressively essential. Adenosine triphosphate (ATP), present in all living cells, can serve as an excellent marker for detecting microbial bioload and its viability. This idea has found numerous commercial applications (e.g. Kikkoman Lumitester PD-20, EnSURE Hygiene Meter, Clean-Trace NG Luminometer UNG2, Charm novaLUM). However, some drawbacks of the ATP-based sampling should be mentioned as well. This method is not sensitive for spore detection as the ATP level is very low in bacterial spores. It is virtually impossible to distinguish the ATP-signals from bacteria, spore, yeast, or even in animal and plant cells. Traces of

Table 1 Currently available microbial sampling techniques for dry surfaces

Sampling method	Common used material/ tool	Process	Typical surface type	Wet or dry	Typical target surface size, cm^2
Swabbing	Swab	Swabbing, brushing	Smooth surfaces	Dry/wet	Localized, <25
Contact plate	Typical/modified plates	Printing, impression	Rough and smooth nonporous	Dry/ rare wet	>25
Sponge	Sponge/spatula	Swabbing, brushing, sweeping	Rough and smooth nonporous	Dry, rare wet	>100
Wipe	Wipe	Wiping	Smooth nonporous	Dry/rare wet	Different, usually >100
Rinse	Water solution	Rinsing, immersion	Rough and smooth nonporous	Dry/wet	Different, usually >100
Bulk	Tweezer	Excision	Rough and smooth porous and nonporous	Dry/wet	>20
Scraping	Scraper, scoop	Scraping, grinding	Rough and smooth nonporous	Wet, rare dry	Different, usually >100
Vacuum	Vacuum filter	Vacuuming	Rough porous	Dry	Different, >1000
Sonication	US generator/transducer	Sonication	Rough porous	Wet/rare dry	Localized, <25
Brushing	Brush	Brushing sweeping	Rough nonporous	Mostly dry	>100

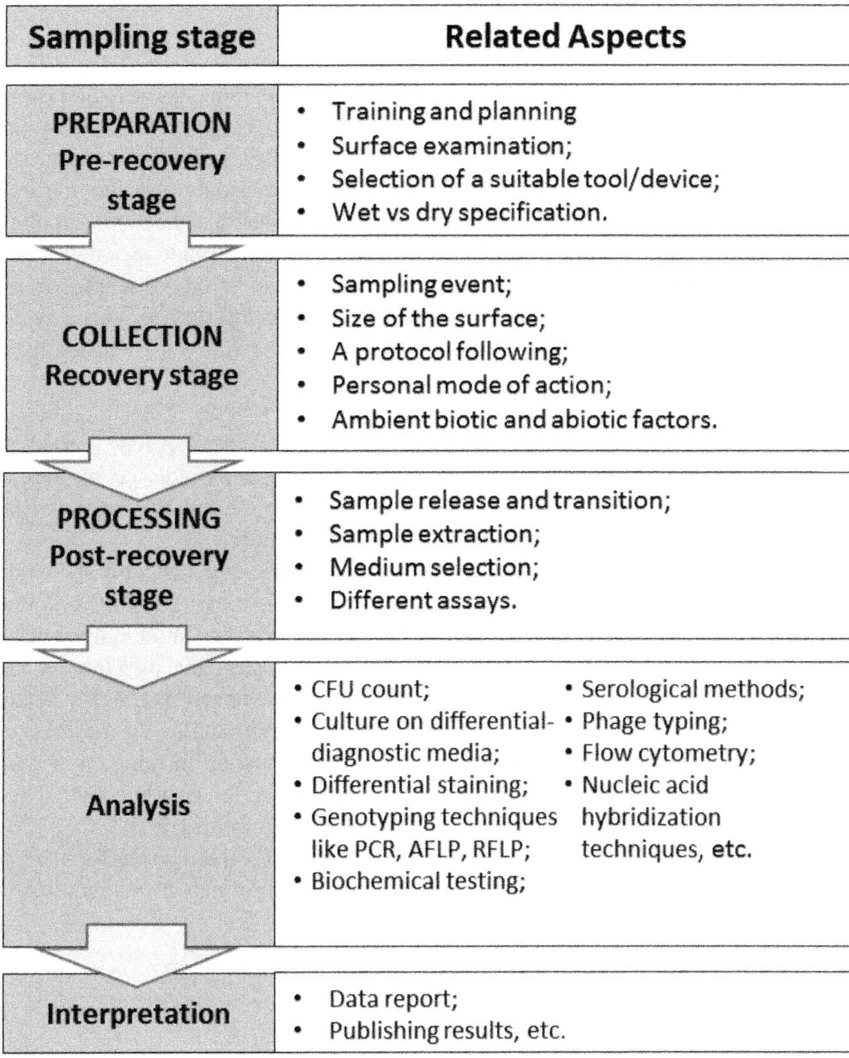

Fig. 4 Main stages of the microbiological sampling and some considerations related to their application

other organics can seriously affect the actual ATP bioluminescence readings as shown by Navid Omidbakhsh and colleagues [57].

Below we will give a brief description of some of the key sampling methods, paying attention to their peculiarities as well as pointing out their strong and weak aspects.

4.1 Swabbing

Using swabs to sample surfaces for microbial contamination has a long history dating back to the beginning of the 20th century. Swab sampling is probably the most used method to recover microbial cells from the target surfaces. In microbiological environmental monitoring programs, swabs are widely used for hygiene monitoring of surfaces and operators. Swabbing is considered as the main sampling standard in the space product assurance, as mentioned in the guidelines developed by The European Space Agency for microbiological examination of flight hardware and cleanrooms [58]. Sampling is performed by wiping the surface with a swab, usually soaked with water or other desirable solvent, to remove particles/biofilms from a surface.

A typical sampling procedure can be described as follows: First, a sufficient number of sterile swabs and test tubes with sterile water are prepared. A swab is aseptically removed from its container and its head (also called flock) is moistened in a test tube with sterile water. Excess moisture from the swab is removed by pressing the swab head against the interior wall of the tube.

During sampling, the swab is held so that the handle creates about a 30° angle with respect to the surface being sampled (Fig. 5). While moving the swab in one direction, the flock of the swab should be rotated slowly and thoroughly over a previously measured 25 cm^2 surface area. In the next step, the direction of the swabbing motion is changed by 90° and the surface is swabbed thoroughly again. Optionally, a third coverage of the surface can be done by changing the direction of the swabbing motion by 135° and repeating the last step. After finishing the wiping steps, the swab is returned into a tube containing the sterile sampling buffer (for example, phosphate buffered saline + 0.02 v/v % Tween 80, pH 7.2).

Prior to plating, each swab in the tube is vortexed at maximum power for 5–10 s. Depending on the swab type used, an additional sonification step can be required in

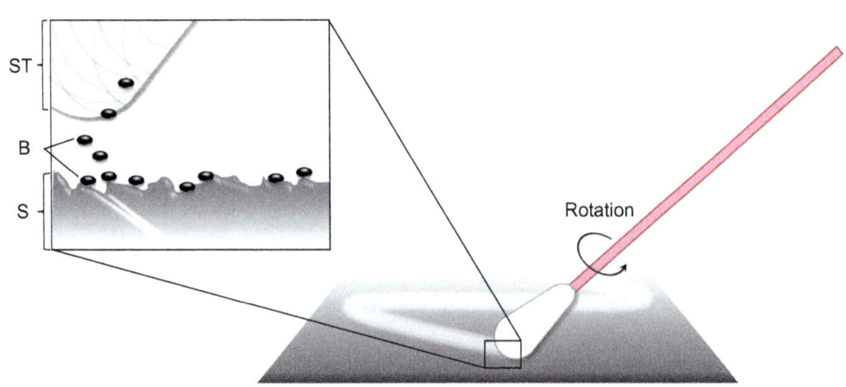

Fig. 5 Swabbing technique. ST head of a swab; B biofilm, S surface to be sampled

order to improve the microorganisms' yield from the swab head material. 24–48 h after plating on petri dishes, the colony forming units (CFU) are examined.

Main benefits of the swabbing methods:

- They are strongly standardized in terms of the sampling procedure and the data evaluation.
- Hard-to-reach and irregular surfaces can be sampled.
- Neither special equipment nor extensive operator training is necessary.
- The materials and reagents are inexpensive.
- Easy disposal.

Depending on the intended customer group and the manufacturer, commercial swabbing kits can be further grouped by several ways, for example as:

(a) direct (e.g. a swab is applied straight on the agar media) or indirect (e.g. a swab is firstly rinsed or dissolved in a buffer);
(b) ready-to-use (e.g. a plastic tube with a swab attached to the lid) or not ready-to-use (e.g. a singly packed dry swabs);
(c) aimed primarily at dry or wet sampling.

It must be recognized that different swabbing methods vary in terms of their selectivity, sensitivity and specificity because they employ different materials, buffers, culture media, etc. Undoubtedly, one type of swab cannot cover all needs and careful selection of a proper swab for the particular surface is additionally very important. Currently, there is a variety of swabs for collection, but their comparative efficiency is not completely known. The material of swab flock has a high influence on recovery efficiency of microbes since physical characteristics of a swab flock can significantly affect the uptake and release of bacteria. Depending on the type of the flock material, swabs can be categorized as **fibrous** (cotton, rayon, nylon, polyester, dacron), **spongeous** (macrofoam, SCRD, polyurethane) or **hydrogel** (calcium alginate).

Traditional rayon swabs are generally used and considered the gold standard in swab collection. Kathryn H Harry et al. tested the efficacy of different swabs (nylon flocked, rayon flocked, HydraFlock®, macrofoam, and nylon flocked) using seven various types of bacteria and highlighted the relevance of surface coatings (fiber) of a swab and its chemical nature. In particular, they showed that the capillary action of a swab is critical for recovery of microorganisms and for protection from dehydration [59].

Nuttada Panpradist et al. demonstrated that polyurethane swabs show the best performance across all swabbing conditions for recovering of *Staphylococcus aureus*. In addition, they also highlighted the importance of the swab size, structure, transfer and fluid capacity of the sample. A similar study by Dalmaso et al. was devoted to development and testing of high-recovery microRheologics flocked spaws [60]. Swabs that most effectively remove bacteria from a surface, however, do not necessarily demonstrate higher bacterial recovery. Some useful guidelines to assist in evaluation of swab choice were formulated by Verdon and coworkers [61].

There are numerous evidences, as well as common sense, that determine that swab-based sampling methods cannot provide a perfect sample collection. Traditional swab-based methods are inaccurate, time consuming and prone to significant variations due to uncontrollable contribution from multiple factors.

Main drawbacks of the swabbing methods:

- Many swabs would be required to sample large areas. Usage of a template is usually required, in order to define the surface to be sampled (typically 25 cm^2). Studies by Mark Buttner et al. demonstrated that the sampling over larger areas provides greater sensitivity [62].
- Swabs definitely do not show same effectivity on surfaces of different roughness. Certain types of roughness (carpets, wood and others) represent a significant problem for sampling using swabs.
- Potential operator contributions to collection efficiency are not well studied. Collection efficacy depends on the pressure applied on the swab, on its speed and the angle to the surface. In general, increasing the level of mechanical energy generated during swabbing would increase the number of bacteria removed from a wet surface. However, it was hypothesized that cellular damage, perhaps caused by the swabbing action itself, may have reduced recoverability from a dry surface [63].
- Microbial recovery is strongly dependent on collection material, wetting agent and extraction conditions. For example, ambient temperature and air humidity can greatly affect the transfer of microorganisms from the sampled surface to the swab (own unpublished data).
- Variability in morphological and physiological properties of microorganisms, such as micro-colonies, biofilm fragments, and connections between single cells represent yet another confounding factor.

The abovementioned points indicate an acute necessity of development of improved swabbing variants and (if possible) combination of swabbing with other sampling methods for the comparison of the obtained data.

4.2 *Wipe Sampling*

Wipe sampling is one of the most frequently used sampling techniques based on friction as it is easy to maintain and readily available. The main idea of this method is the aseptic wipe or fabric is soaked/dipped in sterile distilled water (or other wetting agent) before scrubbing the tested surface, where the bioburden microbial load transfers from the surface to the material (Fig. 6).

A study conducted by Gary S. Brown et al. evaluated the polyester-rayon blend wipes for performance of extraction and collection of powdered *Bacillus atrophaeus* spores from nonporous (painted wallboard and stainless steel) surfaces and revealed a high variability in sampling efficiency rate [12]. Another study by

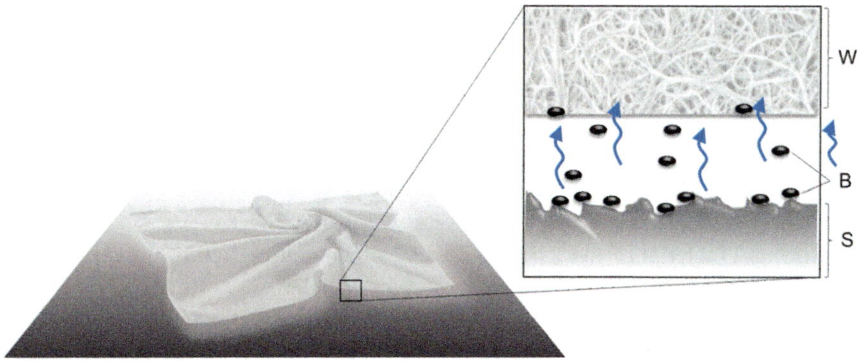

Fig. 6 Performance of the wipe sampling technique. Transfer of bacterial cells (B) from the sampled surface (S) into the wipe tissue (W) is shown (very schematically)

Laura J. Rose et al. brought to light the large variability inherent in sponge wipe environmental sampling as well. Nonetheless, high effectiveness of using cellulose sponge wipe for recovering the *Bacillus* spores from steel surfaces has been demonstrated [64].

An experimental comparison of wipe sampling with HEPA vacuum sock methods upon sampling *B. anthracis* spores from nonporous surfaces by Wayne T. Sanderson and co-workers showed a similar performance of both techniques. However, the authors outlined the superior results of swab sampling method compared to HEPA vacuum and wipe samplings [65].

The collection of microbes from the sampled surface may greatly vary with the characteristics and composition of the surface, the water/solvent used, the mode (style) of sampling etc. [66]. For example, the solutions used to moisten the wipes play a very important role, as demonstrated by N.B. Valentine et al. who observed the highest collection efficiency if 0.3% Tween-80 in phosphate-buffered saline was used [67].

Main benefits of the wiping methods:

- Sampling of larger areas can often provide a better contamination assessment accuracy;
- Suitable for sampling using commercially available cleaning robots;
- Hard-to-reach and irregular surfaces can be sampled;
- They are strongly standardized in terms of the sampling procedure and the data evaluation;
- Neither special equipment nor extensive operator training are necessary;
- The materials and reagents are inexpensive;
- Easy disposal.

Main drawbacks of the wiping methods:

- Labor consuming;
- Since personal protective equipment (gloves etc.) is required to sample, this might increase the contamination risk;
- In spite of personal protective equipment, elevated risk of the exposure of the operator to pathogens;
- Foaming effects are known to interfere with sampling;
- Size and material of the wipes can vary, which causes significant variations in the sampling efficacy.

4.3 *Spatula Sampling*

This group can be considered a hybrid one, since it endeavors to combine the swabbing and the wiping approaches (Fig. 7). The basic structure of a spatula is composed of a bilateral handling stick and with its forefront covered by a piece of a material on both sides. The material can be diverse, resulting in different spatula sampling versions like foam-spatula, nylon-flocked-spatula, etc. In general, before sampling, a spatula is pre-moistened with sterile water or a proper solution and positioned at a 30° angle toward the test surface. After sampling, the spatula is transferred back to its container filled with water solution and the collected cells are extracted by shaking or sonication.

Many studies indicate high sampling effectiveness when using spatula is made of macrofoam (polyurethane). Lewandowski et al. reported that this material could collect $\geq$30% more *Bacillus* spores from steel surfaces than rayon and polyester

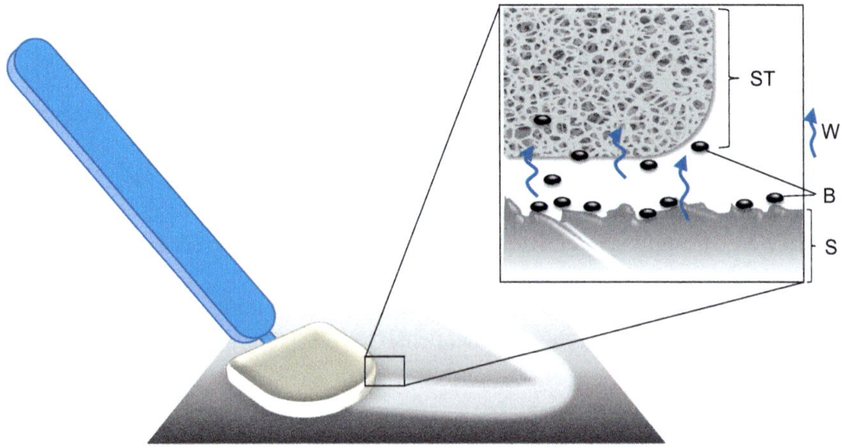

Fig. 7 Implementation of sponge technique. ST tip of a sponge, W water; B biofilm; S surface

ones. The spongy network (pore sizes ranges between 2 and 10 μm) of a macro-foam with many internal cavities makes it more reliable to absorb sampled liquid [68]. Probst et al. claimed foam-spatula assay to be highly effective in microbial load detection on spacecraft-related rough surfaces and superior to NASA wipe assay [69].

Main advantages of the (sponge) spatula methods can be described as follows:

- Relatively high microbial uptake;
- Suitable for many surfaces (wet and dry);
- Convenient for HACCP (Hazard Analysis Critical Control Point);
- Easy handling.

However, an insufficiently sterilized sampling matrix (sponge spatula) can serve itself as the source of contamination, as indicated in the study by Rossi et al. where twenty-four synthetic polyurethane sponges used in food services were found to be highly contaminated and could relocate vast majority of microorganisms to surfaces [70].

Some other drawbacks of the (sponge) spatula methods:

- Low mechanical and chemical stability of sponge material may lead to its disintegration and fragmentation during sampling if the surface of interest is irregular;
- Spatulas show different microbial recovery depending on the surface roughness. Certain types of roughness (carpets, wood and others) represent a significant problem for sampling using spatula;
- Potential operator contributions to collection efficiency are not well studied. Collection efficacy depends on the pressure applied on the spatula, on its speed and the angle to the surface.

4.4 Contact Plating and Tape-Lifting Methods

The techniques using agar syringes, RODAC (Replicate Organism Detection And Counting) plates, agar sausages and membrane filter blots can alse be included in this group. This approach is quite commonly used for dry surface sampling and is often termed as "press plate" or "direct-contact plate" group of methods. Usually, plates and paddles are impregnated with nutrition agar in the form of a dome. This medium is pushed against a horizontal surface so that the microbial load on the surface would stick to the medium and therefore transfer to the agar. In case of a petri dish, an approximately 90 mm diameter surface area is tested and simply converted to the number of microbes per square cm (Fig. 8).

There are two main approaches of contact plates to sample surfaces for microbial contamination: direct agar contact plate (round surface sampling) and dipslides (flat

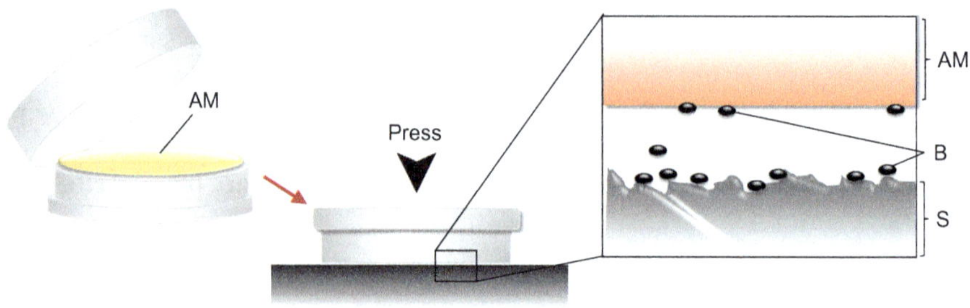

Fig. 8 Implementation of contact plate-based method. AM agar medium; B biofilm; S surface

paddle sampling). Recently, some other related methods (Petrifilms, roller sampler, tape-lifting) have been successfully used to recover microbial cells. The use of **tape-lifting** for collection of touch DNA from fabrics has become routine [71].

The **direct agar contact plate assay** (ACP) involves direct printing/sampling which is based on a petri dish with overfilled an agar medium. This agar is placed against the surface to be tested and pressed at a uniform downward pressure making the microorganisms-exchange between the two surfaces possible, as shown in Fig. 8.

Dipslides (paddle testers, flat paddles) are intended for prompt detection of microbial load on the surfaces of different materials (Fig. 9). Dipslides have been assessed as having high sensitivity and specificity, particularly for dry surfaces [72]. They are also commonly used for the monitoring of microbes in liquid-based systems by dipping or immersing the slide tip in water, this is where the name originally comes from. Dipslides have two sides, one (front) usually composed of non-selective medium (NA, TSA or malt extract), and the other side (rear) made up of selective agar (McConkey, VRBG, Rose Bengal, pseudomonas agar etc.). Dipslides coated with selective agar are recommended for recovering Methicillin-resistant *Staphylococcus aureus* (MRSA) from flat environmental surfaces [73].

Dipslides are produced by numerous manufacturers, like Hygicult, 3M, Hycheck, BD-Difco, Millipore, Orion, Merck, Becton, Dickinson, Oxoid, Biotest, Bode etc. and have several advantages over conventional contact-based sampling techniques. Most dipslides contain special chemical agents that neutralize the most common disinfectants and solvents present on the surface in order to improve bacterial growth as well as recovery.

Roller sampler

J. K. Lutz et al. in 2013 reported a novel sampling tool for recovery and detection of *Staphylococcus aureus* from a stainless steel surface [74]. Because of the prototype rolling design, this contact-based sampler or so-called "roller sampler" has two main benefits over conventional contact-based sampling techniques: it is flexible and suitable for testing large surface area. However, in a direct comparison study by

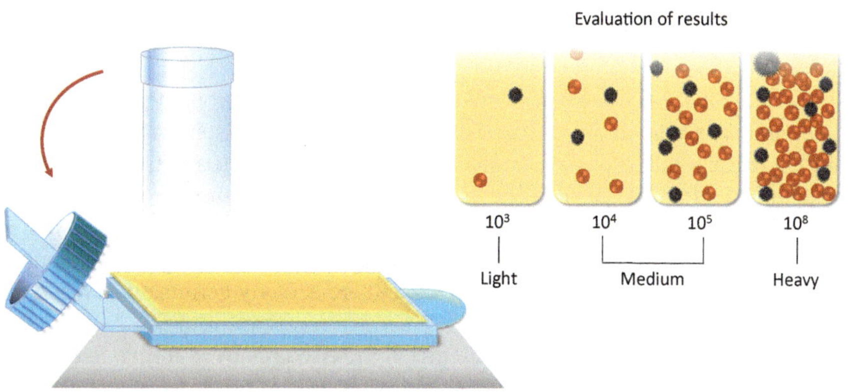

Fig. 9 Surface sampling using dipslides and the quantitation of microbial load

the authors, wipe- and swab-based sampling techniques remained preferred versus both contact-based sampling methods used.

The advantages of using contact plating methods include:

- Easy to use and simple procedure;
- Low cost if made in the lab;
- No further procedures after sampling, like dilution; this saves labor and time.
- Compact and precise area (20 cm^2) is tested;
- Specifically, the advantages of dipslides are better sample security, easier sampling in confined spaces and longer shelf life that may speak in favor of choosing these for bacteria environmental sampling in hospitals, depending on the task. The factorial square design of dipslides allows sampling from different spots. The contact plates can be sealed in a plastic container that prevents risk of contamination and extents shelf life by protecting it from drying out the medium [72].

The disadvantages are:

- Limited use on rough, uneven and curved surfaces;
- In case of high contamination, colonies may overgrow which makes counting and distinguishing individual CFUs troublesome or impossible;
- Microbial recovery may be poor from sticky surfaces;
- Plates are prone to sporadic contamination and cross-contamination;
- Limited and strictly defined sampling area is not always desired;
- Some variants of the contact plating methods are rather expensive;
- Excessive pressure may cause the sample to disperse outside the allocated area, resulting in incorrect counts;
- As for tape-lifting, there is still a paucity of data relating to its effectiveness and the amount of tape-lifting required to generate a useful microbiological profile.

4.5 Bulk Sampling or Direct Extraction Methods

A very difficult challenge upon sampling is to be able to appropriately detect those microorganisms that colonize material's pores, cavities and subsurfaces. The bulk sampling means that a portion/fragment of the sampled material itself is used for extraction of microorganisms. The purpose of the bulk sampling is to collect a representative portion of the material in a suitable amount that should be transported to the laboratory under preserving conditions. The method can be primarily applied to detect whether microbes have colonized and are actively germinating within the **volume of the tested material**.

The fragments to be sampled are aseptically excised/removed from a material of interest via a sterile tweezer (forceps, gloved hands or other suitable tolls) and allocated into a sterilized container or plastic bags (Fig. 10). The size of a sample to pick-up, the mode of its collection and transportation depends on the sample nature and the corresponding protocols to be used.

Bulk sampling is a commonly accepted method for assessment of fungal contamination [75, 76]. Different variants of the bulk sampling method are widely used in soil microbiology as well as in meat-, poultry- and cereal-processing plants for microbial sampling of cereals, carcasses and different anatomical parts/sites. Many researches have extensively studied the applicability of this method for the biological control in slaughterhouses, reporting its great potential [77, 78].

R.A. Pearce et al. compared excision with sponge swabbing for sampling *Enterobacteriaceae* from pork, beef and lamb carcasses and concluded that bulk sampling is the preferred method as it yields the highest microbial recovery [79]. However, swab based sampling of a comparably large area might be superior to bulk sampling of a smaller area for detecting *Enterobacteriaceae* and related bacteria on swine carcass surfaces [80]. Therefore, many authors believe that if an

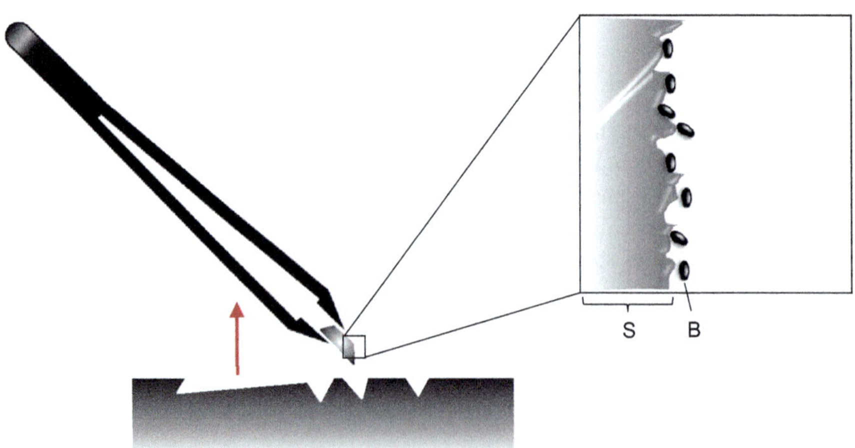

Fig. 10 Bulk sampling technique. S surface; B biofilm

appropriate material (especially, mildly abrasive and absorbent) is used for swabbing, the recovery rate by bulk and swab sampling would be quite similar.

A study conducted by Alexander Probst et al. compared bulk sampling with swab and wipe based protocols to recover *Bacillus* spores from various spacecraft-related rough surfaces. The highest recovery rate for the spores was reached by using the bulk method (15.4%) as compared with other techniques (approx. 13.2%) [13]. However, the authors mentioned poor applicability of the bulk methods for testing of spacecraft surfaces.

Summarizing, the advantages of the bulk sampling methods can be described as follows:

- Enables assessment of both "on surface" and "in volume" contamination;
- Rapid;
- High recovery yield due to extensive extraction;
- Negligible cross-contamination.

Disadvantages:

- Destructive;
- Less suitable for rough and rigid surfaces;
- Inconvenient for flat surfaces;
- Collected samples may be limited in number and size;
- Sensitive to the choice of the sampling site.

4.6 *Scraping*

Like other destructive techniques, scraping based methods provided the highest recovery efficiency for irregular or uneven surfaces. A conventional scraping technique is accomplished by mechanically removing microbial biofilm with different tools such as scrapers, spatulas, curettes, chisels, scalpels, blades etc. In order to pick up and remove a bio-contaminant containing upper layer, a combination of

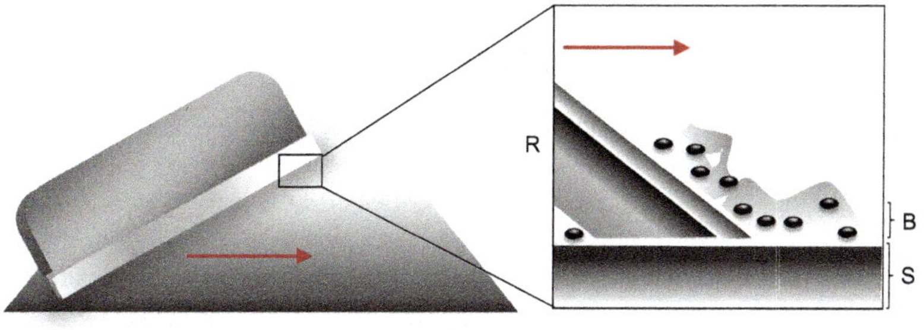

Fig. 11 Performance of the scraping method. R razor; B biofilm; S surface

various scraping/grinding procedures can be used, which usually were followed by an ultrasonic or vacuum treatment (Fig. 11).

The scraping method seems to be useful when analyzing water pipe utilities. B. H. Packard and M.J. Kupferle compared brushing and scraping techniques to recover *Bacillus atrophaeus* (a surrogate for etiologic agent *Bacillus anthracis*) from various pipe materials (polyvinyl chloride, cement-lined ductile iron, and ductile iron pipe). Mean recoveries for all materials ranged from 37 ± 30% to 43 ± 20% for brushing versus 24 ± 10% to 51 ± 29% for scraping [81]. Moreover, this method can be extended and feasibly used in oral therapy to dislodge and detect surface associated microbiota from prostheses and teeth [82].

Sampling by scraping has found wide application in livestock and poultry meat-producing industry to detect hazardous bio-objects from contaminated carcass/meat surfaces. Geir Bjerkan et al. in their comparative study from 2009 described scraping based protocols (scraping with direct culture, scraping followed by vortex mix and scraping followed by sonication) to dislodge different bacteria from titanium and steel surfaces. The authors pointed the simplicity and flexibility of scraping techniques but, considering their low recovery efficiency, they did not recommend scraping as a technique for diagnosis of biofouling-related infection [83].

Benefits: Allows extensive extraction on irregular surfaces, especially corroded and tuberculated surfaces.

Drawbacks: Being intrinsically destructive and highly operator-dependent, scraping/grinding procedures have limited applicability for sampling on solid surfaces in general.

4.7 Vacuuming (Suction) Sampling Techniques

Bio-contaminants, accumulated on various surfaces (especially porous) can be collected with vacuum (-sock) sampling. Initially, the intended use of the method was to sample clean surfaces in laminar flow clean rooms [84]. Later, the vacuuming devices, especially those equipped with HEPA (high efficiency particulate air) were successfully used for sampling surfaces in other environments. W.N. Sanderson and colleagues demonstrated high effectiveness of HEPA vacuum sock sampler when performed on various nonporous surfaces for recovering *Bacillus* spores [65].

This technique primarily allows collection of surface associated dust and other powdery materials from large areas of contamination. Particulate objects are removed and entrained in an air stream and collected in a sock, which is applied to trap these particulate dust substances (Fig. 12).

Currently, there are many studies devoted on recovery of microorganisms and spores on different surfaces using vacuum sampling systems. Vacuuming obviously has several advantages over conventional sampling techniques: it covers large, porous and dusty surfaces, like cloth sets, carpets, tiles and even filter systems. The

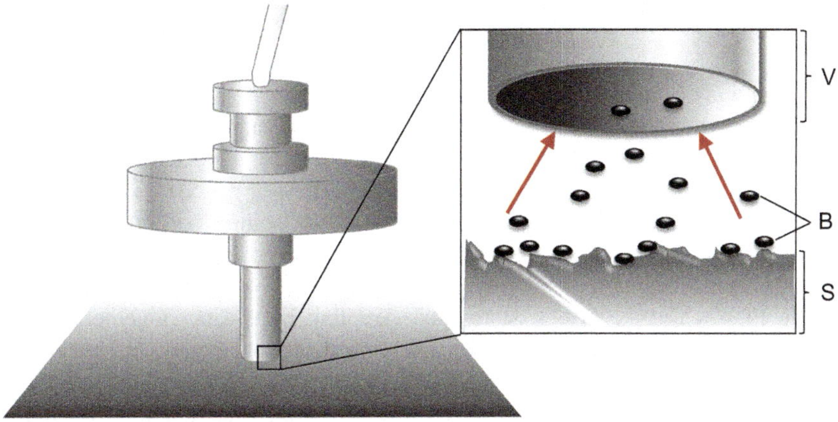

Fig. 12 Implementation of vacuum sampling. V vacuum; B biofilm; S surface

sampling sock is used to seize total dust containing biological objects, which is then evolved to microbial analysis in the laboratory. An especially attractive feature of this method is the possibility of removal (elimination) of dangerous bioloads (like *Bacillus anthracis* spores) from surfaces with readily available equipment, that significantly facilitates the cleaning works following a wide-area biological contamination [85]

In the study by Calfee et al. (2013), *Bacillus* spores settled on different material types (concrete, carpet, and upholstery, stainless steel) were efficiently collected by commonly used vacuuming methods [86].

Combining HEPA-systems with cleaning robots opens a way to commercially available solutions for sampling spores and viable cells on common indoor (carpet, laminate etc.) surfaces. In addition to the known advantages of using robots, such as less risk of contamination and personnel exposure, self-recharging, navigation, large sampling area, S.D. Lee and coauthors reported appreciably high recovery rates [87].

Main benefits of the vacuuming methods can be summarized as follows:

- Great collecting sensitivity due to sampling over large areas;
- Suitable for heavy dust-laden surfaces;
- Suitable for autonomous sampling applications;
- Highly effective on irregular, porous and soft surfaces.

Disadvantages:

- High quantities of collected dust/waste may mask hazardous biological agents;
- Cumbersome sampling equipment;
- Vulnerable to cross-contamination due to the complex construction;
- Large number of factors influence the recovery efficiency: vacuuming time, suction pressure, sampled area, direction, etc.

4.8 Sampling by Rinsing

This technique is sometimes erroneously termed "rinse sampling", which would rather mean examination of the last rinse water for contamination. "Sampling by rinsing" refers to bioload collection from tested surfaces **directly by** rinse water. The surface or object to be examined is sprayed, shower, rinsed or immersed with proper solution, drained through the funnel or just poured and collected into a sterile pocket/bottle. The amount of the liquid is chosen correspondingly to the size of object or surface area of interest (Fig. 13).

The FDA guidance (1993) states that sampling by rinsing, together with the direct sampling methods (swabbing) are considered as mostly acceptable and desirable for validation of cleaning [88].

In 1975 L.C. Blankenship et al. devised spray-rinse method for sampling of *Salmonella typhimurium* from internal cavity of broiler carcasses and discovered its superiority over swabbing and excision methods [89]. Nowadays, spray-rinse sampling method is used in meat and poultry processing industry and allows collecting microbial biomass of the entire cavity of meat, fish, broiler carcasses, hatching eggs and fresh produces [90, 91]. In an extensive study conducted by Qiu Quin Zhang et al. rinsing was recommended as a desirable method in poultry industry to recover *Pseudomonas* spp., *Brochothrix thermosphacta* and *Enterobacteriaceae* [92].

Advantages:

- Collecting large numbers of samples;
- Large surface area may be sampled;
- Systems, generally "inaccessible" for example, due to their geometry, can be sampled;
- High recovery rate.

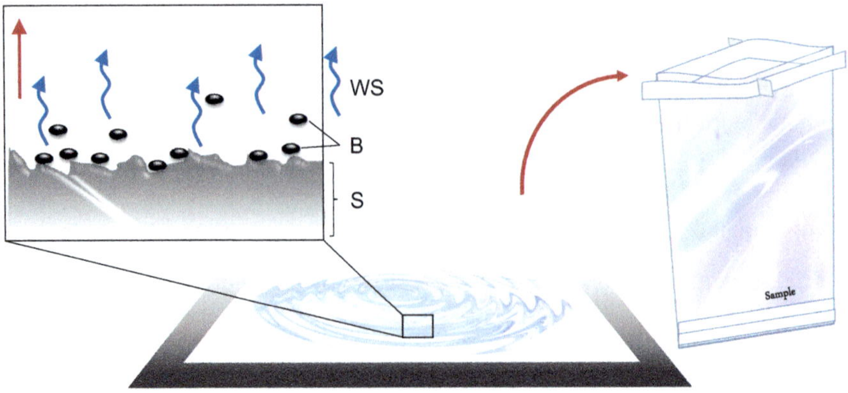

Fig. 13 Performance of rinsing method. WS water solution; B biofilm; S surface

Disadvantages:

- Not suitable for many surfaces;
- Highly depends on operator skills.

4.9 Sonication

Initially, ultrasonic sampling techniques have been developed for surface cleaning and decontamination procedures. To date, ultrasonic engineering achieved great advancement in the processes like extraction, emulsification, changing viscosity, disintegration, decontamination, etc. [93]. It is important to emphasize that ultrasonication cannot be directly applied on dry surfaces because of significant acoustic impedance mismatch as well as insufficient ultrasound propagation distance in the air. Therefore, the surface to be sampled must be submerged in a liquid medium prior to the ultrasound-assisted sampling.

Physical effects accompanying ultrasound, like cavitation and acoustic streaming are best suited to inactivate or remove microorganisms due to the generation and collapse of bubbles. High-intensity ultrasound at low frequencies (about 40 kHz) is frequently used for this purpose [94].

I. J. Seymour et al. studied ultrasound potency for the biological decontamination of raw plant materials and reported that stable ultracavitation increases the uptake of entrapped *Salmonella* spp. on the fruit and vegetable surfaces via shifting/loosening particles through a shearing activity. However, they also reported relatively low decontamination efficiency ($P > 0.69$) that depended on ultrasound frequency (25, 32–40 and 62–70 were used in the study) [95]. In contrary to that, in a study by Kang et al. (2007), where several methods for quantitative recovery of *Listeria monocytogenes* from stainless steel surfaces were compared, the three different sonication methods yielded higher recovery than the other three methods. The sonication methods used in that study were (a) in an ultrasonic water bath (40 kHz); (b) contact with the bristles of a sonicating brush head for 1 min; and (c) indirect contact (2–4 mm distance) with a sonicating brush head for 1 min [96].

N. Oulahal-Lagsir et al. presented the ultrasound-based methodology concerning the removal of fouling in food processing equipment. The ultrasonic apparatus they have developed produces 40 kHz US cavitation allowing the non-destructive and quick detection of a surface bioload showing a four times greater recovery rate compared to a conventional swabbing assay [97] (Fig. 14).

Ultrasonic technologies in oral and skin hygiene practices for removing plagues and wound treatment are growing fast because of health and economic reasons. Ultrasonic waves can dislodge the wound-associated bacteria from human skin and safely extract them into the liquid solution without harming the skin [98]. Pilot experiments for construction of an automatic oral cleaning tool carried out by Takeshi Nishikawa et al. demonstrated dislodgement of *Streptococcus mutans* by sonication exposure (280 kHz, 1 MHz, or 2 MHz; 1–3 min) [99]. Their results also

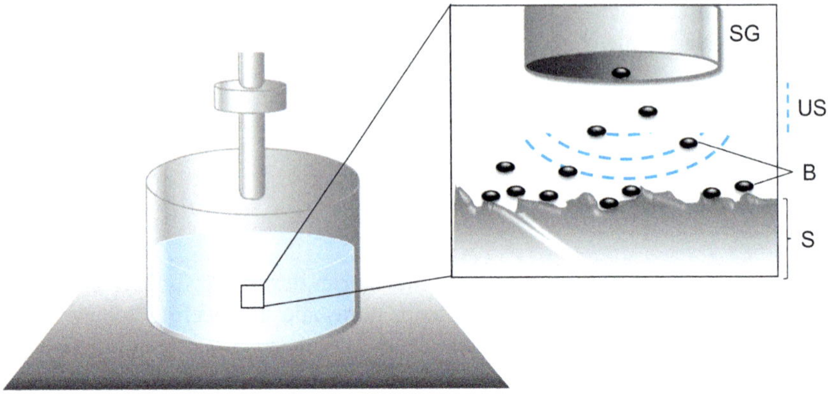

Fig. 14 The sonication system. SG sonication generator; US ultrasound; B biofilm; S surface

suggested that adding a contrast medium, such as micro bubbles (Sonazoid) further enhanced the biofilm removal.

Advantages

- Highly effective on irregular, porous and soft surfaces;
- Systems, generally "inaccessible" for example, due to their geometry, can be sampled;
- Potentially distractive, but without physical risks and chemical damage;
- Suitable for autonomous sampling applications.

Disadvantages:

- Cumbersome sampling equipment;
- The recovery efficiency is sometimes surprisingly low;
- Foam development upon sampling is a further known problem;
- The impact of the interactions of ultrasound waves with material structures (cavities, inner boundaries, etc.) on the sampling outcome needs to be studied;
- High costs related to further development and optimization.

4.10 Sampling by Brushing and Sweeping

Only a limited number of reports are available on brushing and sweeping sampling methods [81]. These techniques are relatively insignificant when compared to conventional methods for the application of bioload recovery on target surfaces. Due to certain drawbacks (such as vastly cross-contamination risk, law safety, direct personal exposure, highly operator dependent, etc.) they are often unable to fulfill the basic quality requirements.

5 Conclusions

Nowadays, many environmental dry surfaces are implicated as a source of healthcare-associated infections presenting a need for reliable, quick and efficient environmental sampling to assess cleanliness, and, improve infection prevention and control. Microbiological sampling should not only be conducted on the "critical" surfaces of direct contact with the patient, water or food, but almost equally importantly on "indirect contact" surfaces in which microorganisms can easily migrate to the "critical" ones. General microbiological monitoring as well as pathogen specific environmental sampling should be undertaken carefully to evaluate the microbial bioload precisely.

Here we have attempted to give a brief overview of the factors contributing to the strength of microbial attachment to dry solid surfaces in order to facilitate rational decision making concerning the approaches to sampling surfaces. Further, we have tried to critically address the existing sampling techniques in terms of suitable surfaces, sampling conditions and potential risks. The choice of decent a sampling method is obviously determined by a number of criteria and factors. For each sampling method, its advantages and disadvantages have been briefly summarized.

This scoping review is to be used firstly, to enhance theoretical understanding of microbial attachment to dry surfaces and secondly, to guide prospective microbial collecting efforts and selection of sampling approaches. We hope that this review may provide helpful information for assessment of a range of sampling situations and strategies.

Acknowledgements Our research for this chapter was partially financially supported by the K2-Commission of FH-Aachen University of Applied Sciences. We have also received very helpful input from Peter Kayser, Dariusz Porst, Alexandra Lösch, Konstantin Kotliar, Bernd Dachwald, Stephan Neumann, Gerhard Artmann, Shachriar Dantism. We would like to thank Ms. Danielle Hillebrecht for the valuable comments and help in the manuscript preparation.

References

1. Ismaïl, R., Aviat, F., Michel, V., Le Bayon, I., Gay-Perret, P., Kutnik, M., et al. (2013). Methods for recovering microorganisms from solid surfaces used in the food industry: A review of the literature. *International Journal of Environmental Research and Public Health, 10*(11), 6169–6183. https://doi.org/10.3390/ijerph10116169.
2. Dunne, W. M. (2002). Bacterial adhesion: Seen any good biofilms lately? *Clinical Microbiology Reviews, 15*(2), 155–166. https://doi.org/10.1128/cmr.15.2.155-166.2002.
3. Watnick, P., & Kolter, R. (2000). Biofilm, city of microbes. *Journal of Bacteriology, 182*(10), 2675–2679.
4. Parsek, M. R., & Fuqua, C. (2004). Biofilms 2003: Emerging themes and challenges in studies of surface-associated microbial life. *Journal of Bacteriology, 186*(14), 4427–4440. https://doi.org/10.1128/jb.186.14.4427-4440.2004.

5. Schmedes, S. E., Sajantila, A., & Budowle, B. (2016). Expansion of microbial forensics. *Journal of Clinical Microbiology, 54*(8), 1964–1974. https://doi.org/10.1128/jcm.00046-16.
6. Smith, D., Martin, D., & Novossiolova, T. (2017). Microorganisms: Good or evil. MIRRI provides biosecurity awareness. *Current Microbiology, 74*(3), 299–308. https://doi.org/10.1007/s00284-016-1181-y.
7. Hess, B. M., Amidan, B. G., Anderson, K. K., Hutchison, J. R. (2016). Evaluating composite sampling methods of bacillus spores at low concentrations. *PLoS One 11*(10). https://doi.org/10.1371/journal.pone.0164582.
8. Priyanka, B., Patil, R. K., & Dwarakanath, S. (2016). A review on detection methods used for foodborne pathogens. *Indian Journal of Medical Research, 144*(3), 327–338. https://doi.org/10.4103/0971-5916.198677.
9. Treaty on Principles Governing the Activities of States in the Exploration and Use of Outer Space, including the Moon and Other Celestial Bodies. http://disarmament.un.org/treaties/t/outer_space/text. Accessed 13 Jun 2017.
10. Pinna, D. (2014). Biofilms and lichens on stone monuments: Do they damage or protect? *Frontiers in Microbiology, 5*. https://doi.org/10.3389/fmicb.2014.00133.
11. Tan, M. S. F., Moore, S. C. Tabor, R. F., et al. (2016). Attachment of *Salmonella* strains to a plant cell wall model is modulated by surface characteristics and not by specific carbohydrate interactions. *BMC Microbiology, 16*. https://doi.org/10.1186/s12866-016-0832-2.
12. Brown, G. S., Betty, R. G., Brockmann, J. E., et al. (2007). Evaluation of vacuum filter sock surface sample collection method for *Bacillus* spores from porous and non-porous surfaces. *Journal of Environmental Monitoring, 9*(7), 666–671. https://doi.org/10.1039/b700163k.
13. Probst, A., Facius, R., Wirth, R., et al. (2010). Validation of a nylon-flocked-swab protocol for efficient recovery of bacterial spores from smooth and rough surfaces. *Applied and Environment Microbiology, 76*(15), 5148–5158. https://doi.org/10.1128/aem.00399-10.
14. Tang, G., Yip, H.-K., Samaranayake, L. P., et al. (2004). Direct detection of cell surface interactive forces of sessile, fimbriated and non-fimbriated Actinomyces spp. using atomic force microscopy. *Archives of Oral Biology, 49*(9), 727–738. https://doi.org/10.1016/j.archoralbio.2004.04.003.
15. Herman, P., El-Kirat-Chatel, S., Beaussart, A., et al. (2014). The binding force of the staphylococcal adhesin SdrG is remarkably strong. *Molecular Microbiology, 93*(2), 356–368. https://doi.org/10.1111/mmi.12663.
16. Dufrêne, Y. F. (2015). Sticky microbes: Forces in microbial cell adhesion. *Trends in Microbiology, 23*(6), 376–382. https://doi.org/10.1016/j.tim.2015.01.011.
17. Marshall, K. C. (1986). Adsorption and adhesion processes in microbial growth at interfaces. *Advances in Colloid and Interface Science, 25*(1), 59–86.
18. Vigeant, M. A.-S., Ford, R. M., Wagner, M., et al. (2002). Reversible and irreversible adhesion of motile escherichia coli cells analyzed by total internal reflection aqueous fluorescence microscopy. *Applied and Environment Microbiology, 68*(6), 2794–2801. https://doi.org/10.1128/aem.68.6.2794-2801.2002.
19. Petrova, O. E., & Sauer, K. (2012). Sticky situations: Key components that control bacterial surface attachment. *Journal of Bacteriology, 194*(10), 2413–2425. https://doi.org/10.1128/jb.00003-12.
20. Hoffman, M. D., Zucker, L. I., Brown, P. J. B., et al. (2015). Timescales and frequencies of reversible and irreversible adhesion events of single bacterial cells. *Analytical Chemistry, 87*(24), 12032–12039. https://doi.org/10.1021/acs.analchem.5b02087.
21. Busscher, H. J., Poortinga, A. T., & Bos, R. (1998). Lateral and perpendicular interaction forces involved in mobile and immobile adhesion of microorganisms on model solid surfaces. *Current Microbiology, 37*(5), 319–323.
22. Kennedy, J. F., & Cabral, J. M. S. (1990). Use of titanium species for the immobilization of cells. *Transition Metal Chemistry, 15*(3), 197–207. https://doi.org/10.1007/bf01038375.
23. Bodenmiller, D., Toh, E., & Brun, Y. V. (2004). Development of surface adhesion in Caulobacter crescentus. *Journal of Bacteriology, 186*(5), 1438–1447.

24. Gottenbos, B., Busscher, H. J., van der Mei, H. C., et al. (2002). Pathogenesis and prevention of biomaterial centered infections. *Journal of Materials Science. Materials in Medicine, 13* (8), 717–722.
25. Donlan, R. M., & Costerton, J. W. (2002). Biofilms: Survival mechanisms of clinically relevant microorganisms. *Clinical Microbiology Reviews, 15*(2), 167–193. https://doi.org/10.1128/cmr.15.2.167-193.2002.
26. Daeschel, M. A., & McGuire, J. (1998). Interrelationships between protein surface adsorption and bacterial adhesion. *Biotechnology and Genetic Engineering Reviews, 15*(1), 413–438. https://doi.org/10.1080/02648725.1998.10647964.
27. Krasowska, A., & Sigler, K. (2014). How microorganisms use hydrophobicity and what does this mean for human needs? *Frontiers in Cellular and Infection Microbiology, 4,* 112. https://doi.org/10.3389/fcimb.2014.00112.
28. Bos, R., van der Mei, H. C., Gold, J., et al. (2000). Retention of bacteria on a substratum surface with micro-patterned hydrophobicity. *FEMS Microbiology Letters, 189*(2), 311–315.
29. Ebersbach, G., & Jacobs-Wagner, C. (2007). Exploration into the spatial and temporal mechanisms of bacterial polarity. *Trends in Microbiology, 15*(3), 101–108. https://doi.org/10.1016/j.tim.2007.01.004.
30. Kaiser, D., & Yu, R. (2005). Reversing cell polarity: evidence and hypothesis. *Current Opinion in Microbiology, 8*(2), 216–221. https://doi.org/10.1016/j.mib.2005.02.002.
31. Klemm, P., & Schembri, M. A. (2000). Bacterial adhesins: Function and structure. *International Journal of Medical Microbiology, 290*(1), 27–35. https://doi.org/10.1016/s1438-4221(00)80102-2.
32. Klemm, P., & Schembri, M. A. (2000). Fimbrial surface display systems in bacteria: from vaccines to random libraries. *Microbiology, 146*(Pt 12), 3025–3032. https://doi.org/10.1099/00221287-146-12-3025.
33. Donlan, R. M. (2008). Biofilms on central venous catheters: is eradication possible? *Current Topics in Microbiology and Immunology, 322,* 133–161.
34. Goller, C. C., & Romeo, T. (2008). Environmental influences on biofilm development. *Current Topics in Microbiology and Immunology, 322,* 37–66.
35. Davey, M. E., & O'Toole, G. A. (2000). Microbial biofilms: From ecology to molecular genetics. *Microbiology and Molecular Biology Reviews, 64*(4), 847–867. https://doi.org/10.1128/mmbr.64.4.847-867.2000.
36. Costerton, J. (1999). Introduction to biofilm. *International Journal of Antimicrobial Agents, 11*(3–4), 217–221. https://doi.org/10.1016/s0924-8579(99)00018-7.
37. Korber, D. R., Choi, A., Wolfaardt, G. M., et al. (1997). Substratum topography influences susceptibility of Salmonella enteritidis biofilms to trisodium phosphate. *Applied and Environment Microbiology, 63*(9), 3352–3358.
38. Rasmussen, T. B., & Givskov, M. (2006). Quorum-sensing inhibitors as anti-pathogenic drugs. *International Journal of Medical Microbiology, 296*(2–3), 149–161. https://doi.org/10.1016/j.ijmm.2006.02.005.
39. An, Y. H., & Friedman, R. J. (2010). *Handbook of bacterial adhesion: Principles, methods, and applications*. Totowa, NJ: Humana Press.
40. Busscher, H. J., & van der Mei, H. C. (2006). Microbial adhesion in flow displacement systems. *Clinical Microbiology Reviews, 19*(1), 127–141. https://doi.org/10.1128/cmr.19.1.127-141.2006.
41. Limoli, D. H, Jones, C. J., Wozniak, D. J. (2015). Bacterial extracellular polysaccharides in biofilm formation and function. *Microbiology Spectrum, 3*(3). https://doi.org/10.1128/microbiolspec.mb-0011-2014.
42. Renner, L. D., & Weibel, D. B. (2011). Physicochemical regulation of biofilm formation. *MRS Bulletin, 36*(5), 347–355. https://doi.org/10.1557/mrs.2011.65.
43. Walden, M., Edwards, J. M., Dziewulska, A. M., et al. (2015). An internal thioester in a pathogen surface protein mediates covalent host binding. *Elife, 4*. https://doi.org/10.7554/elife.06638.

44. Flint, S., Palmer, J., Bloemen, K., et al. (2001). The growth of *Bacillus stearothermophilus* on stainless steel. *Journal of Applied Microbiology, 90*(2), 151–157. https://doi.org/10.1046/j.1365-2672.2001.01215.x.
45. Peng, J.-S., Tsai, W.-C., & Chou, C.-C. (2002). Inactivation and removal of Bacillus cereus by sanitizer and detergent. *International Journal of Food Microbiology, 77*(1–2), 11–18.
46. Berne, C., Ducret, A., Hardy, G. G., et al. (2015). Adhesins involved in attachment to abiotic surfaces by gram-negative bacteria. *Microbiology Spectrum, 3*(4). https://doi.org/10.1128/microbiolspec.mb-0018-2015.
47. Webster, P., Wu, S., Gomez, G., et al. (2006). Distribution of bacterial proteins in biofilms formed by non-typeable *Haemophilus influenzae. Journal of Histochemistry and Cytochemistry, 54*(7), 829–842. https://doi.org/10.1369/jhc.6a6922.2006.
48. Wagner, C., & Hensel, M. (2011). Adhesive mechanisms of *Salmonella enterica. Advances in Experimental Medicine and Biology, 715,* 17–34. https://doi.org/10.1007/978-94-007-0940-9_2.
49. Giltner, C. L., van Schaik, E. J., Audette, G. F., et al. (2006). The *Pseudomonas aeruginosa* type IV pilin receptor binding domain functions as an adhesin for both biotic and abiotic surfaces. *Molecular Microbiology, 59*(4), 1083–1096. https://doi.org/10.1111/j.1365-2958.2005.05002.x.
50. Pratt, L. A., & Kolter, R. (1998). Genetic analysis of *Escherichia coli* biofilm formation: Roles of flagella, motility, chemotaxis and type I pili. *Molecular Microbiology, 30*(2), 285–293.
51. Davey, M. E., Caiazza, N. C., & O'Toole, G. A. (2003). Rhamnolipid surfactant production affects biofilm architecture *in Pseudomonas aeruginosa* PAO1. *Journal of Bacteriology, 185* (3), 1027–1036. https://doi.org/10.1128/jb.185.3.1027-1036.2003.
52. Kuchma, S. L., & O'Toole, G. A. (2000). Surface-induced and biofilm-induced changes in gene expression. *Current Opinion in Biotechnology, 11*(5), 429–433. https://doi.org/10.1016/s0958-1669(00)00123-3.
53. Isenberg, H. D. (1988). Pathogenicity and virulence: Another view. *Clinical Microbiology Reviews, 1*(1), 40–53.
54. Arciola, C. R., Campoccia, D., Ravaioli, S., et al. (2015). Polysaccharide intercellular adhesin in biofilm: Structural and regulatory aspects. *Frontiers in Cellular and Infection Microbiology, 5,* 7. https://doi.org/10.3389/fcimb.2015.00007.
55. Silvestri, E. E., Feldhake, D., Griffin, D., et al. (2016). Optimization of a sample processing protocol for recovery of *Bacillus anthracis* spores from soil. *Journal of Microbiological Methods, 130,* 6–13. https://doi.org/10.1016/j.mimet.2016.08.013.
56. Silvestri, E. E., Yund, C., Taft, S., et al. (2017). Considerations for estimating microbial environmental data concentrations collected from a field setting. *Journal of Exposure Science and Environmental Epidemiology, 27*(2), 141–151. https://doi.org/10.1038/jes.2016.3.
57. Omidbakhsh, N., Ahmadpour, F., Kenny, N. (2014). How reliable are ATP bioluminescence meters in assessing decontamination of environmental surfaces in healthcare settings? *PLoS One, 9*(6). https://doi.org/10.1371/journal.pone.0099951.
58. ECSS Executive Secretariat ECSS-Q-ST-70-55C.
59. Harry, K., Turner, J., & Madhusudhan, K. (2013). Comparison of physical characteristics and collection and elution performance of clinical swab. *African Journal of Microbiology Research, 7*(31), 4039–4048. https://doi.org/10.5897/ajmr.
60. Dalmaso, G., Bini, M., Paroni, R., et al. (2008). Qualification of high-recovery, flocked swabs as compared to traditional rayon swabs for microbiological environmental monitoring of surfaces. *PDA Journal of Pharmaceutical Science and Technology, 62*(3), 191–199.
61. Verdon, T. J., Mitchell, R. J., & van Oorschot, R. A. H. (2014). Swabs as DNA collection devices for sampling different biological materials from different substrates. *Journal of Forensic Sciences, 59*(4), 1080–1089. https://doi.org/10.1111/1556-4029.12427.
62. Buttner, M. P., Cruz, P., Stetzenbach, L. D., et al. (2007). Evaluation of two surface sampling methods for detection of *Erwinia herbicola* on a variety of materials by culture and

quantitative PCR. *Applied and Environment Microbiology, 73*(11), 3505–3510. https://doi.org/10.1128/aem.01825-06.
63. Moore, G., & Griffith, C. (2007). Problems associated with traditional hygiene swabbing: the need for in-house standardization. *Journal of Applied Microbiology, 103*(4), 1090–1103. https://doi.org/10.1111/j.1365-2672.2007.03330.x.
64. Rose, L. J., Hodges, L., O'Connell, H., et al. (2011). National validation study of a cellulose sponge wipe-processing method for use after sampling *Bacillus anthracis* spores from surfaces. *Applied and Environment Microbiology, 77*(23), 8355–8359. https://doi.org/10.1128/aem.05377-11.
65. Sanderson, W. T., Hein, M. J., Taylor, L., et al. (2002). Surface Sampling methods for *Bacillus anthracis* spore contamination. *Emerging Infectious Disease Journal, 8*(10), 1145. https://doi.org/10.3201/eid0810.020382.
66. Stam, C., Behar, A., Cooper, M. (2016). Sampling of microbiological samples. In: M. Micic (Ed) *Sample preparation techniques for soil, plant, and animal samples* (1st ed., pp. 25–39). New York, NY: Humana Press.
67. Valentine, N. B., Butcher, M. G., Su, Y.-F., et al. (2008). Evaluation of sampling tools for environmental sampling of bacterial endospores from porous and nonporous surfaces. *Journal of Applied Microbiology, 105*(4), 1107–1113. https://doi.org/10.1111/j.1365-2672.2008.03840.x.
68. Lewandowski, R., Kozlowska, K., Szpakowska, M., et al. (2010). Use of a foam spatula for sampling surfaces after bioaerosol deposition. *Applied and Environment Microbiology, 76*(3), 688–694. https://doi.org/10.1128/aem.01849-09.
69. Probst, A., Facius, R., Wirth, R., et al. (2011). Recovery of bacillus spore contaminants from rough surfaces: A challenge to space mission cleanliness control. *Applied and Environment Microbiology, 77*(5), 1628–1637. https://doi.org/10.1128/aem.02037-10.
70. Rossi, E. M., Scapin, D., & Tondo, E. C. (2013). Survival and transfer of microorganisms from kitchen sponges to surfaces of stainless steel and polyethylene. *The Journal of Infection in Developing Countries, 7*(3), 229–234. https://doi.org/10.3855/jidc.2472.
71. Verdon, T. J., Mitchell, R. J., & van Oorschot, R. A. H. (2014). Evaluation of tapelifting as a collection method for touch DNA. *Forensic Science International: Genetics, 8*(1), 179–186. https://doi.org/10.1016/j.fsigen.2013.09.005.
72. Ibfelt, T., Foged, C., & Andersen, L. P. (2014). Validation of dipslides as a tool for environmental sampling in a real-life hospital setting. *European Journal of Clinical Microbiology and Infectious Diseases, 33*(5), 809–813. https://doi.org/10.1007/s10096-013-2018-2.
73. Obee, P., Griffith, C. J., Cooper, R. A., et al. (2007). An evaluation of different methods for the recovery of meticillin-resistant Staphylococcus aureus from environmental surfaces. *Journal of Hospital Infection, 65*(1), 35–41. https://doi.org/10.1016/j.jhin.2006.09.010.
74. Lutz, J. K., Crawford, J., Hoet, A. E., et al. (2013). Comparative performance of contact plates, electrostatic wipes, swabs and a novel sampling device for the detection of *Staphylococcus aureus* on environmental surfaces. *Journal of Applied Microbiology, 115*(1), 171–178. https://doi.org/10.1111/jam.12230.
75. Niemeier, R. T., Sivasubramani, S. K., Reponen, T., et al. (2006). Assessment of fungal contamination in moldy homes: comparison of different methods. *Journal of Occupational and Environmental Hygiene, 3*(5), 262–273. https://doi.org/10.1080/15459620600637333.
76. Dillon, H. K., Miller, J. D., Sorenson, W. G., et al. (1999). Review of methods applicable to the assessment of mold exposure to children. *Environmental Health Perspectives, 107*(Suppl 3), 473–480.
77. Tittlemier, S. A., Varga, E., Scott, P. M., et al. (2011). Sampling of cereals and cereal-based foods for the determination of ochratoxin A: An overview. *Food Additives and Contaminants: Part A: Chemistry, Analysis, Control, Exposure & Risk Assessment, 28*(6), 775–785. https://doi.org/10.1080/19440049.2011.559278.

78. Wuest, S. B. (2009). Correction of bulk density and sampling method biases using soil mass per unit area. *Soil Science Society of America Journal, 73*(1), 312. https://doi.org/10.2136/sssaj2008.0063.
79. Pearce, R. A., & Bolton, D. J. (2005). Excision vs sponge swabbing—A comparison of methods for the microbiological sampling of beef, pork and lamb carcasses. *Journal of Applied Microbiology, 98*(4), 896–900. https://doi.org/10.1111/j.1365-2672.2004.02525.x.
80. Lindblad, M. (2007). Microbiological sampling of swine carcasses: a comparison of data obtained by swabbing with medical gauze and data collected routinely by excision at Swedish abattoirs. *International Journal of Food Microbiology, 118*(2), 180–185. https://doi.org/10.1016/j.ijfoodmicro.2007.07.009.
81. Packard, B. H., & Kupferle, M. J. (2010). Evaluation of surface sampling techniques for collection of Bacillus spores on common drinking water pipe materials. *Journal of Environmental Monitoring, 12*(1), 361–368. https://doi.org/10.1039/b917570a.
82. Drago, L., Romanò, C. L., Mattina, R., et al. (2012). Does dithiothreitol improve bacterial detection from infected prostheses? A pilot study. *Clinical Orthopaedics and Related Research, 470*(10), 2915–2925. https://doi.org/10.1007/s11999-012-2415-3.
83. Bjerkan, G., Witsø, E., & Bergh, K. (2009). Sonication is superior to scraping for retrieval of bacteria in biofilm on titanium and steel surfaces in vitro. *Acta Orthopaedica, 80*(2), 245–250. https://doi.org/10.3109/17453670902947457.
84. Whitfield, W. J., Beakley, J. W., Dugan, V. L., et al. (1969). Vacuum probe: New approach to the microbiological sampling of surfaces. *Applied Microbiology, 17*(1), 164–168.
85. Edmonds, J. M., Sabol, J. P., & Rastogi, V. K. (2014). Decontamination efficacy of three commercial-off-the-shelf (COTS) sporicidal disinfectants on medium-sized panels contaminated with surrogate spores of Bacillus anthracis. *PLoS ONE, 9*(6), e99827. https://doi.org/10.1371/journal.pone.0099827.
86. Calfee, M. W., Rose, L. J., Morse, S., et al. (2013). Comparative evaluation of vacuum-based surface sampling methods for collection of Bacillus spores. *Journal of Microbiol Methods, 95*(3), 389–396. https://doi.org/10.1016/j.mimet.2013.10.015.
87. Lee, S. D., Calfee, M. W., Mickelsen, L., et al. (2013). Evaluation of surface sampling for *Bacillus spores* using commercially available cleaning robots. *Environmental Science and Technology, 47*(6), 2595–2601. https://doi.org/10.1021/es4000356.
88. U.S. Food and drug Administration. (1993). Validation of cleaning processes (7/93): Guide to inspections validation of cleaning processes. https://www.fda.gov/iceci/inspections/inspectionguides/ucm074922.htm.
89. Blankenship, L. C., Cox, N. A., Craven, S. E., et al. (1975). Total rinse method for microbiological sampling of the internal cavity of eviscerated broiler carcasses. *Applied Microbiology, 30*(2), 290–292.
90. Sarlin, L. L., Barnhart, E. T., Caldwell, D. J., et al. (1998). Evaluation of alternative sampling methods for *Salmonella* critical control point determination at broiler processing. *Poultry Science, 77*(8), 1253–1257.
91. Fletcher, D. L., Russell, S. M., Walker, J. M., et al. (1993). An evaluation of a rinse procedure using sodium bicarbonate and hydrogen peroxide on the recovery of bacteria from broiler carcasses. *Poultry Science, 72*(11), 2152–2156.
92. Zhang, Q., Qini, U., Ye, K., et al. (2012). Comparison of excision, swabbing and rinsing sampling methods to determine the microbiological quality of broiler carcasses. *Journal of Food Safety, 32*(1), 134–139. https://doi.org/10.1111/j.1745-4565.2011.00360.x.
93. Ashokkumar, M. (2016*). Handbook of ultrasonics and sonochemistry*. Singapore: Springer Science + Business Media.
94. Sanglay, G. C., Eifert, J. D., & Sumner, S. S. (2004). Recovery of *Salmonella* spp. from raw produce surfaces using ultrasonication. *Foodborne Pathogens Diseases, 1*(4), 295–299. https://doi.org/10.1089/fpd.2004.1.295.
95. Seymour, I. J., Burfoot, D., Smith, R. L., et al. (2002). Ultrasound decontamination of minimally processed fruits and vegetables. *International Journal of Food Science & Technology, 37*(5), 547–557. https://doi.org/10.1046/j.1365-2621.2002.00613.x.

96. Kang, D., Eifert, J. D., Williams, R. C., et al. (2007). Evaluation of quantitative recovery methods for *Listeria monocytogenes* applied to stainless steel. *Journal of AOAC International, 90*(3), 810–816.
97. Oulahal-Lagsir, N., Martial-Gros, A., Boistier, E., et al. (2000). The development of an ultrasonic apparatus for the noninvasive and repeatable removal of fouling in food processing equipment. *Letters in Applied Microbiology, 30*(1), 47–52.
98. McDonald, W. S., & Nichter, L. S. (1994). Debridement of bacterial and particulate-contaminated wounds. *Annals of Plastic Surgery, 33*(2), 142–147.
99. Nishikawa, T., Yoshida, A., Khanal, A., et al. (2010). A study of the efficacy of ultrasonic waves in removing biofilms. *Gerodontology, 27*(3), 199–206. https://doi.org/10.1111/j.1741-2358.2009.00325.x.

Author Biographies

Professor Ilya Digel (Germany) Book Co-Editor, Ph.D. He received the Dr. (Biology) degree from the al-Farabi Kazakh National State University in 1995. Since 2002 he has been researcher and lecturer on cell biology and biophysics at the FH Aachen University of Applied Sciences, Germany. Being the head of the Laboratory of Cell- and Microbiology in the Institute of Bioengineering, he conducts ongoing research on molecular biophysics of proteins and water as well as on development of diagnostic and therapeutic approaches in medicine.

Professor Nuraly Sh. Akimbekov (Kazakhstan) Ph.D. He completed his Ph.D. in Biotechnology. Currently he is employed as Assistant Professor at the al-Farabi Kazakh National University and, a Senior Researcher at the Institute of Biology and Biotechnology Issues, Almaty, Kazakhstan. His main research interests are focused on bio-composite materials, heterogeneous structures and multifunctional bio-products for various purposes.

Mrs. Professor Kistaubayeva (Kazakhstan) Ph.D. She is Associated Professor and Head of Department of biotechnology of al-Farabi Kazakh National University, Almaty, Kazakhstan. From 1999 till 2005 she received her bachelor and then master degree in biology/biotechnology at al-Farabi Kazakh National University. He completed her Ph.D. in biotechnology in 2009. Mrs. has completed numerous scientific internships and trainings abroad (Slovaks Republic, Germany) and has more than 120 scientific publications. She received the Kazakhstan Young Scientist Awards and numerous scholarships.

Professor Azhar A. Zhubanova (Kazakhstan) Book Co-Editor, Ph.D. She is a member of the Kazakhstan's National Academy of Natural Sciences, Doctor of Biology, Professor of the Department of Biotechnology at the al-Farabi Kazakh National University. She is well known for her contributions to the fields of microbiology, biochemistry, environmental sciences and biotechnology. She is the author or co-author of over 600 publications, edited 10 books and wrote four monographs. She was invited to numerous guest lectures in the USA, Germany, Russia, India and China.

The Search for Signatures of Life and Habitability on Planets and Moons of Our Solar System

Oliver Funke and Gerda Horneck

1 Introduction

1.1 Life and Habitability, in the Solar System and Beyond

With the advent of space technology, our knowledge about the world beyond our Earth has largely increased: Space probes sent out into our Solar System to visit and explore its planets and their moons have provided striking impressions of those strange worlds; and with space telescopes we are detecting more and more extra-solar planetary systems in our galactic neighborhood. Searching for signatures of life on those remote bodies is one of the major drivers of the space activities.

So far, the Earth is the only planet known to harbor life. Life's history spans at least 3.5 billion years, as witnessed by microfossils detected in the oldest sedimentary rocks [47]. Since then, life has evolved to the present rich biodiversity and thereby it has significantly shaped its environment. Assuming that the basic properties of life are universal and that life is a cosmic imperative, as suggested by the Nobel laureate Christian de Duve [6], the general properties and requirements of terrestrial life should be guiding any search for extra-terrestrial habitats and life (reviewed in [14]).

Despite the diversity of life on Earth, all organisms share a certain set of requirements, such as the availability of elements (in particular C, H, N, O, P, S, trace metals, etc.) needed for macromolecular synthesis, liquid water, energy, and appropriate physical-chemical conditions (such as temperature). To assess the

O. Funke (✉)
German Aerospace Center DLR, Space Administration, Navigation, Bonn, Germany
e-mail: oliver.funke@dlr.de

G. Horneck
German Aerospace Center DLR, Institute of Aerospace Medicine, Cologne, Germany

G. M. Artmann et al. (eds.), *Biological, Physical and Technical Basics of Cell Engineering*, https://doi.org/10.1007/978-981-10-7904-7_20

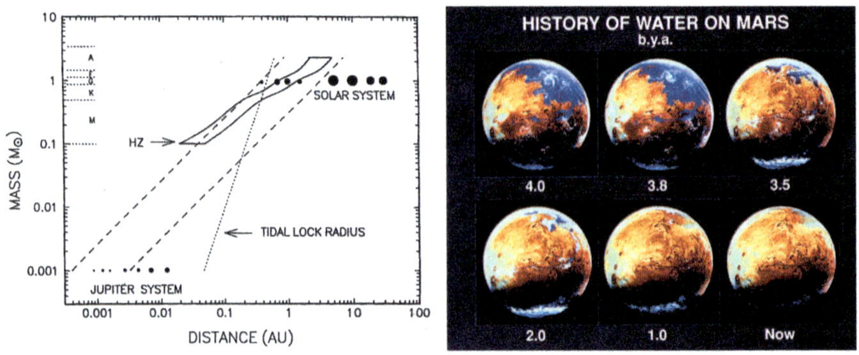

Fig. 1 Left: Habitable zone as a function of the distance from the star and its mass [16]. Right: artist's conceptions of past water coverage on Mars. *Credit* NASA

ability of another celestial body to sustain life, two alternative concepts of habitability have been applied:

(i) the astrophysical approach considers liquid water over extended periods of time on the surface of a planet as prerequisite for habitability, as reflected in the definition of a circumstellar habitable zone (Fig. 1) [16], and
(ii) the ecological approach considers a planet as habitable, if the physical and chemical conditions of its environment match with the physiological requirements of at least one type of microorganism on Earth [3]. This latter approach relies on the versatility and adaptability of extremophile microorganisms on Earth.

1.2 Habitable Candidates in Our Solar System

The habitable zone of our Solar System encompasses orbits from 0.84 to 1.77 AU (Fig. 1 left) [16]. Its range has been estimated on the base of solar luminosity and the orbital distance from the sun, which allows planets to maintain liquid surface water over geological time scales. Venus, Earth and Mars are situated in this habitable zone or in its close neighborhood, whereas the outer planets Jupiter, Saturn, Uranus and Neptune are located far off this zone.

The Earth is the prime example of a habitable planet. It has the right orbit around the Sun to be partially covered by oceans, it sustains an atmosphere that is dense enough to stabilize its surface temperature, and it continuously supports life since about 3.5 billion years. Such planets, where life, if originated, can evolve to multicellular complex life forms which can substantially modify the planets lithosphere, hydrosphere and atmosphere are termed Class I habitats [23].

Venus is located at the inner edge of the habitable zone. It is an extremely dry and hot planet containing very little water vapor in its atmosphere [21]. However, data sent from spacecraft, e.g. Venus Express of ESA, suggest that—about 4 billion years ago—Venus hosted large amounts of water on its surface, thereby providing habitable conditions. As the planet heated up, much of this water evaporated into the atmosphere, where it could then be ripped apart by electric forces in its ionosphere and subsequently was lost to space [4]. Such planets, where life once may have originated but hostile surface conditions may have prevented further evolution to complex life forms are termed Class II habitats [23].

Mars, which is located at the outer border of the habitable zone, is also classified as Class II habitat [23]. The extensive fluvial features seen on its surface attest to a stable flow of water during the first billion years [43], thereby providing habitable conditions on its surface at some time in the past [22, 48] (Fig. 1 right). If life started during this early period on Mars, the gradually decreasing pressure and temperature might have forced the emerging biota to retreat to some protective oases, where it might still exist [13]. But even today, transient liquid water formations might exist near the Mars surface: NASA's Mars Curiosity rover observed a night-time occurrence of transient liquid brines in the uppermost 5 cm of the subsurface at the Gale crater that then evaporated after sunrise [29]. The so called "recurrent slope lineae" (RSL), discovered first in 2011 in images obtained with the Mars Reconnaissance Orbiter High Resolution Imaging Science Experiment, might hint to the presence of brine, containing liquid water. However, this is subject to an ongoing controversial discussion (see Sect. 2.1.2).

Because the definition of the habitability of Class I and II planets is based on the availability of surface liquid water, they are termed "*surface liquid water worlds*". In addition, there are also celestial bodies located outside of the habitable zone, which harbor liquid water, but only in their interiors. They are termed "*interior liquid water worlds*" [2] and are represented by the satellites of the outer planets of our Solar System. The moons Europa, Ganymede and Callisto of Jupiter and Titan and Enceladus of Saturn contain a liquid water reservoir below a km-thick icy crust. Their habitability is assessed on the base of the ecological approach, i.e. the ability of the environment to support the activity of at least one known terrestrial organism [3]. Europa and Enceladus are unique in that their ocean is in contact with the silicate crust below. Interaction of liquid water with the rock material provides copious amounts of molecular hydrogen for organic synthesis and hence for metabolic activity [37]. Similar processes are taking place at the hydrothermal vents at the bottom of Earth's ocean and might also occur within the oceans of the two moons. Such "*interior liquid water worlds*", where the bottom of the ocean is in contact with the lithosphere are classified as Class III habitats (Fig. 2) [23]. They fulfill the following habitability requirements:

(i) the presence of liquid water,
(ii) an energy source adequate to fulfill the metabolic processes,
(iii) the presence of chemical elements necessary for life, and
(iv) relevant physical environmental conditions, above all pressure and temperature.

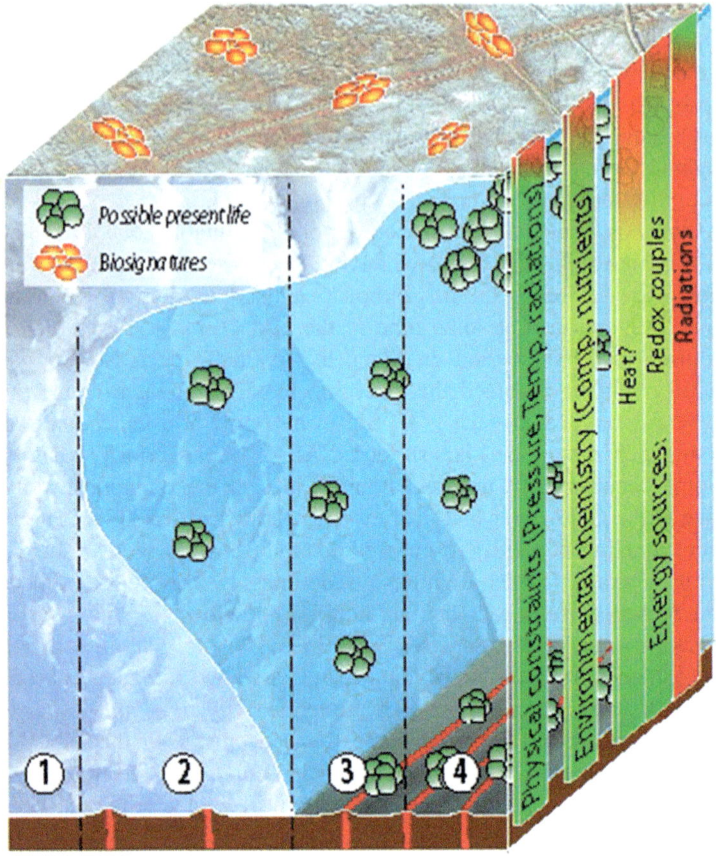

Fig. 2 Class III habitat, where the subsurface ocean is in contact with the lithosphere (from Lammer et al. [23])

Less chance for habitability is given to Class IV bodies, where the subsurface ocean is located between 2 ice layers, because they lack the geological reducing power provided by a rock-water interaction. Jupiter's moons Ganymede and Callisto as well as Saturn's moon Titan belong to those Class IV bodies.

In the following we summarize current activities in Germany to develop key technologies required for future robotic space missions to explore unique and hard to access areas of high astrobiological interest. These areas are located on Mars, the Jovian moon Europa and Saturn's moon Enceladus. The Space Administration of the German Aerospace Center (DLR) initiated and funds some of the technological developments required to demonstrate the feasibility of such space missions. The capability of a computer based artificial intelligence (AI) to autonomously conduct the complete mission or at least major parts of it is a prerequisite [33]. Today's AI needs to be raised significantly to enable such an autonomous mission. This includes also adaptation of individual mission procedures, if changed environmental conditions, failure of technical system components and other unforeseen issues require this.

Different universities and research institutes in Germany collaborate within the EnEx- and VaMEx-Initiative created for this purpose and funded by DLR Space Administration. The developments are focusing mainly on autonomous navigation capability. Once this key technology is successful validated in terrestrial analog scenarios, the correspondent mission could be suggested to ESA for realization in a large scale space mission.

2 Water in the Focus of Future Robotic Exploration of the Solar System

2.1 Mars

2.1.1 Environmental Conditions

Mars is about a half as big as our Earth and disposes only a tenth of the mass. Its gravity amounts to only a little bit more than one third of that of Earth. The temperatures on the Martian surface vary between −143 °C in winter at the Poles and +35 °C in summer at noon near the equatorial region. In the annual average, temperatures are about to −60 °C. Nevertheless, an essential difference to Earth is the absence of a magnetic field today: the solar wind hits unhindered on the atmosphere of Mars, and lighter atmospheric constituents like O_2 and N_2 are volatilized into space due to solar wind pressure and weak bondage by Martian gravity. As a result, the heavier CO_2 is the dominant constituent of the remaining thin Mars atmosphere. On the planet's surface the atmospheric pressure today is about 6 mbar. This corresponds to the triple-point of water: at this low surrounding atmospheric pressure ice which is warmed up by solar irradiation cannot transfer into the liquid phase, but evaporates immediately into the atmosphere. Therefore, liquid water cannot exist at the surface due to the physics of water. In addition to that, the sterilizing UV radiation from the sun reaches the Martian surface practically unfiltered by low atmospheric water vapor content. Subsequently Mars resembles at very first sight even more a hostile to life dry desert, caught in an ice age climate, than a habitable world on which life could have developed.

However, and on the contrary to that, Mars is considered as earth-most similar among the other planets of our solar system. The reason for this can be found in the geologic history of the red planet, which can be distinguished into three main periods (Fig. 1 right side):

During the "Noachian period", approximately 4.5–3.5 billion years ago, the oldest extant surfaces of Mars formed, including the Tharsis region, a volcanic upland. Also, due to volcanic activity the atmosphere was much denser than today, and in

the late time of this period large amounts of liquid water covered the plains of the surface. For the first ~500 Myr in this period Mars possessed a periodically reversing magnetic field due to early plate tectonics [34].
The "Hesperian period" followed from 3.5 billion years ago and lasted for about 200–600 million years. During this period, the formation of extensive lava plains took place by an even increased volcanism.
The "Amazonian period" was the next geological phase that still lasts today. With the decrease of tectonic activity, the magnetic field weakened more and more. The atmosphere thinned out, temperatures dropped, and surface regions formerly covered by liquid water decreased progressively in size. While most of the water is now bound as ice at the polar caps (mainly at the southern pole), recent results of the Mars Reconnaissance Orbiter (MRO) showed evidence that also large amounts of water can be found as frozen ground water ice within the subsurface of the Utopia Planitia region.

This very brief history of the water on Mars shows:

In the early time of the Solar System Mars possessed a much thicker atmosphere than today, and large amounts of water were prominent on the surface. Wide lakes and even seas formed which were fed by rivers and precipitation. The temperatures were more moderate than today and the environmental surroundings resembled those of the young Earth. This period corresponds to that on Earth, when first life evolved. Therefore, it cannot be excluded that like on Earth life has also begun to evolve on Mars at that time. And if life has evolved, the question must be addressed whether it could still exist there today, adapted to and sheltered in habitable niches from today's harsh environmental conditions given on the planetary surface. Such niches could be e.g. the subsurface, the bottom side of rocks, lava tunnels and caves. Sites that show in addition the presence of liquid water for at least some time of the Martian year would be ideal to harbor microorganism life and are therefore favorite places to explore.

2.1.2 Surface Liquid Water in Valles Marineris?

Dohm et al. [7] discussed the origin of the formation process of Valles Marineris as being a result of the Tharsis volcanic processes and thus lava flow. Leone [26] supported this finding on basis of combined use of data obtained by Mars Reconnaissance Orbiter (CTX camera, CRISM spectrometer, and HiRISE camera imagery), and profiles obtained with the Mars Orbiter Laser Altimeter (MOLA) aboard the Mars Global Surveyor (MGS) spacecraft. The gigantic canyon system could therefore still exhibit some habitable niches like former lava tunnels and caves at the slopes that correspond to best locations to search for biosignatures [27].

Aside from the large dimensions with 4000 km in length and up to 200 km in width, Valles Marineris is also of special interest because its ground is located about 7 km below the surrounding planetary surface (in some regions up to 10 km). In consequence, the atmospheric pressure at the ground of this large canyon system is accordingly higher:

As one of the results of the SKAD mission analysis (Phase 0 study, see Sect. 2.1.4), recently conducted by OHB System AG on behalf of DLR Space Administration, the atmospheric pressure within Valles Marineris was calculated to be about 13 mbar—which is well above the triple point of water. Thus, from a physical point of view, the presence of liquid water there would be possible. As brine, it can stay liquid even at temperatures below the freezing point of water.
Recurrent slope lineae (RSL) were discovered first in 2011 in images obtained with the Mars Reconnaissance Orbiter High Resolution Imaging Science Experiment. The highest density of known RSL can be found in Valles Marineris (Fig. 3).

They appear as narrow dark markings of 0.5–5 m in width and several hundred meters in length and occurring on steep (25°–40°) sunlit slopes. They incrementally lengthen down the slopes, fade in colder seasons, and recur annually. A briny aquifer was suggested by several authors [30, 42]. The origin of the suggested water in this process is not clear, absorption of atmospheric water vapor by presence of sufficient deliquescent salts was proposed by McEwan et al. [31] and Ojha et al. [35]. On basis of results from orbital spectroscopy data and the Thermal Emission Imaging System (THEMIS) instrument aboard the Mars Odyssey Orbiter Edwards and Piqueux [8] concluded that the water content within RSL is only at most ~0.5–3 wt%, and a circulating briny fluid is not present.

The discussion about the processes involved in RSL is still ongoing and the physical and chemical driving mechanism is not completely understood yet. So far, however, it appears unlikely that RSL could harbor or even sustain life. Exploration of RSL remains a topic of high scientific interest.

Even more relevant with respect to astrobiology is the finding of Gourronc et al. [11], who concluded from landforms in Valles Marineris in comparison with similar landforms on Earth that during the late Noachian to the early Hesperian period the canyon was entirely glaciated, yielding to a vast abundance of frozen water there, accumulated from the Martian atmosphere.

Okubo [36] interpreted other landforms in Candor and Coprates Chasma as former mud volcanoes: prominent surface characteristics in these regions of the Valles Marineris appear as distinctive knobs, rings and lobate structures. Okubo infers "the source of sediment for this mud volcanism to be the Hesperian eolian deposits that occur within these chasmata". Further, he suggests that "groundwater upwelling during the Hesperian to possibly the Early Amazonian facilitated the mobilization of these sediments within the subsurface and thereby contributed to the ensuing mud volcanism".

Today, groundwater might still be present in Valles Marineris, making this canyon a site of special interest for future exploration: groundwater, lava tunnels and caves, and in situ analysis of RSL should be addressed in such a mission.

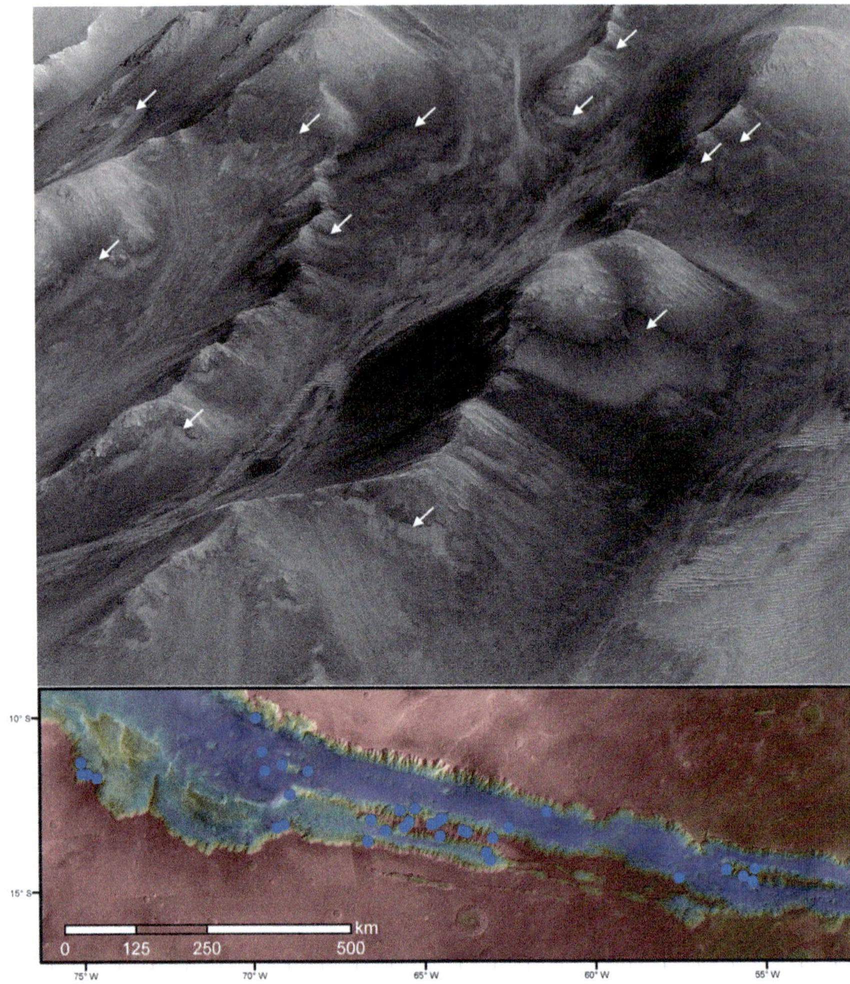

Fig. 3 Top: Recurring slope lineae (RSL) identified in the Coprates Montes area of Mars' Valles Marineris. Image obtained by NASA's Mars Reconnaissance Orbiter with repeated observations from orbit. The scene covers an area approximately 2.5 km wide, RSL appear as faint dark streaks. Bottom: Map of confirmed RSL sites, marked as blue dots, in part of the Valles Marineris. Each marked site includes a few to more than 1000 individual "lineae". *Credit* NASA/JPL-Caltech/Univ. of Arizona

2.1.3 Astrobiological Aspects

As mentioned above, the basic requirements for a planet to be habitable are the existence of liquid water, of an appropriate energy source and the right elements to build up life (Fig. 4). All three requirements were available on early Mars during the end of the Noachian period, at the same time when life started on early Earth.

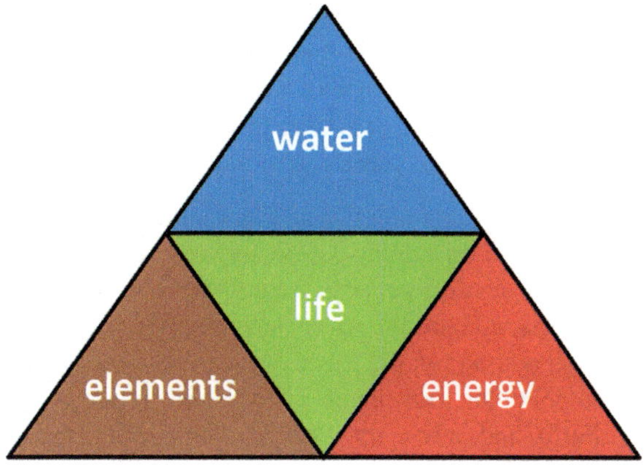

Fig. 4 The 3 basic requirements for life (adopted from Hand [12])

Under the assumption, that life emerges at a certain stage of planetary evolution if the right environmental physical and chemical requirements are provided [6], it is legitimate to assume that the conditions on early Mars were as favorable for life to emerge as on early Earth. However, both planets, although quite similar during their infancy, developed quite differently: Towards a fully blooming planet Earth and a devastated planet Mars scarce of any signs of life. The question arises, whether—if life once started on Mars—

(i) it became extinct later on when conditions became less favorable, but we can still find relicts of these early forms of life as fossils or other biosignatures, or
(ii) it adapted to the changing dry and cold surface conditions by retreating to more favorable protected niches, where it still might vegetate or even flourish.

Therefore it is the paramount next step in the exploration of Mars, to search for areas, where all three requirements for life prevail simultaneously and over extended periods of time [2]. Certain regions in Valles Marineris, as described above, might be suitable places to start with.

2.1.4 Past and Present Rover Missions

As pointed out in Sect. 2.1.2 the huge Valles Marineris canyon offers many interesting sites to search for extraterrestrial microbes. Most of the technologies required for the exploration are already available, such as probing techniques, sensors for navigation, energy supply units, sample analysis and more. Since the Mars Pathfinder Mission in 1997 with its Sojourner, the rover technology has made further progress, and the dimensions of rovers increased as well as the scientific instrumentation onboard. NASA's Mars Exploration Rovers Spirit, Opportunity and Curiosity showed a very well performance and demonstrated the progress achieved during the past two decades. It can be expected that this ongoing technological progress can also be demonstrated by the ESA/ROSCOMOS rover mission ExoMars that is planned for 2020.

Although these missions were designed to be mainly remotely controlled by receiving sequences of orders from an operations center on Earth, the computers onboard the rovers possess also software for autonomous hazard avoidance:

Due to the distance between Mars and Earth telecommands sent from earth will always face a time delay before they can be received by the rover on Mars. To avoid hazardous areas like large rocks, steep crater slopes or fine grained sand, the computers onboard the rovers are equipped with special software that allows at least a partial autonomy: the rover can react to a hazard being apparent in the sensor data by stopping the movement. Then a status report is sent back to the command station on Earth, and new commands are waited for. On Earth, solutions like a change of the formerly chosen rover pathway to avoid the identified obstacle can then be evaluated and sent as new orders to the rover to continue the mission safely.
Accompanying the technological progress also the autonomy software was developed further and shows an increasing degree of (partial) autonomy, too. A fully autonomous Mars rover mission is currently neither given nor in direct planning, because for the required high level artificial intelligence (AI) much development work is still to be done.

The persecuted concept of a "safety first" design for past and present Mars rover missions is principally comprehensible and well-founded, since the financial costs for realization of such missions are of the order of many 100 M$. However, this has also a severe implication on the mission performance:

Mars rovers move rather slowly, typically at a highest velocity of just a few centimeters per second. Covered distances are of the order of up to some ten kilometers (e.g. Opportunity: 40.25 km as of July 28, 2014, announced by NASA) within the duration of the rover mission. Landing sites, that are appropriate to the scientific aims of the mission, have therefore to be chosen very carefully, since the area of exploration is limited to a radius of less than 50 km.
NASA's Curiosity rover is the youngest rover mission and currently still active in Gale Crater. Curiosity is with its car-size and mass of about 900 kg the largest rover ever sent to Mars. The power demand of the numerous sensors and scientific instruments as well as for the movement of the rover is significant and cannot be generated by solar panels alone. Curiosity is therefore powered by a Multi-Mission Radioisotope Thermoelectric Generator (MMRTG) that provides approximately 125 W of electric energy, stored in rechargeable lithium-ion batteries. The thermal heat of the Generator also warms the rover during the nights, especially in winter time on Mars, when temperatures in the investigated area are dropping below −120 °C.

2.1.5 VaMEx—Valles Marineris Explorer

The VaMEx approach offers a basic concept for a possible future large space mission to explore the Valles Marineris intensively. In a very first step, we define this canyon system as a principal area of interest (AoI) for the mission. Regions of interest (RoI) like RSL and apparent lava tubes can be chosen within the AoI from orbital observation data. Due to the large dimensions of the canyon, the RoI are supposed to be at relatively large distances to each other. The landing site of the robotic exploration system will be chosen either in or nearby to a preselected RoI. Exploration starts with the task of reconnaissance to find possible sites of interest (PoI), which could be e.g. places with wet soil, caves and RSL that appear to be accessible by the exploration system starting from the canyon ground.

Thus, the PoI determination can be done best during the running mission by ground control on Earth on basis of obtained reconnaissance data: For this decisive point, there is rather a lot of time, as far as the reconnaissance operations on Mars are running autonomously, governed by a high level of AI ensuring system security. The chosen PoIs are then subject to exploration: the navigation and the pathway of the explorer system towards the PoI can also be planned by ground control.

Considering this scenario expiry, it is apparent, that a fully autonomous exploration system is not necessarily required. However, due to the large distances the system must cross, it must be moving rather fast at a speed of meters per second and if possible even faster—this is much faster than any other rover mission before. The safety aspect can thus only be fulfilled, if the robot disposes of an autonomous, quick and reliable security system that is independent from mission control on Earth.

In VaMEx the concept of a multi agent robotic system (MARS) is pursued: Instead of one single and sophisticated rover, the Valles Marineris Explorer ground segment consists of small rovers, and walking/crawling and flying (UAV: unpiloted aerial vehicle) robots. The elements of this heterogeneous drone swarm are equipped with instruments and/or sensors corresponding to their individual tasks within the mission. For redundancy, the number of each type of drone is at least two, hence the swarm consists of at least six elements. One of the prerequisites to utilize such a swarm for VaMEx is a small size of the drones, since they got to be transported together within the lander. The current size of the VaMEx-drones, as used for terrestrial testing (rover, UAV and hominid robot are shown in Fig. 5), still has to be miniaturized further.

The lander itself acts as a shuttle and shall be designed as a mobile station, following and supporting the swarm during the operations, e.g. with energy supply for recharging batteries. Also, telecommunication with the mission control on Earth can be realized by utilization of the large lander. The lander then acts as a relay station for the swarm elements. The swarm elements in turn can explore the Valles Marineris within a radius of up to ~ 15 km (corresponding of an area of about 1400 km^2) with the lander as central point.

The complete VaMEx scenario consists further of an orbiter, which can be used scientifically for detection of wet soil within the AoI/RoI, and to support the swarm navigation on the ground. For this purpose, the orbiter must be capable of resolving each swarm element at the ground and of referencing its position.

Fig. 5 Daniel Kühn, DFKI system engineer who created the hominid drone Charlie, presents the VaMEx swarm drones at ILA Berlin Air Show 2016: Charlie (right), a rover (left), and a hexacopter UAV drone (top right) as used for terrestrial validation tests of the swarm behavior. *Credit* DFKI/VaMEx-Team

All swarm elements must cooperate closely and share their knowledge about the environment (including hazards, obstacles etc.), thus building a collective awareness and intelligence. Included in the AI are the (more powerful) mobile lander, and —if visible from the lander—the computational power of the orbiter, too. A high-level AI can be realized this way, that must be capable to react in real time to hazardous situations.

The exploration of the Valles Marineris requires a mission duration of at least one year, within that many different RoI should be visited and their PoIs investigated closely. In the past years, robotic drone technology has made large progress, e.g. the impressive works by Boston Dynamics should be mentioned here (although not space proofed so far). It can be assumed, that further progress in miniaturization on the one hand, and robust performance on the other hand will be achieved in the next years. For VaMEx, the most critically technologies are the AI and the UAV.

2.1.6 VaMEx-Initiative

In 2012, the department Navigation of DLR Space Administration initiated and funded the first VaMEx joint research project, which was led by Technische Universität Braunschweig. Project partners were the Technical University of

Munich (TUM), and the DLR Institute of Communications and Navigation. Scientific contents were autonomous navigation techniques, cooperative swarm concepts, and automatic fault detection and recovery for UAV drones. Results were published by Sand et al. [39], Falconí and Holzapfel [9] and Falconí et al. [10]. The project ended in 2015 with a final field test, where the developed algorithms and techniques were successfully demonstrated. The final report "VaMEx—Valles Marineris Explorer" of this project is available (in German language only) at the Technische Informationsbibliothek (TIB) Hannover.

The VaMEx project marks the first step towards development of an autonomous, cooperating drone swarm, and justifies a continuation of the started developments. By the end of 2014 the DLR Explorer Initiatives were established at DLR Space Administration, supported jointly by the following listed four departments:

- Navigation,
- Microgravity Research and Life Sciences,
- Human Spaceflight, ISS and Exploration, and
- General Technologies and Robotics.

In 2015 three new joint research projects were started within the VaMEx-Initiative:

- "VaMEx-CoSMiC: Cooperative Swarm Navigation, Mission and Control" is the direct continuation of the former VaMEx project, with the same project partners.
- "VaMEx-VIPE: Visual and Proprioceptive Exploration" adds with visual based navigation techniques and the hominid robot Charlie two new aspects to VaMEx. The project is led by TUM with the German Research Center for Artificial Intelligence (DFKI GmbH) and the NavVis GmbH as partners. Adding a Charlie-type drone as a VaMEx swarm element is considered as a benefit, due to the robot's ability to move in rocky terrain, where rovers are not suited, to climb slopes and to support/aid other swarm elements being in distress. Charlie could be a technological nucleus for creating a small multi-purpose hominid drone.
- VaMEx-LAOLa: Lokales Ad hoc Ortungs- und Landesystem (local ad hoc localization and landing system) supports further the navigation capability of the MARS (see Sect. 2.1.5). The project is led by University of Würzburg, partners are University Erlangen-Nürnberg and ANavS GmbH.

With "SKAD", acronym for "Satellitenunterstützter Kooperativ Autonomer Drohnenschwarm" (engl.: satellite supported cooperative autonomous drone swarm)" a terrestrial feasibility demonstration of the principal VaMEx set up and scenario was proposed as a candidate for Germany's next large scale national space mission. The proposal was part of a DLR Space Administration internal idea competition, in which a total of 11 different missions were suggested by its departments. Starting in 2016, for each of the proposed missions a mission analysis (Phase 0) study was started. For SKAD, the final Mission Definition Review (MDR) was on March 14, 2017, marking the completion of the Phase 0 analysis.

Objective of the SKAD mission would be the development of all required VaMEx hard- and software for a long duration field test in the Atacama Desert in Chile [25]. The field test shall demonstrate the feasibility of a VaMEx mission on Mars, as described in Sect. 2.1.5, by deploying a heterogeneous drone swarm in the Atacama. The swarm shall consist of (at least)

- two fast driving, but small rovers, each one equipped with a sample analysis laboratory, and capable to carry a crawler drone on its back.
- two crawler drones, capable of crawling slopes, to take samples and return them to a rover for analysis.
- one hominid drone for multi-purpose utilization.
- two vertical take-off and landing (VTOL) UAV-drones, specially designed to fly in the very thin Martian atmosphere (Fig. 6).

Each drone is further equipped with sensors, antennas and everything needed for navigation, communication etc. The drone swarm shall operate autonomously for the duration of approximately one year in the Atacama, supported by a LEO-satellite, providing the swarm with additional information about RoI and snapshots of the actual positions of the swarm elements for reference. As in VaMEx, PoI have to be found and investigated in SKAD with a high degree in autonomy.

On basis of the obtained Phase 0 results, up to two missions will be selected in 2018 for a continuation in phase A/B1. After this, another selection process will be performed to decide, which mission shall be completed in phases C/D, followed by a mission start in 2025/26.

If SKAD will be realized, a TRL 6 will be reached before starting the Atacama long duration field test, which in turn would mark a large step towards a future international VaMEx mission.

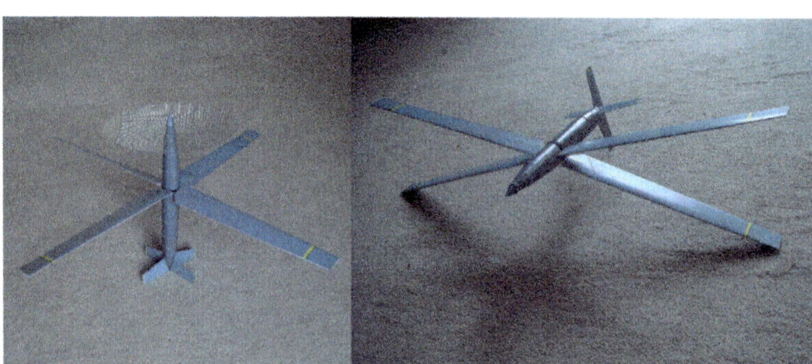

Fig. 6 The SKAD VTOL UAV-concept for operations in the thin Mars' atmosphere. Left: Starting and landing position, the lift-off is supported by a four blade rotor implemented at the center. Right: For airborne operations the center rotor will be fixed in an x-wing constellation, yielding lift, while flight movements are driven by a smaller propeller at the tip of the UAV. *Courtesy* OHB System AG, 2017

2.2 *Icy Water Worlds: Europa and Enceladus*

2.2.1 Global Water Ocean Beneath a Thick Ice Sheet

Oceans trapped under ice appear to be pretty common in the outer regions of our Solar System, e.g., in the moons Europa, Ganymede and Callisto of Jupiter, and Titan and Enceladus of Saturn. At those outer regions of our Solar System, the energy provided by the Sun is too weak to maintain a liquid ocean. Instead, tidal forces, exerted by the gravitational pull of the planet and the neighbor moons, have taken over. Jupiter's gravity is immense and raises huge tides. In addition, the other satellites whizzing by raise also tides. This creates a constant flexing of those moons that heats them up, driving their geology. The closer the moon orbits its planet, the greater is the tidal heating.

The first, although indirect, evidence of a subsurface ocean on Europa was provided by NASA's Galileo mission, which explored the Jupiter system from 1995 to 2003. The images of Europa, sent to the Earth, revealed strange pits, domes and broken landscapes at the surface that suggested a slight turn-over of the icy crust, due to heat from Fig. 7a [1]. However, the most important observation came from the magnetometer on board of Galileo: It detected that Jupiter's magnetic field was disrupted in the space around Europa. The best explanation for this disturbance is the induction of a magnetic field in a salty liquid-water ocean beneath the icy crust of Europa, as the moon passes through its planet's magnetic field. This ocean should have a thickness of at least 10 km [17] (Fig. 7b), and probably it is about 100 km deep. In addition, plumes of water vapor rising at the southern hemisphere over 160 km above Europa's icy surface were recently spectrophotometrically detected by the Hubble Space Telescope [38], another indication of a subsurface water ocean. At its floor the ocean is in contact with the silicate core, a feature that is unique among the Jovian moons and a strong indicator for habitability [40].

Saturn's tiny moon Enceladus—it is only about 500 km in diameter—is the other body in the outer solar system that harbors a global subsurface ocean in contact with its silicate core. Observations by the Voyager 2 probe in the 1980s showed already that its icy surface is remarkably smooth in some places, and bright white all over. The lack of impact craters around the southern area points to a relatively young age of this area. Images taken by the Cassini spacecraft during the flyby in 2005 revealed a distinctive, tectonically deformed region surrounding Enceladus' South Pole. This area is covered by fractures and ridges, and jets of icy particles laced with a brew of water and simple organic chemicals continuously gush out into space, released from the sub-ice ocean (Fig. 8). These ejected icy particles on one hand provide fresh coatings for Enceladus' bright surface and on the other hand they are the dominant source of Saturn's E ring [41]. It has been estimated that in view of Enceladus' small size and slightly higher amount of rock than water the floor of the ocean would be in contact with the silicate core as is the case for Europa [28].

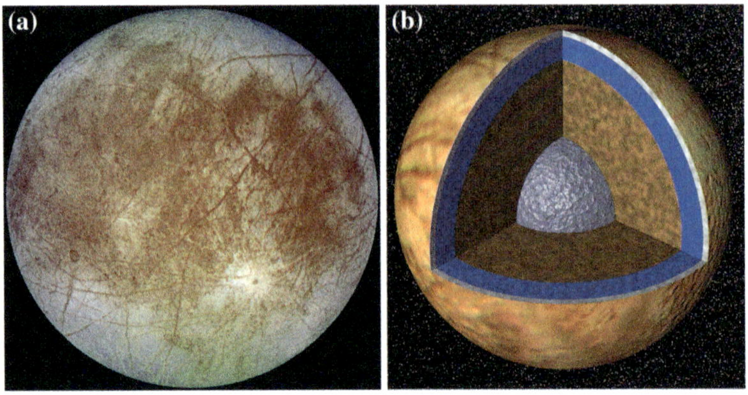

Fig. 7 Jupiter's moon Europa; **a** Image taken on September 7, 1996, at a distance of 677,000 km by the camera onboard NASA's Galileo; picture processed by DLR (*credit* NASA and DLR). **b** Interior of Europa with a metallic (iron, nickel) core (shown in gray), surrounded by a rock shell (shown in brown), surrounded by a shell of liquid water (shown in blue) and covered by ice (shown in white) (*credit* NASA)

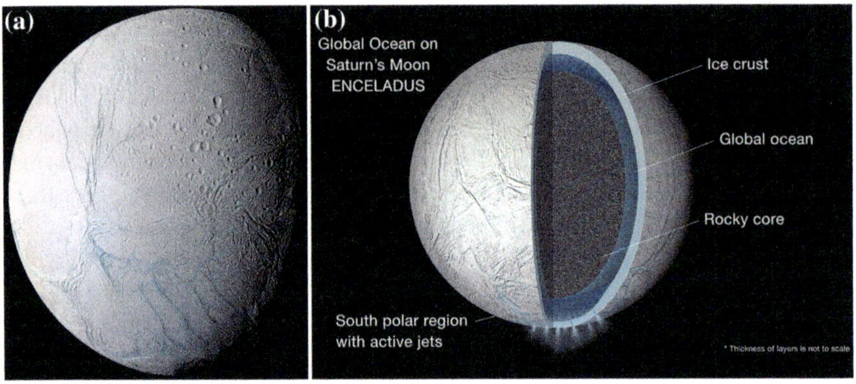

Fig. 8 Saturn's moon Enceladus; **a** High resolution image from a close flyby of the Cassini spacecraft, showing the so-called tiger stripes (in false-color blue), which spew ice from the moon's icy interior into space, creating a cloud of fine ice particles over the moon's south pole; published Jan. 2011 (*credit* NASA). **b** Sketch of the assumed interior of Saturn's moon Enceladus with a global liquid water ocean between its rocky core and icy crust. The thickness of layers shown here is not to scale (*Source* NASA/JPL-Caltec; Published: 15 September 2015)

2.2.2 Astrobiological Aspects

The subsurface-oceans of the moons Europa and Enceladus are the best candidates for habitability in our Solar System, because they fulfill all three basic requirements for life (Fig. 4). Liquid water is available by definition, thermal energy may arise from serpentinization, i.e. rock-water-carbon interactions at the bottom of the oceans [45], and the elements required for life, above all carbon and nitrogen, have been detected in the plumes of Enceladus [46]. They also include organic carbon molecules, such as CH_4 ($\sim 1\%$), HCN ($\sim 1\%$), and a suite of other more complex compounds up to those with six carbon atoms. Therefore, these moons are the most promising candidates for looking for signatures of extant extraterrestrial life in our Solar System [32].

2.2.3 IceMole

A proven way to explore the interior of a terrestrial glacier is by utilization of a melting probe. Classic melting probes are of cylindrical shape with an electrically heated metallic head (e.g. copper). The energy source remains on the surface of the glacier, the needed power (a few kW) is supplied to the melting probe via a permanent cable connection, that is also used for communication and data transfer. For application on a glacier, the probe is placed in vertical position with the melting head facing the ice surface. Once the head gets heated, the melting process starts and the probe melts its way vertically through the glacier. Typical penetration velocities are of the order of 1–3 m/h, depending on the heating power and the cross-section area of the probe.

For these classic probes, the direction of the melting is driven just by gravity. If the heating is switched off, the melting probe in its current position begins to freeze to the glacier.

The utilization of melting probes for accessing the global ocean of Jupiter's icy moon Europa was proposed soon after its existence became obvious. A concept of a radioisotope powered cryobot was proposed by Zimmerman et al. [50] and the melting probe approach was discussed intensively in the scientific community (a review is given by Ulamec et al. [44].

In 2010, Bernd Dachwald and a team of students at FH Aachen University of Applied Sciences developed the first fully maneuverable melting probe, called IceMole [5]. The difference to classic melting probes is the cuboid design and an ice drill, installed on top of the melting head. In addition, the four side walls of the cuboid were equipped with additional semiconductor thermoelements. These at first sight rather simple adaptations had a significant influence on the mobility of the IceMole: the newly gained capability for differential heating of the side walls in combination with the ice drill at the melting head enabled the IceMole to be maneuverable in three dimensions through the glacier. Even an upward movement can be achieved due to the ice drill that enables also a continuously given very close contact of the melting head with the ice sheet.

At the same time the feasibility of an in situ decontamination of a melting probe was investigated by Leimena et al. [24]: for utilization of a steerable melting probe to access e.g., subglacial lakes that are hermetically sealed from the environment by the glacier, it is required to avoid any forward contamination caused by the melting probe. This is also requested by the COSPAR Planetary Protection Policy [18]. Leimena et al. concluded that because the maximum surface temperature of a melting probe (~50 °C) is not sufficient for a sterilization of the probe, additional in situ sterilization must be applied by usage of 30% (v/v) H_2O_2 and 3% (v/v) NaClO. The team of Leimena could thus successfully "reach the threshold of 0 colony forming unit per mL for in situ drilling for both time-course and depth-course experiments".

Requirements for utilization of a melting probe for microbiological sampling can thus be summarized as follows:

- a proved concept for in situ decontamination
- full maneuverability (IceMole concept)
- capability of self-localization and navigation
- detection capability of obstacles (cavities, rocks) and potential target zones (e.g. an aquifer).

The trajectory of the probe can then be planned accordingly. Such a melting probe was build and tested within the joint research project EnEx.

2.2.4 EnEx—Enceladus Explorer

Exploration of the subglacial oceans of Europa and Enceladus is of highest interest for astrobiologists. For Europa, it is supposed that the ocean lies beneath a > ~10 km thick ice crust, that must be penetrated by a probe to get access to the ocean water. For Enceladus, the access could be easier, because ocean water is upwelling through cracks in the overlaying ice sheet: close to the surface and thus the vacuum of space the upwelling water is vaporized immediately, forming a fountain of ice particles (see Sect. 2.2.1). The occurrence of such ice fountains is referred to cryovolcanism, the base points of the fountains are called cryovolcanoes. If life has evolved within the Enceladus ocean, the upwelling water would contain microbes. However, they burst as soon as they reach the surface. Thus, a desirable mission target for astrobiologists could be to obtain a sample of liquid water that was taken well before the point of vacuum exposure. This point of interest is supposed to be located just a few 100 m beneath the surface—compared with the penetration of a many kilometers thick ice sheet, this significantly reduced distance is for time reasons much more appreciable.

This mission scenario is considered in EnEx [19], a project that started in 2012 and was led by FH Aachen. Partners were the Universität der Bundeswehr (University of Armed Forces) Munich, TU Braunschweig, University of Bremen, RWTH Aachen University, and Bergische Universität Wuppertal. In total, the consortia consisted of eight participating institutes. Scientific and technological goal of the project was primarily the development of an appropriate navigation system for the IceMole-concept of FH Aachen. This goal was achieved by a combined navigation system consisting of inertial measurement unit (IMU), magnetometer, and an acoustic pinger network, deployed close to the surface of the glacier [20].

For detection of obstacles, an ultrasonic array was implemented into the melting head.

Secondary goals of the EnEx project were a redesign of the IceMole, so that the navigation system could be implemented as payload, the development of a sampling system for integration into the probe, accompanied by development of an in situ decontamination procedure.

The technologies were tested in successive field campaigns on glaciers in the Alpes and in Antarctica. Finally, the EnEx-IceMole was given the chance to demonstrate its full potential in November/December 2014 during a finalizing field test, that was done in collaboration with a science team led by Jill Mikucki from Department of Microbiology, University of Tennessee: the technology was successfully deployed at Taylor glacier, Anarctica, and returned a water sample (brine) from a subglacial lake there, that feeds the so called "Blood Falls" at the end of the glacier. Prior to the sampling an in situ decontamination of the probe in ice was done. In this field test the EnEx-IceMole proved its potential as a unique tool for future space exploration.

The project resulted in the assembly of the EnEx-IceMole, the first fully maneuverable melting probe for clean sampling and sample return (Fig. 9).

For realization of a future EnEx space mission, however, still much work has to be done. The joint research project EnEx was a first step towards it. Work is continued within the EnEx-Initiative, established in late 2014 after the successful final field test in Antarctica.

Within the second phase of development, started in 2015, the degree of autonomy shall be raised, and the navigation capability as well as the maneuverability of

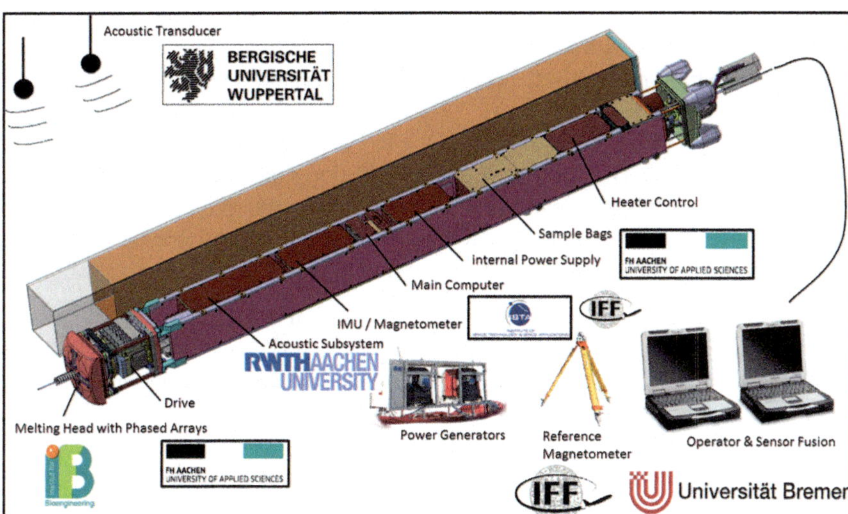

Fig. 9 Concept of the EnEx-IceMole, as deployed for the final field test at the Blood Falls in Antarctica. Sampling is made by inserting a hollow needle through the ice drill mounted at the melting head (bottom left side in the picture). *Credit* FH Aachen/EnEx-Team

the probe shall be improved further. The project partners in EnEx are now working within the EnEx-Initiative in individual projects, to raise the technology readiness level (TRL) of the EnEx-IceMole further. Environmental conditions on Enceladus are considered in more detail, to push the technological development towards a space mission. As a new potential target for a near future field test, investigation of the second highest volcano in Antarctica, Mt. Erebus, is envisaged. At its slopes, numerous ice fumaroles can be found. A task of the EnEx-IceMole could be melting into the ice and down to the bedrock, autonomous detection of a fumarole ahead as a target, planning of the trajectory, approaching and taking a gas sample. Then, the sample should be returned to the surface in its original composition.

The Mt. Erebus field test resembles the EnEx mission scenario, in which a water sample shall be taken from the interior of a cryovolcano on Enceladus. For the EnEx mission a very high degree of autonomy is needed. The Mt. Erebus field test can be envisaged for 2022, and if successful, EnEx may have reached TRL 4 or even TRL 5 then. It is considered as necessary to obtain a TRL 6 before proposing EnEx to ESA for realization as a large space mission. Considering the long duration flight to Enceladus (about 7 years for the Cassini spacecraft), the year 2045 is considered as a realistic time horizon for an EnEx mission starting its operations on Enceladus.

2.2.5 EurEx—Europa Explorer

With a cryobot-concept (e.g. Zimmerman et al. [50]) the thick ice sheet of Europa can be penetrated. For exploration of Europa's ocean, an autonomous underwater vehicle (AUV) named hydrobot was proposed [15], acting as a payload of a cryobot. This approach was investigated further in the EurEx project funded by DLR Space Administration. In this project the DFKI evaluated a very first prototype of such a melting probe shuttle, named "Teredo", carrying aboard the AUV "Leng" [47] (Fig. 10). Once Teredo has reached the ice/water interface, it pins itself in the ice and releases the AUV from its interior. A docking mechanism was developed, enabling the Leng to dock with the Teredo automatically. Within the DFKI large sea water basin a first laboratory "exploration" test successfully demonstrated the basically addressed technologies. The autonomous navigation was guided by

- Acoustic pinger signals, generated by a secondary payload of Leng named MicroGliders, that are released primary to the AUV. Task of the MicroGliders is to spread out and anchor at the bottom side of the ice sheet in some distance from their starting point, the position of the Leng. From that new position, they span an acoustic pinger network. Energy harvesting yields the energy needed to generate the acoustic signals, which can be used by the AUV Teredo for navigation.
- A camera implemented in the front of Leng is used for exploration tasks as well as for the re-docking process with Teredo.

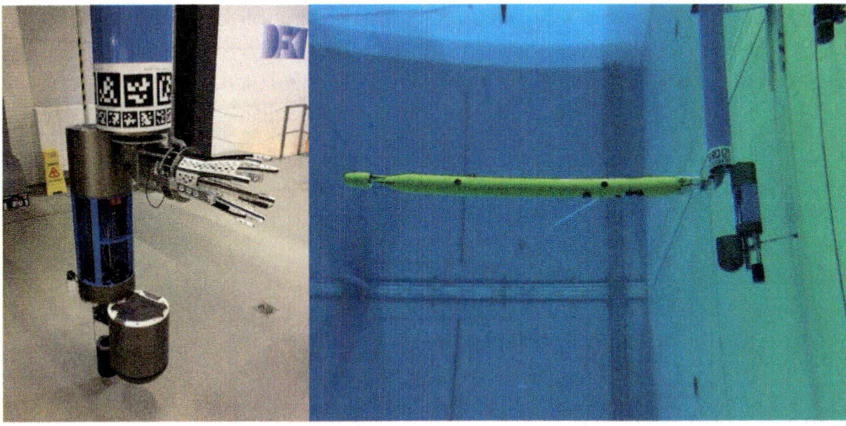

Fig. 10 Left: Docking mechanism of the Ice Shuttle "Teredo" (blue cylinder). Right: AUV "Leng" (yellow craft) after automated docking to Teredo during final demonstration in the EurEx-project. *Credit* DFKI

It is planned to continue the started technology development within the follow-up project EurEx-SiLaNa in 2017. The development of a Europa Explorer mission is a big endeavor, since fully autonomy of the AUV-system is inevitable. As a time horizon for an EurEx mission on Europa we estimate the year 2050. However, all the efforts will be worth, for Europa can be considered as the top candidate to look for extraterrestrial life in our solar system.

3 Outlook and Perspectives

With the Valles Marineris on Mars and the subglacial oceans on Enceladus and Europa three candidates for potential life carrying worlds have been discussed. With VaMEx, EnEx and EurEx three potential future space missions for the exploration of the named candidates have been presented. All three missions differ mainly in the required degree of AI:

In comparison to past and present rover missions on Mars, VaMEx needs an improved degree in autonomy, but assistance from mission control on Earth is possible. The EnEx melting probe needs an even higher degree in autonomy, and for EurEx, a fully autonomous AUV got to be developed.
Within the DLR Explorer Initiatives the mainly required technologies (e.g. autonomous navigation capability) are developed with the goal, to demonstrate the feasibility of these missions first in terrestrial analog field tests. A VaMEx mission could be realized by 2035, while an EurEx mission needs more efforts and therefore more time: a realization not before 2050 has to be expected.

Results of other space missions currently in preparation or planning, like e.g. NASA's Europa Clipper and ESA's JUICE mission, will give valuable input to the development of an EurEx-like mission. As Kevin Hand said, for him as an astrobiologist the exploration of Europa's ocean would clearly be the *"mission of missions"* [12].

References

1. Carr, M. H., Belton, M. J. S., Chapman, C. R., Davies, M. E., Geissler, P., Greenberg, R., et al. (1998). Evidence for a subsurface ocean on Europa. *Nature, 391,* 363–365.
2. Cockell, C. S., Bush, T., Bryce, C., Direito, S., Fox-Powell, M., Harrison, J. P., et al. (2016). Habitability: A review. *Astrobiology, 16,* 89–117.
3. Cockell, C. S., & Westall, F. (2004). A postulate to assess 'habitability'. *International Journal of Astrobiology, 3,* 157–163.
4. Collinson, G. A., Frahm, R. A., Glocer, A., Coates, A. J., Grebowsky, J. M., & Barabash, S. (2016). The electric wind of Venus: A global and persistent 'polar wind'-like ambipolar electric field sufficient for the direct escape of heavy ionospheric ions. *Geophysical Research Letters, 43,* 5926–5934.
5. Dachwald, B., Xu, C., Feldmann, M., & Plescher, E. (2011). Development of a novel subsurface ice probe and testing of the first prototype on the Morteratsch Glacier. *Geophysical Research Abstracts, 13,* 4943.
6. De Duve, C. (2011). Life as a cosmic imperative? *Philosophical Transactions of the Royal Society of London A: Mathematical, Physical and Engineering Sciences, 369,* 620–623.
7. Dohm, J. M., Williams, J. P., Anderson, R. C., Ruiz, J., McGuire, P. C., Komatsu, G., et al. (2009). New evidence for a magmatic influence on the origin of Valles Marineris, Mars. *Journal of Volcanology and Geothermal Research, 185,* 12–27.
8. Edwards, C. S., & Piqueux, S. (2016). The water content of recurring slope lineae on Mars. *Geophysical Research Letters, 43,* 8912–8919.
9. Falconí, G. P., & Holzapfel, F. (2013). Adaptive fault tolerant control allocation for a hexacopter system. In *Proceedings of the American Control Conference*, 2016 (pp. 6760–6766).
10. Falconí, G. P., Schatz, S. P., & Holzapfel, F. (2016). Fault tolerant control of a hexarotor using a command governor augmentation. In *24th Mediterranean Conference on Control and Automation (MED)*, 2016 (pp. 182–187).
11. Gourronc, M., Bourgeois, O., Mège, D., Pochat, S., Bultel, B., Massé, M., et al. (2014). One million cubic kilometers of fossil ice in Valles Marineris: Relicts of a 3.5 Gy old glacial landsystem along the Martian equator. *Geomorphology, 204,* 235–255.
12. Hand, K. (2017). *The search for life in Oceans beyond Earth, Space science week public lecture*. Washington D.C., USA: National Academy of Sciences.
13. Horneck, G. (2000). The microbial world and the case for Mars. *Planetary and Space Science, 48,* 1053–1063.
14. Horneck, G., Walter, N., Westall, F., Grenfell, J. L., Martin, W. F., Gomez, F., et al. (2016). AstRoMap European Astrobiology Roadmap. *Astrobiology, 16*(3), 201–243. (Special Issue).

15. Horvath, J. C., Carsey, F. D., Cutts, J. A., Jones, J. A., Johnson, E. D., Landry, B. M., … & Jeng, T. W. (1997, July). Searching for ice and ocean biogenic activity on Europa and Earth. In *Optical Science, Engineering and Instrumentation '97* (pp. 490–500). International Society for Optics and Photonics.
16. Kasting, J. F., Whitmire, D. P., & Reynolds, R. T. (1993). Habitable zones around main sequence stars. *Icarus, 101,* 108–128.
17. Khurana, K. K., Kivelson, M. G., Stevenson, D. J., Schubert, G., Russell, C. T., Walker, R. J., et al. (1998). Induced magnetic fields as evidence for subsurface oceans in Europa and Callisto. *Nature, 395,* 777–780.
18. Kminek, G., & Rummel, J. (2015). COSPAR's Planetary Protection Policy. *Space Research Today, 193,* 1–14. (COSPAR's information bulletin).
19. Konstantinidis, K., Martinez, C. L. F., Dachwald, B., Ohndorf, A., Dykta, P., Bowitz, P., … & Förstner, R. (2015). A lander mission to probe subglacial water on Saturn's moon Enceladus for life. *Acta astronautica, 106,* 63–89.
20. Kowalski, J., Linder, P., Zierke, S., von Wulfen, B., Clemens, J., Konstantinidis, K., et al. (2016). Navigation technology for exploration of glacier ice with maneuverable melting probes. *Cold Regions Science and Technology, 123,* 53–70.
21. Kulikov, Yu N, Lammer, H., Lichtenegger, H. I. M., Terada, N., Ribas, I., Kolb, C., et al. (2006). Atmospheric and water loss from early Venus. *Planetary and Space Science, 54,* 1425–1444.
22. Lammer, H., Selsis, F., Penz, T., Amerstorfer, U. V., Lichtenegger, H. I. M., Kolb, C., et al. (2005). Atmospheric evolution and history of water on Mars. In T. Tokano (Ed.), *Water on Mars and Life* (pp. 25–44)., Advances in astrobiology and biogeophysics Berlin: Springer.
23. Lammer, H., Bredehöft, J. H., Coustenis, A., Khodachenko, M. L., Kaltenegger, L., Grasset, O., et al. (2009). What makes a planet habitable? *The Astronomy and Astrophysics Review, 17,* 181–249.
24. Leimena, W., Artmann, G. M., Dachwald, B., Artmann, A., Goßmann, M., & Digel, I. (2010). Feasibility of an in-situ microbial decontamination of an ice-melting probe. *Eurasian Chemico-Technological Journal, 12*(2), 145–150.
25. Lemke, M. K., Funke, O., Klein, V., Montenegro, S., Schilling, K., & Buehler, C., et al. (2017). German large national mission candidate SKAD—a satellite-based cooperative autonomous drone swarm for exploration. In *Submitted to 68th International Astronautical Congress 2017.*
26. Leone, G. (2014). A network of lava tubes as the origin of Labyrinthus Noctis and Valles Marineris on Mars. *Journal of Volcanology and Geothermal Research, 277,* 1–8.
27. Léveillé, R. J., & Datta, S. (2009). Lava tubes and basaltic caves as astrobiological targets on Earth and Mars: A review. *Planetary and Space Science, 58*(2010), 592–598.
28. Lunine, J. I. (2017). Ocean worlds exploration. *Acta Astronautica, 131,* 123–130.
29. Martín-Torres, F. J., Zorzano, M.-P., Valentín-Serrano, P., Harri, A.-M., & Genzer, M. (2015). Transient liquid water and water activity at Gale crater on Mars. *Nature Geoscience, 8* (5), 357–361.
30. McEwen, A. S., Ojha, L., Dundas, C. M., Mattson, S. S., Byrne, S., Wray, J. J., et al. (2011). Seasonal Flows on Warm Martian Slopes. *Science, 333,* 740–743.
31. McEwan, A, Chojnacki, M., Dundas, C., Ojha, L., Masse, M., & Schaefer, E., et al. (2015). Recurring slope lineae on Mars: Atmospheric Origin? *EPSC Abstracts, 10,* EPSC2015-786-1.
32. McKay, C. P., Anbar, A. D., Porco, C., & Tsou, P. (2014). Follow the plume: The habitability of enceladus. *Astrobiology, 14,* 352–355.

33. Muscettola, N., Nayak, P. P., Pell, B., & Williams, B. C. (1998). Remote agent: To boldly go where no AI system has gone before. *Artificial Intelligence, 103*(1–2), 5–47.
34. Nimmo, F., & Stevenson, D. J. (2000). Influence of early plate tectonics on the thermal evolution and magnetic field of Mars. *Journal Geophysical Research, 105*(E5), 11969–11979.
35. Ojha, L., Wilhelm, M. B., Murchie, S. L., McEwen, A. S., Wray, J. J., Hanley, J., et al. (2015). Spectral evidence for hydrated salts in recurring slope lineae on Mars. *Nature Geoscience*. https://doi.org/10.1038/ngeo2546.
36. Okubo, C. H. (2016). Morphologic evidence of subsurface sediment mobilization and mud volcanism in Candor and Coprates Chasmata, Valles Marineris, Mars. *Icarus, 269,* 23–37.
37. Proskurowski, G., Lilley, M. D., Seewald, J. S., Früh-Green, G. L., Olson, E. J., Lupton, J. E., et al. (2008). Abiogenic hydrocarbon production at Lost City hydrothermal field. *Science, 319,* 604–607.
38. Roth, L., Saur, J., Retherford, K. D., Strobel, D. F., Feldman, P. D., McGrath, M. A., et al. (2014). Transient water vapor at Europa's south pole. *Science, 343,* 171–174.
39. Sand, S., Zhang, S., Mühlegg, M., Falconi, G., Zhu, C., & Krüger, T., et al. (2013). Swarm exploration and navigation on Mars. In *2013 International Conference on Localization and GNSS (ICL-GNSS).*
40. Sotin, C., & Prieur, D. (2007). Jupiter's Moon Europa: Geology and habitability. In G. Horneck & P. Rettberg (Eds.), *Complete course in astrobiology* (pp. 253–271). Germany: Wiley-VCH, Weinhiem.
41. Spahn, F., Schmidt, J., Albers, N., Hörning, M., Makuch, M., Seiss, M., et al. (2006). Cassini dust measurements at Enceladus and implications for the origin of the E ring. *Science, 311,* 1416–1418.
42. Stillman, D. E., Michaels, T. I., Grimm, R. E., & Hanley, J. (2016). Observations and modeling of northern mid-latitude recurring slope lineae (RSL) suggest recharge by a present-day martian briny aquifer. *Icarus, 265,* 125–138.
43. Tokano, T. (Ed.). (2005). *Water on Mars and Life.*, Advances in astrobiology and biogeophysics Berlin: Springer.
44. Ulamec, S., Biele, J., Funke, O., & Engelhardt, M. (2006). Access to glacial and subglacial environments in the Solar System by melting probe technology. *Reviews in Environmental Science and Bio/Technology, 6*(2006), 71–94.
45. Vance, S., Harnmeijer, J., Kimura, J., Hussmann, H., deMartin, B., & Brown, J. M. (2007). Hydrothermal systems in small Ocean planets. *Astrobiology, 7*(6), 987–1005.
46. Waite, J. H., Lewis, W. S., Magee, B. A., Lunine, J. I., McKinnon, W. B., Glein, C. R., et al. (2009). Liquid water on Enceladus from observations of ammonia and 40 A in the plume. *Nature, 460,* 487–490.
47. Westall, F. (2011). Early life. In: M. Gargaud, P. López-Garcia, & H. Martin (Eds), *Origins and Evolution of Life: An astrobiology perspective* (pp. 391–413). Cambridge University Press.
48. Westall, F., Foucher, F., Bost, N., Bertrand, M., Loizeau, D., Vago, J. L., et al. (2015). Biosignatures on Mars: What, where, and how? implications for the search for martian life. *Astrobiology, 15*(11), 998–1029.
49. Wirtz, M., & Hildebrand, M. (2016). IceShuttle Teredo: An Ice-Penetrating Robotic System to Transport an Exploration AUV into the Ocean of Jupiter's Moon Europa. In *67th International Astronautical Congress (IAC)*, Guadalajara, Mexico, September 26–30, 2016.
50. Zimmerman, W., Bryant, S., Zitzelberger, J., & Nesmith, B. (2001, February). A radioisotope powered cryobot for penetrating the Europan ice shell. In M. S. El-Genk & M. J. Bragg (Eds.), *AIP Conference Proceedings* (Vol. 552, No. 1, pp. 707–715). AIP.

Author Biographies

Oliver Funke (Germany) Dr. rer. nat. is physicist and employee of DLR Space Administration, Department of Navigation. As project manager he is responsible for the department's program line "Innovative Navigation Technologies" and initiates and cares for new projects. Furthermore, he represents the research funders in these projects. His research interests are covering the fields of autonomous robotic exploration, melting probe technologies, mission concepts, remote sensing and astrobiology. He is coordinator of the DLR Explorer Initiatives with several departments of DLR Space Administration involved to study new approaches for future space exploration missions.

Gerda Horneck (Germany) Ph.D. is former Deputy Director of the Institute of Aerospace Medicine of the German Aerospace Center DLR and former Head of its Radiation Biology Section (retired since 2004). She has been involved in radiobiological and astrobiological space experiments since the Apollo era. For her research she was awarded with honors by ESA, NASA, DLR, ISSOL and IAA. She is Honorary President of the European Astrobiology Network Association EANA and Deputy Editor of the journal Astrobiology. She is author of more than 100 scientific publications and has written or edited several books on astrobiology and space biology.

Printed by Printforce, the Netherlands